# Lecture Notes in Computer Science 4003

*Commenced Publication in 1973*
Founding and Former Series Editors:
Gerhard Goos, Juris Hartmanis, and Jan van Leeuwen

Yevgeni Koucheryavy   Jarmo Harju
Villy B. Iversen (Eds.)

# Next Generation Teletraffic and Wired/Wireless Advanced Networking

6th International Conference, NEW2AN 2006
St. Petersburg, Russia, May 29 – June 2, 2006
Proceedings

 Springer

Volume Editors

Yevgeni Koucheryavy
Jarmo Harju
Tampere University of Technology
Institute of Communication Engineering
P.O.Box 553, 33101 Tampere, Finland
E-mail: {yk, harju}@cs.tut.fi

Villy B. Iversen
Technical University of Denmark, Research Center COM
Building 345v, 2800 Kongens Lyngby, Denmark
E-mail: vbi@com.dtu.dk

Library of Congress Control Number: 2006926212

CR Subject Classification (1998): C.2, C.4, H.4, D.2, J.1, K.6, K.4

LNCS Sublibrary: SL 5 – Computer Communication Networks and
Telecommunications

ISSN      0302-9743
ISBN-10   3-540-34429-2 Springer Berlin Heidelberg New York
ISBN-13   978-3-540-34429-2 Springer Berlin Heidelberg New York

Springer is a part of Springer Science+Business Media

springer.com

© Springer-Verlag Berlin Heidelberg 2006
Printed in Germany

Typesetting: Camera-ready by author, data conversion by Scientific Publishing Services, Chennai, India
Printed on acid-free paper      SPIN: 11759355      06/3142      5 4 3 2 1 0

# Preface

We welcome you to the proceedings of 6th NEW2AN 2006 (Next Generation Teletraffic and Wired/Wireless Advanced Networking) held in St. Petersburg, Russia.

This year the scientific program contained significant contributions to next-generation teletraffic as concerns traffic characterization, estimation of traffic parameters, and modeling of new services based on real data and experiments. New methods for designing dynamic optimal systems were presented. In particular, issues of quality of service (QoS) in wireless and IP-based multi-service networks were dealt with, as well economical aspects of future networks.

The papers in this volume also provide new contributions to various aspects of networking. The emphasis is on wireless networks, including cellular networks, wireless local area networks, personal area networks, mobile ad hoc networks and sensor networks. Topics cover several layers from lower layers via routing issues to TCP-related problems. New and innovative developments for enhanced signaling protocols, QoS mechanisms and cross-layer optimization are also well represented within the program.

The call for papers attracted 137 papers from 28 countries, resulting in an acceptance ratio of 35%. With the help of the excellent Technical Program Committee and a number of associated reviewers, the best 49 high-quality papers were selected for publication. The conference was organized in 17 single-track sessions.

The Technical Program of the conference benefited from three distinguished keynote speakers: Ian F. Akyildiz, Georgia Tech, Atlanta, USA; Nikolai Nefedov, NOKIA, Helsinki, Finland; and Nina Bhatti, Hewlett-Packard Laboratories, Palo Alto, California, USA.

We wish to thank the Technical Program Committee members and associated reviewers for their hard work and their important contribution to the conference.

This year the conference was organized in cooperation with ITC (International Teletraffic Congress), IEEE, COST 290, with the support of NOKIA (Finland), BalticIT Ltd. (Russia), and Infosim GmbH. (Germany). The support of these organizations is gratefully acknowledged.

Finally, we wish to thank many people who contributed to the NEW2AN organization. In particular, Dipti Adhikari (TUT) carried a substantial load of submission and review website maintaining, Jakub Jakubiak (TUT) did an excellent job on the compilation of camera-ready papers and interaction with Springer. Sergei Semenov (NOKIA) is to be thanked for his great efforts on conference linkage to industry. Many thanks go to Anna Chachina (Monomax Ltd.) for her excellent local organization efforts and the conference's social program preparation.

We believe that the work done for the 6th NEW2AN conference provided an interesting and up-to-date scientific program. We hope that participants enjoyed the technical and social conference program, Russian hospitality and the beautiful city of St. Petersburg.

March 2006                                                    Yevgeni Koucheryavy
                                                                    Jarmo Harju
                                                               Villy B. Iversen

# Organization

## Steering Committees

| | |
|---|---|
| Ian F. Akyildiz | Georgia Institute of Technology, USA |
| Boris Goldstein | State Univ. of Telecommunications, Russia |
| Igor Faynberg | Lucent Technologies, USA |
| Jarmo Harju | Tampere University of Technology, Finland |
| Andrey Koucheryavy | ZNIIS R&D, Russia |
| Villy B. Iversen | Technical University of Denmark, Denmark |
| Paul Kühn | University of Stuttgart, Germany |
| Kyu Ouk Lee | ETRI, Korea |
| Mohammad Obaidat | Monmouth University, USA |
| Michael Smirnov | Fraunhofer FOKUS, Germany |
| Manfred Sneps-Sneppe | Ventspils University College, Latvia |
| Ioannis Stavrakakis | University of Athens, Greece |
| Sergey Stepanov | Sistema Telecom, Russia |
| Phuoc Tran-Gia | University of Würzburg, Germany |
| Gennady Yanovsky | State Univ. of Telecommunications, Russia |

## Program Committee

| | |
|---|---|
| Ozgur B. Akan | METU, Turkey |
| Khalid Al-Begain | University of Glamorgan, UK |
| Tricha Anjali | Illinois Institute of Technology, USA |
| Konstantin Avrachenkov | INRIA, France |
| Francisco Barcelo | UPC, Spain |
| Torsten Braun | University of Bern, Switzerland |
| Chrysostomos Chrysostomou | University of Cyprus, Cyprus |
| Georg Carle | University of Tübingen, Germany |
| Roman Dunaytsev | Tampere University of Technology, Finland |
| Eylem Ekici | Ohio State University, USA |
| Sergey Gorinsky | Washington University in St. Louis, USA |
| Faramarz Fekri | Georgia Institute of Technology, USA |
| Markus Fidler | NTNU Trondheim, Norway |
| Giovanni Giambene | University of Siena, Italy |
| Stefano Giordano | University of Pisa, Italy |
| Ivan Ganchev | University of Limerick, Ireland |
| Alexander Gurgenidze | Liniya Sviazi, Russia |
| Andrei Gurtov | HIIT, Finland |
| Vitaly Gutin | Popov Society, Russia |
| Martin Karsten | University of Waterloo, Canada |

| | |
|---|---|
| Farid Khafizov | Nortel Networks, USA |
| Yevgeni Koucheryavy | Tampere University of Technology, Finland (Chair) |
| Lemin Li | U. of Electronic Science and Techn. of China, China |
| Piet Van Mieghem | Delft University of Technology, Netherlands |
| Ian Marsh | SICS, Sweden |
| Maja Matijašević | University of Zagreb, FER, Croatia |
| Paulo Mendes | DoCoMo Euro-Labs, Germany |
| Michael Menth | University of Würzburg, Germany |
| Ilka Miloucheva | Salzburg Research, Austria |
| Dmitri Moltchanov | Tampere University of Technology, Finland |
| Edmundo Monteiro | University of Coimbra, Portugal |
| Mairtin O'Droma | University of Limerick, Ireland |
| Jaudelice Cavalcante de Oliveira | Drexel University, USA |
| Evgeni Osipov | University of Basel, Switzerland |
| George Pavlou | University of Surrey, UK |
| Simon Pietro Romano | Universita' degli Studi di Napoli "Federico II", Italy |
| Stoyan Poryazov | Bulgarian Academy of Sciences, Bulgaria |
| Alexander Sayenko | University of Jyväskylä, Finland |
| Sergei Semenov | NOKIA, Finland |
| Burkhard Stiller | University of Zürich and ETH Zürich, Switzerland |
| Weilian Su | Naval Postgraduate School, USA |
| Veselin Rakocevic | City University London, UK |
| Dmitry Tkachenko | IEEE St. Petersburg BT/CE/COM Chapter, Russia |
| Vassilis Tsaoussidis | Demokritos University of Thrace, Greece |
| Christian Tschudin | University of Basel, Switzerland |
| Kurt Tutschku | University of Würzburg, Germany |
| Lars Wolf | Technische Universität Braunschweig, Germany |
| Linda J. Xie | University of North Carolina, USA |

## Additional Reviewers

| | | |
|---|---|---|
| I. Alocci | M. Brogle | M. Fernandez |
| M. Amin | M. Charalambides | P. Flynn |
| S. Avallone | P. Chini | M. Georg |
| T. Bernoulli | S. D'Antonio | S. Georgoulas |
| A. Binzenhoefer | Z. Despotovic | V. Goebel |
| D. Brogan | M. Dick | M. Grgic |

May 29 - June 2 ◆ 2006 ◆ St. Petersburg ◆ RUSSIA

TAMPERE UNIVERSITY OF TECHNOLOGY

NOKIA

# Table of Contents

## Sensor Networks I

## WLAN

## Teletraffic II

# Traffic Characterization and Modeling II

# QoS I

# Sensor Networks II

# MANETs

## Sensor Networks III

## 3G/UMTS II

## Lower Layer Techniques I

# Towards Distributed Communications Systems: Relay-Based Wireless Networks

Nikolai Nefedov

Nokia Research Center, P.O. Box 407 00045, Finland
nikolai.nefedov@nokia.com

**Abstract.** The increasing demand for high data-rate services stimulates growing deployment of wireless local area network technologies. However, the limited communication range of these technologies makes it difficult and expensive to provide high data-rate services at the periphery of service areas and in environments with harsh channel conditions. Furthermore, this problem even more complicated for future 4G wireless broadband networks, which are expected to offer data rates up to 100 Mb/s for mobile users and up to 1 Gb/s for stationary users. To provide a reasonably large coverage for these high data rates the conventional cellular networks appear to be not feasible and it stimulated a search for other architectures.

Recently it is shown that novel concepts such as multi-hop relaying and associated diversity techniques, may significantly enhance the high data rate coverage and increase the throughput beyond-3G networks. In relay-based cellular networks, the mobile terminals which cannot establish (at the required rates) direct links with the base station may communicate to fixed relays or other mobile terminals (e.g., using the unlicensed band) to assist the connection. It allows to substitute a poor-quality single-hop wireless link with a composite multi-hop better-quality link whenever possible. In a more general case, a relay-based wireless network may form a distributed communication system where several multi-hop links may be established and different types of re-transmissions (e.g., amplify-forward, decode-forward) are utilized. Besides, due to the broadcasting nature of wireless communications, transmitted signal from a given source may be processed simultaneously for the further retransmission by several relays and/or mobile terminals. These relays, or its subsets, may utilize different forms of cooperation to form, for example, virtual (distributed) antenna arrays. The relay cooperation may be extended beyond beamforming and include also space-time coding concept, where separated relay antennas interconnected with unreliable radio links form a distributed MIMO relaying structure.

Furthermore, relay nodes may combine several incoming source signals and retransmit a combined (coded) version. This recently proposed novel concept, known as network coding, currently attracts the growing attention.

In this talk we overview the recent advances in wireless relay networks, including distributed MIMO relaying, cooperative strategies and network coding.

Y. Koucheryavy, J. Harju, and V.B. Iversen (Eds.): NEW2AN 2006, LNCS 4003, p. 1, 2006.
© Springer-Verlag Berlin Heidelberg 2006

# Grand Challenges for Wireless Sensor Networks

Ian F. Akyildiz

Broadband and Wireless Networking Lab,
School of Electrical and Computer Engineering,
Georgia Institute of Technology,
Atlanta, GA 30332
USA
ian@ece.gatech.edu

**Abstract.** The technological advances in the micro-electro-mechanical systems (MEMS) and the wireless communications have enabled the deployment of the small intelligent sensor nodes at homes, in workplaces, supermarkets, plantations, oceans, streets, and highways to monitor the environment. The realization of smart environments to improve the efficiency of nearly every aspect of our daily lives by enhancing the human-to-physical world interaction is one of the most exciting potential sensor network applications utilizing these intelligent sensor nodes. However, this objective necessitates the efficient and application specific communication protocols to assure the reliable communication of the sensed event features and hence enable the required actions to be taken by the actors in the smart environment. In this talk, the grand challenges for the design and development of sensor/actor network communication protocols are presented. More specifically, application layer, transport layer, network layer, data link layer, in particular, error control and MAC protocols, and physical layer as well as cross layer issues are explained in detail.

# Dimensioning of Multiservice Links Taking Account of Soft Blocking

V.B. Iversen[1], S.N. Stepanov[2], and A.V. Kostrov[3]

[1] COM · DTU, Technical University of Denmark,
DK-2800 Kgs. Lyngby, Denmark
vbi@com.dtu.dk
[2] Sistema Telecom, 125047 Moscow,
1st Tverskay-Yamskaya 5, Russia
stepanov@sistel.ru
[3] Institute for Problems of Information Transmission RAN
101447, Moscow, Bolshoy Karetniy 19, Russia
A-Kostrov@e-mails.ru

**Abstract.** In CDMA systems the blocking probability depends on both the state of the system and the bandwidth of the request. Due to soft blocking caused by interference from users in same cell and neighboring cells we have to modify the classical teletraffic theory. In this paper we consider a model of a multiservice link taking into account the possibility of soft blocking. An approximate algorithm for estimation of main performance measures is constructed. The error of estimation is numerically studied for different types of soft blocking. The optimal procedure of dimensioning is suggested.

**Keywords:** teletraffic, wireless, CDMA, soft blocking, approximate algorithms, dimensioning, trunk reservation.

## 1 Introduction

New generation mobile networks are based on CDMA technology. Due to peculiarities of radio technology such systems have no hard limit on capacity. In order to maintain the QoS characteristics of already accepted connections, a new connection should not be accepted by the network if this connection increases the noise above a prescribed level for connections already accepted. A theoretical study of this phenomena can be done by means of teletraffic models taking into account the probability of soft blocking [1],[2]. In such models new connections will experience blocking probabilities which depend on both the bandwidth required and the state of the system.

## 2 Model Description and Main Performance Measures

Let us consider a single link traffic model, where the link transmission capacity is represented by $k$ basic bandwidth units (BBU) [3], and let us suppose that we have $n$ incoming Poisson flows of connections with intensities $\lambda_s$,

Y. Koucheryavy, J. Harju, V.B. Iversen (Eds.): NEW2AN 2006, LNCS 4003, pp. 3–10, 2006.

$(s = 1, 2, \ldots, n)$. A connection of $s$'th flow uses $b_s$ bandwidth units for the time of connection. To take into account the possibility of soft blocking we suppose that a demand of $s$'th flow $(s = 1, 2, \ldots, n)$ arriving when exactly $i$ bandwidth units are occupied is blocked with probability $\varphi_{s,i}$. It is obvious that $\varphi_{s,i} = 1$ if $i = k - b_s + 1, k - b_s + 2, \ldots, k$ for all $s = 1, 2, \ldots, n$. We shall assume that the holding times all are exponentially distributed with the same mean value chosen as time unit.

By proper choosing of probabilities $\varphi_{s,i}$ we can consider the variety of situation in servicing demands for bandwidth. For example by taking $\varphi_{s,i} = 0$ if $i = 0, 1, \ldots, \theta_s$ and $\varphi_{s,i} = 1$ when $i = \theta_s + 1, \theta_s + 2, \ldots, k$ for all $s = 1, 2, \ldots, n$ we construct the model of multiservice link with trunk reservation [4].

Let $i_s(t)$ denote the number of connections of the $s$'th flow served at time $t$. The model is described by an $n$-dimensional Markovian process of the type $r(t) = \{i_1(t), i_2(t), \ldots, i_n(t)\}$ with state space $S$ consisting of vectors $(i_1, \ldots, i_n)$, where $i_s$ is the number of connections of the $s$'th flow being served by the link under stationary conditions. The process $r(t)$ is defined on the finite state space $S$ which is a subset of the set $\Omega$. The set of states $\Omega$ is defined as follows:

$$(i_1, \ldots, i_n) \in \Omega, \ i_s \geq 0, \ s = 1, \ldots, n, \ \sum_{s=1}^{n} i_s b_s \leq k.$$

Some of the states $(i_1, \ldots, i_n) \in \Omega$ are excluded from $S$ by choice of probabilities of soft blocking $\varphi_{s,i}$.

Let us by $P(i_1, \ldots, i_n)$ denote the unnormalised values of stationary probabilities of $r(t)$. To simplify the subsequent algebraic transforms we suppose that

$$P(i_1, \ldots, i_n) = 0, \quad (i_1, \ldots, i_n) \in \Omega \setminus S.$$

After normalisation, $p(i_1, \ldots, i_n)$ denotes the mean proportion of time when exactly $\{i_1, \ldots, i_n\}$ connections are established. Let us use small characters to denote the normalised values of state probabilities and performance measures.

The process of transmission of $s$'th flow is described by blocking probabilities $\pi_s, s = 1, \ldots, n$. Their formal definition through values of state probabilities are as follows (here and further on, summations are for all states $(i_1, \ldots, i_n) \in \Omega$ satisfying the formulated condition, and for state $(i_1, \ldots, i_n)$ $i$ denotes the total number of occupied bandwidth units $i = i_1 b_1 + \cdots + i_n b_n$):

$$\pi_s = \sum_{(i_1, \ldots, i_n) \in \Omega} p(i_1, \ldots, i_n) \, \varphi_{s,i}. \tag{1}$$

## 3 Exact and Approximate Estimation of Performance Measures

Because the process $r(t)$ doesn't have any specific properties the exact values of $\pi_s, s = 1, \ldots, n$ can be found only by solving the system of state equations by some numerical method. The system of state equations looks as follows

$$P(i_1, i_2, \ldots, i_n) \left\{ \sum_{s=1}^{n} (\lambda_s (1 - \varphi_{s,i}) + i_s \right\} = \tag{2}$$

$$\sum_{s=1}^{n} P(i_1, \ldots, i_s - 1, \ldots, i_n) \lambda_s (1 - \varphi_{s,i-b_s}) I(i_s > 0)$$

$$+ \sum_{s=1}^{n} P(i_1, \ldots, i_s + 1, \ldots, i_n) (i_s + 1) I(i + b_s \leq k),$$

where $(i_1, \ldots, i_n) \in \Omega$, function $I(\cdot)$ equals one if the condition formulated in brackets is true, and otherwise equals zero. The normalization condition is valid for $P(i_1, \ldots, i_n)$:

$$\sum_{(i_1, \ldots, i_n) \in \Omega} P(i_1, \ldots, i_n) = 1.$$

The system of state equations (2) can be solved by Gauss-Seidel iteration algorithm for 2–3 input flows of demands. We will use this opportunity to study the error of approximate evaluation of the considered model. The only way to find the performance measures of large models is to use approximate algorithms. Let us consider the recursive scheme that exploits the fact that the performance measures (1) can be found if we for process $r(t)$ know probabilities $p(i)$ of being in the state where exactly $i$ bandwidth units are occupied:

$$p(i) = \sum_{i_1 b_1 + \cdots + i_n b_n = i} p(i_1, \ldots, i_n).$$

The corresponding formulas are as follows:

$$\pi_s = \sum_{i=0}^{k} p(i) \varphi_{s,i}, \quad s = 1, 2, \ldots, n. \tag{3}$$

For approximate evaluation of unnormalised values of $P(i)$ it is proposed to use the following recurrence:

$$P(i) = \begin{cases} 0 & \text{if } i < 0 \\ a & \text{if } i = 0 \\ \frac{1}{i} \sum_{s=1}^{n} \lambda_s b_s (1 - \varphi_{s,i-b_s}) P(i - b_s) & \text{if } i = 1, 2, \ldots, k \end{cases} \tag{4}$$

This is exact for special cases when blocking probabilities are chosen so that the process is reversible [2]. Let us study the numerical accuracy of the suggested approach. In the Tables 1–3 the results of exact and approximate estimation of $\pi_s$ for three sets of the probabilities of soft blocking are presented.

– The first set is defined by

$$\varphi_{s,i} = \frac{i^3}{k^3}, \quad i = 0, 1, \ldots, k - b_s,$$
$$\varphi_{s,i} = 1, \quad i = k - b_s + 1, k - b_s + 2, \ldots, k, \quad s = 1, 2, 3.$$

- The second set is defined by

$$\varphi_{s,i} = \frac{i^4}{k^4}, \qquad i = 0, 1, \ldots, k - b_s,$$

$$\varphi_{s,i} = 1, \qquad i = k - b_s + 1, k - b_s + 2, \ldots, k, \qquad s = 1, 2, 3.$$

- The third set is defined by

$$\varphi_{s,i} = \frac{i^5}{k^5}, \qquad i = 0, 1, \ldots, k - b_s,$$

$$\varphi_{s,i} = 1, \qquad i = k - b_s + 1, k - b_s + 2, \ldots, k, \qquad s = 1, 2, 3.$$

The forms of the corresponding curves are shown in Fig. 1.

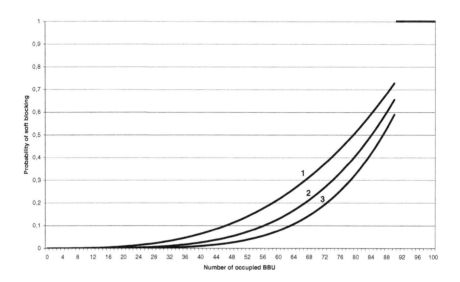

**Fig. 1.** The probabilities of soft blocking for the 3rd flow of demands as a function of the number of occupied basic bandwidth units (BBU). The first set is defined by $\varphi_{3,i} = \frac{i^3}{k^3}$, $i = 0, 1, \ldots, k - b_3$ (curve 1). The second set is defined by $\varphi_{3,i} = \frac{i^4}{k^4}$, $i = 0, 1, \ldots, k - b_3$ (curve 2). The third set is defined by $\varphi_{3,i} = \frac{i^5}{k^5}$, $i = 0, 1, \ldots, k - b_3$ (curve 3).

The model input parameters are as follows: $k = 100$, $n = 3$, $b_1 = 3$, $b_2 = 5$, $b_3 = 10$, $\lambda_s = \frac{\rho k}{n b_s}$, $s = 1, 2, 3$. The value of $\rho$ (offered load per basic bandwidth unit) is varied from 0.1 to 1.

The exact values of blocking probabilities $\pi_s$, $s = 1, 2, 3$ are obtained by solving the system of state equations (2). The approximate values of blocking probabilities $\pi_s$, $s = 1, 2, 3$ are obtained using the recursions (4). Together with exact and approximate values of $\pi_s$ the relative errors of estimation $\pi_s$ are presented in Tables 1–3.

The numerical results presented show that the approximate method has high accuracy. For all numerical results presented the relative error is less than 0.002. It

**Table 1.** The results of exact and approximate estimation of $\pi_s$ for probability of soft blocking defined by $\varphi_{s,i} = \frac{i^3}{k^3}$, $i = 0, 1, \ldots, k - b_s$ and $\varphi_{s,i} = 1$, $i = k - b_s + 1, k - b_s + 2, \ldots, k$, $s = 1, 2, 3$

| | $\pi_1$ | | | $\pi_2$ | | | $\pi_3$ | | |
|---|---|---|---|---|---|---|---|---|---|
| $\rho$ | Exact | Approx. | Error | Exact | Approx. | Error | Exact | Approx. | Error |
| 0.1 | 0.003198 | 0.003198 | 0.0001 | 0.003198 | 0.003198 | 0.0001 | 0.003198 | 0.003198 | 0.0001 |
| 0.2 | 0.015120 | 0.015123 | 0.0002 | 0.015120 | 0.015123 | 0.0002 | 0.015120 | 0.015123 | 0.0002 |
| 0.3 | 0.038432 | 0.038439 | 0.0002 | 0.038432 | 0.038439 | 0.0002 | 0.038433 | 0.038441 | 0.0002 |
| 0.4 | 0.072386 | 0.072395 | 0.0001 | 0.072387 | 0.072396 | 0.0001 | 0.072403 | 0.072415 | 0.0002 |
| 0.5 | 0.113843 | 0.113848 | 0.0000 | 0.113849 | 0.113855 | 0.0001 | 0.113937 | 0.113955 | 0.0002 |
| 0.6 | 0.159089 | 0.159083 | 0.0000 | 0.159109 | 0.159108 | 0.0000 | 0.159407 | 0.159436 | 0.0002 |
| 0.7 | 0.205057 | 0.205037 | 0.0001 | 0.205111 | 0.205102 | 0.0000 | 0.205857 | 0.205902 | 0.0002 |
| 0.8 | 0.249693 | 0.249659 | 0.0001 | 0.249811 | 0.249796 | 0.0001 | 0.251325 | 0.251384 | 0.0002 |
| 0.9 | 0.291834 | 0.291789 | 0.0002 | 0.292051 | 0.292038 | 0.0000 | 0.294699 | 0.294760 | 0.0002 |
| 1.0 | 0.330939 | 0.330892 | 0.0001 | 0.331298 | 0.331296 | 0.0000 | 0.335450 | 0.335493 | 0.0001 |

allows us to use the suggested approximate procedure for solving the problem of dimensioning of the considered model of multiservice link taking the probability of soft blocking into account.

# 4   Optimized Dimensioning of Multiservice Link with Soft Blocking

A main problem that often has to be solved by operators is to determine the volume of telecommunication resources that is sufficient for servicing the given input flows of demands with prescribed characteristics of QoS. Adequacy of volume is determined by comparison of suitably chosen performance measure with prescribed level of system functioning. A traditional dimensioning problem related with multiservice line planning is formulated as follows:

For given input flow, find a minimum value of bandwidth $k$ satisfying the inequality:

$$B \leq \pi. \tag{5}$$

Here $B$ is the performance measure used for dimensioning, and $\pi$ is the prescribed norm of servicing for bandwidth requests. The numerical complexity of this approach is estimated by the numerical complexity of finding the performance measure for given input parameters multiplied by the number of searching. Let us denote this dimensioning scheme described above as traditional.

The implementation of the traditional procedure has at least two negative aspects. The first is the numerical instability of using recursions of type (4) especially for large number of service units. Another negative aspect is the increase of computational efforts. To solve the formulated dimensioning problem we need to perform a run of (4) for each value of $k$ serving as a candidate for desired solution. Each time we need to start calculation from $i = 0$.

**Table 2.** The results of exact and approximate estimation of $\pi_s$ for probability of soft blocking defined by $\varphi_{s,i} = \frac{i^4}{k^4}$, $i = 0, 1, \ldots, k - b_s$ and $\varphi_{s,i} = 1$, $i = k - b_s + 1, k - b_s + 2, \ldots, k$, $s = 1, 2, 3$

| | $\pi_1$ | | | $\pi_2$ | | | $\pi_3$ | | |
|---|---|---|---|---|---|---|---|---|---|
| $\rho$ | Exact | Approx. | Error | Exact | Approx. | Error | Exact | Approx. | Error |
| 0.1 | 0.000779 | 0.000779 | 0.0002 | 0.000779 | 0.000779 | 0.0002 | 0.000779 | 0.000779 | 0.0002 |
| 0.2 | 0.005470 | 0.005473 | 0.0004 | 0.005470 | 0.005473 | 0.0004 | 0.005471 | 0.005473 | 0.0004 |
| 0.3 | 0.018224 | 0.018234 | 0.0005 | 0.018225 | 0.018234 | 0.0005 | 0.018229 | 0.018239 | 0.0005 |
| 0.4 | 0.041794 | 0.041814 | 0.0005 | 0.041797 | 0.041818 | 0.0005 | 0.041849 | 0.041874 | 0.0006 |
| 0.5 | 0.075790 | 0.075813 | 0.0003 | 0.075809 | 0.075836 | 0.0004 | 0.076077 | 0.076124 | 0.0006 |
| 0.6 | 0.117220 | 0.117233 | 0.0001 | 0.117288 | 0.117313 | 0.0002 | 0.118162 | 0.118231 | 0.0006 |
| 0.7 | 0.162326 | 0.162322 | 0.0000 | 0.162501 | 0.162521 | 0.0001 | 0.164587 | 0.164664 | 0.0005 |
| 0.8 | 0.207965 | 0.207947 | 0.0001 | 0.208326 | 0.208348 | 0.0001 | 0.212350 | 0.212405 | 0.0003 |
| 0.9 | 0.252053 | 0.252037 | 0.0001 | 0.252692 | 0.252731 | 0.0002 | 0.259379 | 0.259368 | 0.0000 |
| 1.0 | 0.293448 | 0.293456 | 0.0000 | 0.294455 | 0.294532 | 0.0003 | 0.304440 | 0.304315 | 0.0004 |

We suggest an optimized procedure of dimensioning a multiservice link with soft blocking based on results of [5]. Let us suppose that there exists a $b > 0$ so that for all $s = 1, 2, \ldots, n$, $\varphi_{s,i} = 0$ for $i = 0, 1, \ldots, k - b$. The value of blocking probability (3) for such a case will be calculated according to the expression

$$\pi_s = \sum_{i=k-b+1}^{k} p(i)\,\varphi_{s,i}, \quad s = 1, 2, \ldots, n. \tag{6}$$

Using the general framework formulated in [5] we obtain a one–run algorithm that at each step gives the normalised values of state probabilities which are necessary for solving the problem of dimensioning based on the form of function $B$ specified by (6). We will indicate when necessary the number of available service units by lower index for the corresponding set of probabilities. The one–run three–step algorithm that at each step gives the normalised values of the model's characteristics that are necessary for solving the dimensioning problem based on performance measure (6) looks as follows:

- *Step 1:* Let $p_0(0) = 1$.
- *Step 2:* Let us suppose that $b > \max_{0 \leq s \leq n}(b_s)$. For fixed $k = 1, 2, \ldots$, find normalised value of $p_k(i)$:

$$p_k(k) = \frac{\frac{1}{k} \sum_{s=1}^{n} \lambda_s\, b_s\, (1 - \varphi_{s,k-b_s})\, p_{k-1}(k - b_s)}{1 + \frac{1}{k} \sum_{s=1}^{n} \lambda_s\, b_s\, (1 - \varphi_{s,k-b_s})\, p_{k-1}(k - b_s)}; \tag{7}$$

$$p_k(i) = \frac{p_{k-1}(i)}{1 + \frac{1}{k} \sum_{s=1}^{n} \lambda_s\, b_s\, (1 - \varphi_{s,k-b_s})\, p_{k-1}(k - b_s)},$$

**Table 3.** The results of exact and approximate estimation of $\pi_s$ for probability of soft blocking defined by $\varphi_{s,i} = \frac{i^5}{k^5}$, $i = 0, 1, \ldots, k - b_s$ and $\varphi_{s,i} = 1$, $i = k - b_s + 1, k - b_s + 2, \ldots, k$, $s = 1, 2, 3$

| | $\pi_1$ | | | $\pi_2$ | | | $\pi_3$ | | |
|---|---|---|---|---|---|---|---|---|---|
| $\rho$ | Exact | Approx. | Error | Exact | Approx. | Error | Exact | Approx. | Error |
| 0.1 | 0.000217 | 0.000217 | 0.0001 | 0.000217 | 0.000217 | 0.0001 | 0.000217 | 0.000217 | 0.0001 |
| 0.2 | 0.002190 | 0.002192 | 0.0005 | 0.002190 | 0.002192 | 0.0005 | 0.002191 | 0.002192 | 0.0005 |
| 0.3 | 0.009356 | 0.009364 | 0.0009 | 0.009357 | 0.009365 | 0.0009 | 0.009366 | 0.009375 | 0.0010 |
| 0.4 | 0.025760 | 0.025784 | 0.0009 | 0.025768 | 0.025793 | 0.0010 | 0.025875 | 0.025907 | 0.0013 |
| 0.5 | 0.053337 | 0.053375 | 0.0007 | 0.053381 | 0.053425 | 0.0008 | 0.053935 | 0.054004 | 0.0013 |
| 0.6 | 0.090577 | 0.090615 | 0.0004 | 0.090731 | 0.090788 | 0.0006 | 0.092498 | 0.092591 | 0.0010 |
| 0.7 | 0.133811 | 0.133840 | 0.0002 | 0.134197 | 0.134262 | 0.0005 | 0.138285 | 0.138356 | 0.0005 |
| 0.8 | 0.179235 | 0.179267 | 0.0002 | 0.180006 | 0.180092 | 0.0005 | 0.187627 | 0.187603 | 0.0001 |
| 0.9 | 0.224035 | 0.224100 | 0.0003 | 0.225352 | 0.225482 | 0.0006 | 0.237586 | 0.237388 | 0.0008 |
| 1.0 | 0.266544 | 0.266679 | 0.0005 | 0.268556 | 0.268757 | 0.0007 | 0.286217 | 0.285775 | 0.0015 |

$$\text{when}\quad i = k-1, \; k-2, \; \ldots, \max(k - b + 1, 0)$$

- *Step 3:* Here we calculate the performance measures defined by (6), check the dimensioning criteria (for example, $B = \max_{1 \leq s \leq n} \pi_s$) and either stop or continue the process of estimating the number of service units needed from step 2 by increasing number of channels by one.

When implementing this version of the recurrence algorithm we need to keep a vector of size $O\{b\}$ in computer memory. Computational efforts are estimated by $O\{(n + b)k\}$.

## 5   Conclusion

The model of a multiservice link taking the possibility of soft blocking into account is considered. An approximate algorithm for estimation of main performance measures is constructed. The error of estimation is numerically studied for different types of soft blocking. The optimal procedure of dimensioning is suggested. The suggested approach is numerically stable because it deals with normalised values of global state probabilities used for estimation of main stationary performance measures. The model is applicable to CDMA traffic models and systems with trunk reservation. In future work the model will be extended to Engset and Pascal traffic models.

## References

1. Iversen, V.B. & Benetis, V. & Ha, N.T. & Stepanov, S.N.: Evaluation of Multiservice CDMA Networks with Soft Blocking. ITC 16th Specialist Seminar on Performance Evaluation of Mobile and Wireless Sytems, Antwerp, Belgium, August 31–September 2, 2004. 8 pp.

2. Iversen, V.: Modelling restricted accessibility for wireless multi-service systems. Euro-NGI workshop on Wireless and Mobility, Lake Como, Italy, July 2005. Springer Lecture Notes on Computer Science, LNCS 3883, pp. 93–102.
3. Hui, J.Y.: Resource Allocation for Broadband Networks. IEEE Journal on Selected Areas in Communications, Vol. 6 (1988) : 9, 1598–1608.
4. *Broadband network traffic. Performance evaluation and design of broadband multiservice networks.* Final report of action COST 242. J. Roberts & U. Mocci & J. Virtamo (editors). Springer, 1996. 584 pp.
5. Iversen, V.B. & Stepanov, S.N.: The unified approach for teletraffic models to convert recursions for global state probabilities into stable form. ITC19, 19th International Teletraffic Congress, August 29 – September 2, 2005. Beijing, China. Proceedings pp. 1559–1570.

# On the Packet Loss Process at a Network Node

Dieter Fiems and Herwig Bruneel

SMACS Research Group, Department TELIN (IR07), Ghent University
St-Pietersnieuwstraat 41, 9000 Gent, Belgium
{df, hb}@telin.UGent.be

**Abstract.** It is a well known fact that the packet loss ratio is an important but insufficient measure to assess the influence of packet loss on user perceived quality of service in telecommunication networks. In this paper we therefore assess other loss process characteristics of finite capacity Markov-modulated $M/M/1$-type buffers. Combining a probability generating functions approach with matrix techniques, we derive an expression for the joint probability generating function of the time and the number of accepted packets – packets that are not lost – between packet losses. We then illustrate our approach by means of some numerical examples.

## 1 Introduction

In wired communication networks, packet loss is almost solely caused by buffer overflow in intermediate network nodes. Therefore, the packet loss ratio – the fraction of packets that get lost – is well studied in queueing literature. See a.o. Kim and Schroff [1] and Pihlsgard [2]. However, in the case of multimedia communications, the packet loss ratio is not the only loss characteristic of interest. Multimedia applications can typically tolerate a reasonable amount of packet loss, as long as there are no bursts of packet loss. For example, the perceived visual quality of variable bit rate video streaming heavily depends on the burstiness of the packet loss process [3]. Forward error correction (FEC) techniques may further mitigate the effects of packet loss and increase the tolerable packet loss ratio but again require that packet loss is well spread in time [4].

Some authors have studied other characteristics of the loss process in a finite buffer. Loss characteristics under investigation include a.o. the probability to have a certain $k$ packets that are lost in a block of $n$ consecutive packets, [3,5,6,7,8,9], the spectrum of the packet loss process [10] and the conditional loss probability [11]. Although loss process characteristics have been studied predominantly for the $M/M/1/N$ queueing system [5,6,7,8], some authors also allow correlation in the arrival process [3,10,11].

This contribution addresses the evaluation of packet loss characteristics for discrete-time and continuous-time Markov-modulated $M/M/1$-type queueing systems. In particular, the packet loss characteristics under investigation are the time and the number of packet arrivals that are not lost between consecutive packet losses. By combining a generating functions approach with matrix techniques, we obtain expressions for the various moments of these random variables.

Y. Koucheryavy, J. Harju, and V.B. Iversen (Eds.): NEW2AN 2006, LNCS 4003, pp. 11–20, 2006.

The remainder of this contribution is organised as follows. In the next section, the discrete-time queueing model is introduced and analysed. In Section 3, we then map the equivalent continuous-time system to its discrete-time counterpart by means of uniformisation. Some numerical results then illustrate our approach in section 4. Finally, conclusions are drawn in Section 5.

## 2    Discrete-Time Buffer Model

We first consider a synchronised or discrete-time queueing system. That is, we assume that time is divided into fixed length intervals called slots and that both arrivals and departures are synchronised with respect to slot boundaries. During the consecutive slots, packets arrive at the system, are stored in a finite capacity buffer and are transmitted on a first-in-first-out basis. The buffer can store up to $N$ packets simultaneously (including the packet that is being transmitted) and at most one packet can enter/leave the buffer at a slot boundary. Arrivals and departures are scheduled according to the "late arrival, arrivals first" waiting-room management policy. I.e., both arrivals and departures are scheduled immediately before slot boundaries and arrivals precede departures as depicted in Figure 1. Therefore, it is possible that an arriving packet is rejected at a slot boundary where another packet leaves the queue.

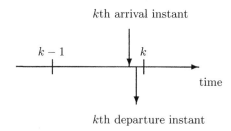

**Fig. 1.** Late-arrival, arrivals first waiting room management

The queueing system under consideration operates in a Markovian environment in the sense that arrival and departure probabilities at the consecutive slot boundaries depend on the state of an underlying finite-state Markov chain with state-space $\{1, 2, \ldots, M\}$. We characterise the arrival and departure processes by means of the $M \times M$ transition matrices $A(n, m) = [a_{ij}(n, m)]$ and $\tilde{A}(n) = \tilde{a}_{ij}(n)$ $(n, m \in \{0, 1\})$. These transition probabilities are defined as,

$$
\begin{aligned}
a_{ij}(n, m) &= \Pr[A_{k+1} = n \wedge D_{k+1} = m \wedge Q_{k+1} = j | Q_k = i \wedge U_k > 0], \\
\tilde{a}_{ij}(n) &= \Pr[A_{k+1} = n \wedge Q_{k+1} = j | Q_k = i \wedge U_k = 0].
\end{aligned}
\tag{1}
$$

Here $k$ is a random slot boundary and $A_{k+1}$ and $D_{k+1}$ denote the number of arrivals and departures at the $(k + 1)$th slot boundary. $Q_k$ denotes the state of the Markovian environment at the $k$th slot boundary and $U_k$ denotes the queue content – the number of packets in the system – at this slot boundary. As such, the transition matrices $A(n,m)$ and $\tilde{A}(n)$ correspond to the case that there are $n$ arrivals and $m$ departures during the transition when the queue is non-empty and the case that there are $n$ arrivals during the transition when the queue is empty respectively.

Notice that the discrete-time queueing model under investigation extends a.o. the discrete-time $PH/PH/1/N$ queueing system and the $DMAP/D/1/N$ system with service interruptions.

## 2.1 Steady-State Analysis

Consider a random (tagged) slot boundary, say slot $k$, and let $U_k$ and $Q_k$ denote the queue content and the state of the Markovian environment at this boundary. Further, let $T_k$ denote the number of slots since the last slot boundary where a packet got lost and let $X_k$ denote the number of packet arrivals since the last slot boundary where a packet got lost.

We may now relate the triple $(U_{k+1}, T_{k+1}, X_{k+1})$ to the triple $(U_k, T_k, X_k)$. When the queue is not full at the $k$th slot boundary ($0 \le U_k < N$) or when the queue is full at this boundary but there are no arrivals at the $(k+1)$th boundary ($U_k = N, A_{k+1} = 0$), no packets are lost. As such, we find,

$$
\begin{aligned}
U_{k+1} &= (U_k - D_{k+1})^+ + A_{k+1}, \\
T_{k+1} &= T_k + 1, \\
X_{k+1} &= X_k + A_{k+1}.
\end{aligned}
\tag{2}
$$

Here $(\cdot)^+$ is the usual shorthand notation for $\max(0, \cdot)$. There is packet loss at the $(k+1)$th slot boundary if there is an arrival at the $(k+1)$th slot boundary and the buffer is full at the $k$th boundary ($U_k = N, A_{k+1} = 1$). As such, we find,

$$
\begin{aligned}
U_{k+1} &= U_k - D_{k+1} \\
T_{k+1} &= 0, \\
X_{k+1} &= 0.
\end{aligned}
\tag{3}
$$

Let $P(x, z; n, i)$ denote the partial joint probability generating function of $T_k$ and $X_k$ at a random slot boundary $k$, given the queue content $U_k = n$ and the state of the underlying Markov process $Q_k = i$,

$$
P(x, z; n, i) = \mathrm{E}[x^{T_k} z^{X_k} | U_k = n, Q_k = i] \Pr[U_k = n, Q_k = i].
\tag{4}
$$

Further, let $P(x, z; n)$ denote the row vector with elements $P(x, z; n, i)$, $i = 1 \dots M$. Conditioning on the queue content and the state of the underlying Markov process at a random slot boundary and on the number of arrivals and departures at the following slot boundary then leads to,

$$
\begin{aligned}
P(x, z; 0) &= P(x, z; 0)x\tilde{A}(0) + P(x, z; 1)xA(0, 1), \\
P(x, z; 1) &= P(x, z; 0)xz\tilde{A}(1) + P(x, z; 1)x\left[A(0, 0) + zA(1, 1)\right] \\
&\quad + P(x, z; 2)xA(0, 1),
\end{aligned}
$$

$$\begin{aligned}
P(x,z;n) &= P(x,z;n-1)xzA(1,0) + P(x,z;n)x\left[A(0,0)+zA(1,1)\right] \\
&\quad + P(x,z;n+1)xA(0,1)\,, \\
P(x,z;N-1) &= P(x,z;N-2)xzA(1,0) + P(x,z;N)xA(0,1) \\
&\quad + P(x,z;N-1)x\left[A(0,0)+zA(1,1)\right] + P(1,1;N)A(1,1)\,, \\
P(x,z;N) &= P(x,z;N-1)xzA(1,0) + P(x,z;N)xA(0,0) \\
&\quad + P(1,1;N)A(1,0)\,,
\end{aligned}$$
(5)

for $n = 2,\ldots N-2$.

Let $P(x,z)$ denote the row vector with elements $P(x,z;n)$, $n = 0,\ldots,N$. The former set of equations then leads to the following matrix equation,

$$P(x,z) = P(x,z)(\Theta_1 x + \Theta_2 xz) + P(1,1)\Theta_3\,,$$
(6)

where the matrices $\Theta_1$, $\Theta_2$ and $\Theta_3$ are of size $M(N+1) \times M(N+1)$ and have the following block representations,

$$\Theta_1 = \begin{bmatrix}
\tilde{A}(0) & 0 & 0 & \cdots & 0 & 0 \\
A(0,1) & A(0,0) & 0 & \cdots & 0 & 0 \\
0 & A(0,1) & A(0,0) & \cdots & 0 & 0 \\
\vdots & \vdots & \vdots & \ddots & \vdots & \vdots \\
0 & 0 & 0 & \cdots & A(0,1) & A(0,0)
\end{bmatrix}\,,$$
(7)

$$\Theta_2 = \begin{bmatrix}
0 & \tilde{A}(1) & 0 & 0 & \cdots & 0 & 0 \\
0 & A(1,1) & A(1,0) & 0 & \cdots & 0 & 0 \\
0 & 0 & A(1,1) & A(1,0) & \cdots & 0 & 0 \\
\vdots & \vdots & \vdots & \vdots & \ddots & \vdots & \vdots \\
0 & 0 & 0 & 0 & \cdots & A(1,1) & A(1,0) \\
0 & 0 & 0 & 0 & \cdots & 0 & 0
\end{bmatrix}\,,$$
(8)

$$\Theta_3 = \begin{bmatrix}
0 \cdots 0 & 0 & 0 \\
\vdots \ddots \vdots & \vdots & \vdots \\
0 \cdots 0 & A(1,1) & A(1,0)
\end{bmatrix}\,.$$
(9)

Plugging in $x = z = 1$ in equation (6) then leads to $\pi = P(1,1) = \pi\sum_i \Theta_i$. Combining the latter equation with the normalisation condition $\pi e^T = 1$ allows us to solve for the vector $\pi$. Here $e^T$ denotes a column vector with all elements equal to 1. Notice that the equation for $\pi$ is the equation of a finite quasi birth death (QBD) process and can therefore be solved efficiently. See, a.o. Latouche and Ramaswami [12, chapter 10].

Once the vector $\pi$ is known, the vector $P(x,z)$ follows from the equation (6),

$$P(x,z) = P(1,1)\Theta_3 \left(I - \Theta_1 x - \Theta_2 xz\right)^{-1}\,.$$
(10)

For large $N$ and $M$, it is however not trivial to perform the matrix inversion in the former equation symbolically. We may however use the moment generating property of probability generating functions to obtain a similar set of equations for the various moments of $X$ and $T$ which one can easily solve numerically.

For example, let $\mu_Y^{(r)}(n, i)$ $(n = 0 \ldots N, i = 1 \ldots M, Y \in \{X, T\})$ denote the $r$th moment of the time $T$ or the number of arrivals $X$ since the last loss at a random slot boundary $k$ while the queue content equals $n$ at this boundary and while the underlying Markov chain is in state $i$,

$$\mu_Y^{(r)}(n, i) = \mathrm{E}[(Y_k)^r | U_k = n, Q_k = i] \Pr[U_k = n, Q_k = i]. \tag{11}$$

Further, let $\mu_Y^{(r)}(n)$ denote the row vector with elements $\mu_Y^{(r)}(n, i)$, $i = 1 \ldots M$ and let $\mu_Y^{(r)}$ denote the row vector with elements $\mu_Y^{(r)}(n)$. In view of expression (6), we then find,

$$\mu_T^{(1)} = P(1, 1)(\Theta_1 + \Theta_2)(I - \Theta_1 - \Theta_2)^{-1}, \tag{12}$$

$$\mu_X^{(1)} = P(1, 1)\Theta_2(I - \Theta_1 - \Theta_2)^{-1}, \tag{13}$$

$$\mu_T^{(2)} = 2\mu_T^{(1)}(\Theta_1 + \Theta_2)(I - \Theta_1 - \Theta_2)^{-1} - \mu_T^{(1)}, \tag{14}$$

$$\mu_X^{(2)} = 2\mu_X^{(1)}\Theta_2(I - \Theta_1 - \Theta_2)^{-1} - \mu_X^{(1)}. \tag{15}$$

Similar expressions may be obtained for higher moments of $T_k$ and $X_k$ as well as for the covariance of $T_k$ and $X_k$.

## 2.2 Loss Characteristics

Let $\tilde{T}$ and $\tilde{X}$ denote the number of slots between consecutive losses and the number of arrivals between consecutive losses (excluding lost packets) respectively and let $Q(x, z)$ denote the joint probability generating function of these random variables,

$$Q(x, z) = \mathrm{E}\left[x^{\tilde{T}} z^{\tilde{X}}\right]. \tag{16}$$

Conditioning on the state of the underlying Markov chain at the boundary where the packet is lost and at the boundary preceding this boundary, on the queue content at the slot boundary preceding the packet loss and on the number of arrivals and departures at the boundary where the packet is lost, then leads to,

$$Q(x, z) = \frac{xP(x, z; N)[A(1, 0) + A(1, 1)]e^T}{\pi(N)[A(1, 0) + A(1, 1)]e^T}. \tag{17}$$

By means of the moment generating property of probability generating functions, we may retrieve expressions for the various moments of $\tilde{T}$ and $\tilde{X}$ in terms of the moments of $T_k$ and $X_k$. For example, the mean time $\mu_{\tilde{T}}^{(1)}$ and the mean number of packet arrivals $\mu_{\tilde{X}}^{(1)}$ between consecutive losses are given by,

$$\mu_{\tilde{T}}^{(1)} = 1 + \frac{\mu_T^{(1)}(N)[A(1, 0) + A(1, 1)]e^T}{\pi(N)[A(1, 0) + A(1, 1)]e^T}, \tag{18}$$

$$\mu_{\tilde{X}}^{(1)} = \frac{\mu_X^{(1)}(N)\,[A(1,0)+A(1,1)]\,e^T}{\pi(N)\,[A(1,0)+A(1,1)]\,e^T}\,, \tag{19}$$

respectively. Here $\mu_X^{(1)}(N) = \lim_{z\to 1}\frac{d}{dz}P(1,z;N)$ is the mean number of arrivals since the last packet loss while the buffer is full. Finally, notice that $\mu_{\tilde{X}}$ and $\mu_{\tilde{T}}$ relate to the packet loss ratio PLR as follows,

$$\text{PLR} = \frac{1}{1+\mu_{\tilde{X}}^{(1)}} = \frac{1}{\rho\mu_{\tilde{T}}^{(1)}}\,. \tag{20}$$

Here $\rho$ denotes the packet arrival intensity.

## 3    Continuous-Time Buffer Model

We now introduce the equivalent continuous-time buffer model and relate this model to its discrete-time equivalent by means of uniformisation [12, Section 2.8]. The continuous-time model under investigation is the $M/M/1$ queue in a random environment, see a.o. Neuts [13, Chapter 6] and Latouche and Ramaswami [12, Section 1.2].

As before, let $N$ denote the maximal number of packets that can be stored in the buffer, including the one being transmitted. The queueing system operates in a Markovian environment with state space $\{1,2,\ldots,M\}$ and generator matrix $\mathcal{M} = [m_{ij}]_{i,j=1\ldots M}$. Here $m_{ij}$ $(i \neq j)$ denotes the transition rate from state $i$ to state $j$. Further, whenever the environment is in state $i$ and the queue is non-empty, the arrival and departure rates equal $\lambda_i$ and $\mu_i$ respectively $(i = 1,\ldots,M)$. Similarly, let $\tilde{\lambda}_i$ denote the arrival rate when the queue is empty and the environment is in state $i$ $(i = 1,\ldots,M)$.

### 3.1    Uniformisation

Let $c_i = \sum_{i\neq j} m_{ij} + \lambda_i + \mu_i$ and $\tilde{c}_i = \sum_{i\neq j} m_{ij} + \tilde{\lambda}_i$ denote the rate at which an event occurs when the environment is in state $i$ and when the buffer is non-empty and empty respectively. Here an event is either a state change, an arrival or a departure. Further, choose $c$ such that $c_i, \tilde{c}_i \leq c < \infty$.

We now take a Poisson process $Y(t)$ with rate $c$ and let $t_0, t_1, \ldots$ denote the Poisson event times. Further we consider an independent marked Markov chain $\{(Q_n, A_n, D_n), n \geq 0\}$ with transition matrices $A(m,n)$ and $\tilde{A}(m)$ (as defined in (1)) given by,

$$
\begin{aligned}
A(0,0) &= \frac{1}{c}(\mathcal{M} - \text{diag}(\lambda_i) - \text{diag}(\mu_i)) + I\,, & A(0,1) &= \frac{1}{c}\,\text{diag}(\mu_i)\,, \\
A(1,0) &= \frac{1}{c}\,\text{diag}(\lambda_i)\,, & A(1,1) &= 0\,, \\
\tilde{A}(0) &= \frac{1}{c}(\mathcal{M} - \text{diag}(\tilde{\lambda}_i)) + I\,, & \tilde{A}(1) &= \frac{1}{c}\,\text{diag}(\tilde{\lambda}_i)\,. \tag{21}
\end{aligned}
$$

It is then easy to verify that the process that evolves on the Poisson event times according to the former transition matrices has the same generator matrices as the original continuous-time process.

The former construction now shows that the continuous-time queueing system can be modelled by means of a discrete-time equivalent system. The slot lengths however do no longer have a fixed length. The slot lengths are now exponentially distributed with mean $1/c$.

## 3.2   Loss Characteristics

Let $\hat{T}$ and $\hat{X}$ denote the mean time between consecutive packet losses and the mean number of packet arrivals between these losses for the continuous time system. Further, let $\tilde{T}$ and $\tilde{X}$ denote the mean time – expressed as a number of slots – between consecutive packet losses and the mean number of packet arrivals between these losses for the discrete-time system with transition matrices given in (21). We then easily find,

$$\hat{T} = \sum_{i=1}^{\tilde{T}} E_i, \quad \hat{X} = \tilde{X}. \tag{22}$$

Here $E_i$ denotes an independent and identically exponentially distributed series of random variables with mean $1/c$. Let $\hat{Q}(s, z)$ denote the joint transform of $\hat{T}$ and $\hat{X}$. In view of the former expressions we find,

$$\hat{Q}(s, z) = \mathrm{E}\left[e^{-s\hat{T}}z^{\hat{X}}\right] = \mathrm{E}\left[\left(\frac{c}{c+s}\right)^{\tilde{T}}z^{\hat{X}}\right] = Q\left(\frac{c}{c+s}, z\right). \tag{23}$$

Here $Q(x, z)$ is given by equation (17) under the assumption that the arrival and transmission processes are characterised by the matrices (21). The moment generating property of characteristic functions and probability generating functions then again lead to expressions for the various moments of $\hat{X}$ and $\hat{T}$.

## 4   Numerical Example

We now illustrate our approach by means of a numerical example.

We consider a discrete-time queueing system. Let the arrivals be governed by a discrete Markovian arrival process (DMAP). At the consecutive slot boundaries, the arrival process is in one of two possible states (say, state $a$ en $b$) and there is a packet arrival with probability $p_a$ whenever the arrival process is in state $a$ at a boundary. There are no arrivals when the process is in state $b$. As such, the arrival process is completely characterised by the transition probabilities $\alpha_{ab}$ and $\alpha_{ba}$ from state $a$ to $b$ and $b$ to $a$ respectively and by the probability $p_a$. Alternatively, we may characterise the arrival process by,

$$\sigma_i = \frac{\alpha_{ba}}{\alpha_{ab} + \alpha_{ba}}, \quad K_i = \frac{1}{\alpha_{ab} + \alpha_{ba}}, \quad \rho_i = p_a\sigma_i. \tag{24}$$

The parameter $\sigma_i$ denotes the fraction of time that the arrival process is in state $a$. The parameter $K_i$ takes values between $\max(\sigma_i, 1 - \sigma_i)$ and $\infty$ and is a measure for the absolute lengths of the $a$-periods and $b$-periods. Finally, $\rho_i$ denotes the arrival load.

Packet transmission times equal 1 slot but we assume that the transmission line is not always available. E.g. the transmission line is shared between multiple buffers and is occupied by traffic from the other buffers from time to time. The transmission line alternates between two states (say, state $c$ and $d$) and is available in state $c$ and unavailable in state $d$. Transition probabilities are denoted by $\beta_{cd}$ and $\beta_{dc}$ from state $c$ to $d$ and from state $d$ to $c$ respectively. Alternatively we may characterise the interruption process by,

$$\sigma_o = \frac{\beta_{dc}}{\beta_{cd} + \beta_{dc}}, \quad K_o = \frac{1}{\beta_{cd} + \beta_{dc}}. \tag{25}$$

Here, $\sigma_o$ denotes the fraction of time that the transmission line is available and $K_o$ is again a measure for the absolute lengths that the transmission line is (un)available.

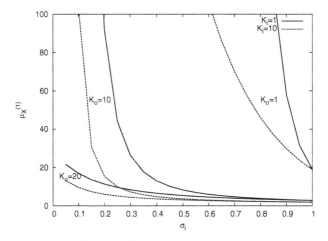

**Fig. 2.** Mean of the number of arrivals between consecutive losses

In Figures 2 and 3, the mean and the coefficient of variation of the number of arrivals between consecutive losses are depicted versus the fraction of time $\sigma_i$ that the arrival process is in state $a$. The buffer can store up to 10 packets simultaneously and there is an arrival with probability $p_i = 1/2$ whenever the arrival process is in state $a$. Further, the transmission line is available during a fraction $\sigma_o = 1/2$ of the time and different values of the input and output burstiness factors are considered as indicated.

Increasing the arrival load implies a decrease of the quality of service: the mean number of packet arrivals between losses decreases and the coefficient of variance increases. Further, the presence of burstiness in the arrival and interruption

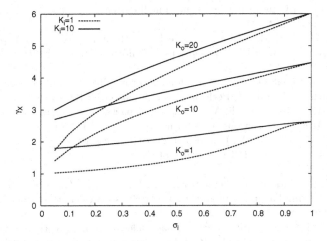

**Fig. 3.** Coefficient of variation of the number of arrivals between consecutive losses

processes $(K_i, K_o > 1)$ implies a decrease of the quality of service as well. Finally notice that for $\sigma_i = 1$ the performance measures do not depend on $K_i$ as in this case the arrival process is constantly in state $a$.

## 5  Conclusions

We studied loss characteristics of discrete-time and continuous-time $M/M/1/N$ queues in a Markovian environment. In particular, we obtained an expression for the joint probability generating function of the time and the number of packet arrivals between consecutive losses. During analysis, we heavily relied on both matrix techniques and probability generating functions. The moment generating property of probability generating functions enabled us to obtain various performance measures. We illustrated our approach by means of various numerical examples.

### Acknowledgement

This work is supported by the Interuniversity Attraction Poles Program - Belgian Science Policy.

### References

1. Kim, H., Schroff, N.: Loss probability calculations and asymptotic analysis for finite buffer multiplexers. IEEE/ACM Transactions on Networking **9** (2001) 755–768
2. Pihlsgard, M.: Loss rate asymptotics in a GI/G/1 queue with finite buffer. Stochastic Models **21** (2005) 913–931
3. Dán, G., Fodor, V., Karlsson, G.: Analysis of the packet loss process for multimedia traffic. In: Proceedings of the 12th International Conference on Telecommunication Systems, Modeling and Analysis, Monterey,CA (2004) 83–84

4. Frossard, P.: FEC performance in multimedia streaming. IEEE Communications Letters **5** (2001) 122–124

5. Gurewitz, O., Sidi, M., Cidon, I.: The ballot theorem strikes again: packet loss process distribution. IEEE Transactions on Information Theory **46** (2000) 2588–2595

6. Dube, P., Ait-Hellal, O., Altman, E.: On loss probabilities in presence of redundant packets with random drop. Performance Evaluation **52** (2003) 147–167

7. Cidon, I., Khamisy, A., Sidi, M.: Analysis of packet loss processes in high speed networks. IEEE Transactions on Information Theory **IT-39** (1993) 98–108

8. Altman, E., Jean-Marie, A.: Loss probabilities for messages with redundant packets feeding a finite buffer. IEEE Journal of Selected Areas in Communications **16** (1998) 778–787

9. Ait-Hellal, O., Altman, E., Jean-Marie, A., Kurkova, I.: On loss probabilities in presence of redundant packets and several traffic sources. Performance Evaluation **36–37** (1999) 485–518

10. Sheng, H., Li, S.: Spectral analysis of packet loss rate at a statistical multiplexer for multimedia services. IEEE/ACM Transactions on Networking **2** (1994) 53–65

11. Schulzrinne, H., Kurose, J., Towsley, D.: Loss correlation for queues with burtsty input streams. In: Proceedings of IEEE ICC. (1992) 219–224

12. Latouche, G., Ramaswami, V.: Introduction to matrix analytic methods in stochastic modeling. Series on statistics and applied probability. ASA-SIAM (1999)

13. Neuts, M.: Matrix-geometric solutions in stochastic models: An algorithmic approach. Dover Publications (1994)

# Transient Analysis of a Queuing System with Matrix-Geometric Methods

Péter Vaderna and Tamás Éltető

Traffic Laboratory, Ericsson Research, H-1117, Budapest, Irinyi J. u. 4-20, Hungary
Peter.Vaderna@ericsson.com
Tamas.Elteto@ericsson.com

**Abstract.** This paper investigates a queuing system with infinite number of servers where the arrival process is given by a Markov Arrival Process (MAP) and the service time follows a Phase-type (PH) distribution. They were chosen since they are simple enough to describe the model by exact methods. Moreover, highly correlated arrival processes and heavy-tailed service time distributions can be approximated by these tools on a wide range of time-scales. The transient behaviour of the system is analysed and the time-dependent moments of the queue length is computed explicitly by solving a set of differential equations. The results can be applied to models where performance of parallel processing is important. The applicability of the model is illustrated by dimensioning a WEB-based content provider.

## 1 Introduction

Designers of mobile or data networks and network elements often go for the means of mathematics and statistics in order to understand system behaviour. Due to the complexity of the system exact mathematical methods are often replaced by simulations, numerical estimations or simple rules-of-thumbs, each having pros and cons.

This paper investigates the MAP/PH/$\infty$ queuing model where MAP stands for the Markov Arrival Process and PH means Phase-type distributions. Numerical methods exist to approximate the moments of the queue-length for the PH/G/$\infty$ system [2] which can be extended to the more general MAP/G/$\infty$ system. However, these solutions rely on the numerical solution of a system of differential equations. Further, authors in [9] derive numerically tractable formulas of the moments of BMAP/PH/$\infty$ system, which still contain elements that can only be obtained approximately. In the following we would like to present a different computational method where the time-dependent moments of the queue length of MAP/PH/$\infty$ system can be obtained exactly. Knowing the moments, discrete and finite support distribution functions (i.e. the range of the random variable is finite) can be calculated without using approximations. This way the number of moments to determine is the number of different values of the random variable.

The model can be applied on systems where demands arrive according to a non-stationary point process, the service is parallel and transient behaviour is

Y. Koucheryavy, J. Harju, and V.B. Iversen (Eds.): NEW2AN 2006, LNCS 4003, pp. 21–33, 2006.

crucial to find out the limiting factors. An application example is shown where the number of parallel working processes in a WEB-based content provider is the measure of interest. The requests arriving at the server can be well modelled by MAP and the service time follows PH distribution. The distribution of the number of parallel processes is evaluated at the time instant when the expected value is the largest. The outcome of the model is the number of processing units (e.g. servers or processing capacity) needed to serve all demands with large probability so that the chance of blocking remain under a certain value.

## 2    Mathematical Background

In this section a short introduction of the MAP is given. More details can be found e.g. in [2]. Consider a continuous-time stochastic process $\{(N(x), J(x)) : x \geq 0\}$ where $N(x)$ is the number of arrivals in time $(0, x]$ while $J(x)$ is the phase at time $x$. The arrival process $N(x)$ is modulated by a phase process $J(x)$ whose states govern the arrival rates. Let us partition the state space $\{(n, j) : n \geq 0, 1 \leq j \leq M\}$ into subsets

$$l(k) = \{(k, 1), \ldots, (k, M)\} \quad \text{for } k \geq 0.$$

The infinitesimal generator matrix $\mathbf{Q}$ of the continuous-time Markov-process $\{(N(x), J(x)) : x \geq 0\}$ has the following form:

$$\mathbf{Q} = \begin{bmatrix} \mathbf{D}_0 & \mathbf{D}_1 & \mathbf{0} & \mathbf{0} & \cdots \\ \mathbf{0} & \mathbf{D}_0 & \mathbf{D}_1 & \mathbf{0} & \cdots \\ \mathbf{0} & \mathbf{0} & \mathbf{D}_0 & \mathbf{D}_1 & \cdots \\ \mathbf{0} & \mathbf{0} & \mathbf{0} & \mathbf{D}_0 & \cdots \\ \vdots & \vdots & \vdots & \vdots & \ddots \end{bmatrix},$$

where $\mathbf{D}_0, \mathbf{D}_1$ are matrices of order $M$ with $\mathbf{D}_1 \geq 0$, $[\mathbf{D}_0]_{i,j} \geq 0$ for $1 \leq i \neq j \leq M$, $[\mathbf{D}_0]_{i,i} < 0$ for $1 \leq i \leq M$ and the matrix $\mathbf{D} = \mathbf{D}_0 + \mathbf{D}_1$ is stochastic, that is, $\mathbf{De} = \mathbf{0}$, where $\mathbf{e}' = (1, \cdots, 1)$. Matrices $\mathbf{D}_0$ and $\mathbf{D}_1$ filter those parts of the Markov process which correspond to non-arrival and arrival transitions respectively.

A natural generalization of MAP is to allow more than one arrivals which requires the definition of $\mathbf{D}_m$ accordingly, for m-sized batches.

A special case of MAP can be derived by letting only the diagonal elements of $\mathbf{D}_1$ be nonzero, which basically means that no state transitions occur at the time of arrivals. This results in the Markov Modulated Poisson Process (MMPP), which is a widely used arrival model [5].

Another special case of MAP is the Phase-type (PH) renewal process. In this case, the arrivals occur according to a renewal process where the time between the arrivals has PH distribution represented by the $(\alpha, \mathbf{T})$ pair [2]. The vector $\alpha$ is the initial probability vector of the PH distribution while $\mathbf{T}$ is a non-singular matrix describing the phase transitions until the absorption such that $[\mathbf{T}]_{i,j} \geq 0$ for $1 \leq i \neq j \leq M$, $[\mathbf{T}]_{i,i} < 0$, for $1 \leq i \leq M$ and $\mathbf{T}\mathbf{e} \leq 0$. In this case the phase process $J(x)$ is restarted after an arrival according to the initial probability vector $\boldsymbol{\alpha}$ and the arrival rate that depends on the phase is described by vector $\mathbf{t}$:

$$\mathbf{t} = -\mathbf{T}\mathbf{e}.$$

The MAP representation in this case is

$$\mathbf{D}_0 = \mathbf{T}, \quad \mathbf{D}_1 = \mathbf{t}\boldsymbol{\alpha}.$$

# 3    Moments of an Infinite-Server Queuing System

In the following an infinite-server queuing system is introduced with MAP arrivals and PH service time distribution. The moments of the queue length are computed, where the queue length stands for the parallel demands being served in the system.

Let $X(t)$ denote the queue length and $J(t)$ the phase of the arrival process at time $t+$ and let $\boldsymbol{\mu}^{(K)}(t)$ denote the $M$-vector whose $i$th element is $\mu_i^{(K)}(t)$ $(K \geq 1)$, where

$$\mu_i^{(K)}(t) = E[X^{(K)}(t)|X(0) = 0, J(0) = i]$$

$$K \geq 1, 1 \leq i \leq M$$

$X^{(K)}(t)$ denotes the factorial product $X(t)[X(t)-1]\cdots[X(t)-K+1]$ and $\mu_i^{(K)}(t)$ denotes the $K$th factorial moments of the number of demands being served if the system is started from state $i$.

The main purpose is to calculate the time-dependence of the moments of the above queuing system. According to the calculations in [1], the following system of differential equations can be written for each factorial moments of the queue length:

$$\frac{d}{dt}\boldsymbol{\mu}^{(1)}(t) = \mathbf{D}\boldsymbol{\mu}^{(1)}(t) + \{1 - H(t)\}\mathbf{D}_1\mathbf{e}$$

$$\boldsymbol{\mu}^{(1)}(0) = \mathbf{0} \tag{1}$$

and for $K \geq 2$

$$\frac{d}{dt}\boldsymbol{\mu}^{(K)}(t) = \mathbf{D}\boldsymbol{\mu}^{(K)}(t) + K\{1 - H(t)\}\mathbf{D}_1\boldsymbol{\mu}^{(K-1)}(t)$$

$$\boldsymbol{\mu}^{(K)}(0) = \mathbf{0}. \tag{2}$$

$H(t)$ denotes the cumulative distribution function (c.d.f.) of the service time and $\mathbf{e}$ is the $M$-vector whose each elements is 1.

Starting from solving Equation 1, the time dependence of each moment of the queue length can be iteratively calculated by using the preceding moment in Equation 2.

The calculation used in this paper is based on solving the above system of differential equations with PH service time distributions. PH distribution can also be represented by a mixed Erlang distribution [3]:

$$H(t) = 1 - \sum_{i=1}^{I} e^{-\beta_i t} \sum_{j=0}^{J_i} \gamma_{ij} t^j, \tag{3}$$

where $\beta_i$s and $\gamma_{ij}$s are the exponents and the coefficients, respectively, $I$ is the number of different exponents and $J_i$ is the maximal $t$-power belonging to the $i$th exponent. Since $H(t)$ is a c.d.f., $\beta_i > 0$ for all $1 \leq i \leq I$.

The solution method is based on standard techniques of solving first order linear inhomogeneous ordinary differential equation systems with constant coefficients, which can be found e.g. in [10] and [11]. For details see the Appendix.

## 4   Possible Application

### 4.1   Application Description

In this section the usage of the model is illustrated by a technical application. The system under investigation consists of processing modules of a WEB-based content provider. The service operates via a central file-server containing the news items and different multimedia objects (pictures, videos, animations). The server can identify the type of browser a certain request is sent from. The appearance of the article depends on the terminal type so the server has to optimize it according to the type of the browser. The aim of the operator is to send the messages in a format that appears in as good quality as possible (e.g. the size and resolution of the pictures or the rendering of the text). Web-servers may adapt to the limited capabilities of mobile equipments as well in order to improve the performance and the quality of browsing.

The content provider consists of a central server, converter units and a storage unit. The operation steps of the service is illustrated in Figure 1. The timing sequence of the events is represented by the numbers next to the arrows.

1. When a new article arrives from the news agency at the server a basic version is generated.
2. A message arrives from a user requesting for the article.
3. The server converts the basic version according to the browser type.
4. Since there may be another user with the same browser type requesting this article the converted version is stored temporarily so that the conversion need not be performed once again.
5. The storage unit sends the requested version of the article to the server.
6. The server forwards the article to the user.

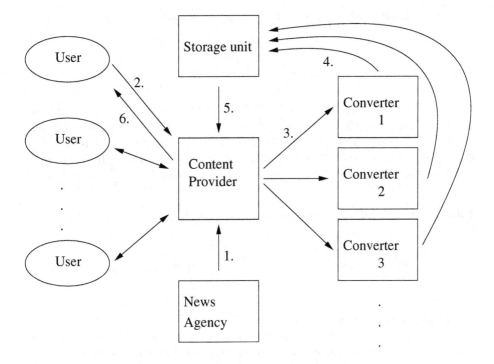

**Fig. 1.** Outline of the news service

When a request arrives from a browser of type that has been processed so far, the server can turn to the storage unit. For a new browser type the conversion has to be made. Note that the converters are not necessarily separate processing units, they can be different program threads on the same processor.

The purpose is to investigate the functionality of the conversion from the publication of a new article by the agency to the state where the article is converted to all possible forms. In order for the fast service more than one different conversion can be processed at the same time. If the number of converters is not large enough all the converters may be occupied resulting in rejection of a request coming from a terminal which has a new type. Increasing the number of converters may meet financial limits. The task is to fix an upper limit for the processing converters where the probability of being all of them occupied is very small.

## 4.2    Mathematical Modeling

Matrix-geometric methods can be applied to the above model in the following way. If there are $n$ types of browser, matrices $\mathbf{D}$, $\mathbf{D}_0$ and $\mathbf{D}_1$ can be built from the relative frequency of the occurrence of each browser types and the arrival intensity of the requests from the users. Table 1 shows the relative frequencies of $n$ browser types.

**Table 1.** Share of the different types of browsers (frequency)

| Type | 1 | 2 | ... | n |
|------|---|---|-----|---|
| Frequency | $F_1$ | $F_2$ | ... | $F_n$ |

Since $F_i$s are relative frequencies, $\sum_{i=1}^{n} F_i = 1$. If all $F_i$s are different, the number of states in the MAP is $S = 2^n$ thus, the $S \times S$ matrix $\mathbf{D}$ describing the model becomes rather large. However, the size of the matrix can be reasonably decreased by letting the frequency of some types the same thus, $n_i$ browser types have the same frequency. Table 2 shows such a scenario, here $\sum_{i=1}^{k} n_i F_i = 1$.

**Table 2.** Share of the different types of browsers with reduced number of different frequencies

| Type | 1 | 2 | ... | $n_1$ | $n_1 + 1$ | $n_1 + 2$ | ... | $n_1 + n_2$ | ... | $n_1 + ... + n_k$ |
|------|---|---|-----|-------|-----------|-----------|-----|-------------|-----|-------------------|
| Frequency | $F_1$ | $F_1$ | ... | $F_1$ | $F_2$ | $F_2$ | ... | $F_2$ | ... | $F_k$ |

The resulting number of states is $S = \prod_{i=1}^{k} n_k$.

The states represent the number of messages of certain browser types that have arrived so far. State transitions may only occur if a new type of request arrives. If the requests have Poisson arrival with rate $\lambda$, the elements of the MAP representation matrix $\mathbf{D}$ can be built from $\lambda$ multiplied by the proper frequency values. For details see Section 4.3.

### 4.3 Numerical Example

An example is shown where the time-dependent moments of the queue-length of a $MAP/M/\infty$ is computed. Let's assume that the requests of the users arrive according to Poisson process with intensity 8 requests per sec. In our example 10 different types of terminals are known with different converting procedures. The time of conversion is exponentially distributed with average 5 seconds (though the model can handle more complex distributions).

$$H(t) = 1 - e^{-\frac{t}{5}}.$$

The share of the 10 different types and their average number of requests in a second is summarized in Table 3.

One can see from Table 4 that the state space of the underlying MAP can be described by 4-tuples. These vectors point out which terminal types have a properly converted version of the latest article in the storage unit. The meaning of the elements of the 4-tuples is:

**1.** If the article is not converted for type 1 then its value is 0 otherwise 1.
**2.** If the article is not converted for type 2 then its value is 0 otherwise 1.

**Table 3.** Share of the different types of browsers (frequency) and the number of requests in a second generated by them (intensity)

| Type | Frequency | Intensity |
|---|---|---|
| 1. type | 21% | 1.68 |
| 2. type | 20% | 1.6 |
| 3. type | 13% | 1.04 |
| 4. type | 13% | 1.04 |
| 5. type | 13% | 1.04 |
| 6. type | 4% | 0.32 |
| 7. type | 4% | 0.32 |
| 8. type | 4% | 0.32 |
| 9. type | 4% | 0.32 |
| 10. type | 4% | 0.32 |

**Table 4.** Different types of browsers sorted by their intensities

| Frequency | Types | Intensity |
|---|---|---|
| 21% | 1. | 1.68 |
| 20% | 2. | 1.6 |
| 13% | 3., 4., 5. | 1.04 |
| 4% | 6., 7., 8., 9., 10. | 0.32 |

3. Conversions of types 3-5 are counted here. Its value can be between 0 and 3. The arrival order does not matter since types 3-5 have the same frequency.
4. Conversions of types 6-10 are counted here. Its value can be between 0 and 5. The arrival order does not matter since types 6-10 have the same frequency.

The number of states is $2 \cdot 2 \cdot 4 \cdot 6 = 96$. State transition is allowed only between states whose 4-vector representation differs in only one digit. The initial state is $(0, 0, 0, 0)$, i.e. the storage unit is still empty. The state transitions are given in a $96 \times 96$ matrix $\mathbf{D_1}$ in the following way (see Figure 2).

If the state transition corresponds to the conversion of terminal type 1 then the value of the matrix element is $8 \ 1/s \cdot 0.21 = 1.68 \ 1/s$ as it can be seen in Table 3 in column "Intensity". The matrix element of the state transition corresponding the conversion of terminal type 2 is $1.6 \ 1/s$.

In case of types 3-5 the corresponding state transition changes the 3rd vector element. In this case the number of conversions needs to be maintained as well since the first request is expected to arrive with intensity $3 \cdot 1.04 \ 1/s = 3.12 \ 1/s$ from one of types 3-5. However, after the first request, only the remaining two types can generate new requests so the intensity decreases to $2 \cdot 1.04 \ 1/s = 2.08 \ 1/s$. If 2 types are already processed from types 3-5, the arrival intensity of the remaining request is $1.04 \ 1/s$.

The case of types 6-10 is similar to the above case of types 3-5. Here the initial intensity is $5 \cdot 0.32 \ 1/s = 1.6 \ 1/s$ and it decreases to $0.32 \ 1/s$ if one type is left to be processed.

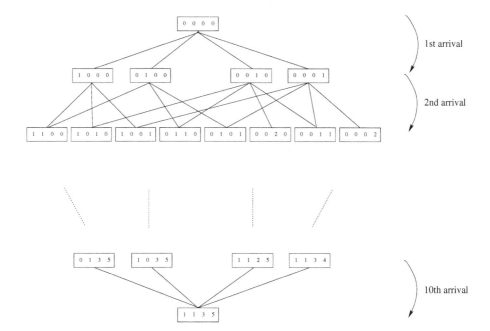

**Fig. 2.** State transition upon arrivals

From the publishment of the article the requests arrive continuously to the server. After the first types are processed, the number of needed conversions is smaller. If 10 converters are available, all demands can be served. However, we will see that the service can be completed with a smaller number of converters.

Since every request from a new type generates a state transition, $D_0$ has only diagonal elements, each of them assigned so that $D = D_0 + D_1$ is stochastic, i.e. the sum of the elements in a row is 0. By replacing the $D$ and $D_1$ matrices and the $H(t)$ function into Equations 1 and 2 and solving them following the steps shown in the Appendix, the time-dependent moments can be obtained. The formulae are evaluated by Matlab, for the computation example see [13].

In Figure 3 the time evolution of the average number of busy converters is depicted. It can be seen clearly that the largest number of working converters is expected to operate 0.75 seconds after the news publishment.

In order to find a reasonable limit for the number of available converters where the probability of saturation is small at the maximum utilization, the moments of the number of busy converters are evaluated 0.75 seconds after the publishment of the new article. Assuming that we have unlimited number of available converters, Table 5 contains the results.

Since the number of conversions is at most 10, the distribution of the number of parallel conversions can be computed from the moments with the help of a

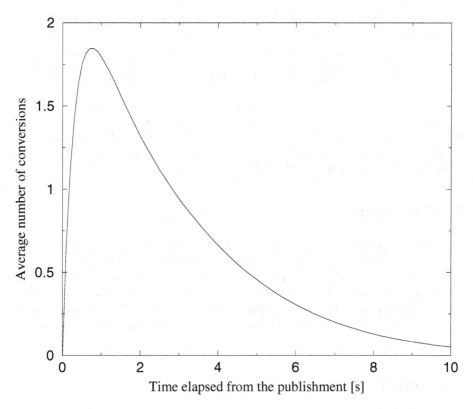

**Fig. 3.** Time evolution of the average number of busy converters after the publishment of a new article

**Table 5.** Moments of the number of parallel conversions 0.75 seconds after the publishment of the new article

|     | Factorial moment | Moment      |
| --- | ---------------- | ----------- |
| 1.  | 1.848            | 1.848       |
| 2.  | 3.063            | 4.911       |
| 3.  | 4.499            | 15.537      |
| 4.  | 5.763            | 56.048      |
| 5.  | 6.307            | 224.208     |
| 6.  | 5.734            | 976.641     |
| 7.  | 4.157            | 4573.540    |
| 8.  | 2.253            | 22805.388   |
| 9.  | 0.812            | 120181.397  |
| 10. | 0.146            | 665327.954  |

Vandermonde-type matrix [12]. Table 6 shows the probability of the number of parallel conversions ($N$) exceeding a given limit ($n$).

**Table 6.** Distribution function of the number of parallel conversions 0.75 seconds after the publishment of the new article

| $n$ | 0. | 1. | 2. | 3. | 4. | 5. | 6. | 7. | 8. | 9. | 10. |
|---|---|---|---|---|---|---|---|---|---|---|---|
| $P(N > n)$ | 0.871 | 0.577 | 0.276 | 0.095 | 0.023 | 0.004 | $5 \cdot 10^{-4}$ | $4 \cdot 10^{-5}$ | $2 \cdot 10^{-6}$ | $3 \cdot 10^{-8}$ | 0 |

If the system was designed so that the probability of reaching the capacity limit should be less than 0.1%, from Table 6 it can be deduced that instead of 10 converters, 5 converter units or processing capacity corresponding to 5 parallel processing threads would be sufficient.

## 5   Conclusions

The paper introduces a mathematical model and shows one possible application. The transient behaviour of the moments of a MAP/PH/∞ queuing system is determined exactly, which can be used to obtain the exact distribution. This model can be used to describe transient behaviour of systems with parallel servers, general arrival and general processing times. The applicability of the computational method is illustrated by solving a dimensioning problem of content and multimedia servers.

## Acknowledgements

This paper describes work partially undertaken in the context of the E-NEXT - IST FP6-506869 project, which is partially funded by the Commission of the European Union. The views contained herein are those of the authors and do not necessarily represent those of the E-NEXT project.

## References

1. V. Ramaswami and M. F. Neuts. Some explicit formulas and computational methods for infinite-server queues with phase-type arrivals. *J. Appl. Prob. 17*, pages 498-514, 1980.
2. G. Latouche, V. Ramaswami. Introduction to Matrix Analytic Methods in Stochastic Modeling. *ASA-SIAM series on statistics and applied probability*, 1999.
3. A. Pfening and M. Telek. Evaluation of task completion time of phase type work requirement. In Proceedings of *Relectronic'95* (Budapest), pp. 73-78, 1995.
4. S. Rácz, A. Tari, and M. Telek. MRMSolve: Distribution estimation of large Markov reward models. in *Tools 2002* (London, England), pp. 72-81, Springer, LNCS 2324, April 2002.
5. W. Fischer, K. Meier-Hellstern. The Markov-modulated Poisson process (MMPP) cookbook. *Performance Evaluation* 18, 149-171, 1993.
6. D. Burman, R. Smith. An Asymptotic Analysis of a Queuing System with Markov-Modulated Arrivals. *Operations Research* Vol 34. 1986.

7. L. Takács. On Erlang's Formula. *The Annals of Mathematical Statistics*, Vol. 40, No. 1, 71-78, 1969.
8. B. L. Nelson and M. R. Taaffe. The $Ph_t/Ph_t/\infty$ Queueing System: Part I — The Single Node. To appear in *INFORMS JOC* in 2004.
9. H. Masuyama and T. Takine. Analysis of an Infinite-Server Queue with Batch Markovian Arrival Streams *Queueing Systems* vol.42. no.3, pp.269-296, 2002.
10. E. W. Weisstein. Ordinary Differential Equation. *MathWorld–A Wolfram Web Resource*.    http://mathworld.wolfram.com/OrdinaryDifferentialEquation  SystemwithConstantCoefficients.html
11. I. N. Bronstein, K. A. Semendjajew, G. Musiol and H. Mühlig. *Taschenbuch der Mathematik*. Verlag Harri Deutsch, Frankfurt am Main, Thun 1999.
12. J. A. Shohat and J. D. Tamarkin. *The Problem of Moments*. American Mathematical Society, Providence, Rhode Island, 1943.
13. http://gawain.elte.hu/~vpeti/map_ph_inf.m

# Appendix

The solution for the factorial moments in Equations 1 and 2 can be expressed as the sum of the general solution of the homogeneous part and a particular solution of the inhomogeneous part of the differential equation system. Since $D_1$ and $D_0$ matrices are constant, the solution of the homogeneous part remains the same for all moments. For $K \geq 2$, the solution of the inhomogeneous part of the differential equation system for the $K$th moment depends on the solution of the differential equation system corresponding to the $K - 1$st moment. We start with computing $\mu^{(1)}(t)$, then the same method can be used for the higher moments, accordingly.

The first step of solving Equation 1 is to obtain the solution of the homogeneous part of the equation:

$$\frac{d}{dt}\mu^{(1)(H)}(t) = D\mu^{(1)(H)}(t) \tag{4}$$

The general solution is the linear combination of exponentials:

$$\mu_n^{(1)(H)}(t) = \sum_{m=1}^{M} V_{nm} c_m e^{r_m t} \tag{5}$$

The $M \times M$ matrix $V$ is the similarity transformation matrix generating Jordan form from the constant coefficient matrix $D$: $VDV^{-1} = J$. If $D$ has single eigenvalues the columns of $V$ are the eigenvectors of $D$, $r_m$s are the corresponding eigenvalues, $c_m$s are unknown variables.

Since $D$ is a stochastic matrix it has an eigenvalue equal to 0, and the real part of all other eigenvalues are negative.

*Remark 1.* Since one of the exponents is 0, the above solution of the homogeneous part of the differential equation system has a term independent of $t$. This term plays an important role, when we take the limit $t \to \infty$. We may choose $r_1 = 0$ without breaking the generality.

*Remark 2.* In case of complex eigenvalues, complex conjugate pairs occur in the set of roots of the characteristic polynomial of $\mathbf{D}$ and also in the exponents, resulting in real numbers in the solution.

*Remark 3.* In case of multiple eigenvalues, extra polynomials should be taken into account but the basic solution methodology remains the same.

For the general solution of the differential equation system, one should also get one of the particular solutions. For this purpose, the method of undetermined coefficients can be applied, i.e. we consider $c_m$ ($1 \leq m \leq M$) as a function of $t$.

$$\sum_{m=1}^{M} V_{km} \dot{c}_m(t) e^{r_m t} = f_k(t), \tag{6}$$

where

$$f_k(t) = (1 - H(t)) \sum_{i=1}^{M} D_{1ki}$$

and it can be written the following general form:

$$f_k(t) = \sum_{i=1}^{I} e^{-\beta_{ik} t} \sum_{j=0}^{J_i} \gamma_{ij}^{(k)} t^j \tag{7}$$

The derivatives of the $c_m(t)$ functions ($1 \leq m \leq M$) can be expressed by $f_k(t)$ functions and the inverse of $\mathbf{V}$. Let the $M \times M$ matrix $\mathbf{A}$ be the inverse of $\mathbf{V}$, $\mathbf{A} = \mathbf{V}^{-1}$. Solving Equation 6 for $\dot{\mathbf{c}}$, the following equations hold for the derivatives of the coefficients:

$$\dot{c}_m(t) = \sum_{k=1}^{M} A_{mk} f_k(t) e^{-r_m t} = \sum_{k=1}^{M} A_{mk} \sum_{i=1}^{I} e^{-(\beta_{ik} + r_m)t} \sum_{j=0}^{J_i} \gamma_{ij}^{(k)} t^j,$$

$$1 \leq m \leq M \tag{8}$$

After the integration of both sides of Equation 8, we get the following formula for $c_m$ (the constant part of the integration is taken as 0 for simplicity because a particular solution is sufficient):

$$c_m^{\diamond}(t) e^{r_m t} = \sum_{(\diamond)k=1}^{M} \sum_{(\diamond)i=1}^{I} \sum_{j=0}^{J_i} \sum_{l=0}^{j} A_{mk} \gamma_{ij}^{(k)} \frac{\Gamma(j+1)}{\delta_{ik}^{(m)\,(j+1)}} \frac{e^{-\beta_{ik} t} \delta_{ik}^{(m)\,l} t^l}{l!},$$

$$1 \leq m \leq M$$

where $\delta_{ik}^{(m)} = \beta_{ik} + r_m$. The square sign indicates that the above formula is evaluated only for those $i$ and $k$ indexes for that $\delta_{ik}^{(m)} \neq 0$. If $\delta_{ik}^{(m)} = 0$, those $i$ and $k$ indexes should be treated separately:

$$c_m^{*}(t) e^{r_m t} = \sum_{(*)k=1}^{M} \sum_{(*)i=1}^{I} \sum_{j=0}^{J_i} A_{mk} \gamma_{ij}^{(k)} \frac{t^{j+1}}{j+1} e^{r_m t}$$

where the star sign indicates that the summation is performed for only those $i$ and $k$ indexes for that $\delta_{ik}^{(m)} = 0$. The $n$th component of the particular solution can be written as

$$\mu_n^{\diamond(1)(P)}(t) = \sum_{m=1}^{M} V_{nm} c_m^{\diamond}(t) e^{r_m t}$$

$$\mu_n^{*(1)(P)}(t) = \sum_{m=1}^{M} V_{nm} c_m^{*}(t) e^{r_m t}$$

$$1 \leq n \leq M \tag{9}$$

The particular solution of the Equation system 1 is the sum of the two different cases.

$$\mu_n^{(1)(P)}(t) = \mu_n^{\diamond(1)(P)}(t) + \mu_n^{*(1)(P)}(t) \tag{10}$$

Note, that both $\mu_n^{\diamond(1)(P)}(t)$ and $\mu_n^{*(1)(P)}(t)$ converge with $t \to \infty$ to zero since $\beta_{ik} > 0$ for all $i$ and $k$ in the corresponding interval, $r_m > 0$ for all $2 \leq m \leq M$ and $r_1 = 0$ can not occur in $\mu_n^{*(1)(P)}(t)$ since in that case $\delta_{ik}^{(m)} = 0$ does not hold.

The general solution of the first factorial moment can then be derived by replacing Equations 5 and 10 in the following formula:

$$\mu_n^{(1)}(t) = \mu_n^{(1)(H)}(t) + \mu_n^{(1)(P)}(t) \tag{11}$$

In order to calculate $c_m$ coefficients in Equation 5 the initial condition of Equation 1 can be used, thus

$$\mu_n^{(1)}(0) = \mu_n^{(1)(P)}(0) + \sum_{m=1}^{M} V_{nm} c_m = 0$$

where $c_m$ is considered as constant. Using the notation $\mathbf{A} = \mathbf{V}^{-1}$ again, the vector of coefficients can be calculated by

$$\mathbf{c} = -\mathbf{A} * \boldsymbol{\mu}^{(1)(P)}(0) \tag{12}$$

The time-dependent solution of Equation 1 is thus described in Equation 11 where $\mu_n^{(1)(H)}(t)$ is specified in Equation 5 where the $c_m$ expressed in Equation 12 should be replaced and $\mu_n^{(1)(P)}(t)$ is specified in Equation 10.

# HEAF: A Novel Estimator for Long-Range Dependent Self-similar Network Traffic

Karim Mohammed Rezaul[1], Algirdas Pakstas[1], R. Gilchrist[1], and Thomas M. Chen[2]

[1] Department of Computing, Communications Technology and Mathematics,
London Metropolitan University, England
morekba786@yahoo.co.uk, a.pakstas@londonmet.ac.uk,
r.gilchrist@londonmet.ac.uk
[2] Department of Electrical Engineering, Southern Methodist University, USA
tchen@engr.smu.edu

**Abstract.** Long-range dependent (LRD) self-similar chaotic behaviour has been found to be present in internet traffic by many researchers. The 'Hurst exponent', H, is used as a measure of the degree of long-range dependence. A variety of techniques exist for estimating the Hurst exponent; these deliver a variable efficacy of estimation. Whilst ways of exploiting these techniques for control and optimization of traffic in real systems are still to be discovered, there is need for a reliable estimator which will characterise the traffic. This paper uses simulation to compare established estimators and introduces a new estimator, HEAF, a 'Hurst Exponent Autocorrelation Function' estimator. It is demonstrated that HEAF(2), based on the sample autocorrelation of lag2, yields an estimator which behaves well in terms of bias and mean square error, for both fractional Gaussian and FARIMA processes. Properties of various estimators are investigated and HEAF(2) is shown to have promising results. The performance of the estimators is illustrated by experiments with MPEG/Video traces.

## 1 Introduction

Self-similar and long-range dependent (LRD) characteristics of internet traffic have attracted the attention of researchers since 1994 [1, 2]. It is especially important to understand the link between self-similar and long-range dependence of traffic and performance of the networks. Thus, in [3] it was observed that the performance of networks degrades gradually with increasing self-similarity, which results in queuing delay and packet loss. The more self-similar the traffic, the longer the average queue size. The queue length distribution is caused by self-similar traffic. The tail of the queue length distribution tends to be higher when the traffic is self-similar and thus resulting in a higher probability of buffer overflow (packet loss). The performance results [4] prove that the degree of self-similarity in the traffic increases as the cell loss and cell delay increase for certain output port buffer size.

The LRD property of the traffic fluctuations has important implications on the performance, design and dimensioning of the network. Self-similarity in packetised data networks can be caused by the distribution of file size and by human interactions such as teleconferences, voice chat, online video and games etc.

Y. Koucheryavy, J. Harju, and V.B. Iversen (Eds.): NEW2AN 2006, LNCS 4003, pp. 34–45, 2006.

Estimation of the Hurst parameter, H, has been carried out in biophysics, hydrology, stock markets, meteorology, geophysics, etc. [5]. The Hurst exponent is also used to measure the degree of LRD in internet traffic. Having a reliable estimator can yield a good insight into traffic behaviour and eventually may lead to improved traffic engineering.

A number of methods have been proposed to estimate the Hurst parameter. Some of the most popular include: aggregated variance time (V/T), Rescaled-range (R/S), Higuchi's method and wavelet-based methods. The various methods demonstrate variable performance. We here discuss the properties of these estimators.

This paper compares the established estimators by simulation experiments, and introduces a new estimator which we call a HEAF estimator; i.e. a 'Hurst Exponent by Autocorrelation Function' estimator. This estimator, HEAF, estimates H by a process which is simple, quick and reliable. In order to investigate the properties of HEAF and to compare it with other estimators, two different simulation studies were performed. The first one uses fractional Gaussian noise (fGn) sequences generated by the Dietrich-Newsam algorithm [6], generating exact self-similar sequences. The second set of simulations uses a fractional autoregressive moving average (FARIMA) process [7]. In [8], it is shown that the FARIMA process can be effectively used to model network traffic. The fGN process is a simpler model which is convenient to use for comparison.

The paper is organized as follows. Section 2 describes the definitions of self-similarity and long-range dependence. Section 3 shows the relationship between the autocorrelation function and LRD. The established methods for estimating Hurst parameter are highlighted in section 4. The algorithm for generating self-similar sequences is described in section 5. Section 6 introduces the new estimator, HEAF. Finally the results are presented in section 7.

## 2  Self-similarity and Long-Range Dependence

A phenomenon which is self-similar looks the same or behaves the same when viewed at different degree of magnification. Self-similarity is the property of a series of data points to retain a pattern or appearance regardless of the level of granularity used and can be the result of long-range dependence (LRD) in the data series. If a self-similar process is bursty at a wide range of timescales, it may often exhibit long-range-dependence.

Long-range-dependence means that all the values at any time are correlated in a positive and non-negligible way with values at all future instants. The auto-correlation function is a measure of how similar a time series $X(t)$ is to itself shifted in time by $k$, creating the new series $X(t+k)$. A continuous time process $Y = \{Y(t),\ t \geq 0\}$ is self-similar if it satisfies the following condition [9]:

$$Y(t) \overset{d}{=} a^{-H} Y(at), \quad \forall a \geq 0, \ \forall a > 0, \quad for \ 0 < H < 1, \tag{2.1}$$

where $H$ is the index of self-similarity, called the Hurst parameter and the equality is in the sense of finite-dimensional distributions.

Consider a stationary series of length $n$ and let $M$ be the integral part of $n/m$, where the original series $X$ has been divided into non-overlapping blocks of size $m$. Then define the corresponding aggregated sequence with level of aggregation $m$ by

$$X^{(m)}(k) = \frac{1}{m} \sum_{i=(k-1)m+1}^{km} X(i) \; ; \quad k = 1,2,3,...,M. \text{ (The index, } k \text{ labels the block)} \quad (2.2)$$

If $X$ is the increment process of a self-similar process $Y$ defined in (2.1), i.e.,

$$X(i) = Y(i+1) - Y(i), \text{ then } X \stackrel{d}{=} m^{1-H} X^{(m)} \; ; \text{ for all integer m .} \quad (2.3)$$

A stationary sequence $X = \{X(i), \; i \geq 1\}$ is called exactly self-similar if it satisfies (2.3) for all aggregation levels $m$.

## 3  Relationship Between LRD and Auto-Correlation Function (ACF)

The stationary process X is said to be a long-range dependent process if its ACF is non-summable [10], meaning that $\sum_{k=-\infty}^{\infty} \rho_k = \infty$ .

The details of how the ACF decays with $k$ are of interest because the behaviour of the tail of ACF completely determines its summability. According to [1], $X$ is said to exhibit long-range-dependence if $\rho_k \sim L(t) k^{-(2-2H)}$, as $k \rightarrow \infty$, \quad (3.1)

Where $1/2 < H < 1$ and $L(.)$ slowly varies at infinity, i.e., $\lim_{t \to \infty} L(xt)/L(t) = 1$,

for all $x > 0$. Equation (3.1) implies that the LRD is characterized by an autocorrelation function that decays hyperbolically rather than exponentially fast.

## 4  Estimation of the Hurst Parameter

In this paper, we use four different methods in conjunction with HEAF to estimate the Hurst exponent, $H$, namely: (a) Aggregated Variance Time (V/T) analysis, [11, 12] (b) Rescaled-range (R/S) analysis for different block sizes, [1, 11] (c) Higuchi's method, [13] and (d) the wavelet method [14, 15].

## 5  Generation of Self-similar Sequences

With the intention of understanding the long-range dependence, we started our simulation study with fractional Gaussian noise (fGN) [6] and with FARIMA (0,d,0), which is the simplest and the most fundamental of the fractionally differenced ARIMA processes [7]. For the fGN process, we used Dietrich and Newsam algorithm [6]. The algorithm is implemented in Matlab which is publicly available in [16]. A Matlab implementation is given in [16] for the simulation of FARIMA process.

# 6 HEAF: A 'Hurst Exponent by Autocorrelation Function' Estimator

We here introduce a new estimator by extending the approach of Kettani and Gubner [17]. As in [17], for a given observed data $X_i$ (i.e. $X_1,.........,X_n$ ), the sample autocorrelation function can be calculated by the following method:

Let
$$\hat{\mu}_n = \frac{1}{n}\sum_{i=1}^{n} X_i \qquad (6.1)$$

and $\hat{\gamma}_n(k) = \frac{1}{n}\sum_{i=1}^{n-k}(X_i - \hat{\mu}_n)(X_{i+k} - \hat{\mu}_n)$ , where $k = 0, 1, 2, ....., n,$ (6.2)

with
$$\hat{\sigma}_n^2 = \hat{\gamma}_n(0). \qquad (6.3)$$

Then the sample autocorrelations of lag k are given by

$$\hat{\rho}_k = \hat{\gamma}_n(k)/\hat{\sigma}_n^2 \qquad (6.4)$$

(Equations (6.1), (6.2), (6.3) and (6.4) denote the sample mean, the sample covariance, the sample variance and the sample autocorrelation, respectively). A second-order stationary process is said to be exactly second-order self-similar with Hurst exponent $1/2 < H < 1$  if

$$\rho_k = [(k+1)^{2H} - 2k^{2H} + (k-1)^{2H}]/2 \qquad (6.5)$$

From equation (6.5), Kettani and Gubner suggest a moment estimator of H. They consider the case where $k = 1$ and replace $\rho_1$ by its sample estimate $\hat{\rho}_1$, as defined in equation (6.4). This gives an estimate for H of the form

$$\hat{H} = 0.5 + (0.5/\log_e 2) \ \log_e(1 + \hat{\rho}_1) \qquad (6.6)$$

Clearly, this estimate is straightforward to evaluate, requiring no iterative calculations. For more details of the properties of this estimator, see Kettani and Gubner [17].

We now propose an alternative estimator of H based upon equation (6.5), by considering the cases where $k>1$. Note that the sample equivalent of equation (6.5) can be expressed as

$$f(H) = \hat{\rho}_k - 0.5\{(k+1)^{2H} - 2k^{2H} + (k-1)^{2H}\} = 0. \qquad (6.7)$$

Thus, for a given observed $\hat{\rho}_k$, $k>1$, we can use a suitable numerical procedure to solve this equation, and find an estimate of H. We call this a HEAF(k) estimate of H.

To solve equation (6.7) for H, we use the well-known Newton-Raphson (N-R) method. This requires the derivative of f(H).

The algorithm to estimate HEAF($k$), for any lag $k$, consists of the following steps:

1. Compute the sample autocorrelations for lag $k$ of a given data set by equation (6.4). (Note that $X_i$ can be denoted as the number of bits, bytes, packets or bit rates observed during the $i$ th interval. If $X_i$ is a Gaussian process, it is known as fractional Gaussian noise).
2. Make an initial guess of $H$, e.g. $H_1 = 0.6$, then calculate $H_2$, $H_3$, $H_4$, ..., successively using $H_{r+1} = H_r - f(H_r) / f'(H_r)$, until convergence, to find the estimate $\hat{H}$ for the given lag $k$. Our initial consideration is of the case where $k = 2$ in equation (6.2); i.e. we first consider HEAF(2).

One of the major advantages of the HEAF estimator is speed, as the NR-method converges very quickly to a root. The form of the equation (6.7) appears to give quick converge (within at most four iterations) for any initial guess in our range of interest, namely $H$ in (0.2, 1). If an iteration value, $H_r$ is such that $f'(H_r) \cong 0$, then one can face "division by zero" or a near-zero number [18]. This will give a large magnitude for the next value, $H_{r+1}$ which in turn stops the iteration. This problem can be resolved by increasing the tolerance parameter in the N-R program. We have considered HEAF($k$), for $k = 2$, ...,11, and have encountered no difficulty in finding the root in (0.5, 1). Clearly, HEAF($k$) is a moment based estimator. We will consider aspects of its properties which arise from this in a subsequent paper.

# 7   Results and Discussion

We compared the estimators for sample sizes $n = 2^{13}$, $2^{14}$, $2^{15}$ by generating 20 different realisations for each of several values of $H$. The value of $H$ was estimated applying variety of methods, including HEAF(2).

   The results are presented in the Tables 1(a) and 1(b). The (absolute) bias, mean square error (mse), and variance (var) of the estimators are shown. A confidence interval was found for the Hurst parameter (and hence effectively for the bias), to give an indication of the accuracy of our simulated biases. HEAF(2) generally exhibits better performance (for $H$=0.6, 0.7, 0.8) than the other estimators as the length of the sample varies. There is a little difference observed between the size of the bias and mse of the wavelet and HEAF(2) estimators for $H$=0.5 and $H$=0.9. Table 2 describes the simulation results for samples from the FARIMA (0, d, 0) process. We generated 100 realizations for each $H$, each of length 16384. For $H = 0.6$, HEAF(2) is less biased than other estimators, except $R/S$. Also HEAF(2) exhibits less mean square error (mse) than others for all cases (i.e., for $H = 0.6, 0.7, 0.8, 0.9$).

## 7.1   Why Lag 2 (i.e. k = 2) in HEAF ?

To investigate a suitable lag ($k$) of the ACF for this estimator, we generated 50 realizations of varying sample lengths (For reasons of space, we here show the results for $n$=2048 and $n$=16384. Results for interim $n$ are similar) from the fGN process and the FARIMA(0,d,0) process.

Figure 1 shows the bias of the HEAF($k$) estimator based on the  ACF of lag $k$, $k=1,\ldots,11$. It is clear from the figure that HEAF(1) shows the least absolute bias compared to others and as a second choice, HEAF(2) is better than HEAF($k$), for most higher $k$. Figure 2 represents the mean square error (mse) of the HEAF($k$) estimators. HEAF(1) shows less mse than others. As an alternative choice, HEAF(2)  is better than the other HEAF estimators based on higher lags.

**Table 1(a).** Estimation of H for 20 independent realisations of various lengths (for the fGN process)

| Estimators | Sample size | | H = 0.5 | H = 0.6 | H = 0.7 | H = 0.8 | H = 0.9 |
|---|---|---|---|---|---|---|---|
| HEAF(2) | $2^{13}$ | $\hat{H}$ | 0.4937 | 0.5953 | 0.6939 | 0.7875 | 0.8698 |
| | | \|bias\| | 0.0063 | 0.0047 | 0.0061 | 0.0125 | 0.0302 |
| | | mse | 0.0005 | 0.0001 | 0.0001 | 0.0002 | 0.001 |
| | | var | 0.000022 | 0.000006 | 0.000003 | 0.000003 | 0.000005 |
| | | CI | 0.484, 0.503 | 0.59, 0.6 | 0.69, 0.697 | 0.784, 0.791 | 0.866, 0.874 |
| | $2^{14}$ | $\hat{H}$ | 0.5004 | 0.5993 | 0.6972 | 0.7916 | 0.8752 |
| | | \|bias\| | 0.0004 | 0.0007 | 0.0028 | 0.0084 | 0.0248 |
| | | mse | 0.0001 | 0 | 0 | 0.0001 | 0.0007 |
| | | var | 0.000007 | 0.000002 | 0.000001 | 0.000002 | 0.000003 |
| | | CI | 0.495, 0.506 | 0.596, 0.602 | 0.695, 0.699 | 0.789, 0.794 | 0.872, 0.879 |
| | $2^{15}$ | $\hat{H}$ | 0.5044 | 0.6007 | 0.6996 | 0.7946 | 0.8797 |
| | | \|bias\| | 0.0044 | 0.0007 | 0.0004 | 0.0054 | 0.0203 |
| | | mse | 0.0001 | 0.0001 | 0 | 0 | 0.0004 |
| | | var | 0.000005 | 0.000003 | 0.000001 | 0.000001 | 0.000002 |
| | | CI | 0.5, 0.509 | 0.598, 0.604 | 0.697, 0.702 | 0.793, 0.797 | 0.877, 0.882 |
| Wavelet | $2^{13}$ | $\hat{H}$ | 0.4982 | 0.6076 | 0.7143 | 0.8194 | 0.9231 |
| | | \|bias\| | 0.0018 | 0.0076 | 0.0143 | 0.0194 | 0.0231 |
| | | mse | 0 | 0.0001 | 0.0002 | 0.0004 | 0.0006 |
| | | var | 0.000002 | 0.000002 | 0.000002 | 0.000002 | 0.000003 |
| | | CI | 0.495, 0.501 | 0.605, 0.611 | 0.711, 0.717 | 0.816, 0.823 | 0.92, 0.926 |
| | $2^{14}$ | $\hat{H}$ | 0.4997 | 0.6089 | 0.7156 | 0.8205 | 0.9245 |
| | | \|bias\| | 0.0003 | 0.0089 | 0.0156 | 0.0205 | 0.0245 |
| | | mse | 0 | 0.0001 | 0.0003 | 0.0004 | 0.0006 |
| | | var | 0.000002 | 0.000002 | 0.000001 | 0.000001 | 0.000001 |
| | | CI | 0.497, 0.502 | 0.606, 0.611 | 0.713, 0.718 | 0.818, 0.823 | 0.922, 0.927 |
| | $2^{15}$ | $\hat{H}$ | 0.5013 | 0.6104 | 0.717 | 0.8218 | 0.9255 |
| | | \|bias\| | 0.0013 | 0.0104 | 0.017 | 0.0218 | 0.0255 |
| | | mse | 0 | 0.0001 | 0.0003 | 0.0005 | 0.0007 |
| | | var | 0.000001 | 0.000001 | 0.000001 | 0.000001 | 0.000001 |
| | | CI | 0.499, 0.504 | 0.608, 0.613 | 0.715, 0.719 | 0.82, 0.824 | 0.923, 0.928 |
| Higuchi | $2^{13}$ | $\hat{H}$ | 0.4995 | 0.603 | 0.7116 | 0.8232 | 0.9469 |
| | | \|bias\| | 0.0005 | 0.003 | 0.0117 | 0.0232 | 0.0469 |
| | | mse | 0.002 | 0.0021 | 0.0026 | 0.0029 | 0.0034 |
| | | var | 0.000103 | 0.000111 | 0.000132 | 0.000126 | 0.000063 |
| | | CI | 0.479, 0.52 | 0.582, 0.624 | 0.689, 0.735 | 0.801, 0.846 | 0.931, 0.963 |
| | $2^{14}$ | $\hat{H}$ | 0.4994 | 0.6028 | 0.7113 | 0.8272 | 0.9461 |
| | | \|bias\| | 0.0006 | 0.0028 | 0.0113 | 0.0272 | 0.0461 |
| | | mse | 0.0018 | 0.002 | 0.0025 | 0.0034 | 0.0033 |
| | | var | 0.000094 | 0.000103 | 0.000122 | 0.000141 | 0.00006 |
| | | CI | 0.48, 0.519 | 0.582, 0.623 | 0.689, 0.733 | 0.803, 0.851 | 0.931, 0.962 |
| | $2^{15}$ | $\hat{H}$ | 0.4993 | 0.6027 | 0.711 | 0.8268 | 0.9316 |
| | | \|bias\| | 0.0007 | 0.0027 | 0.0111 | 0.0268 | 0.0315 |
| | | mse | 0.0017 | 0.0019 | 0.0023 | 0.0033 | 0.0044 |
| | | var | 0.000088 | 0.000098 | 0.000117 | 0.000135 | 0.000181 |
| | | CI | 0.481, 0.518 | 0.583, 0.622 | 0.689, 0.733 | 0.804, 0.85 | 0.905, 0.958 |

**Table 1(b).** Estimation of H for 20 independent realisations of various lengths (for the fGN process)

| Estimators | Sample size | | H = 0.5 | H = 0.6 | H = 0.7 | H = 0.8 | H = 0.9 |
|---|---|---|---|---|---|---|---|
| Variance | $2^{13}$ | $\hat{H}$ | 0.5352 | 0.6211 | 0.7106 | 0.8099 | 0.9426 |
| | | \|bias\| | 0.0352 | 0.0211 | 0.0106 | 0.0099 | 0.0425 |
| | | mse | 0.0136 | 0.0122 | 0.0112 | 0.0106 | 0.0117 |
| | | var | 0.000651 | 0.00062 | 0.000585 | 0.000554 | 0.00052 |
| | | CI | 0.484, 0.586 | 0.571, 0.671 | 0.662, 0.759 | 0.763, 0.857 | 0.897, 0.988 |
| | $2^{14}$ | $\hat{H}$ | 0.546 | 0.6363 | 0.729 | 0.8307 | 0.9644 |
| | | \|bias\| | 0.046 | 0.0363 | 0.029 | 0.0307 | 0.0644 |
| | | mse | 0.015 | 0.0139 | 0.0124 | 0.0128 | 0.0156 |
| | | var | 0.000679 | 0.000662 | 0.000578 | 0.000623 | 0.000602 |
| | | CI | 0.494, 0.598 | 0.585, 0.688 | 0.681, 0.777 | 0.781, 0.881 | 0.915, 1.014 |
| | $2^{15}$ | $\hat{H}$ | 0.5618 | 0.6527 | 0.746 | 0.8483 | 0.9822 |
| | | \|bias\| | 0.0618 | 0.0527 | 0.046 | 0.0483 | 0.0822 |
| | | mse | 0.018 | 0.0167 | 0.0155 | 0.0152 | 0.0188 |
| | | var | 0.000746 | 0.000733 | 0.000706 | 0.000675 | 0.000632 |
| | | CI | 0.507, 0.616 | 0.599, 0.707 | 0.693, 0.799 | 0.796, 0.9 | 0.932, 1.033 |
| R/S | $2^{13}$ | $\hat{H}$ | 0.4993 | 0.5866 | 0.6687 | 0.7555 | 0.8278 |
| | | \|bias\| | 0.0007 | 0.0133 | 0.0313 | 0.0445 | 0.0722 |
| | | mse | 0.0036 | 0.0045 | 0.0056 | 0.0073 | 0.0106 |
| | | var | 0.000191 | 0.000227 | 0.000243 | 0.00028 | 0.000285 |
| | | CI | 0.472, 0.527 | 0.557, 0.617 | 0.638, 0.7 | 0.722, 0.789 | 0.794, 0.862 |
| | $2^{14}$ | $\hat{H}$ | 0.4955 | 0.5845 | 0.6728 | 0.7571 | 0.832 |
| | | \|bias\| | 0.0045 | 0.0155 | 0.0272 | 0.0429 | 0.068 |
| | | mse | 0.0036 | 0.0045 | 0.0056 | 0.0072 | 0.0101 |
| | | var | 0.000186 | 0.000224 | 0.000256 | 0.000282 | 0.00029 |
| | | CI | 0.468, 0.523 | 0.555, 0.614 | 0.641, 0.705 | 0.724, 0.791 | 0.798, 0.866 |
| | $2^{15}$ | $\hat{H}$ | 0.493 | 0.5826 | 0.6722 | 0.7586 | 0.8494 |
| | | \|bias\| | 0.007 | 0.0174 | 0.0278 | 0.0415 | 0.0506 |
| | | mse | 0.0037 | 0.0045 | 0.0056 | 0.0071 | 0.0085 |
| | | var | 0.000193 | 0.000222 | 0.000256 | 0.000283 | 0.000311 |
| | | CI | 0.465, 0.521 | 0.553, 0.612 | 0.64, 0.704 | 0.725, 0.792 | 0.814, 0.885 |

Figure 3 and Figure 4 depict the size of the bias and mse of HEAF(k) for $k=1,\ldots,11$, for the estimates for a FARIMA series. From Figure 3, it is seen that HEAF(1) yields the most biased estimator of $H$. For $n=$ 2048, HEAF(3), HEAF(7) and HEAF(9) give less bias for $H = 0.9, 0.7, 0.6$, respectively. For $n=$ 16384, the bias of the estimators fluctuates for all lags except lag1.

In Figure 4, HEAF(1) generally has higher mse than for HEAF($k$), $k>1$. For $n=$ 2048, HEAF(2) has least mse for $H = 0.6, 0.8, 0.9$. For $n=16384$, HEAF(3) has least mse for $H = 0.8$ and $H= 0.9$. HEAF(2) and HEAF(10) have smallest mse for $H = 0.6$ and $H = 0.7$, respectively. The difference between the mse's of the estimators of HEAF(2) and HEAF(3) is less than the difference between the mse's of HEAF(1) and HEAF(2). A similar scenario can be observed in case of bias of the estimators, which is shown in Figure 3.

In sharp contrast, HEAF(1) is an estimator with greater bias, with more mse, compared to HEAF(2) and HEAF(3), for the simulations from the FARIMA process. On the other hand, HEAF(1) is suitable for the fGN process (which is not surprising as it is based upon the ACF of that process). But the difference between the mse (or bias) of HEAF(1) and HEAF(2) is far less for the simulated fGN process than the difference for the simulated FARIMA process.

Thus, on the basis of bias and mse, HEAF(2) may be considered to give an estimator which is suitable for both fGN and FARIMA process. However, since the FARIMA process is more appropriate in simulating network traffic than fGN [19], and is also more flexible than fGN in terms of simultaneously modelling both short-range dependence (SRD) and LRD [20, 21], we suggest that HEAF(2) might be used to estimate the Hurst parameter as it can more or less adapt to both processes.

**Table 2.** Simulation of properties of estimators of the Hurst parameters for 100 realizations of sample length 16384 (for the FARIMA (0, d, 0) process)

| Estimators | | H = 0.6 | H = 0.7 | H = 0.8 | H = 0.9 |
|---|---|---|---|---|---|
| HEAF | $\hat{H}$ | 0.5929 | 0.6815 | 0.7737 | 0.86 |
| | \|bias\| | 0.0071 | 0.0185 | 0.0263 | 0.04 |
| | mse | 0.0002 | 0.0004 | 0.0008 | 0.0017 |
| | var | 0.0000011 | 0.0000006 | 0.0000007 | 0.000001 |
| | CI | 0.591,0.595 | 0.68,0.683 | 0.772,0.775 | 0.858,0.862 |
| Wavelet | $\hat{H}$ | 0.5889 | 0.6772 | 0.767 | 0.8559 |
| | \|bias\| | 0.0111 | 0.0228 | 0.033 | 0.0441 |
| | mse | 0.0002 | 0.0006 | 0.0011 | 0.002 |
| | var | 0.0000003 | 0.0000003 | 0.0000004 | 0.0000003 |
| | CI | 0.588,0.59 | 0.676,0.678 | 0.766,0.768 | 0.855,0.857 |
| Higuchi | $\hat{H}$ | 0.5841 | 0.6877 | 0.7697 | 0.8583 |
| | \|bias\| | 0.0159 | 0.0123 | 0.0303 | 0.0417 |
| | mse | 0.0045 | 0.006 | 0.0063 | 0.0065 |
| | var | 0.0000433 | 0.0000594 | 0.0000544 | 0.0000476 |
| | CI | 0.571,0.597 | 0.672,0.703 | 0.755,0.784 | 0.844,0.872 |
| Variance | $\hat{H}$ | 0.6799 | 0.7738 | 0.8251 | 0.9078 |
| | \|bias\| | 0.0799 | 0.0738 | 0.0251 | 0.0078 |
| | mse | 0.0342 | 0.0242 | 0.0264 | 0.0245 |
| | var | 0.0002808 | 0.000189 | 0.0002604 | 0.0002466 |
| | CI | 0.646,0.713 | 0.746,0.801 | 0.793,0.857 | 0.876,0.939 |
| R/S | $\hat{H}$ | 0.598 | 0.6874 | 0.7713 | 0.8531 |
| | \|bias\| | 0.002 | 0.0126 | 0.0287 | 0.0469 |
| | mse | 0.0061 | 0.0067 | 0.0097 | 0.0138 |
| | var | 0.0000614 | 0.0000657 | 0.0000893 | 0.0001167 |
| | CI | 0.582,0.614 | 0.671,0.704 | 0.752,0.79 | 0.831,0.875 |

## 7.2  Performance of the Estimators

The estimated Hurst parameters from the wavelet analysis, Aggregated variance and Whittle method have been presented in [22] for real data and synthetic data, where it is observed that the Whittle and wavelet methods overestimate the degree of self-similarity (i.e., produce $H > 1$). It seems that the wavelet method can be used to obtain relatively well performing estimators of the Hurst parameter for stationary traffic traces [23], which accords with our results. The study [23] explores the advantages and limitations of the wavelet estimators. Here the authors found that a traffic trace with a number of deterministic shifts in the mean rate results in steep wavelet spectrum, which leads to overestimating the Hurst parameter.

In this section, Hurst parameters are estimated for MPEG/video trace files (for example, frame size of MPEG video files, music video and movies) that are publicly available in [24, 25]. In Table 3, Hurst parameters are calculated for frame size (byte) of different types of frames. The frame type I, P, B denote Intracoded, Predictive and

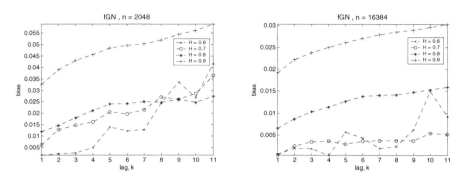

**Fig. 1.** The (absolute) bias of HEAF estimators for the ACF of different lags, investigated 50 realizations by fGN process

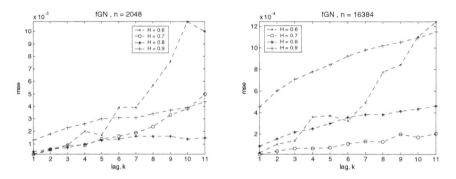

**Fig. 2.** The mean square error (mse) of HEAF estimators for the ACF of different lags, investigated 50 realisations by fGN process

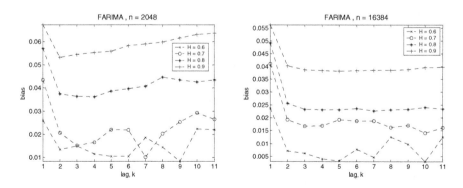

**Fig. 3.** The (absolute) bias of HEAF estimators for the ACF of different lags, investigated 50 realisations by FARIMA process

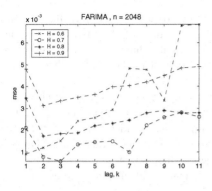
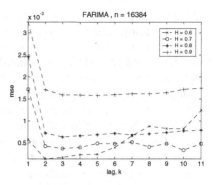

**Fig. 4.** The mean square error (mse) of HEAF estimator for the ACF of different lags, investigated 50 realisations by FARIMA process

**Table 3.** Estimated Hurst parameters for MPEG trace files [25]

| MPEG files | Hurst parameter for frame type | | | | Estimators |
|---|---|---|---|---|---|
| | I | P | B | IPB | |
| ATP | 1.151 | 0.994 | 1.032 | 0.412 | Wavelet |
| | 0.905 | 0.917 | 0.956 | 0.689 | HEAF(2) |
| | 0.661 | 0.765 | 0.730 | 0.756 | R/S |
| | 1.417 | 1.247 | 1.284 | 1.302 | V/T |
| | 0.997 | 0.995 | 0.996 | 0.996 | Higuchi |
| MrBean | 1.304 | 0.938 | 0.954 | 0.360 | Wavelet |
| | 0.967 | 0.953 | 0.979 | 0.563 | HEAF(2) |
| | 0.954 | 0.936 | 0.920 | 1.027 | R/S |
| | 1.284 | 0.945 | 1.081 | 1.128 | V/T |
| | 1.000 | 0.985 | 0.991 | 0.993 | Higuchi |
| Lambs | 1.229 | 0.861 | 1.043 | 0.373 | Wavelet |
| | 0.949 | 0.916 | 0.968 | 0.525 | HEAF(2) |
| | 0.831 | 0.921 | 0.895 | 1.003 | R/S |
| | 1.382 | 1.354 | 1.400 | 1.468 | V/T |
| | 0.996 | 0.998 | 0.997 | 0.997 | Higuchi |
| Bond | 1.189 | 1.003 | 1.026 | 0.241 | Wavelet |
| | 0.951 | 0.951 | 0.965 | 0.631 | HEAF(2) |
| | 0.951 | 0.903 | 0.865 | 0.976 | R/S |
| | 1.393 | 1.337 | 1.384 | 1.410 | V/T |
| | 1.006 | 1.007 | 1.005 | 1.006 | Higuchi |

**Table 4.** Hurst parameters for Movies [24]

| Files (Movies) Frame-type: I, Length (byte) | Estimators | |
|---|---|---|
| | Wavelet | HEAF(2) |
| Star Wars I CD1 | 0.437 | 0.944 |
| Star Wars I CD2 | 0.441 | 0.948 |
| Hackers | 0.490 | 0.555 |
| KissoftheDragon | 0.810 | 0.928 |
| LordOfTheRingsII-CD1 | 0.617 | 0.872 |
| LordOfTheRingsII-CD2 | 0.635 | 0.893 |
| RobinHoodDisney | 0.398 | 0.885 |
| TombRaider | 0.760 | 0.886 |

**Table 5.** Hurst parameters for Music Videos [24]

| Files (Music Videos) Frame-type: I, Length (byte) | Estimators | | | | |
|---|---|---|---|---|---|
| | Wavelet | HEAF(2) | R/S | V/T | Higuchi |
| Alanis Morissette - Ironic | 0.406 | 0.675 | 0.950 | 1.638 | 0.997 |
| Alizee - J'en Ai Mare | 0.066 | 0.580 | 0.110 | 1.651 | 1.000 |
| Crazy Town - Butterfly | 0.422 | 0.666 | 0.817 | 1.056 | 1.001 |
| DMX - Ruff Ryderz Anthem | 0.014 | 0.907 | 0.702 | 1.463 | 0.998 |
| Eminem - My Name Is | 0.455 | 0.559 | 0.586 | 1.231 | 1.000 |
| Jennifer Lopez - If You Had My Love | 0.540 | 0.613 | 0.691 | 1.225 | 0.999 |
| Metallica - Nothing Else Matters | 0.455 | 0.613 | 0.738 | 1.624 | 0.998 |
| Offspring - Self Esteem | 0.364 | 0.679 | 0.731 | 1.593 | 0.997 |
| Pink - Family Portrait | 0.520 | 0.608 | 0.720 | 1.158 | 1.002 |
| Shania Twain - Im Gonna Getcha Good | 0.535 | 0.615 | 0.637 | 1.599 | 0.998 |

Bi-directional frames respectively. For I-frames of all MPEG files, the wavelet based estimator produces a Hurst parameter of greater than 1.0. Similarly, the wavelet based estimator produces an estimate of $H$ that is greater than 1.0 for B-frame except "Mr. Bean" sequence. The $H$ calculated by the wavelet method is less than 0.5 for IPB frame of movies ATP, Lambs and Bond.

Table 4 illustrates the Hurst parameters estimated by the wavelet method and by HEAF(2), for the frame size (byte) of I-type in different movies. Table 5 depicts estimated Hurst parameters for frame size (byte) of music videos. In most cases, the wavelet-based estimate of $H$ is found to be less than 0.5. For "Alizee" and "DMX", the results seem strange as $H$ is estimated to be less than 0.015. It seems that the estimated $H$ is extremely underestimated. The V/T analysis always gave an estimate of $H> 1$, for all music videos.

## 8  Conclusions

It has been found that the established estimators for the Hurst parameter (except the wavelet method) can give poor estimates, as they sometimes underestimate or overestimate the degree of self-similarity. For example, for the simulation of $H = 0.6$ and $H = 0.8$, the estimated $H$ by R/S analysis were found to be 0.38 and 1.059 respectively. Because of space limitations we cannot provide all our simulation results here. This paper proposes a new estimator, HEAF, to estimate the Hurst parameter, $H$, of self-similar LRD network traffic. In particular, our results show that HEAF(2) is an estimator of $H$ with relatively good bias and mse, when estimating fractional Gaussian or FARIMA processes.

## References

1. Leland Will E. Taqqu M. S., Willinger W. and Wilson D. V., On the Self-similar nature of Ethernet Traffic (Extended version), IEEE/ACM Transactions on Networking, February 1994, Vol. 2, No. 1, pp. 1-15.
2. Thomas Karagiannis, Mart Molle, Michalis Faloutsos, Long-range dependence: Ten years of Internet traffic modeling, IEEE Internet Computing, 2004, Vol. 8, No. 5, pp. 57-64.
3. K. Park , G. Kim and M.E. Crovella, On the Relationship Between File Sizes Transport Protocols and Self-Similar Network Traffic, Int'l Conf. Network Protocols, IEEE CS Press, 1996, pp. 171-180.
4. Chen Y., Z. Deng, and C. Williamson, A Model for Self-Similar Ethernet LAN Traffic: Design, Implementation, and Performance Implications, In Proceedings of the 1995 Summer Computer Simulation Conference, July 24-26, 1995, pp. 831-837, Ottawa, Canada.
5. Hurst, H. E., Blank, R. P. and Simaika, Y. M. Long Term Storage: An Experimental Study. London, Constable, 1965.
6. C.R. Dietrich and G.N. Newsam, Fast and exact simulation of stationary Gaussian processes through circulant embedding of the covariance matrix, SIAM Journal on Scientific Computing, 1997, vol.18, pp.1088-1107.
7. Kokoszka, P. S. and Taqqu, M. S., Fractional ARIMA with stable innovations, Stochastic Processes and their Applications, 1995, vol.60, pp.19–47.

8. Fei Xue, T. T. Lee, Modeling and Predicting Long-range Dependent Traffic with FARIMA Processes, Proc. International symposium on communication, Kaohsiung, Taiwan, November 7-10,1999.

9. Walter Willinger, Vern Paxson, and Murad Taqqu, Self-similarity and Heavy Tails: Structural Modeling of Network Traffic. In A Practical Guide to Heavy Tails: Statistical Techniques and Applications, Adler, R., Feldman, R., and Taqqu, M.S., editors, Birkhauser, 1998, pp. 27-53.

10. Cox D., Long-Range Dependence: a Review. Statistics: An Appraisal, Iowa State Statistical Library, The Iowa State University Press, H. A. David and H. T. David (eds.), 1984, pp. 55-74.

11. Hurst H. E., Long-term storage capacity of reservoirs, Transactions of the American Society of Civil Engineers, 1951, vol.116, pp. 770-808.

12. Ton Dieker, Simulation of fractional Brownian motion, Masters Thesis, Department of Mathematical Sciences, University of Twente, The Netherlands, 2004.

13. T. Higuchi, Approach to an irregular time series on the basis of the fractal theory, Physica D, 1988, vol.31, pp. 277-283.

14. P. Abry, P. Flandrin, M. S. Taqqu and D. Veitch, Wavelets for the Analysis, Estimation, and Synthesis of Scaling Data, Self-Similar Network Traffic and Performance Evaluation. K. Park and W. Willinger (editors), John Wiley & Sons, New York, 2000, pp. 39-88.

15. P. Abry and D. Veitch, Wavelet Analysis of Long-Range Dependent traffic. IEEE Transactions on Information Theory, 1998, vol.44, No 1, pp. 2-15.

16. Stilian Stoev, Free MATLAB code for FARIMA, Boston University, Department of Mathematics and Statistics, 2004, http://math.bu.edu/people/sstoev/  (visited on September 03, 2005)

17. H. Kettani and J. A. Gubner, A Novel Approach to the Estimation of the Hurst Parameter in Self-Similar Traffic, Proceedings of the 27th Annual IEEE Conference on Local Computer Networks (LCN 2002), Tampa, Florida, November, 2002, pp. 160-165.

18. John H. Mathews, Numerical Methods for mathematics, science and Engineering, Prentice-Hall Intl, 1992.

19. Hong Zhao, Nirwan Ansari, Yun Q. Shi: Self-similar Traffic Prediction Using Least Mean Kurtosis, International Conference on Information Technology: Coding and Computing (ITCC), 2003, pp. 352-355.

20. Abdelnaser Adas, Traffic Models in Broadband Networks, IEEE Communications Magazine, July 1997, pp. 82-89.

21. Huaguang Feng, Thomas R. Willemain and Nong Shang, Wavelet-Based Bootstrap for Time Series Analysis, Communications in Statistics: Simulation and Computation, 2005, vol.34, No 2, pp 393-413.

22. UNC Network Data Analysis Study Group, University of North Carolina, http://www-dirt.cs.unc.edu/net_lrd/ (visited on May 23, 2005).

23. Stilian Stoev, Murad Taqqu, Cheolwoo Park and J.S. Marron, Strengths and Limitations of the Wavelet Spectrum Method in the Analysis of Internet Traffic, SAMSI, Technical Report #2004-8, March 26, 2004.

24. Frank H.P. Fitzek, Video and Audio Measurements for Pre-encoded Content, http://trace.kom.aau.dk/preencoded/index.html (visited on August 12, 2005).

25. MPEG traces, University of Würzburg Am Hubland, http://www-info3.informatik.uni-wuerzburg.de/MPEG/traces/ (visited on September 10, 2005).

# On Modeling Video Traffic from Multiplexed MPEG-4 Videoconference Streams

A. Lazaris, P. Koutsakis, and M. Paterakis

Dept. of Electronic and Computer Engineering
Information & Computer Networks Laboratory
Technical University of Crete, Chania, Greece
{alazaris, polk, pateraki}@telecom.tuc.gr

**Abstract.** Due to the burstiness of video traffic, video modeling is very important in order to evaluate the performance of future wired and wireless networks. In this paper, we investigate the possibility of modeling this type of traffic with well-known distributions. Our results regarding the behavior of single videoconference traces provide significant insight and help to build a Discrete Auto-regressive (DAR(1)) model to capture the behavior of *multiplexed     MPEG-4 videoconference movies* from VBR coders.

## 1 Introduction

As traffic from video services is expected to be a substantial portion of the traffic carried by emerging wired and wireless networks, statistical source models are needed for Variable Bit Rate (VBR) coded video in order to design networks which are able to guarantee the strict Quality of Service (QoS) requirements of the video traffic. Video packet delay requirements are strict, because delays are annoying to a viewer; whenever the delay experienced by a video packet exceeds the corresponding maximum delay, the packet is dropped, and the video packet dropping requirements are equally strict.

Hence, the problem of modeling video traffic, in general, and videoconferencing, in particular, has been extensively studied in the literature. VBR video models which have been proposed in the literature include first-order autoregressive (AR) models [2], discrete AR (DAR) models [1, 3], Markov renewal processes (MRP) [4], MRP transform-expand-sample (TES) [5], finite-state Markov chain [6, 7], and Gamma-beta-auto-regression (GBAR) models [8, 9]. The GBAR model, being an autoregressive model with Gamma-distributed marginals and geometric autocorrelation, captures data-rate dynamics of VBR video conferences well; however, it is not suitable for general MPEG video sources [9].

In [3] the authors show that H.261 videoconference sequences generated by different hardware coders, using different coding algorithms, have gamma marginal distributions (this result was also employed by [10], which proposes an Autoregressive Model of order one for sequences of H.261 encoding) and use this result to build a Discrete Autoregressive (DAR) model of order one, which works well when several sources are multiplexed.

Y. Koucheryavy, J. Harju, and V.B. Iversen (Eds.): NEW2AN 2006, LNCS 4003, pp. 46–57, 2006.

In [11-13], different approaches are proposed for MPEG-1 traffic, based on the lognormal, Gamma, and a hybrid Gamma/lognormal distribution model, respectively. Standard MPEG encoders generate three types of video frames: *I* (intracoded), *P* (predictive) and *B* (bidirectionally predictive); i.e., while *I* frames are intra-coded, the generation of *P* and *B* frames involves, in addition to intra-coding, the use of motion prediction and interpolation techniques. *I* frames are, on average, the largest in size, followed by *P* and then by *B* frames.

An important feature of common MPEG encoders (both hardware and software) is the manner in which frame types are generated. Typical encoders use a fixed Group-of-Pictures (GOP) pattern when compressing a video sequence; the GOP pattern specifies the number and temporal order of *P* and *B* frames between two successive *I* frames. A GOP pattern is defined by the distance N between *I* frames and the distance M between *P* frames. In practice, the most frequent value of M is 3 (two successive B frames) while the most frequent values of N are 6, 12, and 15, depending on the required video quality and the transmission rate.

In this work, we focus on the problem of modeling videoconference traffic from MPEG-4 encoders (the MPEG-4 standard is particularly designed for video streaming over wireless networks [14]), which is a relatively new and yet open issue in the relevant literature.

## 2   Videoconference Traffic Model

### A. Frame-Size Histograms

We use four different long sequences of MPEG-4 encoded videos (from [15]) with low or moderate motion (i.e, traces with very similar characteristics to the ones of actual videoconference traffic), in order to derive a statistical model which fits well the real data. The length of the videos varies from 45 to 60 minutes and the data for each trace consists of a sequence of the number of cells per video frame and the type of video frame, i.e., *I, P,* or *B*. We use packets of ATM cell size throughout this work, but our modeling mechanism can be used equally well with packets of other sizes. We have investigated the possibility of modeling the four videoconference videos with quite a few well-known distributions and our results show that the best fit among these distributions is achieved for all the traces studied with the use of the Pearson type V distribution. The Pearson type V distribution (also known as the "inverted Gamma" distribution) is generally used to model the time required to perform some tasks (e.g., customer service time in a bank); other distributions which have the same general use are the exponential, gamma, weibull and lognormal distributions [20]. Since all these distributions have been often used for video traffic modeling in the literature, they have been chosen as fitting candidates in order to compare their modeling results in the case of MPEG-4 videoconferencing.

The four traces under study are, respectively, a video stream extracted and analyzed from a camera showing the events happening within an office (Video Name: "Office Cam"); a video stream extracted and analyzed from a camera showing a lecture (Video Name: "Lecture Room Cam"); a video stream extracted and analyzed from a talk-show (Video Name: "N3 Talk"); a video stream extracted and analyzed from another talk-show (Video Name: "ARD Talk"). For each one of these movies we have used the high quality coding version, in which new video frames arrive every 40 msecs, in Quarter Common Intermediate Format (QCIF) resolution. The compression pattern used to encode all the examined video streams is IBBPBBPBBPBB, i.e., N=12, M=3, according to the definitions used in Section 1.

The frame-size histogram based on the complete VBR streams is shown, for all four sequences, to have the general shape of a Pearson type V distribution (this is shown in Figure 1, which presents indicatively the histogram for the lecture sequence; the other three traces have similar histograms).

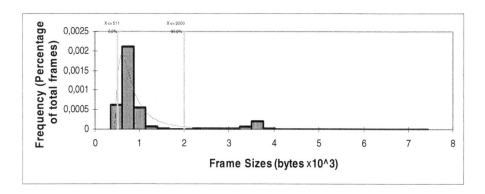

**Fig. 1.** Histogram for the frame size of the lecture camera trace

## B. Statistical Tests and Autocorrelations

The statistical test was made with the use of Q-Q plots. The Q-Q plot is a powerful goodness-of-fit test [3, 20], which graphically compares two data sets in order to determine whether the data sets come from populations with a common distribution (if they do, the points of the plot should fall approximately along a 45-degree reference line). More specifically, a Q-Q plot is a plot of the quantiles of the data versus the quantiles of the fitted distribution (a z-quantile of $X$ is any value $x$ such that $P((X \leq x) = z)$.

In Figure 2, we have plotted the 0.01-, 0.02-, 0.03-,... quantiles of the actual trace versus the respective quantiles of the various distribution fits for the ARD Talk trace (the results are similar for all the traces).

The Pearson V distribution fit is shown to be the best in comparison to the gamma, weibull, lognormal and exponential distributions, which are presented here (comparisons were also made with the negative binomial and Pareto distributions, which were also worse fits than the Pearson V). However, as already mentioned, although the Pearson V was shown to be the better fit among all distributions, the fit is not perfectly accurate. This was expected, as the gross differences in the number of bits required to represent $I$, $P$ and $B$ frames impose a degree of periodicity on MPEG-encoded streams, based on the cyclic GOP formats. Any model which purports to

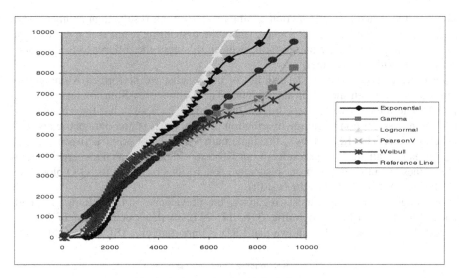

**Fig. 2.** Q-Q plot for the ARD TALK trace

reflect the frame-by-frame correlations of an MPEG-encoded video stream must account for GOP cyclicity, otherwise the model could produce biased estimates of cell loss rate for a network with some given traffic policing mechanism [9, 13].

Hence, we proceeded to study the frame size distribution for each of the three different video frame types (*I, P, B*), in the same way we studied the frame size distribution for the whole trace. This approach was also used in [9, 19].

Another approach, similar to the above, was proposed in [11]. This scheme uses again lognormal distributions and assumes that the change of a scene alters the average size of *I* frames, but not the sizes of *P* and *B* frames. However, it is shown in [4, 12] that the average sizes of *P* and *B* frames can vary 20% and 30% (often more than that), respectively, in subsequent scenes, therefore the size changes are statistically significant.

As it will be shown from our results, none of the above choices of distribution fits are relevant to the case of *I, P* and *B* frames of MPEG-4 videoconference traffic.

The mean, peak and variance of the video frame sizes for each video frame type (*I, P* and *B*) of each movie were taken again from [15] and the Pearson type V parameters are calculated based on the formulas for the mean and variance of Pearson V (the parameters for the other fitting distributions are similarly obtained based on their respective formulas). The autocorrelation coefficient of lag-1 was also calculated for all types of video frames of all four movies, as it shows the very high degree of correlation between successive frames of the same type (it was larger than 0.7 in all the cases, and in most of the cases it was larger than 0.9). The autocorrelation coefficient of lag-1 will be used in the following Sections of this work, in order to build a Discrete Autoregressive Model for each video frame type.

From the five distributions examined (Pearson V, exponential, gamma, lognormal, weibull) the Pearson V distribution once again provided the best fitting results for 11 of the 12 cases examined, i.e., for all video frame types of the office, N3 Talk and lecture camera traces, and for the *P, B* frame types of the ARD Talk trace. The only case in which the Pearson V distribution exhibits worse fitting results than another distribution is that of the N3 Talk *P* frames, where the best fitting result is derived with the use of the lognormal distribution (still, even in this case the difference in the goodness-of-fit results is marginal).

In order to further verify the validity of our results, we performed Kolmogorov-Smirnov tests for all the 12 fitting attempts. The Kolmogorov-Smirnov test (KS-test) tries to determine if two datasets differ significantly. The KS-test has the advantage of making no assumption about the distribution of data, i.e, it is non-parametric and distribution free. The KS-test uses the maximum vertical deviation between the two curves as its statistic *D*. The results of our KS-tests (Figure 3 is a characteristic example of these results), confirm our respective conclusions based on the Q-Q plots (i.e., the Pearson V distribution is the best fit).

Although controversy persists regarding the prevalence of Long Range Dependence (LRD) in VBR video traffic ([16, 17, 23]), in the specific case of MPEG-encoded video, research has shown that LRD is important [11, 18]. The results of our

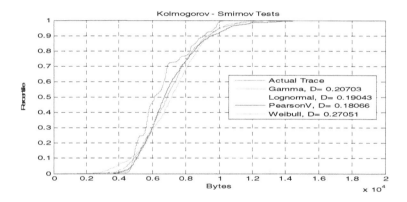

**Fig. 3.** KS-test (Comparison Percentile Plot) for the ARD TALK I-frames

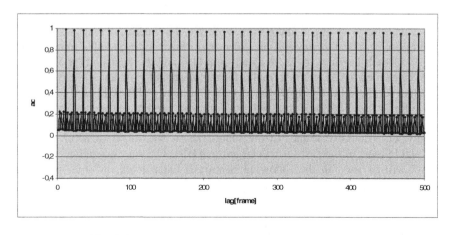

**Fig. 4.** Autocorrelation function of the lecture camera trace

study on single MPEG-4 videoconferencing agree with this conclusion. The autocorrelation function for the lecture camera trace is shown in Figure 4 (the respective Figures for the other three traces are similar). Two apparent periodic components are observed, one containing lags with low autocorrelation and the other lags with high autocorrelation. We observe that autocorrelation remains high even for large numbers of lags and that both components decay very slowly; both these facts are a clear indication of the importance of LRD. The existence of strong autocorrelation coefficients is due to the periodic recurrence of I, B and P frames.

Although the fitting results when modeling each video frame type separately, with the use of the Pearson V distribution, are clearly better than the results produced by modeling the whole sequence uniformly, the high autocorrelation shown in the Figure above can never be perfectly "captured" by a distribution generating frame sizes independently, according to a declared mean and standard deviation, and therefore none of the fitting attempts (including the Pearson V), as good as they might be, can achieve perfect accuracy. However, these results lead us to extend our work in order to build DAR models which inherently use the autocorrelation coefficient of lag-1 in their estimations and which will be shown to capture well the behavior of *multiplexed MPEG-4 videoconference movies*, by generating frame sizes independently for *I, P* and *B* frames.

## 3   The DAR (1) Model

A Discrete Autoregressive model of order $p$, denoted as DAR($p$) [21], generates a stationary sequence of discrete random variables with an arbitrary probability distribution and with an autocorrelation structure similar to that of an Autoregressive model. DAR(1) is a special case of a DAR(p) process and it is defined as follows: let $\{V_n\}$ and $\{Y_n\}$ be two sequences of independent random variables. The random variable $V_n$ can take two values, 0 and 1, with probabilities 1-$\rho$ and $\rho$, respectively. The random variable $Y_n$ has a discrete state space $S$ and $P\{Y_n = i\} = \pi(i)$. The sequence of random variables $\{X_n\}$ which is formed according to the linear model:

$$X_n = V_n X_{n-1} + (1- V_n) Y_n \tag{1}$$

is a DAR(1) process.

A DAR(1) process is a Markov chain with discrete state space $S$ and a transition matrix:

$$P = \rho I + (1-\rho) Q \tag{2}$$

where $\rho$ is the autocorrelation coefficient, $I$ is the identity matrix and $Q$ is a matrix with $Q_{ij} = \pi(j)$ for i, j $\in$ S.

Autocorrelations are usually plotted for a range W of lags. The autocorrelation can be calculated by the formula:

$$\rho(W) = E[(X_i - \mu)(X_{i+w} - \mu)]/\sigma^2 \tag{3}$$

where $\mu$ is the mean and $\sigma^2$ the variance of the frame size for a specific video trace.

## 4   DAR(1) Modeling Results and Discussion

As in [3], where a DAR(1) model with negative binomial distribution was used to
model the number of cells per frame of VBR teleconferencing video, we want to build
a model based only on parameters which are either known at call set-up time or can
be measured without introducing much complexity in the network. DAR(1) provides
an easy and practical method to compute the transition matrix and gives us a model
based only on four physically meaningful parameters, i.e., the mean, peak, variance
and the lag-1 autocorrelation coefficient $\rho$ of the offered traffic (these correlations, as
already explained, are typically very high for videoconference sources). According to
[22], the DAR(1) model can be used with any marginal distribution.

As explained in our work on modeling a single MPEG-4 videoconference trace, the
lag-1 autocorrelation coefficient for the $I$, $P$ and $B$ frames of each trace is very high in
all the studied cases. Therefore, we proceeded to build a DAR(1) model for each
video frame type for each one of the four traces under study. More specifically, in our
model the rows of the $Q$ matrix consist of the Pearson type V probabilities ($f_0$, $f_1$, …
$f_k$, $F_K$), where $F_K = \Sigma_{k > K} f_k$, and $K$ is the peak rate. Each $k$, for $k<K$, corresponds to
possible source rates less than the peak rate of K.

From the transition matrix in (2) it is evident that if the current frame has, for example, i cells, then the next frame will have i cells with probability $\rho+(1-\rho)*f_i$, and
will have k cells, $k \neq i$, with probability $(1-\rho)*f_k$. Therefore the number of cells per
video frame stays constant from one ($I$, $P$ or $B$) video frame to the next ($I$, $P$ or $B$)
video frame, respectively, in our model with a probability slightly larger than $\rho$ (for
example, in the ARD Talk trace, with probability slightly larger than 93.23%,
77.81%, 94.49% for the $I$, $P$ and $B$ frames of the trace, respectively). This is evident
in Figure 5, where we compare the actual $I$ frames of the ARD Talk trace and their respective DAR(1) model and it is shown that the DAR(1) model's data produce a
"pseudo-trace" with a periodically constant number of cells for a number of video
frames. This causes a significant difference when comparing a segment of the

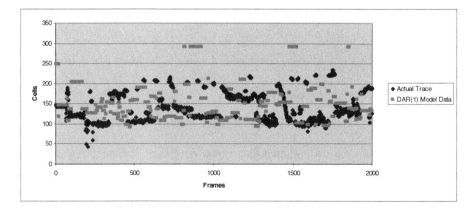

**Fig. 5.** Comparison for a single trace between a 2000 frame sequence of the actual $I$ frames
sequence of the ARD Talk trace and the respective DAR(1) model in number of cells/frame
(Y-axis)

sequence of *I, P, or B* frames of the actual ARD Talk video trace and a sequence of the same length produced by our DAR(1) model. The same vast differences also appeared when we plotted the DAR(1) models versus the actual *I, P* and *B* video frames of the actual N3Talk, office camera and lecture camera traces for a single movie.

However, our results have shown that the differences presented above become small for all types of video frames and for all the examined traces *for a superposition of 5 or more sources,* and are almost completely smoothed out in most cases, as the number of sources increases (the authors in [3] have reached similar conclusions for their own DAR(1) model and they present results for a superposition of 20 traces). This is clear in Figures 6-8, which present the comparison between our DAR(1) model and the actual *I, P, B* frames' sequences of the ARD Talk video, for *a superposition* of 20 traces (the results were perfectly similar for all video frame types of the other three traces; we have used the initial trace sequences to generate traffic for 20 sources, by using different starting points in the trace). The common property of all these results (derived by using a queue to model multiplexing and processing frames in a FIFO manner) is that the DAR(1) model seems to provide very accurate fitting results for *P* and *B* frames, and relatively accurate for *I* frames.

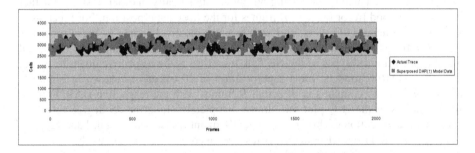

**Fig. 6.** Comparison for 20 superposed sources between a 2000 I frame sequence of the actual ARD Talk trace and the respective DAR(1) model in number of cells/frame (Y-axis)

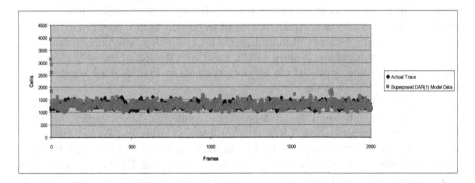

**Fig. 7.** Comparison for 20 superposed sources between a 2000 P frame sequence of the actual ARD Talk trace and the respective DAR(1) model in number of cells/frame (Y-axis)

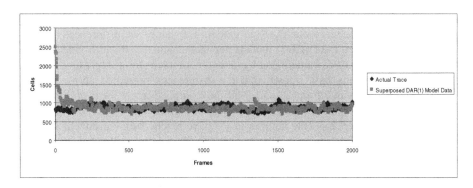

**Fig. 8.** Comparison for 20 superposed sources between a 2000 B frame sequence of the actual ARD Talk trace and the respective DAR(1) model in number of cells/frame (Y-axis)

However, although Figures 6-8 suggest that the DAR(1) model captures very well the behavior of the multiplexed actual traces, they do not suffice as a result. Therefore, we proceeded again with testing our model statistically in order to study whether it produces a good fit for the *I, P, B* frames for the trace superposition. For this reason we have used again Q-Q plots, and we present indicatively some of these results in Figures 9-10, where we have plotted the 0.01-, 0.02-, 0.03-,... quantiles of the actual *B* and *I* video frames' types of the N3 Talk trace versus the respective quantiles of the respective DAR(1) models, for a superposition of 20 traces. As shown in Figure 9, which presents the comparison of actual *B* frames with the respective DAR(1) models, the points of the Q-Q plot fall almost completely along the 45-degree reference line, with the exception of the first and last 3% quantiles (left- and right-hand tail), for which the DAR(1) model underestimates and overestimates, respectively, the probability of frames with a very small (large) number of cells. The very good fit shows that the superposition of the *B* frames of the actual traces can be modeled very well by a respective superposition of data produced by the DAR(1) model (similar results were derived for the superposition of *P* frames), as it was suggested in Figures 7, 8. Figure 10 presents the comparison of actual *I* frames with the respective DAR(1) model, for the N3 Talk trace. Again, the result suggested from Figure 6, i.e., that our method for modeling *I* frames of multiplexed MPEG-4 videoconference streams provides only relative accuracy, is shown to be valid with the use of the Q-Q plots. The results for all the other cases which are not presented in Figures 9-10 are similar in nature to the ones shown in the Figures.

One problem which could arise with the use of DAR(1) models is that such models take into account only short range dependence, while, as shown earlier, MPEG-4 videoconference streams show LRD. This problem is overcome by our choice of modeling *I, P* and *B* frames separately. This is shown in Figure 11. It is clear from the Figure that, even for a small number of lags, (e.g., larger than 10) the autocorrelation of the superposition of frames decreases quickly, for all the traces. Therefore, although in some cases the DAR(1) model exhibits a quicker decrease than that of the actual traces' video frames sequence, this has minimal impact on the fitting quality of the DAR(1) model. This result further supports our choice of using a first-order model.

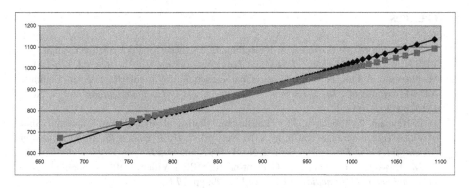

**Fig. 9.** Q-Q plot of the DAR(1) model versus the actual video for the B frames of the N3 Talk trace, for 20 superposed sources

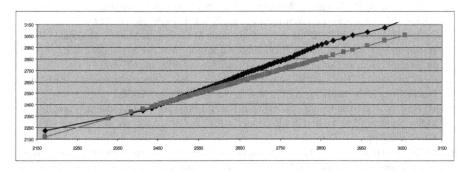

**Fig. 10.** Q-Q plot of the DAR(1) model versus the actual video for the I frames of the N3 Talk trace, for 20 superposed sources

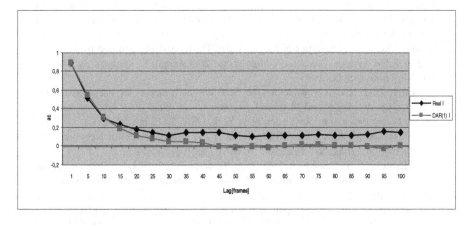

**Fig. 11.** Autocorrelation vs. number of lags for the I frames of the actual N3 Talk trace and the DAR(1) model, for 20 superposed sources

# References

1. D. M. Lucantoni, M. F. Neuts, A. R. Reibman, "Methods for Performance Evaluation of VBR Video Traffic Models," *IEEE/ACM Trans.Networking,* Vol. 2, 1994, pp. 176-180.
2. M. Nomura, T. Fuji, N. Ohta, "Basic Characteristics of Variable Rate Video Coding in ATM Environment", *IEEE Journal on Selected Areas in Communications,* Vol. 7, No. 5, 1989, pp. 752-760.
3. D. P. Heyman, A. Tabatabai, T. V. Lakshman, "Statistical Analysis and Simulation Study of Video Teleconference Traffic in ATM Networks", *IEEE Transactions on Circuits and Systems for Video Technology,* Vol. 2, No. 1, 1992, pp. 49-59.
4. A. M. Dawood, M. Ghanbari, "Content-based MPEG Video Traffic Modeling", *IEEE Transactions on Multimedia,* Vol. 1, No. 1, 1999, pp. 77-87.
5. B. Melamed and D. E. Pendarakis, "Modeling Full-Length VBR Video Using Markov-Renewal Modulated TES Models", *IEEE Journal on Selected Areas in Communications,* Vol. 16, No. 5, 1998, pp. 600-611.
6. K. Chandra, A. R. Reibman, "Modeling One- and Two-Layer Variable Bit Rate Video", *IEEE/ACM Transactions on Networking*, Vol. 7, No. 3, 1999, pp. 398-413.
7. Q. Ren, H. Kobayashi, "Diffusion Approximation Modeling for Markov Modulated Bursty Traffic and its Applications to Bandwidth Allocation in ATM Networks", *IEEE Journal on Selected Areas in Communications,* Vol. 16, No. 5, 1998, pp. 679-691.
8. D. P. Heyman, "The GBAR Source Model for VBR Videoconferences", IEEE/ACM Transactions on Networking (1997), Vol. 5, No. 4, 1997, pp. 554-560.
9. M. Frey, S. Ngyuyen-Quang, "A Gamma-Based Framework for Modeling Variable-Rate Video Sources: The GOP GBAR Model", *IEEE/ACM Trans. on Networking*, Vol. 8, No. 6, 2000, pp. 710-719.
10. S. Xu, Z. Huang, "A Gamma autoregressive video model on ATM networks", *IEEE Transactions on Circuits and Systems Video Technology,* Vol. 8, No. 2, 1998, pp. 138–142.
11. M. Krunz, S. K. Tripathi, "On the Characterization of VBR MPEG Streams," *in Proceedings of ACM SIGMETRICS*, Vol. 25, June 1997.
12. U. K. Sarkar, S. Ramakrishnan, D. Sarkar, "Modeling Full-Length Video Using Markov-Modulated Gamma-Based Framework," *IEEE/ACM Trans. on Networking*, Vol. 11, No.4, 2003, pp. 638-649.
13. O. Rose, "Statistical Properties of MPEG Video Traffic and Their Impact on Traffic Modeling in ATM Systems," *in Proceedings of the 20th Annual Conference on Local Computer Networks*, October 1995.
14. D. Arifler, B. L. Evans, "Modeling the Self-Similar Behavior of Packetized MPEG-4 Video Using Wavelet-Based Methods", *in Proceedings of the IEEE International Conference on Image Processing,* Rochester, New York, USA, 2002, pp. 848-851.
15. F. H. P. Fitzek, M. Reisslein, "MPEG-4 and H.263 Video Traces for Network Performance Evaluation", *IEEE Network,* Vol. 15, No. 6, 2001, pp. 40-54.
16. D. P. Heyman, T. V. Lakshman, "What are the Implications of Long-Range Dependence for VBR-Video Traffic Engineering", *IEEE/ACM Transactions on Networking*, Vol. 4, No.3, 1996, pp. 301-317.
17. B. K. Ryu, A. Elwalid, "The Importance of Long-Range Dependence of VBR Video Traffic in ATM Traffic Engineering: Myths and Realities", *in Proceedings of the ACM SIGCOMM 1996,* Stanford, CA, USA, pp. 3-14.

18. J. Beran, R. Sherman, M. S. Taqqu, W. Willinger, "Long-Range Dependence in Variable Bit-Rate Video Traffic", *IEEE Transactions on Communications*, Vol. 43, No. 2/3/4, 1995, pp. 1566-1579.
19. M. Krunz, H. Hughes, "A Traffic Model for MPEG-coded VBR Streams", *in Proceedings of the ACM SIGMETRICS*, Ottawa, Canada, 1995, pp. 47-55.
20. A. M. Law, W. D. Kelton, "Simulation Modeling & Analysis", $2^{nd}$ Ed., McGraw Hill Inc., 1991.
21. P. A. Jacobs, P. A. W. Lewis, "Time Series Generated by Mixtures", *Journal of Time Series Analysis*, Vol. 4, No. 1, 1983, pp. 19-36.
22. T. V. Lakshman, A. Ortega and A. R. Reibman, "VBR Video: Trade-offs and potentials", *Proceedings of the IEEE*, Vol. 86, No. 5, 1998, pp. 952-973.
23. K. Park, W. Willinger (editors), "Self-Similar Network Traffic and Performance Evaluation", John Wiley & Sons, Inc., 2000.
24. R. Schafer, "MPEG-4: A Multimedia Compression Standard for Interactive Applications and Services", *IEE Electronics and Communications Engineering Journal*, Vol. 10, No. 6, 1998, pp. 253-262.

# A New MAC Protocol Based on Multimedia Traffic Prediction in Satellite Systems

P. Koutsakis[*] and A. Lazaris

Dept. of Electronic and Computer Engineering
Information & Computer Networks Laboratory
Technical University of Crete, Chania, Greece
{polk, alazaris}@telecom.tuc.gr

**Abstract.** The provision of acceptable Quality-of-Service (QoS) for integrated multimedia traffic over a geosynchronous earth orbit (GEO) satellite network demands the existence of a well-designed Medium Access Control (MAC) protocol. This paper proposes a new dynamic satellite bandwidth allocation technique which is based on accurate videoconference traffic prediction. Our work is combined with another work on data traffic modeling and prediction and is shown to provide very good throughput and delay results.

## 1 Introduction

In recent years, broadband satellite networks have attracted significant attention as a part of the global communications infrastructure. Satellite networks have specific characteristics which make them an attractive solution in order to complement terrestrial networks in providing worldwide access to the present and future generation multimedia services. More specifically, satellite networks include global coverage, providing access for remote users to terrestrial wide area networks from any location within the satellite coverage area, and broadcast/multicasting.

However, at the same time other characteristics of satellite networks, such as long propagation delays (270 ms round-trip delay for a GEO satellite network, i.e., at least 540 ms from the time the signaling information is sent from the station to the satellite until it reaches the receiver terminal), the limited bandwidth to be shared among many users, and limitation in power (which implies stringent use of buffer memory, transponder capacity and processing power) create the need for a well-designed Medium Access Control (MAC) protocol, especially in today's networks which are expected to handle bursty multimedia traffic. Among the above mentioned problems, the propagation delay is the most significant one, as it makes bursty users' (such as video users') current traffic profile rather useless for bandwidth allocation, since the profile will probably have changed significantly by the time the bandwidth allocation is made by the Network Control Center (NCC), which we consider to be integrated in the satellite on-board device.

---

[*] This research was supported by the Operational Program for Education and Initial Vocational Training "PYTHAGORAS: Funding of research groups in Technical University of Crete", M 2.2 MIS, 89192, co-funded by the EU through the Third Community Support Framework.

Y. Koucheryavy, J. Harju, and V.B. Iversen (Eds.): NEW2AN 2006, LNCS 4003, pp. 58 – 69, 2006.

Videoconference traffic is expected to be a substantial portion of the traffic carried by emerging networks, hence presenting a challenge for satellite system designers. For Variable Bit Rate (VBR) coded video, statistical models are needed to design networks which are able to guarantee the strict Quality of Service (QoS) requirements of video traffic. Video packet delay requirements are strict, because delays are annoying to a viewer; whenever the delay experienced by a video packet exceeds the corresponding maximum delay, the packet is dropped, and the video packet dropping requirements are equally strict. Therefore, a good statistical model can be very useful in evaluating network performance under various videoconferencing loads.

The most well-known and used video standards for this application today are H.263 and MPEG-4. In this work, we first build a model which accurately captures the behavior of multiplexed H.263 videoconference movies, and we proceed to use this model for predicting the behavior of videoconference traffic in our new MAC protocol proposal for a Digital Video Broadcasting Return Channel Satellite (DVB-RCS) system. Our proposed MAC protocol is shown, via a simulation study, to achieve video packet delays and video packet dropping probability only slightly worse than the ideal scenario in which the actual video user requirements for each video frame were a priori known to the NCC.

Finally, our modeling scheme is combined with the modeling approach used in [10] for predicting self-similar data traffic and the combination is shown to provide once again very satisfying results in terms of minimizing bandwidth waste.

## 2  H.263 Video Traffic Modeling

H.263 is a video standard that can be used for compressing the moving picture component of audio-visual services at low bit rates. It adopts the idea of PB frames, i.e., two pictures being coded as a unit. Thus a PB-frame consists of one P-picture which is predicted from the previous decoded P-picture and one B-picture which is predicted from both the previous decoded P-picture and the P-picture currently being decoded. The name B-picture was chosen because parts of B-pictures may be bidirectionally predicted from the past and future pictures. With this coding option, the picture rate can be increased considerably without increasing the bit rate much [1].

Our work focuses on the accurate fitting of the marginal (stationary) distribution of video frame sizes of single videoconference traces. More specifically, our work follows the steps of the work presented in [2], where Heyman et al. analyzed three videoconference sequences coded with a modified version of the H.261 video coding standard and two other coding schemes, similar to the H.261. The authors in [2] found that the marginal distributions for all the sequences could be described by a gamma (or equivalently negative binomial) distribution.

In our work, we have studied three different long sequences of H.263 VBR encoded videos, from the publicly available library of frame size traces of long MPEG-4 and H.263 encoded videos provided by the Telecommunication Networks Group at the Technical University of Berlin [3]. We have investigated the possibility of modeling the traces with a number of well-known distributions and our results have shown that the best fit among these distributions for modeling a single movie is achieved for all traces examined with the use of the Pearson type V distribution (also known as the inverted gamma distribution).

The three traces are, respectively:

1. A video stream extracted and analyzed from a camera showing the events happening within an office ("Office Cam").
2. A video stream extracted and analyzed from a talk-show ("ARD Talk").
3. A video stream extracted and analyzed from another talk-show ("N3 Talk").

All three of these traces are movies with low or moderate motion.

In the work presented in this paper, we will show that, based on the good fit of the Pearson V distribution for modeling a single movie, the behavior of *both single and multiplexed H.263 videoconference movies* from VBR coders can be accurately captured and used for efficient proactive resource management in satellite systems *(the case of modeling multiplexed videoconference streams is especially significant for our study, since numerous sources are multiplexed in the uplink channel)*. To do this, we will first briefly explain our results on modeling a single movie.

For each one of the three videos under study we have used the VBR coding version, in which new video frames arrive every 80 msecs. The length of the videos varies from 45 to 60 minutes and the data for each trace consists of a sequence of the number of cells per video frame. Table 1 presents the trace statistics for each trace (packet size=48 information bytes; we use packets of ATM cell size throughout this work, but our mechanism can be used equally well with packets of other sizes, as the nature of our modeling results would not be altered at all), as well as the parameters $(\alpha, \beta)$ of the Pearson type V distribution fit for each movie. The Probability Density Function (PDF) of a Pearson type V distribution with parameters $(\alpha, \beta)$ is:

$$f(x)= [x^{-(\alpha+1)} e^{-\beta/x} ]/ [\beta^{-\alpha} \Gamma(\alpha)], \text{ for all } x>0 \qquad (1)$$

and zero otherwise.

The mean and variance are given by the equations:

$$\text{Mean}=\beta/(\alpha-1) \qquad (2)$$

$$\text{Variance}=\beta^2/[(\alpha-1)^2(\alpha-2)] \qquad (3)$$

**Table 1.** Trace Statistics

| Movie | Mean (packets) | Peak (packets) | Standard Deviation (packets) | Pearson type V parameters $(\alpha,\beta)$ |
|---|---|---|---|---|
| Office | 18.8 | 109 | 6.8 | (9.64,162.43) |
| ARD Talk | 49.5 | 277 | 27 | (5.36,215.82) |
| N3 Talk | 53.1 | 291 | 30.3 | (5.07,216.11) |

Autoregressive models have been used in the past to model the output bit rate of VBR encoders, e.g. [4, 5]. A Discrete Autoregressive model of order $p$, denoted as DAR($p$) [6], generates a stationary sequence of discrete random variables with an arbitrary probability distribution and with an autocorrelation structure similar to that of an Autoregressive model. DAR(1) is a special case of a DAR(p) process and it is defined as follows: let $\{V_n\}$ and $\{Y_n\}$ be two sequences of independent random variables. The random variable $V_n$ can take two values, 0 and 1, with probabilities $1-\rho$ and $\rho$, respectively. The random variable $Y_n$ has a discrete state space $S$ and $P\{Y_n = i\} = \pi(i)$. The sequence of random variables $\{X_n\}$ which is formed according to the linear model:

$$X_n = V_n X_{n-1} + (1 - V_n) Y_n \qquad (4)$$

is a DAR(1) process.

A DAR(1) process is a Markov chain with discrete state space $S$ and a transition matrix:

$$P = \rho I + (1-\rho) Q \qquad (5)$$

where $\rho$ is the autocorrelation coefficient, $I$ is the identity matrix and $Q$ is a matrix with $Q_{ij} = \pi(j)$ for $i, j \in S$.

Autocorrelations are usually plotted for a range W of lags. The autocorrelation can be calculated by the formula:

$$\rho(W) = E[(X_i - \mu)(X_{i+w} - \mu)]/\sigma^2 \qquad (6)$$

where $\mu$ is the mean and $\sigma^2$ the variance of the frame size for a specific video trace.

As in [2], where a DAR(1) model with negative binomial distribution was used to model the number of cells per frame of VBR teleconferencing video, we want to build a model based only on parameters which are either known at call set-up time or can be measured without introducing much complexity in the network. DAR(1) provides an easy and practical method to compute the transition matrix and gives us a model based only on four physically meaningful parameters, i.e., the mean, peak, variance and the lag-1 autocorrelation coefficient $\rho$ of the offered traffic (these correlations are typically very high for videoconference sources). According to [7], the DAR(1) model can be used with any marginal distribution.

More specifically, in our model the rows of the $Q$ matrix consist of the Pearson type V probabilities $(f_0, f_1, \dots f_k, F_K)$, where $F_K = \Sigma_{k > K} f_k$, and $K$ is the peak rate. Each $k$, for $k<K$, corresponds to possible source rates less than the peak rate of K.

The lag-1 autocorrelation coefficient $\rho$ is estimated by Equation (2) to be equal to 0.943 for the office camera trace, 0.867 for the ARD Talk trace and 0.872 for the N3 Talk trace.

We proceeded with testing our model statistically in order to study whether it produces a good fit for the trace superposition. For this reason we have used Q-Q plots. The Q-Q plot is a powerful goodness-of-fit test [2, 8], which graphically compares two data sets in order to determine whether the data sets come from populations with a common distribution (if they do, the points of the plot should fall approximately along a 45-degree reference line). More specifically, a Q-Q plot is a plot of the quantiles of the data versus the quantiles of the fitted distribution (a z-quantile of $X$ is any value $x$ such that $Pr(X \leq x) = z$).

In Figure 1, we have plotted the 0.025-, 0.05-, 0.075-,... quantiles of the actual office camera trace versus the respective quantiles of the DAR(1) model for the superposition of the 15 traces. As shown in the Figure, the points of the Q-Q plot fall almost completely along the 45-degree reference line, with the exception of the last

2.5% quantile (right-hand tail), for which the DAR(1) model greatly overestimates the probability of frames with a very large number of cells. The very good fit shows that the superposition of the actual traces can be modeled very well by a respective superposition of data produced by the DAR(1) model.

The same conclusions were deducted by our results for the superposition of various number of sources transmitting the other two traces. The small differences observed in our results were that, in the case of the ARD Talk trace, the overestimation made by the DAR(1) model was respectively smaller than that shown in Figure 1, and that, in the case of the N3 Talk trace, the overestimation starts "earlier", i.e., it covers the last 5% quantile.

Nevertheless, all our results show that our model captures with great accuracy the behavior of H.263 videoconference traffic.

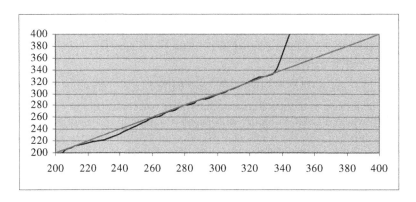

**Fig. 1.** Q-Q plot of DAR(1) model versus the actual office camera trace for 15 superposed sources

## 3  Our MAC Protocol Proposal

The Digital Video Broadcasting Return Channel Satellite (DVB-RCS) standard [12] develops a communication system for the return channel (uplink channel), i.e., the link from the user terminal to the network gateway. Due to the expected services features and large delay-bandwidth product of satellite networks, DVB-RCS represents a proper test bed for the proposed resource allocation schemes. The most relevant elements of the DVB-RCS Network are: a) RCSTs (Return Channel Satellite Terminals), i.e., a generic access terminal, b) the NCC (Networks Control Center), a device in charge of managing the access and bandwidth allocation for RCSTs, c) Gateways and Feeders, which are the elements that receive and transmit information outside the network [10].

As in [10], our proposed satellite medium access scheme is based on a Multi-Frequency Time Division Multiple Access (MF-TDMA) approach, according to which a carrier is divided in timeslots (grouped in frames and superframes). MF-TDMA schemes are capable of providing efficient and flexible bandwidth utilization [10, 14]. The system parameters are taken from [10] and are presented in Table 2.

**Table 2.** System Parameters

| Frame Duration | 26.5 ms |
|---|---|
| Carriers | 4 |
| Slots/frame/carrier | 128 |
| Bytes/slot | 53 |
| System global rate | 8 Mbps |

The NCC allocates to each active RCST a set of timeslots, each characterized by a frequency, bandwidth, start time and duration time. The DVB-RCS standard provides five allocation request types [12, 13] which can be joined in order to satisfy the QoS requirements:

- Continuous Rate Assignment (CRA) is a fixed capacity negotiated between the RCST and the NCC. It is maintained across frames until a new negotiation.
- Rate Based Dynamic Capacity (RBDC) is a capacity allocated to the RCST based on its rate request (bytes/frame), and is subject to a maximum rate limit negotiated between the RCST and the NCC. The last request from a RCST overwrites all previous RBDC requests from the same RCST.
- Volume Based Dynamic Capacity (VBDC) is an assignment strategy in which a terminal signals its request in terms of total number of slots required to empty its queue. The request remains effective as long as not all of the requested time slots have been granted. This strategy is especially suited for bursty traffic [14].
- Absolute Volume Based Dynamic Capacity (AVBDC) request is a request similar to VBDC, with the difference that the last request from a RCST overwrites all previous AVBDC requests from the same RCST.
- Free Capacity Assignment is not a true request, allocating the otherwise unused capacity. It is automatic and does not involve signaling from the RCST to the NCC.

The modeling approach proposed in [10] for self-similar data traffic (such as World Wide Web traffic) lets an RBDC request correspond to the data traffic prediction, plus a corrective factor $\zeta_N$. This approach will also be incorporated in the second part of our results, in Section 4, where data traffic is incorporated into our simulations for the satellite system.

The first difference between the work presented in [10] and our work is that, additionally to the MF-TDMA frame structure, we adopt the idea that, after all requests have been satisfied, the bandwidth left is distributed freely following a certain algorithm (this approach is named in the literature as a Combined Free and Demand Assignment Scheme, CFDAMA scheme); the algorithm implemented in our scheme is a simple round-robin assignment algorithm to all RCSTs which are currently active. This idea of a hybrid protocol was first proposed in [15] and is especially useful in allocating slots to video users, as the difficulty in providing them with adequate bandwidth due to their frequent changes in bandwidth needs could be somewhat alleviated by their acquiring the unused channel bandwidth freely in a round-robin manner. We use the Combined Free and Demand Assignment Scheme with Piggybacking (CFDAMA-PB) version of the protocol, which was shown in [16] to be the most efficient way of making reservations. According to the PB strategy, user stations send their capacity requests embedded in the header of their packets. The free capacity distribution performed by the protocol brings the end-to-end delay performance at low loads close to that obtained with random access protocols, while the demand-based bandwidth allocation at the beginning of each frame guarantees the protocol's stability, robustness and efficient utilization of transmission bandwidth at high loads. Free capacity distribution in an MF-TDMA frame structure was also used in [14], with which we will conceptually compare our work in Section 4.

The second and most important difference of our work with [10] is that an approach similar to the one used in [10] for data traffic (letting the RBDC request correspond to the data traffic prediction) is not enough for real time Variable Bit Rate (rt-VBR) videoconference traffic, which has strict QoS requirements in terms of average video packet dropping (set to a maximum of 0.1% in our work) and end-to-end video packet delays (set to a maximum of 0.6 seconds in our work, which is especially strict considering that, for each possible failure of our prediction due to underassignment, the respective packets which would have to wait for a new assignment will have a minimum end-to-end video packet delay of 0.54 seconds). For this reason, we propose the following different approach in our MAC scheme.

As explained in Section 2, the Pearson V fit provides a good (but not perfect) fit for a single H.263 videoconference trace. Also, a logical assumption for next generation networks is that videoconference users will be allowed to adopt one of just a few specific "modes" (each corresponding to a set of traffic parameters). This is especially plausible for videoconference traffic, as the number of variations between source bandwidth requirements is naturally restricted by the type of application (a much larger pool of "modes" would have to be used in the case of video traffic). Therefore, in this work we consider that a videoconference user can adopt one of the three "modes" presented in Table 1 (a slightly larger pool of modes would have to be used in an actual satellite system scenario). Based on the good model for single videoconference traces and the highly accurate model of multiplexed traffic, *we propose that the "burden" of traffic prediction for the RCSTs should fall on the NCC instead of the RCSTs.* More specifically, the NCC should run a real-time simulation, both for single and for multiplexed videoconference sources. Hence, based on the "mode" declared by the RCSTs at call establishment, the NCC does not need to wait for a request from the RCSTs every channel frame (which would arrive with a delay of more than 5 channel frames, due to the propagation delay); instead, it can start allocating resources to the videoconference terminals, by simulating the single source models with the sources' mean rate as a simulation start point, and by computing the free slots in each channel frame (using the DAR(1) models for multiplexed videoconference traffic, and subtracting the estimated used slots from the total number of slots in the system) in order to allocate the estimated number of free slots in a round-robin manner to all active RCSTs. With this slot allocation scheme, the RCST will not need to send frequent requests to the NCC but it will only need to send a "corrective" AVBDC request every superframe (defined in our work as equal to 6 channel frames, to account for the propagation delay). The reason for sending this request will be for the RCST to help the NCC correct any mistakes (due to either slots overassignment or underassignment) of the models produced at the NCC via online simulation. After receiving the AVBDC request, the NCC will resume its simulation with the current RCST state (in terms of bandwidth requirements) as a start point. As it will be shown from our results, this approach, which minimizes the need for signaling among the video RCSTs and the NCC, provides clearly improved results in terms of videoconference users' QoS requirements satisfaction, due to the quality of the prediction made by our video modeling scheme.

## 4 Results and Discussion

### 4.1 Conceptual Comparison with Relevant Work

In [14], the authors propose a satellite MAC protocol based on a CFDAMA MF-TDMA access scheme (i.e., they use the same approach with our work). The protocol, however, proposes the use of a CRA-type assignment for rt-VBR traffic, such as the videoconferencing traffic used in our work, with the difference that the assignment is fixed for the duration of the connection (no new negotiation is needed), and equal to the rt-VBR user's peak transmission rate; the reason for this choice in [14] is that rt-VBR traffic has strict delay and packet dropping constraints, and no accurate traffic prediction mechanism is provided in [14], hence leading to the "defensive" choice of peak cell rate assignment, which leads to significant bandwidth waste, as the assignment is most of the time larger than the rt-VBR user's actual needs (the other problem with this type of assignment is that if the assignment was made for less than the peak transmission rate, it would at times lead to severe packet dropping; the use, in [14], of the CFDAMA policy could probably help to partially alleviate the packet dropping problem, through the use of the free slot assignment, but the authors chose the "safer" solution of peak transmission rate assignment). The inferior performance of such an assignment will also be shown in our simulation results. Finally, it should be noted that video traffic in [14] is generated with the use of a Markov Modulated Poisson Process (MMPP) model, whereas we use actual video traces in our work.

### 4.2 First Implementation Case: Only Video Traffic Present in the Channel

At the start of our simulation study, we let each videoconference user choose one of the three traffic parameter sets ("modes") which are presented in Table 1, with equal probability.

Three MAC schemes will be compared in this Section. The first is our proposed scheme, the second is a scheme conceptually similar to [14], allocating to each video terminal its peak rate (declared at call establishment) and the third is an "ideal" scheme, as we want to compare our protocol with a similar one in which the NCC would "know", without any information exchange (therefore, no contention is necessary among video RCSTs), exactly what the video RCSTs' bandwidth demands for the next video frame will be.

Figure 2 presents our simulation results for the average video packet dropping metric versus the system utilization. Utilization indicates the traffic load normalized to the uplink capacity, e.g., a traffic load equal to 40% represents 40% of the 8 Mbps uplink capacity, i.e., 3.2 Mbps system throughput. As it is shown in the Figure, the difference in video packet dropping between our scheme and the "ideal" case is so small that it can be considered almost negligible for all normalized video traffic loads. Our scheme can handle up to 76% system load while at the same time satisfying the strict QoS requirement of maximum video packet dropping equal to 0.1%; the respective maximum system load which the "ideal" scheme can handle is 79%. The reason that none of the two schemes can achieve a higher throughput is the high burstiness of video traffic; in certain channel frames, video bursts from more than one RCST happen to take place simultaneously in the uplink channel. Although our traffic modeling scheme can often predict such bursts, the total amount of requested bandwidth in certain channel frames may surpass the system's available capacity; this will lead to inevitable video packet dropping, as some of the packets may will not be sent within the roughly three channel frames which pass before the arrival of the next video frame (when a new video frame arrives, all packets of the previous video frame which have not yet been sent are discarded).

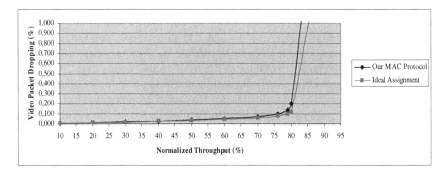

**Fig. 2.** Average Video packet dropping vs. System Utilization

Figure 3 presents our simulation results for the average end-to-end video packet delay versus the system utilization. The results are similar in nature with those of Figure 2, denoting that our scheme's results are again very close to the ones achieved by the "ideal" scheme.

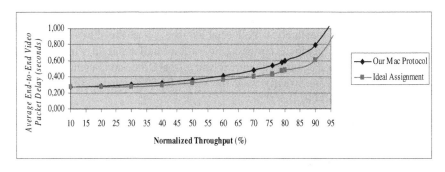

**Fig. 3.** Average End-to-end video packet delay vs. System Utilization

The scheme which allocates the peak rate to each video terminal is not shown in Figures 2-3, as it achieves zero packet dropping and the minimum possible end-to-end video packet delay (respectively, constant and equal to 0.274 seconds, i.e., equal to the propagation delay plus a very small additional amount of time in order for the video RCST to gain access to the channel). However, these two exceptionally good results come at a high cost. As explained earlier and shown in Table 1, all the video-conference traces used in our study are bursty. More specifically, all three traces used in our study have a peak-to-mean ratio larger than 5.4. This means that the constant allocation of the peak rate to all active video sources causes a double disadvantage to the MAC scheme:

1. it causes significant loss of valuable bandwidth resources, as many slots are left unused,
2. the number of free slots in each channel frame, which could be used for other types of traffic (such as data traffic) is heavily decreased, hence restricting them from accessing the system and failing to efficiently satisfy their QoS requirements (this will be shown through our simulation results in Section 4.3).

For these reasons, the maximum channel throughput achieved by allocating the peak rate to each video RCST is very low, equal to 32%, i.e., less than half the achieved throughput from our scheme.

### 4.3 Second Implementation Case: Combining Video and Data Traffic Models

Self-Similar traffic modeling has been shown to fit the data traffic of both a typical Ethernet network [17] and of World Wide Web (WWW) applications [18]. For this reason, self-similarity has been widely used in the literature either simply for modeling data traffic (e.g., [14]), or for data traffic prediction [9, 10]. As already explained earlier, in this work we adopt the approach of [10] for data traffic prediction, and we let an RBDC request correspond to the data traffic prediction, plus a corrective factor $\zeta_N$ (as in [10], we use the corrective factor $\zeta_1$ in our simulations). Data traffic has lower priority than video traffic, as it is much more delay-tolerant. For this reason, we set an upper end-to-end delay bound of 1 second for data packets.

Figure 4 presents our simulation results for the average end-to-end data packet delay versus the system utilization. The results presented in the Figure are the average of three different "divisions" of the system load: in the first case, 30% of the total load was offered from video traffic (e.g., for a normalized system load of 60%, 18% was offered from video traffic), and 70% from data traffic; in the second case, both types of traffic offered 50% of the total system load; in the third case, 70% of the total load was offered from video traffic and 30% from data traffic. Once again, the results presented in Figure 4 are generally similar in nature with those of Figures 2 and 3, denoting that our scheme's results are very close to the ones achieved by the "ideal" scheme. More specifically, as the system load increases, the "ideal assignment" scheme achieves a lower delay of about 0.1-0.2 seconds for small and medium system loads and more than 0.3 seconds for high loads in comparison to our scheme, due to the lack of contention (and therefore, lack of collisions) in the "ideal assignment" scenario. Regarding the achieved maximum system throughput (maximum throughput for which all the QoS requirements of video and data RCSTs are satisfied), the results are again qualitatively similar with those in Section 4.2: the maximum throughput achieved by our scheme is 83%, by the "ideal assignment scheme" 85% and by the scheme assigning the peak rate to video RCSTs equal to 41%. It is clear that all three schemes achieve a much higher throughput with the addition of data traffic into the system. The reason for this is that a significant portion of the slots left unused in the case when only video traffic exists in the system, are ideally filled with the much less demanding data traffic, which does not have an equally urgent need to be transmitted as video traffic (no data packets are dropped if they are not transmitted within a specified amount of time), therefore it can "compromise" with the use of whichever slots are left unused. Still, the channel throughput results cannot reach beyond 85% even for the ideal assignment scheme; this is once more due to the burstiness of video traffic.

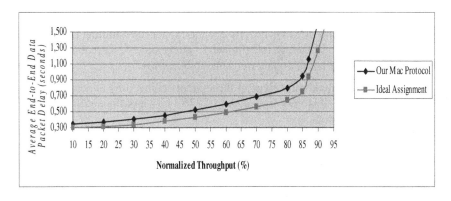

**Fig. 4.** Average End-to-end data packet delay vs. System Utilization

# References

1. ITU-T Recommendation, H.263, 3/1996.
2. D. P. Heyman, A. Tabatabai and T. V. Lakshman, "Statistical Analysis and Simulation Study of Video Teleconference Traffic in ATM Networks", *IEEE Transactions on Circuits and Systems for Video Technology,* Vol. 2, No. 1, pp. 49-59, 1992.
3. http://www-tkn.ee.tu-berlin.de/research/trace/trace.html
4. B. Maglaris, D. Anastassiou, P. Sen, G. Karlsson, and J. D. Robbins, "Performance Models of Statistical Multiplexing in Packet Video Communications", *IEEE Transactions on Communications*, Vol. 36, No.7, pp. 834-844, 1988.
5. C. Shim, I. Ryoo, J. Lee and S. Lee, "Modeling and Call Admission Control Algorithm of Variable Bit Rate Video in ATM networks", *IEEE Journal on Selected Areas in Communications*, vol.12, No.2, pp. 332-344, 1994.
6. P. A. Jacobs and P. A. W. Lewis, "Time Series Generated by Mixtures", *Journal of Time Series Analysis*, Vol. 4, No. 1, pp. 19-36, 1983.
7. T. V. Lakshman, A. Ortega and A. R. Reibman, "VBR Video: Trade-offs and potentials", *Proceedings of the IEEE,* Vol. 86, No. 5, pp. 952-973, 1998.
8. A. M. Law and W. D. Kelton, "Simulation Modeling & Analysis", 2nd Ed., McGraw Hill Inc., 1991.

9. Z. Jiang and V. C. M. Leung, "A Predictive Demand Assignment Multiple Access Protocol for Internet Access over Broadband Satellite Networks", *International Journal of Satellite Communications and Networking*, Vol. 21, No. 4-5, pp. 451-467, 2003.

10. F. Chiti, R. Fantacci and F. Marangoni, "Advanced Dynamic Resource Allocation Schemes for Satellite Systems", *in Proceedings of the IEEE International Conference on Communications (ICC) 2005*, Vol. 3, pp. 1469-1472, Seoul, Korea.

11. H. Peyravi, "Medium Access Control Protocols Performance in Satellite Communications", *IEEE Communications Magazine*, Vol. 37, No. 3, pp. 62-71, 1999.

12. Std., ETSI EN 301 790 V1.3.1 (2003-03).

13. Std., ETSI EN 101 790 V1.2.1 (2003-01).

14. A. Iuoras, P. Takats, C. Black, R. DiGirolamo, E. A. Wibowo, J. Lambadaris and M. Devetsikiotis, "Quality of Service-Oriented Protocols for Resource Management in Packet-Switched Satellites", *International Journal of Satellite Communications*, Vol. 17, No. 2-3, pp. 129-141, 1999.

15. S. V. Krishnamurthy and T. Le-Ngoc, "Performance of CF-DAMA Protocol with Pre-Assigned Request Slots in Integrated Voice/Data Satellite Communications", *in Proceedings of the IEEE International Conference on Communications (ICC) 1995*, Vol. 3, pp. 1572-1576, Seattle, USA.

16. T. Le-Ngoc and S. V. Krishnamurthy, "Performance of Combined Free/Demand Assignment Multiple Access Schemes in Satellite Communications", *International Journal of Satellite Communications*, Vol. 14, No.1, pp. 11-21, 1996.

17. W. E. Leland, M. S. Taqqu, W. Willinger and D. Wilson, "On the Self-Similar Nature of Ethernet Traffic (extended version)", *IEEE Transactions on Networking*, Vol. 2, No. 1, pp. 1-15, 1994.

18. M. E. Crovella and A. Bestavros, "Self-Similarity in World Wide Web Traffic: Evidence and Possible Causes", *IEEE Transactions on Networking*, Vol. 5, No. 6, pp. 835-846, 1997.

# Measurement-Based Optimization of a 3G Core Network: A Case Study

Fabio Ricciato, René Pilz, and Eduard Hasenleithner

Telecommunications Research Center Vienna (ftw.)
ricciato@ftw.at

**Abstract.** We consider the optimization of the Core Network section of a mobile cellular network. While we focus on GPRS the proposed method can be applied to UMTS as well. The problem is to find an optimal assignment of PCUs (Packet Control Units, a module of the BSC) to SGSNs based on measured data. Two concurrent optimization goals apply: balance the number attached Mobile Stations among the available SGSNs, and at the same time minimize the inter-SGSN Routing Area Updates. The input data for the optimization can be extracted from the live traffic signaling by passively monitoring the Gb links between the PCUs and SGSNs. We show how to estimate the mobility matrix and the distribution of attached Mobile Stations for each Routing Area, and how to clean-up the data at hand. A novel ILP formulation is provided for the (re)assignment problem. We present exemplary numerical results for a case study based on real traces from an operational network.

## 1 Introduction

Public wide-area wireless networks are now migrating towards third-generation systems (3G) supporting packet-switched data services. A 3G network includes two main sections: a Packet-Switched Core Network (PS-CN), which is based on IP, and one or more Radio Access Network (RAN). Most operators with legacy GSM networks now maintains two separate RANs for data traffic, GPRS and UMTS, sharing a common PS-CN. Each RAN is divided into *subsystems*, each consisting of one *controller*, called Base Station Controller (BSC) in GPRS and Radio Network Controller (RNC) in UMTS, connected to several Base Transceiver Stations (BTSs). The latter maintain the air interface in the cells, while the BSC/RNC control the radio connections with the Mobile Stations (MS) and the wired interface to the core network. On the PS-CN side each BSC/RNC connects to a Serving GPRS Support Node (SGSN), which performs functions like access control, mobility management, paging, route management etc. Additional details of the GPRS/UMTS network structure can be found in [1]. We focus here on the GPRS section, but the proposed methods can be applied to UMTS as well. The module responsible for handling packet traffic within the BSC is called Packet Control Unit (PCU). The PCUs are connected to the SGSNs via dedicated links (Gb interface). In this work we address one aspect of the PS-CN engineering, namely the optimal (re)assignment of PCUs to SGSNs based on measured data. The PCU-SGSN assignment problem has a practical

Y. Koucheryavy, J. Harju, and V.B. Iversen (Eds.): NEW2AN 2006, LNCS 4003, pp. 70–82, 2006.

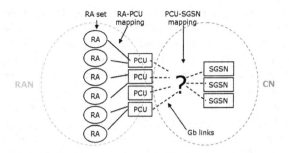

**Fig. 1.** Scheme of the RA-PCU and PCU-SGSN mapping

interest, particularly for those networks where SGSN locations are highly concentrated at one or few physical sites, such that the re-assignment cost is limited to a local rewiring.

The paper is organized as follows. In Section 2 we state the problem, and in Section 3 provide a novel Integer Linear Programming (ILP) formulation for it. In Section 4 we show how to extract the relevant input data (mobility matrix, attached MS) by means of passive traffic monitoring on the Gb links between the PCUs and SGSNs. In Section 5 we present numerical results for a semi-empirical case study based on real traces from an operational network. In Section 6 we discuss some practical issues and constraints found in the practical application of the optimization problem for the engineering of a real network. Finally, in Section 7 we conclude.

*Related Works.* For an overview of 3G network planning and optimization see [2]. Most of the previous literature in cellular network optimization has addressed the Location Area planning (e.g. [3]) or the cell-to-switch assignment problem [4] [5]). These are different from the PCU-SGSN assignment considered here, which does not appear to be addressed in any previous paper. Also, to the best of our knowledge this is the first work presenting numerical results based on a large-scale signaling dataset from an operational GPRS network.

## 2 Problem Statement

A simplified scheme of the GPRS network is given in Figure 1. A Routing Area (RA) is the homologous of a Location Areas in the circuit-switched network (see [1, p. 129]) and represents the basic unit for MS location management. By passively sniffing the Gb interface it is possible to determine the current RA as well as any RA change for each MS, as explained later in Section 4. In the most simple case one PCU covers a single RA, but it is also possible that one RA is split among multiple PCUs, or that a single PCU covers multiple RAs. The design of the radio coverage (i.e. the placement of cells and the assignment of cells to RAs) as well as the association between RAs and PCUs (RA-PCU mapping) are part of the RAN planning process, that is out of the scope of this work. **Here we assume that the configuration of the radio access**

**network is given, and the problem is to find an optimal assignment of PCU to SGSNs.** In other words the problem is to find the optimal PCU-SGSN mapping *given* a certain RA-PCU mapping and the measured mobility patterns. As a side value, the optimization method will output the minimum number of required SGSNs. We do not cover here the problem of *dimensioning* individual Gb links, which is decoupled from the assignment problem.

In order to minimize the resource consumption within each individual SGSN, it is desirable to balance the total workload across the set of available SGSNs. Generally speaking, the SGSN workload involves several dimensions of physical and logical resources (CPU cycles, buffer, bandwidth, memory states etc.) consumed by different types of traffic units (e.g. signaling messages, data packets). To make the problem tractable we will consider a mono-dimensional metric for the SGSN workload, namely the (peak) number of contemporary attached MSs. This is consistent with the fact that the capacity of commercial SGSNs is typically expressed in terms of the maximum number of attachable MSs. Denoting by $\alpha_m$ the peak number of MSs attached to SGSN $m$, the primary optimization objective is to minimize the maximum value $\alpha = \max_m \{\alpha_m\}$. An additional objective is to minimize the incidence of inter-SGSN Routing Area Updates. A MS performs a Routing Area Update (RAU) procedure when moving from RA-$i$ to RA-$j$. If both RAs are attached to the same SGSN, the RAU consumes resources exclusively within the SGSN. If instead the two RAs are attached to different SGSNs, a more complex procedure is triggered, namely the inter-SGSN RAU (iRAU for short). Each iRAU involves four (!) different elements - the two SGSNs, the GGSN and the HLR - consuming resources within *each* element and communication overhead between them (signaling traffic on the Gn network). Therefore, it is desirable to minimize the rate of iRAU messages across the network, hereafter denoted by $\phi$. In general, the two goals of minimizing $\phi$ and $\alpha$ are concurrent and a trade-off is in place between them. The optimization process must be designed to find the optimal trade-off curve between $\phi$ and $\alpha$, based on which the network engineers can select the best operational point by taking into account external informations (e.g. equipment features) to weigh the relative importance of each objective.

We will consider discrete time-series as input data. From the point of view of network planning it is important to minimize the *peak values* of the quantities related to resource consumption. Within the scope of this work we will adopt a static optimization approach, therefore we will preliminarily reduce each time-series to a single scalar value representing its peak (i.e. $a_i(k) \rightarrow \bar{a}_i$ and $b_{ij}(k) \rightarrow \bar{b}_{ij}$). A more sophisticated approach would feed the optimization process with the whole time-profile of each input process $a_i(k)$ rather than just its peak $\bar{a}_i$. In this way, one can achieve a tighter estimation of the required resources, but at the cost of a larger formulation complexity and hence computation time. This approach, referred to as "time-varying optimization" or "multi-period optimization" [6, Chapter 11] has been explored in some past papers [7] [8]. From a mathematical point of view, the problem formulation given in Section 3 could be easily extended towards multi-period optimization. We leave this option to future extensions.

# 3    Problem Formulation

## 3.1    Notation

The following notation is introduced.

- $a_i(k)$ is the number of MS attached to RA-$i$ at time $t_k$.
- $b_{ij}(k)$ is the intensity of the RAU flow from RA-$i$ to RA-$j$ at time $\tau_k$;

We use indices $i, j$ for the RAs and $m$ for the SGSNs. Note that the time axis for the two time-series $a_i(k)$ and $b_{ij}(k)$ are independent from each other, i.e. the sequences $\{t_k\}$ and $\{\tau_k\}$ do not need to be synchronized. The methods for measuring both time-series from passive traces are discussed later in Section 4. The total number of RAs and SGSNs will be denoted by $N$ and $M$ respectively.

- $\overline{a}_i = \max_k a_i(k)$ and $\overline{b}_{ij} = \max_k b_{ij}(k)$ denote the peak values of the series $a_i(k)$ and $b_{ij}(k)$ respectively. the complete set $\{\overline{b}_{ij}\}$ will be referred to as the (peak) "Mobility Matrix".
- $x_{i,m}$ is the binary variable accounting for the RA-SGSN mapping: $x_{i,m} = 1$ means RA-$i$ is connected to the $m$th SGSN.
- $\alpha_m$ is the peak number of MS attached to the $m$th SGSN;
- $\alpha = \max_m \{\alpha_m\}$ is the maximum number of MS attached to a single SGSN;
- $\phi_m^+$ and $\phi_m^-$ are the peak intensities of the outbound / inbound iRAU flow from / to the $m$th SGSN;
- $\phi = \max_m \{\phi_m^+ + \phi_m^-\}$ is the maximum rate of cumulated iRAU flow (inbound and outbound) for a single SGSN
- $y_{i,j,m}^+$ and $y_{i,j,m}^-$ are binary support variables defined for RA pairs with non-null RAU flow (formally $i, j : \overline{b}_{ij} \neq 0$); the value $y_{i,j,m}^+ = 1$ [resp. $y_{i,j,m}^- = 1$] means that the RAU flow from RA-$i$ to RA-$j$ contributes to the outbound [resp. inbound] iRAU flow for the $m$th SGSN.

The choice to minimize the *maximum* iRAU flow per-SGSN ($\phi = \max_m \{\phi_m^+ + \phi_m^-\}$) is based on the implicit assumption that the most critical resources consumed by iRAU messages are located within each SGSN; alternatively, in case that they are deemed to be within the HLR and/or GGSNs, one should consider the minimization of the *total* iRAU over all the SGSNs, i.e. $\phi = \frac{1}{2} \sum_m \{\phi_m^+ + \phi_m^-\}$, and accordingly change the constraint (8) introduced below.

## 3.2    ILP Formulation

The goal of the optimization is to jointly minimize both $\alpha$ and $\phi$ given the set of constraints formulated below. As usual with dual objective optimization, two possible strategies can be adopted. The first option is to use a combined cost function $C = w \cdot \alpha + (1 - w) \cdot \phi$, wherein the relative weight factor $w$ serves as a tuning knob to trade-off between the two minimization objectives. The alternative strategy foresees a two-step optimization. In Step I the primary objective $\alpha$ is minimized and the absolute optimal value is found, denoted by $\alpha_{opt}$. In Step II, the primary variable is constrained within a neighborhood of the absolute optimal value defined by a slack factor $\sigma$ (i.e. $\alpha \leq \alpha_{opt} \cdot \sigma$), and the secondary variable $\phi$ is minimized. With this approach the role of tuning knob is played by the value of $\sigma$. In this work, we adopt the two-step strategy.

The following set of constraints define an Integer Linear Programming (ILP) formulation of the problem:

*minimize:* $\alpha$ (in Step I) or $\phi$ (in Step II)

*subject to:*

$$\alpha \leq \alpha_{opt} \cdot \sigma \quad \text{(only for Step II)} \tag{1}$$

$$\sum_{m=1..M} x_{i,m} = 1, \ i = 1..N \tag{2}$$

$$\sum_{m=1..M} y_{i,j,m}^{+} = \sum_{m=1..M} y_{i,j,m}^{-} \ \forall i,j : \bar{b}_{ij} \neq 0 \tag{3}$$

$$\sum_{m=1..M} y_{i,j,m}^{+} \leq 1, \ \sum_{m=1..M} y_{i,j,m}^{-} \leq 1 \ \forall i,j : \bar{b}_{ij} \neq 0 \tag{4}$$

$$x_{i,m} + y_{i,j,m}^{-} = x_{j,m} + y_{i,j,m}^{+}, \ \forall i,j : \bar{b}_{ij} \neq 0, m = 1..M \tag{5}$$

$$y_{i,j,m}^{-} + y_{i,j,m}^{+} \leq 1, \ \forall i,j : \bar{b}_{ij} \neq 0, m = 1..M \tag{6}$$

$$\phi_m^{-} = \sum_{i,j:\bar{b}_{ij} \neq 0} \bar{b}_{ij} \cdot y_{i,j,m}^{-}, \quad \phi_m^{+} = \sum_{i,j:\bar{b}_{ij} \neq 0} \bar{b}_{ij} \cdot y_{i,j,m}^{+}, \quad m = 1..M \tag{7}$$

$$\phi \geq \bar{\phi}_m^{+} + \bar{\phi}_m^{-}, \forall m = 1..M \tag{8}$$

$$\alpha_m = \sum_{i=1..N} \bar{a}_i \cdot x_{i,m}, \ m = 1..M \tag{9}$$

$$\alpha \geq \alpha_m, \ m = 1..M \tag{10}$$

$$x_{i,m}, y_{i,j,m}^{+}, y_{i,j,m}^{-} \quad \text{binary} \tag{11}$$

Constraint (1) is present only in Step II (recall the value of $\alpha_{opt}$ is the output of Step I). Constraint (2) forces each RA to be attached to one SGSN. Constraints (3)-(4) enforce iRAU flow conservation. The constraints (5)-(6) are the core of the formulation: together with the previous flow conservation constraints they drive the support variable $y_{i,j,m}^{+} = 1$ [resp. $y_{i,j,m}^{-} = 1$ ] iff $x_{i,m} = 1$ and $x_{j,m} = 0$ [resp. $x_{i,m} = 0$ and $x_{j,m} = 1$]. Recall that the binary support variables $y_{i,j,m}^{+}$ and $y_{i,j,m}^{-}$ are key to the formulation: they account for the contribution of the RAU subflow from RA-$i$ to RA-$j$ to the outbound /inbound iRAU flow from / to the $m$th SGSN. Constraints (7) build up the inbound / outbound iRAU flow for each SGSN from the values of the support variables $y_{i,j,m}^{+}$ and $y_{i,j,m}^{-}$. Constraints (9) and (10) define the number of MS attached to each SGSN $\alpha_m$ and its maximum $\alpha$ respectively.

# 4   Input Data Extraction and Preparation

## 4.1   Monitoring Setting

The input data for the optimization can be obtained by passively monitoring the traffic on the Gb links between the PCU and the SGSNs. For this purpose it is required to capture both signaling and data packets, and to access packet fields at different protocol layers. The development of a large-scale passive monitoring system and its deployment in the operational network were accomplished within the METAWIN project [9]. It includes a real-time parser for the whole PS-CN protocol stack, and advanced features like packet-to-MS association and MS location. The latter are of key importance for the extraction of the optimization input data. Frames are captured with DAG cards and recorded with GPS synchronized time-stamps. The packet traces are anonymized on-the-fly by hashing all subscriber related fields (e.g. IMSI, MSISDN) at all protocol layers. In this section we describe the process of extracting the input data $a_i(k)$ and $b_{ij}(k)$ from passive traces captured on Gb. During the analysis of a dataset based on real traces we recognized the need for some "data cleaning" stages that are briefly discussed here, along with exemplary figures from our dataset.

## 4.2   Extraction of the Mobility Matrix $b_{ij}(k)$

Whenever a MS moves from RA $i$ to RA $j$ it sends a RAU message to the SGSN covering $j$. The RAU message seen on Gb contains both the identifiers of the "old" and "new" RAs, $i$ and $j$ respectively. Our code parses all RAU messages and counts the occurrences of $(i, j)$ pairs at time bins of length $T_b$ (we used $T_b = 5$ min). The resulting dataset is then "cleaned-up" across the following stages:

**Non-monitored RAs.** Our monitoring system covers only a fraction of the Gb links, hence of PCUs and RAs. The identifier of a non-monitored RAs (say $z$ in Figure 2) appears in the "old RA" field of a RAU message when the MS moves into a monitored RA (say $x$), thus introducing a non-null element $(z, x)$ and

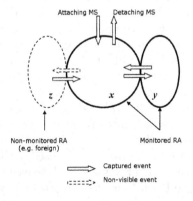

**Fig. 2.** Monitored RAs

hence a new row in the mobility matrix. In order to limit the size of the latter, we merge all the non-monitored RAs into a single virtual "external RA", so that their RAU flows are mapped into a single row of the mobility matrix (see Section 6 for a discussion about the handling the external-RA in the optimization process, and the associated monitoring requirements).

**Invalid identifiers.** For each RAU message seen on Gb, the identifier for the destination RA $j$ is filled in by the BSC, and is therefore fully trustable (assuming no configuration errors in the BSC). Instead, the identifier of the source RA $i$ is advertised directly by the MSs, that could advertise a wrong or invalid identifier (e.g. due to terminal bug). For instance, in our dataset we found a large amount of messages carrying the null value.

**Sporadic elements.** Sporadic elements. It can be expected that the mobility matrix is sparse, i.e. most elements are null, as distant RAs can not exchange RAU flow. This is an advantage for our optimization, since the size of the ILP problem (specifically the number of constraints) depends on the number of non-null elements $\bar{b}_{ij} > 0$. In some cases we found pairs of RAs with very sporadic RAU messages, i.e. $b_{ij}(k)$ holds non-null values only in very few time bins. This is probably the case of distant RAs that occasionally come into proximity (e.g. due to fluctuations in the radio coverage) or to RAs in a region with very low mobility (e.g. rural areas distant from major roads). For our purpose it is convenient to ignore these spurious elements, so as to reduce the size of the problem instance, with a negligible impact on the quality of the final solution. A simple threshold-based filtering suffices to eliminates sporadic elements.

**Anomalous peaks.** The problem formulation given in Section 3 takes as input the peak values of the RAU flow vector $b_{ij}(k)$. However for some $(i, j)$ pairs the time series $b_{ij}(k)$ displays some anomalous high peaks that are considerably higher than the "normal" peaks we want to consider in the network planning. These can be originated by occasional events like BSC failures or pre-planned rebooting. In fact, when RA $i$ is switched off at time $t_1$, the attached MSs migrate immediately to the neighboring RAs, say $j$, generating a high peak in the RAU flow $b_{ij}(k)|_{k \approx t_1}$. Conversely, when RA $i$ is switched on again at time $t_2$, most MS (but not all, due to hysteresis) will return back producing a smaller peak in $b_{ji}(k)|_{k \approx t_2}$. If the peak value $\bar{b}_{ij}$ is computed by simple maximization of the time-series $b_{ij}(k)$, these peaks would inflate its value and lead to large resource over-estimation. To avoid that it is required to identify and filter away outlier peaks before applying the maximization. We used the following simple method based on a sliding window. Given a discrete-time vector $x(k)$ we consider a moving window of length $W$. We denote by $m_W(k)$ and $\sigma_W(k)$ respectively the median and standard deviation of the samples in the range $(k - W/2, k + W/2)$. The generic sample in $k$ is marked as "outlier" if it exceeds the value of $m_W(k) + 3\sigma_W(k)$. In this case its value is set down to $m_W(k)$. We used window length $W=1$ hour. We applied these procedures to our dataset, and derived a "cleaned up" version of the $b_{ij}(k)$ vectors for the 127 RAs in the dataset. From it, we derived the peak

Mobility Matrix $\{\overline{b}_{ij}\}$ by simple maximization. A graphical representation of the matrix is given in Figure 4 (note that index 128 is associated to the external-RA, present only as source RA).

### 4.3   Estimation of Per-RA Attached MSs $a_i(k)$

From Gb traces it is also possible to estimate the number of attached MSs in each RA at each instant. Due to privacy constraints the MS are identified by an arbitrary unique identifier instead of the real IMSI (International Mobile Subscriber Identifier, see [1, p. 44]). Our monitoring system is able to associate signaling and data frames to the corresponding MS. For each MS, tracking key signaling messages (e.g. Attach / Detach, Routing Area Updates) and the data packets exchanged during activity periods enables the exact localization of the MS at the level of RA. This requires the implementation of simplified state machines to track the location and the *state* (Attached / Detached ) of each MS. Following the 3GPP specifications, a generic MS is considered "implicitly detached" after $T_d$ seconds from the last packet (the Routing Area Update Timeout). We have developed an ad-hoc stateful code scalable enough to track the entire population of MS in the network under study. An internal vector of counter indicates the number of attached MSs to each RA. The counter is sampled at regular intervals (1 min) in order to generate the discrete signal $a_i(k)$. Note that the time-series $a_i(k)$ can have different periodicity than $b_{ij}(k)$ (1 min vs. 5 min) as the two are extracted independently. The estimation method is prone to two types of error due to the finiteness of the monitoring scope in time and space. Such "border effects" are discussed in details in Section 6.

Similarly to the RAU flow signal $b_{ij}(k)$, also the signal $a_i(k)$ requires outlier-filtering to eliminate anomalous peaks before maximization. We used the same outlier filtering method described above. Some sample graphs of $a_i(k)$ are given in Figure 3. Both display an anomalous peak around $k \approx 700$, likely due to the (unplanned) rebooting of a neighboring BSC.

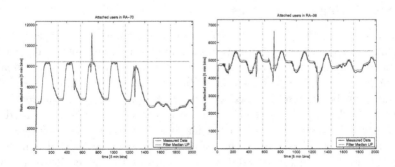

**Fig. 3.** Number of attached MS to some sample RAs (one week in October 2004). The horizontal line denotes the peak value after outlier filtering.

# 5   A Case Study

## 5.1   Input Dataset

The input traces were collected during the METAWIN project [9] from the operational network of mobilkom austria AG&CoKG, the leading mobile communication operator in Austria, EU. The time-series $b_{ij}(k)$ (RAU flow) and $a_i(k)$ (attached MSs) were extracted in post-processing by new modules developed ad-hoc for the estimation procedures described in Section 4.2 and 4.3 respectively.

The dataset used here includes one full week of measurements in October 2004 (from Monday 00:00 to Sunday 24:00) on a subset of the Gb links, specifically those attached to the $K$ SGSNs co-located at a single site. It includes 127 different RAs, representing a fraction $F$ of the total network coverage. *For proprietary reasons we can not disclose the values of $K$ and $F$, nor provide absolute quantitative values like traffic volumes, number of MSs, number of Gb links, etc. Therefore in the following graphs all numerical values have been re-scaled by an arbitrary factor (un-disclosed).* Figure 3 reports the signal $a_i(k)$ before and after the filtering for two sample RAs. A graphical representation of the peak Mobility Matrix is given in Figure 4.

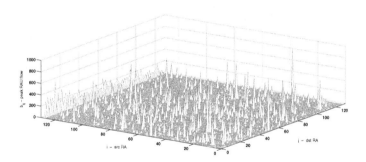

**Fig. 4.** Mobility Matrix (peak values)

## 5.2   Simplifying Assumptions

In a real network there are some constraints that must be considered when designing the PCU-SGSN mapping. For instance a RA can not be attached to multiple SGSNs, therefore if one RA is split among multiple PCUs, all the latter - and hence their attached RAs - must be assigned to the same SGSN. This constraint can be exploited to reduce the resolution time, as explained in the following. All the RAs bound to be assigned to the same SGSN are merged into a single virtual "super-RA", so that a single set of routing variables $\{x_{i,m}\}_m$ can be used for it. This reduce the number of $x_{i,m}$ variables and constraints in the ILP problem instance. Additionally, some PCUs (hence RAs) might be preferably attached to certain SGSNs based on geographical proximity or other deployment-specific factors. This can be reflected in the ILP formulation by fixing *a priori* the value of the routing variables $x_{i,m}$ for these RAs.

For sake of simplicity we take the following simplifiying assumptions for the case study presented below: (1) the mapping between PCUs and RAs is strictly 1:1 and (2) there are no preferential RA-SGSN associations. Furthermore, we have ignored the RAU flows from / towards the external RA (i.e. RA not monitored by our system, including those of other neighboring operators). More details about the requirements and adaptations needed to tackle real-world optimization instances are discussed later in Section 6.

## 5.3 Numerical Results

The optimization problem formulated in Section 3.2 was implemented in AMPL language [10] and solved with CPLEX [11]. Several problem instances were solved with different values of the parameters $M$ (number of SGSNs) and $\sigma$. Recall the latter serves as a tuning knob to trade-off between the minimization of $\alpha$ and $\phi$. For each combination of these parameters, the optimal RA-SGSN assignment is found and the associated values of $\phi$ (Y-axis) and $\alpha$ are reported on a XY plane (Figure 5). The resulting curve represents the optimal trade-off region between $\phi$ and $\alpha$. Clearly, the minimum value of $\alpha$ (computed with $\sigma = 0$, marked by vertical lines in Figure 5) depends on the number $M$ of available SGSNs. With more SGSNs one can achieve lower values of $\alpha$ by distributing the MSs into more subsets, but that comes at the cost of higher iRAU flow $\phi$. For large values of the slack factor $\sigma$ the iRAU flow minimization dominates the overall optimization process, and the solutions for $M = X$ overlap with those for $M = X - h$ ($h$=1,2..), meaning that $h$ out of $M$ SGSN are left unused in the optimal solution. Interestingly, we note that in some cases the availability of one more SGSN allows a better placement of the RAs, yielding a smaller value of $\phi$ for the same value of $\alpha$ (e.g. compare $M = 5$ vs. $M = 4$ for $\alpha = 1.6e5$, and similarly $M = 4$ vs. $M = 3$ for $\alpha = 2.1e5$). The curve in Figure 5 can be used

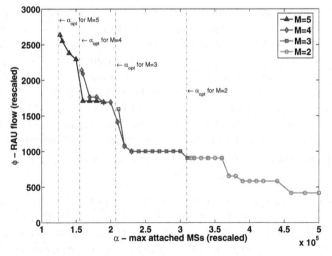

**Fig. 5.** Optimal $\phi - \alpha$ curve for the case-study (rescaled values)

to find the more convenient operating point in the $\alpha - \phi$ plane and to learn the optimal number of SGSN $M$ required to implement the optimal solution.

# 6   Application to Real Networks

The case-study presented in the previous section is semi-empirical: it is based on real input data, but it includes some artificial simplifications, e.g. complete freedom in the assignment of RA to SGSN. These were due in order to skip a number of technical aspects dependent from the detailed structure of the specific network under study. This enables a simpler illustration, allows for a better focus on the problem itself and preserves a certain level of generality. Such approach is well suited for a research paper, which is illustrative in nature. The application to the planning of a *real* network would require taking into account a number of **additional constraints**. Below we discuss how to adapt the optimization process to address such aspects.

**Physical constraints.** Geographical proximity and considerations related to the costs of the wired infrastructure interconnecting the PCU to the SGSN would probably dictate the assignment of certain PCUs (hence RAs) to some pre-defined SGSNs. Such constraints can be easily implemented by fixing the values of the routing variables $x_{i,m}$ associated to these RAs.

**Logical constraints.** The RA-PCU assignment is not necessarily 1:1. It is possible that several RAs are attached to the same BSC/PCU, in which case all such RAs must be assigned to the same SGSN. It is also possible that different cells within the same RA are connected to different PCUs, e.g. for very large RAs. In this case, all such PCUs (sharing the same RA) must be linked to the same SGSN, since the 3GPP specifications forbid that a single RA is split between multiple SGSNs. These aspects introduce mutual constraints between the associations of RA to SGSN, which can be used to reduce the size of the optimization problem. In fact, if $n$ different RAs are forced to be attached to the same SGSN, we can group them under a single virtual entity, called "super-RA", with a single set of routing variables $x_{i,m}$ instead of $n$ (in this case the index $i$ shall refer to the super-RA). In practice, it is required to pre-process the set of RAs and associated PCUs in order to define the minimum set of super-RAs, then the ILP formulation given in Section 3.2 is applied to the super-RA set. The reduction of the (set of) RAs to super-RAs is straightforward. To see that, consider the non-directed graph where RAs and PCUs are mapped to nodes, and RA-PCU associations are mapped to arcs. Finding the minimum set of super-RA is equivalent to find the set of connected components in such simple graph.

**Coverage Requirements.** The estimation mechanism for $a_i(k)$ described in Section 4.3 is prone to two types of error:

**Type I.** Denote by $t_0$ the start of the monitoring period. A generic MS that is attached to the network (say in RA $i$) at time $t_0$ will not be "seen" before it sends the next packet, say at time $t_x$. This will lead to an initial underestimation error in $a_i(k)$ at the beginning of the monitoring period.

Note however that by the time $t_0 + T_d$ all MS must have sent at least one RAU message, therefore this error disappear after $T_d$.

**Type II.** Consider a MS that at time $t_s$ moves from a monitored RA (say $i$) to a non-monitored one ("external RA"). The latter can be for example the foreign RA of another operator, but also a RA of the same network whose Gb link is not tapped. In both cases, the Routing Area Update message will not be "seen" by the monitoring system, which will consider the MS still attached to RA $i$ until $t_s + T_d$, when finally the MS is declared implicitly detached. This effect leads to a continuous overestimation of $a_i(k)$, particularly for those "border RA" neighboring non-monitored areas. Differently from Type I, such error is present in the entire measurement period.

Both estimation errors can be considered as "border effects" due to the finiteness of the monitoring scope in time (Type II) and space (Type II). In order to reduce Type II errors it would be required (1) to monitor *all* the Gb links attached to the SGSNs under optimization and (2) to monitor also the Gn interfaces near the SGSN in order to capture SGSN-to-SGSN signaling. In fact, the transition to a non-monitored RA is notified by the MS to the new (non-monitored) SGSN which will then contact the old (monitored) SGSN. This is in fact part of the iRAU procedure, whose frequency we are trying to minimize (!).

**Partial Subproblem.** In some cases the network engineers might be interested in solving only a partial sub-problem, namely to re-optimize the PCU-SGSN assignment for only a subset of SGSN, typically those co-located in the same physical site. In this case it is not required to monitor the RAs attached to other SGSNs, since they are excluded *a priori* from the assignment problem. However, to achieve a meaningful solution it still required to monitor *all* the RAs attached to the set of SGSN under optimization, plus all other RAs that are candidate to be linked to be them.

**Non-monitored RA.** The iRAU flow from / to external SGSNs (i.e. not considered in the optimization, like foreign SGSNs) contributes to the iRAU load of each monitored SGSN, therefore should be taken into account in the optimization. This can be handled by embedding all non-monitored RAs into an additional virtual RA, the "external-RA", which is not assigned to any SGSNs. Hence, the RAU flows between the monitored RAs and the external RA is always accounted as inter-SGSN RAU. Note that only the inbound iRAU flow from the external-RA can be measured for each monitored RA, not the outbound (towards the external-RA). A reasonable workaround is to assume flow symmetry between the two, and set the external outbound flow equal to the inbound for each RA.

# 7   Conclusions

In this work we introduced the problem of optimal PCU-SGSN mapping in a 3G cellular data network. We provided a novel ILP formulation for the (re)assignment problem and solved an instance of 127 RAs and up to 5 SGSNs. We expect that most of the practical problem found in real-networks will hardly exceed this range, considered that in most cases the optimization will be applied to a subset of SGSNs co-located at a single physical site rather than to the whole-network. Furthermore the logical deployment constraints discussed in the last section are likely to reduce significanly the number of variables (super-RAs). The approach developed in this work allows for a better engineering and optimization of the Gb section of a 3G network: the curve in Figure 5 can be used by the network staff to find the more convenient operating point in the $\alpha - \phi$ plane and learn the minimum number of SGSN required to implement the optimal solution.

The case study presented here was meant primarily as a proof-of-concept of the correctness and consistency of the whole optimization procedure, from the measurement extraction to the ILP resolution. Accordingly, we adopted a semi-empirical approach applying a real dataset to a simplified network instance. It was not in our goal to show the level of improvement that can be achieved by applying this optimization problem to an existing network with a legacy PCU-SGSN wiring. In this case the gain depends on the goodness of the latter. Network engineers can use the proposed approach to periodically validate the goodness of the current setting against an absolute performance bound, and based on that decide about the convenience of rewiring.

The input data required for the optimization can be extracted from packet traces captured by passively monitor the Gb links near the SGSN, without direct access to the RAN. Notably such data could not be derived easily from the network equipments. For example enabling complete RAU logging at the SGSNs or BSCs would introduce a major overload onto these elements and hence performance risks. Therefore such an application represents yet another example of the opportunities enabled by passive network monitoring a 3G network.

# References

1. J. Bannister, P. Mather. *Convergence Technologies for 3G Networks*. Wiley, 2004.
2. Ajay R. Mishra. *Fundamentals of Cellular Network Planning and Optimization*. Wiley, 2004.
3. I. Demirkol et al. Location Area Planning in Cellular Networks Using Simulated Annealing. *Proc. of IEEE INFOCOM'01, Anchorage, AK, USA*, April 2001.
4. A. Merchant, B. Sengupta. Assignment of Cells to Switches in PCS Networks. *IEEE/ACM Trans. on Networking*, 3(5), October 1995.
5. A. Quintero, S. Pierre. Assigning Cells to Switches in cellular mobile networks: a comparative study. *Computer Communications (Elsevier)*, 26(9), June 2003.
6. M. Pioro, D. Medhi. *Routing, flow, and Capacity Design in Communication and Computer Networks*. Morgan Kaufmann, 2004.
7. F. Ricciato, S. Salsano, M. Listanti. Off-line Configuration of a MPLS over WDM Network under Time-Varying Offered Traffic. *IEEE INFOCOM'02*, June 2002.
8. F. Ricciato, U. Monaco. Routing Demands with Time-Varying Bandwidth Profiles on a MPLS Network. *Computer Networks (Elsevier)*, 47(1), January 2005.
9. METAWIN and DARWIN home page. *http://userver.ftw.at/~ricciato/darwin*.
10. AMPL: A Modeling Language for Mathematical Programming. *www.ampl.com*.
11. ILOG CPLEX. *http://www.ilog.com/products/cplex/*.

# A Middleware Approach for Reducing the Network Cost of Location Traffic in Cellular Networks

Israel Martin-Escalona and Francisco Barcelo

Technical University of Catalonia, c/ Jordi Girona 1-3, 08034, Barcelona, Spain
{imartin, barcelo}@entel.upc.edu

**Abstract.** Location is a valuable information for services implemented in wireless networks. Location systems often use the infrastructure of already deployed cellular networks. Accordingly, location systems spend resources from the network they use. This paper proposes a middleware to reduce the consumption of network resources and optimize the location traffic being carried. This middleware, called MILCO (Middleware for Location Cost Optimization), selects the optimum location technique depending on the request, i.e. the location technique that fulfills the required quality of service (QoS) and minimizes the resource operating expense. In addition, MILCO takes advantage from ongoing and carried location processes to reduce the overall expenditure of resources. Results show that MILCO can reduce the location-process failures and improve the figures of latency for the location provisioning and resource usage in cellular networks such as UMTS.

## 1 Introduction

Many operators see in location a key feature for advanced services on wireless networks in the near future. Till now, only a few location-based services (LBS) have been implemented for the mass-market. There are several factors that are delaying the introduction of LBS in the market. The recent deployment of the last generation of 3G networks removed some of them, such as the low-bandwidth channels and the lack of definition on the location system architecture and protocol stacks. However, other issues still remain open. Probably, the main one refers to the mismatch between the QoS offered and demanded.

Nowadays there are several location techniques ready for being deployed: cell identification, terrestrial signal trilateration, satellite navigation, finger-printing, angle of arrival, etc. Each of them provides certain quality of service usually measured in terms of accuracy, response time, availability and consistency [1], where the first two are considered to be the most relevant. On the other hand, there exist a wide variety of LBS, each requiring different QoS depending on the purpose of the service itself. Thus, the capabilities of location systems for carrying location requests coming from different LBS directly relies on the features of the location techniques implemented on them.

Table 1 outlines the quality of service obtained by the most popular location tech zniques. This table shows that none of the location techniques perform excellent in all the conditions. For example, availability of cell identification methods is very

Y. Koucheryavy, J. Harju, and V.B. Iversen (Eds.): NEW2AN 2006, LNCS 4003, pp. 83–95, 2006.

good, but accuracy and consistency are usually poor since they depend on the cell size; GPS techniques on the other hand give accurate positions but availability is poor indoors, etc.

Hybrid techniques are proposed as a way to overcome the shortcomings of using location techniques as standalone. They are based on combining measurements coming from several techniques to take advantage from the strong points of each one [2]. The use of this kind of techniques enhances the QoS offered by the system and allows more LBS to be carried. However, the QoS figures obtained by this kind of techniques are often much higher than necessary for many LBS, which can lead to an inefficient use of the network resources.

This paper proposes a new approach to optimize the use of resources in location systems. The work is structured as follows. The proposed approach is explained in Section 2. Section 3 presents the simulation tool used to carry out the analysis and the scenarios being simulated and Section 4 analyzes the results obtained. Finally, in Section 5 the main conclusions are summarized.

**Table 1.** QoS of the main location techniques

| *Technique* | *Accuracy* | *Response Time* | *Availability* | *Consistency* |
|---|---|---|---|---|
| Cell ID | Fair/Poor | Very good | Very good | Very poor |
| Signal strength | Poor | Good | Very good | Very poor |
| TOA/TDOA | Good | Good | Fair | Good |
| AoA/DoA | Good | Good | Fair/Poor | Fair |
| Fingerprint | Good | Fair | Good | Fair |
| GPS | Very good | Fair | Good | Good |
| Hybrid systems | Very good | Poor | Very good | Good |

## 2  MILCO

### 2.1  System Definition

The network resources consumed by the location system belong to the infrastructure of the underlying cellular network on which the location service is running. As a consequence, the resources used for location purposes are not available for other traffics. This paper proposes to use a middleware to optimize the use of resources in location systems: MILCO (i.e. Middleware for Location Cost Optimization) [3]. MILCO manages all location processes, aiming at reducing the resources usage as long as the QoS requested is fulfilled. Notice that there are several proposals for location middlewares, but they are focused on technology independence and LBS quick development [4], not on efficient use of resources.

MILCO is implemented as a new piece of software inside the location managers, e.g. inside the Serving Mobile Location Centers (SMLCs) in the case of ETSI/3GPP notation [5]. Figure 1 shows a location system architecture including the MILCO. In this figure, ETSI/3GPP notation has been used as reference. Each time a location request reaches the location system through the GMLC, it is delivered to the MILCO

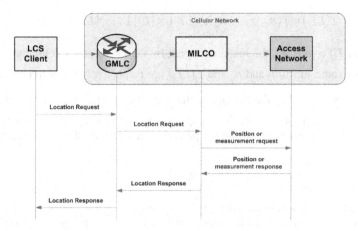

**Fig. 1.** Location system integrating MILCO

entity. Then MILCO selects the location technique that better fits the request, i.e. the one that is expected to achieve the requested QoS and minimizes the spent resources. Finally, MILCO uses the network facilities to get the user's position and forward the result to the LoCation Service (LCS) client that requested it.

Figure 2 shows the steps that define the MILCO's performance for each location request: filtering, location-technique selection and result management. Filtering stage involves all the blocks above the cost function and aims at filtering any location technique that is not suitable for the request. Location techniques can be marked as non-suitable due to three reasons: there is an incompatibility (i.e. either the network or the user terminal do not implement the technique); the location technique is unable to achieve the QoS being requested (e.g. the maximum accuracy achieved by the technique is worse than requested); or there is an input module that can handle the request without running the location technique. The second stage is the location-technique selection. In this stage, MILCO selects the optimum location technique from the remaining set (i.e. after filtering). This is achieved by means of a cost function, which ranks the resource consumption of all the location techniques. Finally the third stage manages the results, i.e. decides the procedure to handle the failures, maintain a database with the previous location measurements and calculations, etc. The default behavior on location failures is to execute another location technique. Notice that in these cases, the requirement for the response time may be much more constrained, since some time has been spent in previous location attempts.

## 2.2 Cost Function

The cost function is the core module of MILCO. It ranks the location techniques suitable for the request according to the use of resources made by each of them, i.e the more resources the technique consumes the lower it is ranked. This rank is further used to select the optimum location technique, i.e. the one that use less resources.

The cost function is composed of several factors, which are used to quantify the network-resource usage. Thus, it is defined as:

$$Z(LT_i) = f(\alpha_1, \alpha_2, ..., \alpha_n; z_1(LT_i), z_2(LT_i), ..., z_n(LT_i)), \qquad (1)$$

where $Z(LTi)$ represents the resources spent by the $i^{th}$ location technique (i.e. $LT_i$), $f$ stands for some function and $\alpha_j$ and $z_j(LT_i)$ are respectively the weight and value of the $j^{th}$ factor applied to the location technique $LT_i$. Several functions ($f$) can be used to calculate the resource usage. This paper proposes to use a simple additive function with $m$ factors in order to evaluate the performance of the module. It is defined as:

$$Z(LT_i) = \sum_{i=1}^{m} \alpha_i \cdot z_i(LT_i). \qquad (2)$$

The factors taken into account in this paper and which use is penalized are the signaling volume, the use of low-bandwidth channels and the power consumption at the user terminal [3]. Signaling volume penalizes those techniques that involve exchanging big amounts of data. Use of low-bandwidth channels favors those techniques that use wide-band channels, e.g. the ones that don't need to cross the radio interface. Finally, power consumption is included to penalize the techniques that quickly drop the user's terminal battery.

### 2.3   Input Modules

Input modules are used to extend the functionalities of the cost function and to improve the performance. Two input modules are in the design: location cache and concurrence manager. Location cache avoids running a location technique whenever the user position can be estimated with enough accuracy. Location cache works on the basis of two hypotheses: an older position is available for the user and the user is close to this position. There are several approaches to verify that the terminal position is close enough to the last stored position [6]. MILCO builds a database with the result of previous location processes and uses the age of the stored positions as constraint for the location cache, i.e. it uses old stored positions only when they are not older than a threshold value. If it is the case, the average speed of the terminal (calculated from the data stored in the database) is used to estimate the current position of the mobile station. Then, depending on the required QoS, this estimation can be enough and help to avoid spending new resources. Notice that the more static the users are the better performance should be expected.

Concurrence aims at avoiding collisions on location technique executions. A collision happens whenever a location request for a specific user is received while another one, with higher QoS requested, is in progress. In such situations, concurrence manager blocks the last received request until the ongoing one finishes. Then, the resulting position is shared by both requests, even though the QoS returned for some of them is better than necessary. Notice that this procedure should perform better as the location traffic (per user) increases.

Location cache and concurrence manager contribute by reducing the number of requests reaching the cost function, hence no location techniques are run for requests that use these modules. As a consequence the overall amount of resources used for location is also reduced.

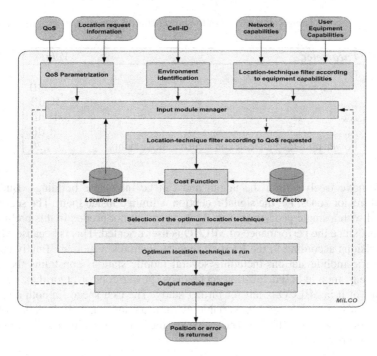

**Fig. 2.** Block diagram of MILCO

## 3   Simulator and Scenarios

The simulator used in this paper has been specifically developed to test MILCO. It allows any kind of network and cellular scenario to be simulated (e.g. channel allocation method, mobility and propagation patterns, admission and power control algorithms, etc). The simulator doubly wraps the simulation area to minimize the impact of the edge-effect on the results. This means that the simulation area is turned into a torus [7], becoming a virtually infinite surface from mobility and propagation points of view.

This paper evaluates MILCO on urban cellular UMTS networks. An admission control algorithm [8] and a power control [9] have been implemented. Control algorithm accepts new users whenever the target SIR of any of the ongoing calls in the cell does not drop more than 1 dB. Power control parameters are detailed on Table 2. A basic scenario is proposed to be simulated, where several location loads ranging from 1 request per 30 seconds to 2 requests per second are applied. Figure 3 shows the resulting layout. The scenario is composed of 100 Node-Bs (NB), which are uniformly spread along the simulation area. Each Node-B is placed in the center of a square-shaped building and achieves a cell coverage of 1135 meters. Notice that an important share of the whole area is overlapping area, i.e. covered by more than one Node-B. This puts the simulation closer to reality and at the same time allows OTDOA, not possible in areas covered by only one or two Nodes-B.

**Table 2.** Parameters of the power control algorithm

| Parameter | Value |
|---|---|
| $\eta_0 = \eta_1 = \eta_{01}$ | 2 |
| $\Delta_0 = \Delta_1 = \Delta_{01}$ | 10 |
| $\Delta_{max}$ | 10 dB |
| $\Delta_{min}$ | 0.01 dB |
| $\Delta_{initial}$ | 1 dB |
| Power updates between consecutive movements | 20 |

Users move freely within the layout and can go into these buildings. Buildings represent indoor zones, i.e. the signal reception is limited inside them. The scenario is populated with a single pedestrian user. More users are not needed in this preliminary evaluation, since the performance of MILCO is user oriented. This is because MILCO takes decisions according to the location requests features, which are finally targeted to a specific mobile station. Including several mobile stations constrains the access network (e.g. mobile-based location techniques, power-control algorithm, etc.) and the impact of it in MILCO is left for further study. The user speed (in both directions $x$ and $y$) follow a normal random variable, with mean and standard deviation of 0.59 m/s and 0.17 m/s respectively. The value of the user speed in both directions is updated once per second.

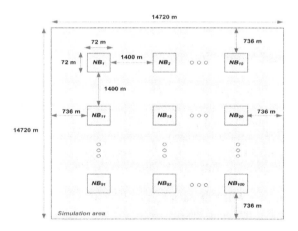

**Fig. 3.** Layout of the scenarios simulated

Propagation pattern follows the Okumura-Hata model, with path-loss slope and zero-meter losses set to 4 and 23dB [10] respectively. SIR is calculated in according to [11], where spreading and orthogonality factors are 10 dB and 0.4 respectively [10]. Handoffs are requested each time the received power or SIR in a Node-B or mobile station (MS) falls bellow a threshold (named handoff-threshold). The handoff request is hold until achieving a new free channel, as long as the received power or SIR are between the handoff-threshold and the sensitivity of the terminal. If SIR or received power falls bellow the sensitivity, the handoff is tried for 15 seconds maximum. If no channel is achieved during this time, the service is interrupted and the user terminal backs off for an exponential time of mean 5 seconds. It must be noted that successful handoffs drop all ongoing requests carried by the mobile station. Table 3 gathers the main parameters of the propagation pattern, which have been extracted from [10, 11].

**Table 3.** Parameters of the propagation pattern

| Parameter | Value |
|---|---|
| Minimum SIR | -9 dB |
| Sensitivity of the stations | -109.2 dBm |
| Maximum MS transmission power | 21 dBm |
| Minimum MS transmission power | -44 dBm |
| Node-B transmission power | 43 dBm |
| Handoff threshold for received power | -106.2 dBm |
| Handoff threshold for SIR at reception | -6 dB |

All location techniques available in UMTS are taken into account in this scenario (i.e. Cell-ID, OTDOA and A-GPS). A hybrid tight-synchronized OTDOA/A-GPS location technique [2] is also included to show the features of the hybridization upgrade. Table 4 characterizes the accuracy and response times achieved by each of them [12]. In this table *mean* is the average value, *range* indicates the set of values that the variable can take and *std* stands for the standard deviation. Availability of OTDOA is computed according to the received power and SIR. This means that three or more BS are expected to be seen in the MS in order to run OTDOA. Otherwise, OTDOA is considered not available at that instant/position. In the case of satellite-based techniques, availability is checked in a different way. The scenario defines a default number of satellites in sight of 5. However, in the indoor zones this figure is uniformly distributed from 1 to 2 satellites.

The position requests are generated by a single service, being the accuracy uniformly distributed from 10 meters (e.g. tracing, tracking and emergency services,

**Table 4.** QoS achieved by the location techniques

| | | Location Techniques | | | |
|---|---|---|---|---|---|
| | | Cell ID | OTDOA | A-GPS | Hybrid |
| Accuracy | Distribution | Deterministic | Uniform | Gaussian | Gaussian |
| | Range | Not applies | [50..250] m | Not applies | Not applies |
| | Mean | 1135 m | 150 m | 3 m | 50 m |
| | Std | 0 m | 57.73 m | 0.90 m | 15 m |
| Response time | Distribution | Deterministic | Exponential | Exponential | Exponential |
| | Mean | 0 s | 7 s | 11 s | 27 s |

etc.) to 2 Km (e.g. location-based information, enhanced call routing, etc.). The response time required is also uniformly distributed between 0 seconds, which means that the location must be provided immediately (e.g. emergency services) and 60 seconds (e.g. push services). Both accuracy and delay constrain the QoS, i.e. not reaching one of them results in a QoS failure (3GPP allows other QoS approaches but the most restrictive one is used in this study).

All the features of MILCO are implemented: cost function and input modules. Positions remain cached for 2 seconds. The cost function used in the scenarios is presented in (2). The quantification of the cost factors can be seen in Table 5, where $N_{NB}$ and $N_{SAT}$ are the number of Node-Bs and satellites involved in the positioning. Two modes are specified in Table 5 for OTDOA and A-GPS: assisted (AS) and not (NAS). Assisted mode involves sending the assistance data to the mobile station, while not assisted assumes that this information has been previously sent. Simulations are made under the assumption that the assistance data (for OTDOA and A-GPS) expires after 30 seconds, i.e. new assistance information is required after 30 seconds from its reception. In the computation of the signaling volume only the topmost protocol in the stack (i.e. RRC for all techniques except Cell-ID) has been taken into account. The quantification of the low-bandwidth channels usage is made by counting the number of times a interface is crossed and dividing this figure by the throughput of the channel used. Power consumption heavily depends on the user terminal performance. Accordingly, the authors propose a quantification for the power consumption based on the number of signals transmitters involved in the location technique.

**Table 5.** Quantification for the factors used in the cost function

| Technique | Signaling Volume (bits) | Low-bandwidth channel usage (ns) | Power Consumption |
|---|---|---|---|
| Cell-ID | 0 | 0 | 0 |
| OTDOA (AS) | $375 + 134 \cdot N_{NB}$ | | $N_{NB}$ |
| OTDOA (NAS) | 268 | | $N_{NB}$ |
| A-GPS (AS) | $473 + 1199 \cdot N_{SAT}$ | $\left[ \dfrac{2 \cdot 10^9}{155\,Mbps} + \dfrac{2 \cdot 10^9}{384\,Kbps} \right]$ | $N_{SAT}$ |
| A-GPS (NAS) | $461 + 647 \cdot N_{SAT}$ | | $N_{SAT}$ |
| Hybrid | $653 + 134 \cdot N_{NB}$ $+ 1254 \cdot N_{SAT}$ | | $N_{NB} + N_{SAT}$ |

The weight of the factors in the cost function is the same and the maximum value of a weighted factor is set to 1. Thus, the maximum value of the cost function is 3. Table 6 gathers the weights $(\alpha_j)$ applied to the factors in order to achieve this behavior. This static assignment of weights is proposed for evaluation purposes. Tunning the weights is out of the scope of this work and it is left for the optimization stage of the MILCO design.

# 4  Performance Analysis

This section analyzes the performance of MILCO, focused on the resources used by each LBS and the location traffic being carried. Location techniques used as standalone are included for comparison.

Figure 4 displays the percentage of unsuccessful LBS in the scenario loaded with 30 requests per second, using only the cost function (i.e. location cache and concurrence manager are disabled). Failures of the mobile-based location techniques due to handoffs are not taken into account. The best results among the standalone location techniques are obtained by A-GPS, with 17.72% of unsuccessful LBS. Using MILCO reduces this figure in 57.85% and better results are obtained if compared with the rest of techniques. Results achieved using only OTDOA and only A-GPS are similar. This is because accuracy and response time constrain both the QoS. Thus, response time constrains the A-GPS results and accuracy does the same for OTDOA. Hybrid technique is also constrained by the long response times, resulting in a poor successful LBS rate.

**Table 6.** Weights of the factors used in the cost function

| *Factor* | *Value* |
|---|---|
| Signaling volume | $1.3651 \cdot 10^{-4}$ |
| Low-bandwidth channel usage | $1.9152 \cdot 10^{-4}$ |
| Power consumption | $1.25 \cdot 10^{-1}$ |

Performance is expected to be improved by input modules. Hereafter, all the measurements are taken implementing all modules (i.e location cache and concurrence manager also enabled). Figure 5 plots the evolution of the unsuccessful LBS requests with the load, taking into account and not the mobility of the terminal (i.e. the LBS failures due to handoff). Inter-arrival time stand for the time elapsed between two consecutive location requests, i.e. the shorter the inter-arrival time the heavier the load. As it can be seen, the unsuccessful LBS rate is higher when mobility is accounted for. This is because handoffs interrupt all ongoing requests, independently of the handoff result [5]. The difference between results considering and not mobility is scarce, since the average speed of the mobile station is very low. However, greater differences are expected with higher average MS speeds. Figure 5 shows that at medium/low request rates (i.e. inter-arrival times longer than 10 seconds), MILCO achieves an unsuccessful rate of LCS between the 7% and 8%. However, the higher the load the lower the unsuccessful LCS rate, since lower time is spent between consecutive location requests and therefore cache and concurrence feature are more likely to be used. This proves that the scalability of the proposed approach is guaranteed: for heavier loads the input modules performance reduce the percentage of unsuccessful requests.

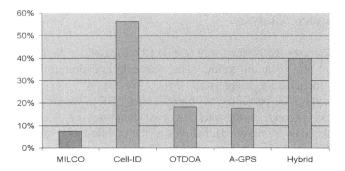

**Fig. 4.** Percentage of unsuccessful location processes

Reducing the use of resources is the other strong point of MILCO. Figure 6 plots the average and standard deviation of the number of location techniques run per location process. This figure shows that at lighter loads (i.e. inter-arrival times longer than 20 seconds), the average number of location techniques remains very close to 1. This is because at these rates concurrence and cache are not useful and only the cost function is used. Therefore, only few processes have enough time to run more than one location techniques (if necessary) and fulfill the response-time requirements. The number of location techniques being used falls as the load is increased, since input modules handle more requests: requests attended by input-modules involve running no techniques. In addition, standard deviation of this variable increases with traffic for rates greater than 1 request per second. This is due to the high difference between requests served by cost function (involves at least 1 location technique) and input modules (no location techniques are run). At the top right of Figure 6, for inter-arrival times lower than 1 second, the standard deviation falls since the number of the location techniques being served by location cache and concurrence manager is much higher than the ones attended by cost function, reaching up to the 88%.

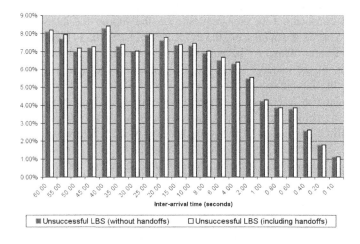

**Fig. 5.** Evolution of unsuccessful LBS with MILCO

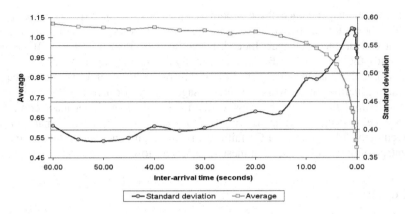

**Fig. 6.** Average number of techniques run per location process

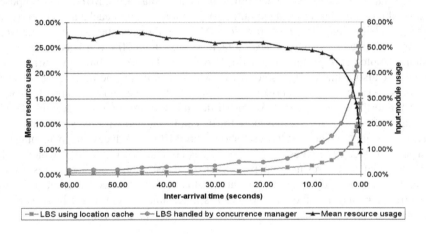

**Fig. 7.** Average resource usage with MILCO

Figure 7 shows the average resources consumed by MILCO according to the definition provided in (2) and the percentage LBS that make use of location cache and concurrent LCS. In this figure, percentage of resource usage is provided relative to the hybrid technique (considered 100%), according to data in Table 5 and Table 6. Therefore, the average resource usage represents the cost of requests handled by MILCO regarding the consumption of resources achieved using only the hybrid technique. At lightly loaded scenarios, location cache and concurrence manager are seldom used. Figure 7 shows that less than 2% and 5% are handled successfully by location cache and concurrence manager in these scenarios. This means that MILCO is reduced to cost function. However, using only this feature reduces the use of resources more than 70%, since the location techniques executed are selected according to the QoS demanded, the features of the network and the capabilities of the base station. Better figures are achieved when the load increases, reaching up to 95% of saving. This improvement is due to the increasing use of location cache and concurrent manager, since these modules attend a location requests at zero cost. Figure 7 shows that at heavier loads, most of the requests are handled by location cache or concurrence manager. Indeed, at the heaviest loaded scenario (10 requests per second), the 88% of the LBS are successfully handled by input modules.

Even though location concurrence algorithm provides several benefits on carried load and reduction on network resource usage, its performance is far from being optimum. Simulations show that most of the LCS failures at higher loads come from concurrence manager (e.g. at 10 requests per second, concurrence manager was involved in the 72.58% of the LCS failures). However, it is not very relevant since the percentage of unsuccessful LCS at these rates is very low.

## 5   Conclusion

This paper presents MILCO, a middleware that aims at reducing the use of network resources and maximize the carried traffic in location platforms. MILCO is composed of several modules, but three stand out: cost function, location cache and concurrence manager. Using only the cost function, MILCO reduces the usage on network resources (defined by means of several factors) more than 70% if it is compared with systems implementing only the hybrid A-GPS/OTDOA technology. Significant savings are also achieved if it is compared with other standalone techniques. MILCO improves thus the system performance just selecting the most suitable technique according to the QoS requested. When all the modules are used together, this saving goes up to 95%. This means that more traffic can be carried with the same amount of network resources. The results also show that MILCO reduces the percentage of unsuccessful location requests a 57.85% if it is compared with A-GPS, which is the technique that better results achieves when used as standalone. Better results are obtained when the middleware is compared with other techniques. Location cache and concurrence manager makes that MILCO is a scalable system, since increasing the load involves reducing unsuccessful LBS rate and the resource usage.

## Acknowledgment

This research has been partially funded by the *EC* through the IST 2000-26040 (Emily project) and by *FEDER* and the *Spanish Government* through project TIC2003-01748.

## References

1. Soliman, S. S., Wheatley, C. E., "Geolocation Technologies and Applications for Third Generation Wireless", Wireless Communications and Mobile Computing. 2, 229-251, 2002.
2. Barcelo F. and Martin-Escalona I., "Coverage of Hybrid Terrestrial-Satellite Location in Mobile Communications", Fifth European Wireless Conference: Mobile and Wireless Systems beyond 3G. 475-479, 2004.
3. Martin-Escalona I., Barcelo F., "Optimization of the Cost of Providing Location Services in Mobile Cellular Networks", 15th IEEE International Symposium on Personal, Indoor and Mobile Radio Communications. 3, 2076-2081, 2004.

4. Spanoudakis, M., Batistakis, A., Priggouris, I., Ioannidis, A., Hadjiefthymiades, S., Merakos, L., "Extensible platform for location based services provisioning", Fourth International Conference on Web Information Systems Engineering Workshops. 72- 79, 2003.
5. 3GPP TS 23.271, "Functional Stage 2 Description of Location Services (LCS).R6", 2004.
6. Biswas P., Han S., Wu J., "Location Caching in the Mobile Middleware Platform", Third International Conference on Mobile Data Management, pp. 172-174, 2002.
7. Zander J., Kim S.L, "Radio Resource Management for Wireless Networks", Artech House, 2001.
8. Capone A., Redana S., "Call Admission Control Techniques for UMTS", IEEE VTS 54th Vehicular Technology Conference, 2001. VTC 2001 Fall, Volume: 2 , pp. 925-929, October 2001.
9. Nuyami L., Lagrange X., Godlewski P., "A Power Control Algorithm for 3G WCDMA System", European Wireless 2002, February 2002.
10. Holma H., Toskala A., "WCDMA for UMTS", John Wiley & Sons, 2000.
11. ETSI TR 125.942, "RF system scenarios", 2004.
12. EMILY.IST-2000-26040, "EMILY system trials report", 2005.

# Location Conveyance in the IP Multimedia Subsystem

Miran Mosmondor, Lea Skorin-Kapov, and Renato Filjar

Research and Development Center, Ericsson Nikola Tesla,
Krapinska 45, HR-10000 Zagreb, Croatia,
{miran.mosmondor, lea.skorin-kapov, renato.filjar}@ericsson.com

**Abstract.** The 3rd Generation Partnership Project (3GPP) has proposed the IP Multimedia Subsystem (IMS) as a key element in the
next-generation network (NGN) converged architecture supporting multimedia services. Extending the IMS towards provisioning support for location based services (LBS) will enable enhanced services and offer new
revenues to the system. However, conveying location information in the
IMS and connecting the IMS with a real positioning system are still open
issues. This paper presents the design and implementation of an IMS Location Server (ILS) integrating IMS with a positioning system. From the
IMS perspective, the ILS serves as a service enabler for LBS. Considerable work has been done by the IETF in the area of location information
transport based on the Session Initiation Protocol (SIP). This paper proposes some improvements in this area. In order to demonstrate proof-of-
concept in enhancing IMS-based services, a Location-aware Push-to-Talk
(LaPoC) prototype service has been developed. The service has been integrated and tested with the Ericsson Mobile Positioning System (MPS).
The paper also gives the results of performance measurements including
traffic load analysis and session establishment time.

## 1   Introduction

Location awareness is an important issue affecting numerous human activities
and even forcing the creation of a special scientific discipline called navigation
in order to develop the means for location and travelling management. In recent
years, location has acquired a completely new dimension through introduction
of a special group of telecommunication services that explore location awareness.
Location-based services (LBS) have become one of the most prosperous groups
of emerging telecommunication services. In their essence, location-based services
successfully integrate three basic building blocks [10]: positioning systems, (mobile) communication systems, and location content.

Positioning systems serve as the entities for determination of an end-user's
position in a suitably chosen reference frame in space. Position determination
is conducted by combining several positioning sensors' outputs (satellite positioning, network positioning, radio positioning, etc.) with required quality of
positioning service [9], which is usually entitled positioning sensor fusion. Mobile

Y. Koucheryavy, J. Harju, and V.B. Iversen (Eds.): NEW2AN 2006, LNCS 4003, pp. 96–107, 2006.

communication systems provide reliable means for position reports and location content exchange between mobile units and the rest of the LBS system.

Location content refers to location-related information presented in various forms (charts and maps, numerical and textual content, multimedia, etc.) delivered either to the mobile unit or some third application (emergency call E112 service, for instance). Since location content is gradually shifting towards a multimedia form of presentation, it seems a natural step forward in LBS development to consider the utilization of the latest related communications technologies, such as the IP Multimedia Subsystem (IMS).

The IMS is the internationally recognized standard for providing end users with advanced multimedia services [1]. A wide range of location applications built around existing and emerging IMS services may be foreseen, including:

- presence and location (friend location visible in address book)
- users sending messages or initiating IMS sessions only with other users located at a defined distance
- context aware adaptation based on user location (e.g. user's communication preference is changed depending on whether the user is at work or at home)
- users sharing their location via shared maps
- location aware multimedia information broadcasting

With the provisioning of user location information considered as a generic and reusable network-provided enabling technology, this paper proposes the introduction of a service enabler called IMS Location Server (ILS) located in the IMS Service Layer. The ILS provides an interface towards a positioning system to retrieve user location information, thus providing the means of making this information available to other IMS application servers. The solution is independent of a particular service and is intended to support the enhancement of existing and emerging IMS-based services.

The basics of the IMS and SIP location conveyance are described in the following section. Section 3 presents the design and implementation of the proposed ILS. The solution is demonstrated with the development of a prototype service called Location-aware Push-to-talk (LaPoC) and through integration with the Ericsson Mobile Positioning System (MPS), as described in Section 4. Section 5 describes tests that were conducted to measure signaling performance.

## 2  Background

This section provides an introduction to the IMS and SIP. Furthermore, the status of standardization in the area of SIP location conveyance is described.

### 2.1  IP Multimedia Subsystem (IMS)

In the move towards a converged network architecture, the IMS represents a key element in the UMTS architecture supporting ubiquitous access to multimedia services. Originally specified by the Third Generation Partnership Project

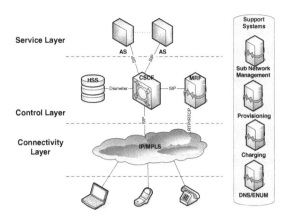

**Fig. 1.** Simplified IP Multimedia Subsystem (IMS) layered architecture

(3GPP), IMS is considered to play a key role in merging the Internet and cellular worlds [8]. Eventually, it will work with any network with packet-switched functions (e.g. GPRS, UMTS, CDMA2000, WLAN, DSL, cable etc.), while interworking with legacy networks will be supported through gateways. Key IMS benefits include: Quality of Service (QoS) support securing enhanced service quality; service integration by defining a standard interfaces over an IP-based infrastructure; and support for flexible charging. Based on a horizontally layered architecture, the IMS provides open call/session control with interfaces to service and connectivity layers in both wireless and wireline industries.

As shown in Fig. 1, the IMS consists of 3 layers: Service Layer, Control Layer and Connectivity Layer. Of the most significant protocols used in the IMS, we point out the Session Initiation Protocol (SIP) chosen by 3GPP as the protocol for session establishment, modification, and release.

The Service Layer comprises application and content servers to execute value-added services. The IMS allows for generic and common functions (implemented as services in SIP Application Servers) to be reused as building blocks for multiple applications and services. This implies the introduction of new services offering rich user experiences, with fast time-to-market and simplified service creation and delivery. Accordingly, the proposed ILS provides location services to other IMS Application Servers (AS) via a standard SIP interface.

The Control Layer comprises network control servers for managing call or session set-up, modification and release. The key IMS entity in the Control Layer is the Call Session Control Function (CSCF) which is responsible for session control and processing of signaling traffic. The Home Subscriber Server (HSS) is a user database, which maintains each end-user's profile. The Media Resource Function (MRF) is responsible for the manipulation of multimedia streams.

The Connectivity Layer comprises routers and switches, both for the backbone and the access network.

## 2.2   SIP Location Conveyance

The Session Initiation Protocol (SIP) [15] has emerged as part of the overall IETF multimedia architecture, providing advanced signaling and control functions for a wide range of multimedia services. SIP is defined as an application-layer control (signaling) protocol for creating, modifying and terminating sessions with one or more participants, and has been adopted by 3GPP as the key session establishment protocol in the IMS.

Conveying location information over SIP is a relatively new issue and is not completely standardized, although there are several Internet Drafts and RFCs [13][12][11], released by the IETF, dealing with this subject. Various SIP methods are applicable to carry location information, in the body or in the header of a message, but no method is pointed out as the preferred standard. More details on the framework and requirements for usage of SIP to convey user location information from one SIP entity to another SIP entity are given in [13].

The IETF proposes a protocol independent object format for conveying such location information [12], extending the XML-based Presence Information Data Format (PIDF) to allow the encapsulation of location information within a presence document. As the baseline location format in PIDF-LO objects, the Geography Markup Language (GML) 3.0 [5] was selected. Conveying static location in PIDF-LO bodies is straightforward. However, the difficult part about asynchronous notification of location information is that many forms of location are measured as a continuous gradient. Unlike notifications using discreet quantities, it is difficult to know when a location change is large enough to warrant notifications. Moreover, different applications require a variety of location resolutions.

Location filters are necessary to specify events that will trigger notifications to subscribers because location information is continuous and not discreet. We can not expect to flood the network (periodically or not) with responses carrying location information. Defined location filters [11] are XML documents which limit location notification to events which are of relevance to the subscriber.

## 3   Design and Implementation of ILS

The ILS functionality proposed in this work has been tested with the development of a prototype service.

### 3.1   IMS Location Server (ILS)

The ILS is designed as a generic SIP Application Server located in the IMS Service Layer. Methods for determining user positions are not implemented within ILS; rather, ILS is responsible for delegating the location request to the positioning system. Using the terminology proposed in [2], the ILS takes on the role of a Location Services (LCS) Client and obtains location information from an LCS Server. All other Application Servers (AS) requiring location data may send requests to the ILS via a SIP interface. Such a concept provides a central location

in the IMS Service Layer that provides location data, rather than having each
AS separately requesting data from a LCS Server.

## 3.2   Location-Aware Push-to-Talk (LaPoC) Service

The proposed solution was demonstrated based on development of a prototype service called Location-aware Push-to-talk (LaPoC). The architecture of
the LaPoC system integrated in IMS and connected with a generic Positioning System is shown in Fig. 2. It consists of two new components; the LaPoC
Application Server (LaPoC AS), and the IMS Location Server (ILS).

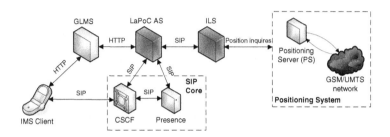

**Fig. 2.** Location-aware Push-to-talk (LaPoC) architecture

Push-to-talk over Cellular (PoC) [3] is one of the first IMS services provided
by numerous network operators. In this walkie-talkie type of service, the user
must press and hold a button when he/she wants to communicate, and can start
talking only when the terminal notifies them. By releasing the button, users
signal the end of their speech. Because Push-to-talk is a half-duplex service, only
one user can speak at a time. In the IMS network, a PoC Server is responsible
for session control functionality.

The LaPoC service implemented in this work extends the functionality of
the PoC service to make it location aware. This means that the PoC Server
is enhanced to establish and modify PoC sessions in the IMS system taking
into account end-user location information. The new proof-of-concept service
demonstrates how to establish and modify a group PoC session only with users
that are at a certain designated distance from the originating user (e.g. 1 km).

With location awareness, the service is even more similar to walkie-talkie.
The difference in comparison with the classic walkie-talkie solution is the possibility for the user to define the coverage area. In LaPoC, this area can range
practically indefinitely, while in the classic walkie-talkie the coverage area is limited by propagation characteristics of radio waves. Example use cases for such
a service include a person wishing to establish a PoC session only with selected
colleagues located within company premises; or a security officer speaking to
officials securing the grounds at a soccer stadium.

The general aim of the latest interdisciplinary activities is to provide a generic
model for successful implementation of a variety of location services within the

IMS. One can find the simplest case in the service of another user location provision to the initiator of the service. Called the immediate location request, this service asks for location response to be delivered immediately after the location request is received [2].

In another case, a location request is sent to a server, but a response is received only when a certain condition stated in the request is fulfilled. Such a service is called a deferred location request. An example is a service initiator requesting to be notified when another user enters a certain area, such as a building or a city. Practical implementation of this service requires the location server to monitor the users' locations. Requirements become even more challenging in the LaPoC service because the designated area is not static, and a group of users is to be monitored instead of a single user.

### 3.3 LaPoC Application Server

The LaPoC AS represents a PoC Server enhancement. Besides implementing classic PoC functionalities, the LaPoC AS is responsible for contacting the ILS to obtain information about which users in the group are inside the designated radius from the originating user, and which are not. It is also responsible for modifying the session in accordance with the location of session members. This means if one user moves outside of range from the originating user, the LaPoC AS will receive a notification from ILS and will terminate the session. In the same way, when one user from the group enters the designated area, ILS sends notification to LaPoC AS which then includes him in the session.

The simplified session establishment sequence diagram is shown in Fig. 3. After receiving a SIP request for a group session, the LaPoC AS contacts the Group List Management Server (GLMS) to retrieve the group member list. For each member of the group the server checks their presence status, whether they are online, offline, or busy, by contacting the Presence Server. Finally, the LaPoC Server retrieves location information from the ILS for available users, to determine whether they are within range from the session originator. Most interfaces of the LaPoC AS are based on SIP, including communication with the ILS.

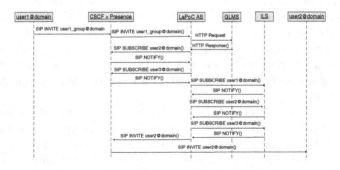

**Fig. 3.** Simplified LaPoC session establishment with *enterOrExit* filters

### 3.4    ILS SIP Interface

As mentioned earlier, the IETF is working towards standardization of SIP location conveyance. The usage of Presence-based GEOPRIV Location Objects (PIDF-LO) [12] carried in the body of a SIP message is proposed. Several SIP methods are applicable to carry the PIDF-LO but none are pointed out as preferred [13]. In the LaPoC prototype, location conveyance using SUBSCRIBE and NOTIFY SIP methods is selected since this model is used for specific event notification in SIP [14].

Furthermore, the usage of location filters to specify events that will trigger notifications to subscribers is also proposed. Several such events and corresponding filters are defined [11], of which the *enterOrExit* filter is most suitable for the LaPoC prototype. The *enterOrExit* filter triggers notification when one of the LaPoC session participants enters or exits a named 2-dimensional or 3-dimensional region or list of regions corresponding to a GML feature.

The problem with this filter definition is that an area is static and that such a filter can be sent for one user only. In the LaPoC service, there is a group of users that needs to be monitored for one dynamically changing location area. If we would like to apply this filter definition, we would first have to send an inquiry for the position of the originating user to form an *enterOrExit* filter and then send a subscription carrying filter for each user in the group (Figure 3). Furthermore, during the session lifetime if or when the originating user changes their position, a new *enterOrExit* filter needs to be formed and again sent for each user in the group. This does not necessarily mean that any of the users have changed their state (entered or exited the defined area). Thereby, in order to reduce signaling between the ILS and LaPoC AS, we propose a new filter definition called *groupInRange*.

### 3.5    *groupInRange* Filter

The idea of a *groupInRange* filter is to encapsulate the solutions of the two main disadvantages of *enterOrExit* filter. First, to avoid re-sending the same filter for each group member, the whole list of users is sent together with one filter definition to the ILS. This principle could be applied for any type of filter and that could significantly reduce initial signaling. Secondly, since the *enterOrExit* filter defines a static area and results in redundant signalization when the originating user changes position, a new event has been defined. This event describes the situation when one resource (user) falls in or out of range from an originating user. This corresponding filter is defined with resource (user) identification and range length only.

The *groupInRange* filter enables the LaPoC AS to send the whole list of users and range length in one SIP message to the ILS and receive back a list of users with information regarding a particular user being in or out of range from the originating user (Fig. 4). Each time a user leaves or enters the range, a notification is sent to the LaPoC AS. A traffic analysis and comparison of these two events is given in Section 5.

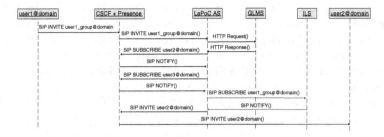

**Fig. 4.** Simplified LaPoC session establishment with *groupInRange* filters

# 4   Case Study

In this section we describe the integration of ILS with a real positioning system, namely the Ericsson Mobile Positioning System (MPS).

## 4.1   Mobile Positioning System (MPS)

Ericsson Mobile Positioning System (MPS) [7] comprises the functionalities of two entities of a 3G network: Gateway Mobile Positioning Centre (GMPC) and Serving Mobile Positioning Centre (SMPC). It collects all available location-related information from the mobile communication network and performs the fusion of two main positioning services when they are available: satellite positioning methods, and network positioning methods (Cell-ID, E-OTD). By combining position mechanisms with location-specific information, MPS can offer customized personal communication services through the mobile phone or other mobile devices. The system is fully scalable and it supports both GSM (MPS-G) and UMTS (MPS-U). The MPS utilizes the Mobile Location Protocol (MLP) for data exchange with the Location Services (LCS) Client [4] (Fig. 5).

The reason for connecting ILS with MPS is interesting because the MPS Software Development Kit (SDK) that was used to emulate MPS functionalities does not support deferred location requests in its currently available version. On one side, the ILS receives location requests through a SIP interface, and on the other, it delegates the request through MLP to the MPS emulator. One major difference between these location requests on different interfaces is that

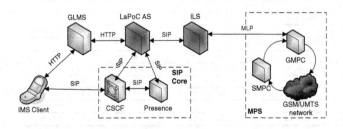

**Fig. 5.** LaPoC with MPS architecture

SIP location requests can also carry location filters [11], hence the ILS SIP interface does have support for deferred location requests. A similar concept is going to be supported in the next version of the MPS SDK with so called spatial triggers. However, due to lack of support for deferred location requests in the currently available version of MPS SDK, the ILS sends location requests for each user periodically to MPS. When the ILS detects that one user has entered or exited the area defined by a location filter, it sends a SIP notification to LaPoC AS. Location signalization between the LaPoC AS and ILS is minimized, as described in the previous section, but periodic signalization with MPS presents a problem. In order to decrease this signalization, we implemented ILS in a way that the time between location requests to MPS depends on the user's velocity. For example, if a user stands still, the location inquiry period is maximized, but as the user starts to move, the period decreases. With this we have slightly improved the signalization amount between ILS and MPS.

## 5   Measurements

This section describes traffic and performance measurements for the developed LaPoC service. Several sets of measurements were conducted. First, measurements were conducted to compare the session establishment time of LaPoC with the classic PoC service for different group sizes. The second set of measurements was performed to compare signalization load between the LaPoC AS and ILS for the LaPoC service implemented with different location filter types, namely *enterOrExit* according to IETF and our proposed *groupInRange* filter. Finally, a comparison of the traffic load between the ILS and MPS implemented with periodic and improved non-periodic location querying was made.

The measurements were performed in the Ericsson Nikola Tesla (ETK) Research & Development Center research lab. The IMS Client, SIP Core and GLMS functionalities were realized with Ericsson PoC Reference Test Suite [6], while the MPS Emulator from the Ericsson MPS SDK was used to emulate MPS functionalities. The whole system was deployed on eleven computers, connected with a 100 Mbit/s Ethernet network switch (Table 1).

**Table 1.** Hardware and software configuration

| COMPUTER | HARDWARE | SOFTWARE |
|---|---|---|
| rlabsrv | Pentium 4, 1.3 GHz, 40 GB HDD, 512 MB RAM | Windows 2000 Server, IMS Clients |
| rlab2, rlab3, rlab5 | Pentium 3, 800 MHz, 10 GB HDD, 512 MB RAM | Windows 2000 Server, IMS Clients |
| rlab4 | Pentium 3, 800 MHz, 40 GB HDD, 512 MB RAM | Windows 2000 Server, LaPoC Server |
| rlab6 | Pentium 3, 866 MHz, 20 GB HDD, 512 MB RAM | Windows 2000 Server, SIP Core Server, MPS emulator |
| rlab7 | Pentium 3, 866 MHz, 20 GB HDD, 512 MB RAM | Windows 2000 Server, GLMS, IMS Location Server (ILS) |
| rlab8 | Pentium 4, 3 GHz, 80 GB HDD, 1 GB RAM | Windows 2000 Server, IMS Clients |
| rlab9 | Pentium 4, 1.3 GHz, 40GB HDD, 256 RAM | Windows Server 2003, IMS Clients |
| rlab10 | Pentium 4, 1.5 GHz, 20GB HDD, 512 RAM | Windows Server 2003, IMS Clients |
| rlab11 | Pentium 4, 1.3 GHz, 80GB HDD, 512 RAM | Windows Server 2003, IMS Clients |

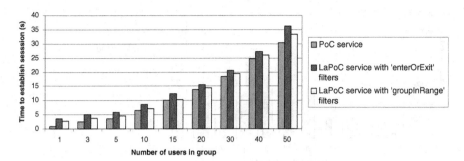

**Fig. 6.** Comparison of PoC , LaPoC with *enterOrExit* filters and LaPoC with *groupIn-Range* filters service session establishment time

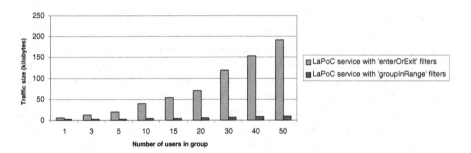

**Fig. 7.** Traffic load between LaPoC AS and ILS at session establishment

The first set of measurements includes a comparison of session establishment time for the LaPoC and PoC services for different group sizes (Fig. 6). The session establishment time prolongation for the LaPoC service is expected and is relatively small if we consider the value added to the PoC service. Furthermore, results illustrate improvements in session establishment time that were made with definition of the *groupInRange* filter (Section III). The reason for this time improvement lies in reduced traffic load between the LaPoC AS and ILS for the LaPoC service for *groupInRange* location filters. Fig. 7 shows that *groupInRange* significantly reduces the amount of signaling in comparison to the *enterOrExit* filter implemented according to IETF recommendations.

The third set of measurements compares traffic load between ILS and MPS implemented with periodic and improved non-periodic location querying. The first test was done to measure the quantity of signalization for various group sizes, where the location of all users was requested periodically every ten seconds. The second test was done to measure the quantity of signalization for various group sizes, where all users were static. In this case, location requests were sent non-periodically. Finally, the last test was done to measure quantity of signalization for various group sizes, where about 30-40% of users (phone routes) were static and the rest changed their position. In this case, location requests were also sent non-periodically. It is clear that aggregated traffic load between ILS and

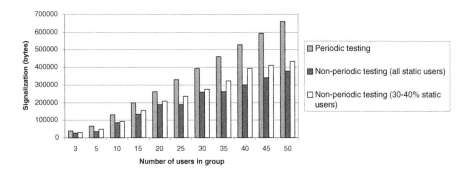

**Fig. 8.** Traffic load between ILS and MPS

MPS increases with time. The results presented in Fig. 8 show the total sum of signalization for a one minute time period, dependant on the number of users in the group. For non-periodic testing where all users were static, the amount of signalization is at its minimum. As the number of users that are randomly changing their movement speed increases, the traffic load increases. In our case, where 30-40% of users remain static, the traffic is still considerably lower than in periodic testing. If all the users in a group are moving with maximum speed, then the traffic load in the case with using non-periodic testing is equal to the worst case scenario when the location of all users is requested periodically.

## 6   Conclusions and Future Work

This paper proposes the introduction of an IMS Location Server (ILS) in the IMS network responsible for retrieving user location information, thus providing the means of making this information available to other IMS application servers. A novel service was developed and integrated with the Ericsson MPS to demonstrate proof-of-concept and to provide a basis for performance measurements related to signaling. Improved SIP location conveyance is presented through definition of a new type of location filter. The emphasis was on location signalization with a positioning system that does not have support for deferred location requests, and on improvements to reduce signalization load.

Instead of using a laboratory environment, in our future work we will consider performing the same measurements in a real 3G network deploying IMS and a real positioning system. Furthermore, privacy related issues that have not been particularly discussed in this article will be also studied in the future.

## References

1. 3GPP TS 23.228: IP Multimedia Subsystem (IMS); Stage 2. Rel. 7, 2005-06 (2005)
2. 3GPP TS 23.271: Functional stage 2 description of Location Services (LCS). Release 7, 2005-03 (2005)

3. -, Open Mobile Alliance OMA-AD_PoC-V1.0-20051104-C: Push to talk over Cellular (PoC) - Architecture. Candidate Version 1.0 , November (2005).

4. -, Open Mobile Alliance OMA-LIF-MLP-V3_1-20040316-C: Mobile Location Protocol (MLP). Candidate Version 3.1, March (2004).

5. OpenGIS: Open Geography Markup Language (GML) Implementation Specification. OGC 02-023r4, January 2003, http://www.opengis.org/techno/implementation.htm (2003)

6. Ericsson Corp: About IMS and Push-to-talk over Cellular. Accessed at: http://www.ericsson.com/mobilityworld/sub/open/index.html, Jan. (2006)

7. Ericsson Corp: Mobile Positioning System. Accessed at: http://www.ericsson.com/mobilityworld/sub/open/index.html, Jan. (2006)

8. G. Camarillo, M. Garcia-Martin: The 3G IP Multimedia Subsystem: Merging the Internet and the Cellular Worlds. John Wiley and Sons, Ltd., England (2004)

9. R. Filjar, D. Huljenic, S. Desic: Distributed Positioning: A Network-based Method for Satellite Positioning Performance Improvement. Journal of Navigation, vol. 55, no. 3, pp. 477-484 (2002)

10. R. Filjar, S. Desic, K. Trzec: LBS Reference Model. Proc. of NAV04 Conference (on CD-ROM), Westminster, London, UK (2004)

11. R. Mahy: A Document Format for Filtering and Reporting Location Notications in PIDF-LO. draft-mahy-geopriv-loc-filters-00.txt, Internet Draft, October 2005, work in progress (2005)

12. J. Peterson: A Presence-based GEOPRIV Location Object Format. IETF RFC 4119, December (2005)

13. J. M. Polk, B. Rosen: Session Initiation Protocol Location Conveyance. draft-ietf-sip-location-conveyance-01.txt, Internet Draft, July 2005, work in progress (2005)

14. A. B. Roach: Session Initiation Protocol (SIP)-Specific Event Notification. IETF RFC 3265, June (2002)

15. J.Rosenberg et al.: SIP: Session Initiation Protocol. IETF RFC 3261, June (2002)

# A Random Backoff Algorithm for Wireless Sensor Networks

Chiwoo Cho[1], Jinsuk Pak[1], Jinnyun Kim[1], Icksoo Lee[1], and Kijun Han[2,*]

Department of Computer Engineering, Kyungpook National University,
1370 Sankyuk-dong, Book-gu, Daegu, 702-701, Korea
{color78, jspak, duritz, islee}@netopia.knu.ac.kr,
kjhan@bh.knu.ac.kr

**Abstract.** Medium Access Control (MAC) protocols employ a backoff algorithm to resolve contention among nodes to acquire channel access. It is desirable to design the backoff algorithm so that the node with lots of remaining energy has a high probability to win in channel contention since the network lifetime can be prolonged by balancing energy consumption over the wireless sensor network. However, most MAC protocols designed for wireless sensor networks have fixed contention period regardless of residual energy, which gives every node the same opportunity to win in the competition. In this paper, we propose a backoff algorithm for wireless MAC which uses dynamic contention period based on the amount of residual energy at each node. Simulation results show that our scheme achieves more power saving and a longer lifetime comparing with the conventional backoff algorithms.

## 1 Introduction

Wireless sensor networking is one of the most essential technologies for implementation of ubiquitous computing. Sensor networks will be applied in variant environments, i.e. health care, military, warehousing, and transportation management. The sensor nodes are usually scatted in a sensor field and data are routed back to the sink by multi-hop. These sensor networks usually share the same communication channel. Sensor nodes have limited in power, computational capacities, memory and short-range radio communication ability. The limited battery life of sensor nodes raises the efficient energy consumption as a key issue in wireless sensor networks. There are four major sources of energy waste; collision, overhearing, control packet overhead and idle listening. Collision of transmitted packets increases energy consumption due to the follow-on retransmissions. Overhearing also spends unnecessary power since a node picks up packets that are destined to other nodes. Sending and receiving control packets consumes energy too. Idle listening meaninglessly consumes battery power by listening to receive possible traffic that is not sent [1].

MAC protocols support nodes to access the communication channels in the networks. Traditional MAC protocols focus on improving fairness, latency, bandwidth utilization

---

* Correspondent author.

Y. Koucheryavy, J. Harju, V.B. Iversen (Eds.): NEW2AN 2006, LNCS 4003, pp. 108 – 117, 2006.
© Springer-Verlag Berlin Heidelberg 2006

and throughput. But, MAC design for wireless sensor networks additional requires energy efficiency as one of its primary concerns due to the specific energy constrained environment. MAC protocols for wireless sensor networks should try to reduce the energy wastage while allocating shared medium among sensor nodes and prevent nodes from transmitting at the same time [2][3]. We focus on the energy efficient MAC protocols for wireless sensor networks.

MAC protocols employ a backoff algorithm to resolve contention among nodes to acquire channel access. The backoff algorithm uniformly chooses a random value from the range [0, CW], where CW is the contention window size. Every node has the same contention window for the backoff algorithm regardless of node status such as node's remaining power. So, the nodes with the low energy level can win in channel contention with the same probability as the nodes with much power. This may lead to the formation of hole in the network since the node with the low energy level can die quickly, which reduces network lifetime substantially [2].

In this paper, we propose a new backoff algorithm for MAC in wireless sensor networks that adaptively determines the contention period of sensor nodes based on their residual energy. The rest of the paper is organized as follows. In Section 2, we review some MAC protocols and backoff algorithm used in wireless sensor networks. In Section 3, our backoff algorithm is introduced in details. Section 4 contains the performance evaluation via simulations. Finally, Section 5 contains the conclusion.

## 2  Related Works

There have been several MAC protocols designed for wireless sensor networks. There are two categories of existing MAC protocol. The first category is a contention-based MAC protocols such as IEEE 802.11 [4]. The main problem of contention-based MAC is that they consume much energy by idle listening. The second category is a contention free MAC protocols such as TDMA. TDMA for wireless sensor networks has two problems that it does not support scalability and it needs centralized control of all nodes [5].

In this section, we briefly review some contention-based MAC protocols. Sensor-MAC (S-MAC) is probably most well known sensor MAC protocol for energy efficiency. It has the following characteristics. S-MAC frame consists of the sleep and the listen periods. S-MAC solves an idle listening problem by putting nodes into periodic sleep state. Each node sleeps for some time, and then wakes up and listens to detect if any other node wants to communicate to it. During sleeping, the nodes turn off radio, and set the wake up time according to the schedule. Before each sensor node starts its periodic listen and sleep, it needs to select a schedule and exchange it with its neighbor nodes. Each sensor node maintains a schedule table, which is composed of neighbor schedules. The schedule is updated periodically to maintain synchronization among the neighboring nodes by SYNK packets. The listen period is divided to receive SYNK packets and data packets [1][6]. The frame structure used in S-MAC is shown in Fig. 1.

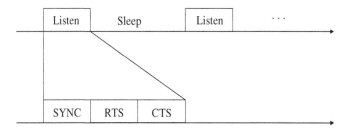

**Fig. 1.** Periodic listen and sleep of S-MAC

S-MAC can reduce the idle listening time, but it is not an optimal solution since it uses a fixed duty cycle. So, S-MAC still has the idle listening problem because sensor nodes waste their energy in active time while there is no traffic.

Several MAC protocols have been developed to resolve the problem of S-MAC. To maintain an optimal active time under variable traffic, Timeout-MAC (T-MAC) dynamically ends the active part when nodes are idle state for a time threshold TA. But T-MAC has a synchronization problem among nodes [7]. DMAC is proposed to achieve very low latency and energy efficiency. DMAC is designed to solve the interruption problem and allow continuous packet forwarding. DMAC adjusts the duty cycle adaptively based on their traffic load in the network [8]. Dynamic Sensor MAC (DSMAC) has been proposed to decrease the latency for delay-sensitive applications. DSMAC is able to dynamically determine the sleeping interval with fixed listen interval and one-hop latency values. Therefore, the duty cycle of sensor nodes is adjusted to adapt to the current traffic condition [3]. Pattern-MAC (PMAC) is proposed to save more power saving than the existing MAC protocols without compromising on the throughput. PMAC adaptively determines the sleep-wake up schedules for a node based on its own traffic, and the traffic patterns of its neighbors. In PMAC, a sensor node gets information about the activity in its neighborhood before exchange through patterns. Based on these patterns, sensor nodes can put itself into a long sleep for several time frames when there is no traffic in the network [9].

In all these contention-based MAC mentioned so far, every node randomly selects a time slot in the fixed contention window to finish its carrier sensing operation. If it has not detected any transmission by the end of that time, it wins the channel contention and acquires transmission opportunity. The channel access mechanism is shown in Fig. 2. The backoff counter is decreased by a slot time as long as the channel is sensed idle, while it is frozen when the channel is sensed busy. When the backoff counter reaches zero, the station starts its data frame transmission. Since every node has the same contention window for the backoff algorithm, it has the same opportunity to win in channel contention regardless of node status. For example, the nodes with the lower energy level can win in channel contention with the same probability as the nodes with much power. So, the node with the low energy level can die quickly. This may lead to the formation of hole in the network, thereby reducing network lifetime substantially [2][10].

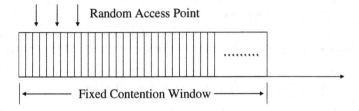

**Fig. 2.** Channel access scheme of a contention-based MAC protocol

## 3  Proposed Backoff Algorithm

In this paper, we propose a backoff algorithm where the contention window is dynamically adjusted based on the amount of remaining energy at each node. In our scheme, the nodes with lots of remaining energy have the higher probability to win in channel contention while the nodes with a little remaining energy have the lower probability to access the channel.

In other words, as the node consumes more energy, it is less likely to win in channel contention. Nodes are initially given the same contention window size (which determines the minimum backoff duration) to have the same opportunity to win in the channel competition. But, when a node consumes more and more energy as times goes on, its contention window becomes longer to have less probability of channel access. Fig. 3 shows the basic concept of our backoff algorithm that uses the contention window size to be determined depending on the remaining energy of each node.

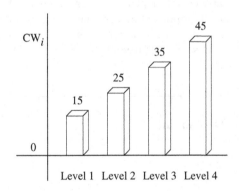

**Fig. 3.** Dynamic contention window in our backoff algorithm

In our algorithm, each node is categorized into 4 levels depending on its residual energy. Each node initially starts from level 1 where its contention window $CW_1$ size is given 15 as shown in Fig. 3. When the node consumes more than 25% of its initial energy, its category is changed to level 2 and its contention window size $CW_2$ is increased to 25. Similarly, if the node consumes more than 50% of its initial energy, its category is changed to level 3 and its contention window size $CW_2$ is again increased to 35. In this way, the contention window size becomes longer as the node consumes more power. The backoff time used by the node in the category i is given by

$$Backoff\ Time\ =\ Random\ (0, CW_i) \times aSlotTime \qquad (1)$$

where aSlotTime means the time duration of a slot.

**Table 1.** Contention window size depending on the residual energy

| Category | Percentage of residual energy relative to initial power | | | Contention window size ($CW_i$) |
|---|---|---|---|---|
| Level 1 | $75\% <$ | Re sidual | energy | $\leq 100\%$ | 15 |
| Level 2 | $50\% <$ | Re sidual | energy | $\leq 75\%$ | 25 |
| Level 3 | $25\% <$ | Re sidual | energy | $\leq 50\%$ | 35 |
| Level 4 | $0\% <$ | Re sidual | energy | $\leq 25\%$ | 45 |

Our backoff algorithm makes all sensor nodes in the network consume their energies uniformly. Balancing the energy consumption among the nodes in the sensor networks avoids the early energy depletion of certain nodes. As a result, the network lifetime can be prolonged by preventing early network disconnection [11]. Note that the contention period should be determined properly considering that too long contention period may cause the idle listening problem. The pseudo-code of proposed algorithm is as follows:

---

**[Channel Access Mechanism]**

$CW_{min} = 15$

*Determine CW size based on the amount of residual energy at each node:*

$CW_i = CW_{min} + \Delta CW(E_{level} - 1)$    *; $\Delta CW$ : Increment of CW size*

*; $E_{level}$ : Energy level* $(1, 2, 3, 4)$

---

*Choose a random value in* $[0, CW]$ *at each node*

*// To acquire transmission opportunity*

*If (Channel is idle)*

    *Decrease the timer*

    *If (Timer is zero)*

        *Send data*

*Else*

    *Freeze the timer*

**Fig. 4.** The pseudo-code of proposed algorithm

## 4  Simulations

We evaluate our backoff algorithm via simulation. We used the Digital wireless LAN module as the energy model where Idle:Rx:Tx ratio is 1:2:2.5 [12]. The sleeping energy consumption is set to 0 (it is usually ignored). Simulation parameters for performance evaluation are shown in Table 2.

**Table 2.** Parameters used in the simulations

| Parameter | Value |
| --- | --- |
| Channel bandwidth | 20 Kbps |
| Control packet length | 10 bytes |
| Data packet length | 200 bytes |
| Slot size | 2 ms |
| Frame size | 1000 ms |
| Transmit energy consumption | 15 mW |
| Receive energy consumption | 12 mW |
| Idle energy consumption | 6 mW |

In the simulation, we use constant bit rate (CBR) model with different time intervals. If the message inter-arrival period is 1 second, each node generates a message every 1 second. We measure the energy consumption at each source node working in different modes: transmitting, receiving and idle modes.

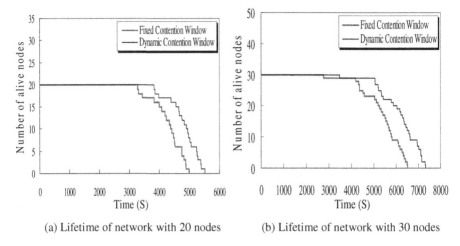

(a) Lifetime of network with 20 nodes          (b) Lifetime of network with 30 nodes

**Fig. 5.** Number of alive nodes

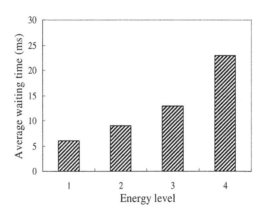

**Fig. 6.** The average waiting time to acquire channel access

The advantage of our backoff algorithm with dynamic contention window can be clearly seen in Fig. 5. Fig. 5 shows the number of alive nodes as time goes on. We observe the elapsed time until the first node (or the last node) fails due to dead batteries. In particular, time for first node to die is very important factor because it gives the time instant when the first node runs out of energy, which reflects the time for network partitioning [13][14]. From this figure, we can see that our backoff algorithm with dynamic contention windows extends the network lifetime comparing with the conventional one using a fixed contention window. This indicates that our scheme is more energy efficient than the conventional scheme. Energy efficiency of our scheme can be more apparently seen when the sensor network becomes large and it has more sensor nodes in the network.

Fig. 6 shows the average waiting time experienced by data packet until it acquires channel access from the time that it is generated. In this test, the 20 source nodes periodically generate data packets. We run simulation for a period of 3000 seconds. In this figure, it can be seen that the node which has more remaining energy has a shorter average waiting time which successfully satisfies the intention of our algorithm.

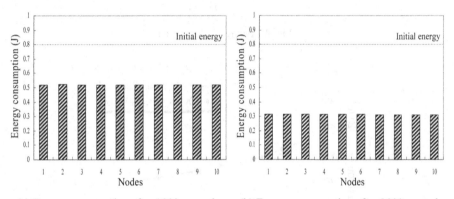

(a) Energy consumption after 1000 seconds     (b) Energy consumption after 2000 seconds

**Fig. 7.** Energy consumption at each node

Fig. 7 shows how uniformly each node consumes its energy over the network. For test, all nodes are given the same initial power, and the consumed energy at each node is measured after some time (e.g. 1000 seconds or 2000 seconds). We can see that our backoff algorithm balances the energy consumption among the nodes in the sensor network. This indicates that our algorithm is more energy efficient than the conventional one since the early energy depletion of certain nodes is prevented, and thus the network lifetime can be extended by our algorithm.

Fig. 8 shows the average residual energy among nodes in the network. In this test, the 10 source nodes periodically generate data packets. For test, all nodes are given the same initial power by 1J, and the average of residual energy at all nodes is measured after 1000, 2000 and 3000 seconds. This result is also coincident with the assurance that our algorithm is more energy efficient than the conventional one.

Finally, we evaluate the data throughput of our backoff algorithm with different message inter-arrival times. In the test, the 10 source nodes periodically generate data packets. Simulations are carried out in single hop environments. We run our simulation for a period of 1000 seconds. Each node generates packets with some

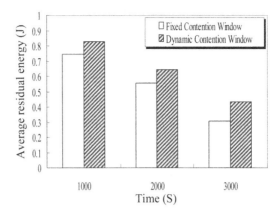

**Fig. 8.** Average residual energy

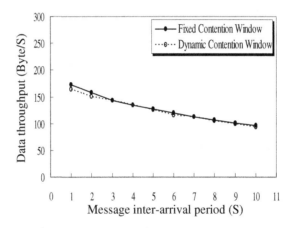

**Fig. 9.** Throughput under different traffic loads

payload (200 bytes) to be sent to their neighbor nodes via contention. Fig. 9 shows that our backoff algorithm with dynamic contention windows offers a similar data throughput to the conventional backoff which uses a fixed contention window. We cannot observe any significant difference in data throughput between them.

## 5   Conclusion

We have proposed a backoff algorithm in which each node adaptively determines its contention window size based on the amount of its residual energy. In the proposed backoff algorithm, as each node consumes more energy, it increases its contention window size to be less likely to win in channel contention. This can balance energy consumption among nodes to prolong network lifetime. Simulation results show that our scheme achieves a more power saving and a longer lifetime when compared with the conventional backoff algorithm.

## Acknowledgement

This research is supported by Program for the Training of Graduate Students for Regional Innovation.

## Reference

[1]  W. Ye, J. Heidemann, and D. Estrin, "An Energy-Efficient MAC protocol for Wireless Sensor Networks," *INFOCOM*, pp. 1567-1576, June 2002.

[2]  R. Kannan, R. Kalidindi, and S. S. Iyengar, "Energy and Rate based MAC Protocol for Wireless Sensor Networks," *SIGMOD Record*, vol. 32, No 4, December 2003.

[3]  P. Lin, C. Qiao, and X. Wang, "Medium Access Control with A Dynamic Duty Cycle for Sensor Networks," *IEEE Global Telecommunication Conference, GLOBECOM'03*, vol. 6, pages: 3547 – 3552, 1-5 December 2003.

[4]  LAN MAN Standards Committee of the IEEE Computer Society, "Wireless LAN Medium Access Control (MAC) and Physical Layer (PHY) specification," *IEEE standard*, 1999.

[5]  S. Biaz and Y. Dai Barowski, ""GANGS": an Energy Efficient MAC Protocol for Sensor Networks," *ACMSE'04*, April 2004, USA.

[6]  W. Ye, J. Heidemann, and D.Estrin, "Medium Access Control with Coordinated, Adaptive Sleeping for Wireless Sensor Networks," *IEEE/ACM Transaction*, vol. 12, pp.493-506, June 2004.

[7]  T.van Dam and K.Langendoen, "An Adaptive Energy-Efficient MAC Protocol for Wireless Sensor Networks," *SenSys*, pp.171-180, November 2003.

[8]  G. Lu, B. Krishnamachari, and C.S. Raghavendra, "An Adaptive Energy-Efficient and Low-Latency MAC for Data Gathering in Wireless Sensor Networks," *Proceedings of 18$^{th}$ International Parallel and Distributed Processing Symposium*, Pages: 224, April 2004.

[9]  T. Zheng, S. Radhakrishnan, and V. Sarangan, "PMAC: An adaptive energy-efficient MAC protocol for wireless Sensor Networks," *Proceedings of 19$^{th}$ IEEE International Parallel and Distributed Processing Symposium*, 2005.

[10] J. Choi, J. Yoo, S. Choi and C. kim, "EBA: An Enhancement of the IEEE 802.11 DCF via Distribute Reservation," *IEEE TRANSACTIONS ON MOBILE COMPUTING*, vol. 4, No 4, July 2005.

[11] I. Chatzigiannakis, A. Kinalis, S. Nikoletseas, "An adaptive power conservation scheme for heterogeneous wireless sensor networks with node redeployment," *Proceedings of the 17$^{th}$ annual ACM Symposium on Parallelism in algorithm and architectures*, pp. 96-105, 2005.

[12] http://www.inf.ethz.ch/personal/kasten/research/bathtub/energy_consumption.html.

[13] M. Younis, M. Bangad and K. Akkaya, "Base-Station Repositioning For Optimized Performance of Sensor   Networks," *in the Proceedings of the IEEE VTC 2003 - Wireless Ad hoc, Sensor, and Wearable Networks*, Orlando, Florida, October 2003.

[14] M. Younis, M. Youssef and K. Akkaya, "Energy-aware management for cluster-based sensor networks," *The International Journal of Computer and Telecommunications Networking*, vol. 43, pp. 649-668, 2003.

# Prolonging the Lifetime of Wireless Sensor Network Via Multihop Clustering

Ying Qian, Jinfang Zhou*, Liping Qian, and Kangsheng Chen

Department of Information Science & Electronics Engineering,
Zhejiang University, Hangzhou 310027, China
zhoujf@zju.edu.cn

**Abstract.** Wireless nodes in sensor network detect surrounding events and then deliver the sensed information to a base station. Organizing these sensors into clusters enables efficient utilization of the limited network resources. Many clustering algorithms have been proposed such as LEACH, HEED, GAF and so on. While LEACH has many excellent features such as highly adaptive, self-configuring cluster formation, application-specific data aggregation, etc., it does not scale well when the network size or coverage increases. In this paper, the Enhanced Multihop Clustering Algorithm (EMCA) is proposed which utilizes multihop links for both intra-cluster and inter-cluster communication. To model the energy consumption more accurately, each cluster is modeled as a Voronoi Cell instead of a circle. The optimal parameter values are determined to minimize the total energy consumption so as to prolonging the lifetime of the whole network. Numerical results show that when both LEACH and EMCA operate with optimal parameter values, the total energy consumption of EMCA is much smaller than that of LEACH. Moreover, EMCA scales much well when the network scale increases, which proves that EMCA is highly scalable and is especially suitable for relatively large-scale wireless sensor networks.

## 1 Introduction

Wireless Sensor network (WSN) is a form of ad hoc network made up of a large number of potentially small and inexpensive sensors. Advances in wireless communication and microelectronics technologies have enabled ordinary nodes to sense environment information, process the collected data and deliver these data to central base stations through wireless channels. Sensor networks are usually used in data gathering applications, such as military surveillance, environmental monitoring, medical treatment, etc [1]. Sensors gather data periodically or when event happens according to the application specification.

Although sensors in a wireless sensor network are almost always connected in ad hoc manner, sensor networks have great difference with ad hoc networks. One of the most prominent characteristics is that, sensor networks often need to be deployed in remote, rough environments autonomously, so that there usually

---

* Corresponding author.

Y. Koucheryavy, J. Harju, V.B. Iversen (Eds.): NEW2AN 2006, LNCS 4003, pp. 118–129, 2006.

have to be huge numbers of densely distributed sensors in a network to guarantee service. From the point view of cost, sensors must be inexpensive which lead to limited power supply and computing resources. Energy is usually supplied by batteries, which are also limited and cannot expect to be replaced or recharged. Hence, energy efficiency is the first design goal in a sensor network [2].

Since sensor networks have the significant energy shortage characteristic as mentioned above, the network protocols employed have to be designed elaborately. We think well tailored sensor network protocols should possess three features. Firstly, the protocols must make the most of limited energy to prolong the lifetime of networks as long as possible. This further includes both energy saving and load balancing. The ideal situation is that all sensors die at the same moment, so that there is no residual energy wasted. Secondly, dynamic and autonomous network operation is required. Sensor applications demand that information be gathered correctly and delivered to the center in time. At the beginning, sensors are deployed randomly and have no idea of the whole network's status. To accomplish the common sensing task, all the nodes have to work cooperatively with consideration of the transmitting/computing ability of each others. Still, all the information needed for collaboration must be communicated wirelessly in ad hoc manner and with restricted resources. Thirdly, scalability is an essential. The number of nodes in a sensor network is expected to be much larger than in a general ad hoc network, and the geography area deployed may be extremely vast, say, a sensor network in wild forests. To be effective in such a large-scale network, the protocols have to be designed so that the performance will not degrade significantly with increment in the scale of the network. Specifically, if the performance degradation is unavoidable, we anticipate the degradation is linear instead of exponential.

Many researches have been carried out in several aspects in wireless sensor networks such as routing, Medium Access Control (MAC), etc., all striving for energy efficient communication. In this paper, we focus on clustered sensor networks as clustering is a better way to save energy and make efficient use of network resources in contrast with flat topology. Moreover, clustering also allows for scalability. In a clustered sensor network, all nodes are classified into clusters according to some clustering algorithm. In each cluster, one node is selected to act as the cluster head (CH) and takes a majority of energy burden in the cluster, e.g., collecting all the information sensed by the cluster, performing data aggregation as necessary and reporting to a base station. All the other nodes only need to communicate within the cluster, or rather, with the cluster head. According to the type of sensor nodes involved, clustered sensor networks can be classified as either homogeneous or heterogeneous. Clustering algorithms in the former usually adopt rotation of cluster heads in order to balance energy dissipation while in the latter use fixed cluster heads which would enable easier hardware implementation to achieve lower cost.

In this paper, we consider homogeneous sensor networks without loss of generality. After extensive research on clustering algorithms proposed for wireless sensor network, we find that single-hop communication is the main reason for unbalanced energy drainage and lacking of scalability. Since LEACH is the most representative clustering algorithm in sensor networks, we adapt LEACH to use multihop communication for information exchange. Our objective is to find the optimal parameter values for this enhanced multihop LEACH-based clustering which we call EMCA (Enhanced Multihop Clustering Algorithm) and prove its better performance compared to LEACH, in terms of both energy consumption and scalability, especially for wireless sensor network of relatively large scale.

The most difference between our algorithm and LEACH is that multihop links are utilized for both intra-cluster and inter-cluster communication. Sensors in the same cluster whereas not within each other's radio range can be scheduled to transmit simultaneously to reduce time delay, while relaying between cluster heads and the base station prevents those remote cluster heads from exhausting quickly. Although many researches have been made in multihop clustering technology, as in [3,4,5,6], our algorithm is quite different from them. E-LEACH [3] uses a chain for communication between cluster heads and the base station while sensors transmit data to their cluster head directly. [4,5,6] deal mainly with the optimization of intra-cluster communication; however, cluster heads transmit data to the base station directly.

Sensors in LEACH are assumed to be able to tune the transmitting power to achieve minimum energy consumption. We, on the other hand, assume a network in which all the sensors transmit at a fixed power level. Communication between two sensors not within each other's radio range uses intermediate nodes to forward data just as in [7]. We also utilize the results provided in [8] to obtain the optimal parameter values used during clustering.

The rest of this paper is organized as follows. In section II, we discuss the related work on clustering algorithms in wireless sensor networks. In section III, the system model for analysis is presented. Next, we evaluate LEACH in detail and give our proposition for EMCA. Numerical analysis results are given in section V to show that EMCA can prolong the network's lifetime and is more scalable than LEACH. Finally, section VI concludes the paper.

## 2    Related Work

During the last few years, a lot of clustering algorithms have been proposed for wireless sensor networks. Grouping a large number of sensors into clusters and keeping them communicate regularly are quite complex. Here, we mention some of the most recent work in different views of clustering.

In [9], Heinzelman et al. developed and analyzed LEACH, an application-specific protocol architecture for microsensor networks. LEACH divides time into rounds. Clusters are organized at the beginning of each round and data are transferred from the nodes to the cluster head and on to the base station after the set-up phase. As LEACH is such a typical clustering protocol, several modifications have been made [10,3,11,4].

The Expellant Self-Organization (ESO) scheme is designed to replace the cluster formation at the beginning of each round in LEACH [12]. One of the advantages of ESO is that it uses only local information. In [13], the Energy Efficient Clustering Scheme (EECS) elects a constant number of candidate nodes and lets them compete for cluster heads according to the nodes' residual energy. This scheme can produce a near uniform distribution of cluster heads. Qin and Zimmermann propose a novel Voting-based Clustering Algorithm (VCA) which lets sensors vote for their neighbors as cluster heads [14]. In the Hybrid Energy-Efficient Distributed (HEED) clustering protocol [15], cluster heads are randomly selected based on their residual energy, and nodes join clusters such that communication cost is minimized. In [7], a distributed, randomized clustering algorithm is proposed which can also be extended to generate a hierarchy of cluster heads. Results show that the energy saving increases with the number of levels in the hierarchy. A new routing and MAC protocol – CuMPE is proposed in [5]. The routing part is cluster-based and adopts a selective flooding algorithm to find the minimum cost path to cluster head. The TDMA scheduling part minimizes transmission delay by first finding the critical path of a route tree and avoiding extending the critical path when scheduling nodes for transmission. The Time-Controlled Clustering Algorithm (TCCA) is developed by Selvakennedy and Sinnappan [6]. TCCA allows multihop clusters using message timestamp and time-to-live(TTL) information to control the cluster formation.

All of the above schemes aim at homogeneous networks. There are also several researches targeted at heterogeneous sensor networks. The Unequal Clustering Size (UCS) model proposed in [16] deals with the problem of unbalanced energy consumption, particularly among the cluster heads, assuming that this type of node is much more expensive than the simpler sensor node. The authors develop a network clustering scheme where unequal sized clusters are formed. Smaragdakis et al. propose Stable Election Protocol (SEP), a heterogeneous-aware protocol to prolong the time interval before the death of the first node [17]. SEP is based on weighted election probability of each node to become cluster head according to the remaining energy in each node. In [18], Energy-Driven Adaptive Clustering (EDCA) protocol, as an improvement over LEACH in heterogeneous networks, uses energy-driven cluster heads rotation, which enables cluster heads change asynchronously and coordinate energy consuming based on the heterogeneous property of nodes' energy. Gupta and Younis investigate the performance of a new clustering algorithm for sensor networks [19]. The proposed approach balances the load among clusters and simultaneously tries to cluster sensor nodes as close to high-energy cluster heads, which are called gateways, as possible. A minimum cost heterogeneous sensor network is considered in [20]. By ensuring certain conditions for connectivity and coverage of the area during the lifetime of the network, it is possible to minimize the overall network cost.

## 3   System Model

In this section, we present the system model for analysis of EMCA, which will be discussed in the next section.

## 3.1  Assumptions

For the development of our model, the following assumptions are made:

1. All the sensor nodes in a wireless sensor network are randomly distributed in a two-dimensional plane according to a homogeneous spatial Poisson process with $\lambda$ intensity.
2. All the sensor nodes in the network are homogeneous. They transmit at the same power level and have the same radio coverage $r$.
3. When two communicating sensor nodes are not within each other's radio range, data is forwarded by other nodes. A distance of $d$ between any two nodes is equivalent to $\lceil d/r \rceil$ hops.
4. The base station is located at the center of the field.
5. A routing and MAC infrastructure is in place. The communication environment is contention- and error-free.

## 3.2  Energy Model

In this paper, the free space channel model is used [9]. To transmit an $l$-bit packet at a distance of $r$, the sensor consumes

$$E_{tx} = lE_{elec} + l\varepsilon_{fs}r^2 \tag{1}$$

and to receive this packet, the sensor consumes

$$E_{rx} = lE_{elec} \tag{2}$$

where $E_{elec}$, the electronics energy, depends on factors such as the digital coding, modulation, filtering and spreading of the signal, whereas the amplifier coefficient, $\varepsilon_{fs}$ depends on the acceptable bit-error rate.

# 4  EMCA

In this section, we will describe EMCA in detail. Before doing this, we present a brief review of LEACH first.

## 4.1  LEACH

LEACH uses a CDMA-TDMA hybrid communication scheme. For inter-cluster communication, each cluster employs a unique spreading code to minimize interference while each cluster head sets up a TDMA schedule for collision-free intra-cluster communication. As shown in Fig. 1, time is divided into rounds. For each round, clusters are formed in set-up phase and data is transmitted to the base station in steady-state phase. Cluster formation is randomized, adaptive and self-configuring. The cluster heads rotate periodically for load balancing. All non-cluster-head nodes choose the closest cluster head to join in. They use power control to set the amount of transmitting power based on the distance to their

**Fig. 1.** Time-line showing LEACH in operation [9]

cluster heads and turn off radios until their allocated transmission time. At the end of each frame, the cluster heads perform data aggregation, which enables considerable energy saving, and forward the aggregated data to the base station.

Although LEACH has above features, we have found several drawbacks in it:

1. Variable number of clusters produced in each round.
2. Uneven size of clusters.
3. The cluster head far away from base station will exhaust its energy quickly while there is still a lot of residual energy in the network. This makes the clustering algorithm less scalable.
4. No time slots allocated to cluster heads. Hence, when cluster head transmits aggregated data to base station and at the same time data collected by one of its members arrives, collision will happen.

Several extensions to LEACH have been proposed as mentioned in section II to deal with the first three drawbacks. Most of our research focuses on scalability of the clustering algorithm. By taking the channel access of cluster heads into consideration, we also avoid the collision problem.

### 4.2   EMCA

**Motivation.** Simulation results show that the performance of LEACH protocol degrades significantly with the increase in simulation region size and network scale. This is mainly because of single-hop communication between cluster heads and the base station. Those cluster heads far away from the base station have to consume most of their energy to compensate multipath fading. Hence, we suggest the employment of multihop links for inter-cluster communication.

On the other hand, LEACH uses CSMA for cluster head to transmit the aggregated data, in which collision is inevitable. If we use multihop inter-cluster communication instead, obviously CSMA will no longer be suitable. Because there is usually only one transceiver in each sensor, it is impossible for cluster heads to transmit while receiving. So we must allocate special time slots for cluster heads.

**Algorithm.** Time-line of EMCA is the same as LEACH, however the operation is different.

The set-up phase is divided into four subphases: cluster-formation, routing-establishment, TDMA-schedule and synchronization.

– Set-up phase: each sensor becomes a cluster head with probability $p$, then advertises this to the other sensors within its radio range. This advertisement is forwarded over limited hops. Any non-cluster-head node that receives such advertisements joins the closest cluster.

- Routing-establishment subphase: In this subphase, we assume that some routing scheme is in place to establish paths from non-cluster-head nodes to their cluster heads and from cluster heads to the base station. Data exchange between two communicating sensors not within each other's radio range is forwarded by intermediate sensors.
- TDMA-schedule subphase: The similar scheme proposed in [5] can be used to set up a TDMA schedule in a cluster. It begins with the outmost nodes and ends with the one-hop neighbor of cluster head. Cluster head only receives packets from its direct neighbors.
- Synchronization subphase: All cluster heads needs to be synchronized for aggregated data transmission. This is done by the base station. It compares the frame lengths of all clusters and chooses the longest one as the frame length. Cluster heads modify their schedules by adding several empty time slots in front of the original ones. Then the base station synchronizes all cluster heads to start steady-state phase at the same time.

The steady-state operation is broken into frames too. Since cluster head only communicates with its direct neighbors, its transceiver is idle at the beginning of each frame. Hence in EMCA, data is aggregated and transmitted to the base station at the beginning of each frame. At the same time, the outmost non-cluster-head nodes start to transmit data to their next hops according to the TDMA schedule. In this way, collision-free communication is guaranteed.

**Energy Consumption.** For the derivation of optimal parameter values in EMCA, a function is defined for the total energy consumption during one frame in steady-state phase. The optimal parameter values are determined so that the energy function is minimized.

As per the assumptions, the sensors are distributed according to a homogeneous spatial Poisson process. Therefore, the number of sensors in a square area of side $M$ is a Poisson random variable, $N$ with mean value of $\lambda M^2$. The probability of becoming a cluster head is $p$; hence, on average, $\lambda M^2 p$ sensors will become cluster heads. Since the locations of all cluster heads are independent of each other, the average total length of segments from a cluster head to the base station is $\frac{0.765M}{2}$ [7].

Now, since a sensor becomes a cluster head with probability $p$, the cluster heads and the non-cluster-head nodes are distributed as per independent homogeneous spatial Poisson processes of intensity $p\lambda$ and $(1-p)\lambda$ respectively. While many papers use circle to model the shape of a cluster such as in [6], each non-cluster-head node joins the losest cluster to form a Voronoi tessellation as illustrated in Fig. 2 [8]. Thus the network is divided into zones called Voronoi Cells, with each cell corresponding to a cluster. According to the results in [8], the average number of non-cluster-head nodes in each cluster is

$$E[N] = \frac{(1-p)\lambda}{p\lambda} = \frac{1-p}{p} \tag{3}$$

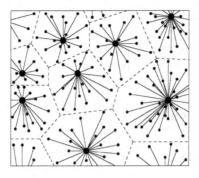

**Fig. 2.** Voronoi Cells [8]

while the average total length of all segments connecting the non-cluster-head nodes to their cluster head is

$$E[L] = \frac{(1-p)\lambda}{2(p\lambda)^{\frac{3}{2}}} = \frac{1-p}{2p^{\frac{3}{2}}\lambda^{\frac{1}{2}}} \qquad (4)$$

Assuming packet size to be $l$, we obtain the total energy consumption required to transmit all the collected data to cluster heads in a network as follows:

$$E[C_1] = \lambda M^2 p * \frac{E[L]}{r} * (E_{rx} + E_{tx}) \qquad (5)$$

where $\frac{E[L]}{r}$ refers to the average total hops needed for every non-cluster-head node to reach the corresponding cluster head, and each hop involves both sending by the transmitting node and reception by the intermediate/destination node.

Let $E_{DA}$ be the energy consumption for aggregating one bit of data. The total energy required for all the aggregation is

$$E[C_2] = E_{DA}\lambda M^2 l \qquad (6)$$

Finally, the energy used by cluster heads to transmit the aggregated data to base station is

$$E[C_3] = \lambda M^2 p * \frac{0.765M}{2r}(E_{rx} + E_{tx}) \qquad (7)$$

Therefore, the total energy consumption in a frame is

$$E[C] = E[C_1] + E[C_2] + E[C_3]$$

$$= \lambda M^2 p * [\frac{1-p}{2p^{\frac{3}{2}}\lambda^{\frac{1}{2}}r}(E_{rx} + E_{tx})] + E_{DA}\lambda M^2 l + \lambda M^2 p * \frac{0.765M}{2r}(E_{rx} + E_{tx})$$

$$= \lambda M^2 l[(\frac{1-p}{2p^{\frac{1}{2}}\lambda^{\frac{1}{2}}} + \frac{0.765Mp}{2})(\frac{2E_{elec}}{r} + \varepsilon_{fs}r) + E_{DA}]$$

$$(8)$$

From the above equation, we can see that the total energy consumption ($C$) can be minimized if we choose the appropriate values of $p$ and $r$.

## 5    Numerical Analysis and Results

This section discusses the numerical experimentation, which includes an analysis on how to determine optimal parameters and the adopted sensor network scenario.

### 5.1    Optimal Parameters Determination

1) $p$

To get the optimal value of $p$, we need to minimize the value of $(\frac{1-p}{2p^{\frac{1}{2}}\lambda^{\frac{1}{2}}} + \frac{0.765Mp}{2})$ in equation (8), which can be represented as follows:

$$\frac{1-p}{2p^{\frac{1}{2}}\lambda^{\frac{1}{2}}} + \frac{0.765Mp}{2} = \frac{1}{2\lambda^{\frac{1}{2}}}p^{-\frac{1}{2}} - \frac{1}{2\lambda^{\frac{1}{2}}}p^{\frac{1}{2}} + \frac{0.765M}{2}p \qquad (9)$$

The right side of the above can be seen as a function of $p^{\frac{1}{2}}$ and we make a derivative of this function $(f)$:

$$\frac{df}{d(p^{\frac{1}{2}})} = \frac{d(\frac{1}{2\lambda^{\frac{1}{2}}}p^{-\frac{1}{2}} - \frac{1}{2\lambda^{\frac{1}{2}}}p^{\frac{1}{2}} + \frac{0.765M}{2}p)}{d(p^{\frac{1}{2}})}$$
$$= \frac{1}{2\lambda^{\frac{1}{2}}}[1.53\sqrt{\lambda M^2}(p^{\frac{1}{2}}) - (p^{\frac{1}{2}})^{-2} - 1] \qquad (10)$$

Define $f_1(x) = 1.53\sqrt{\lambda M^2}x$ and $f_2(x) = x^{-2} + 1$. When $x \in (0,1)$ , $f_1(x)$ monotonically increases from 0 to $1.53\sqrt{\lambda M^2}$ and $f_2(x)$ monotonically decreases from $+\infty$ to 2. Since $\lambda M^2$ represents the number of sensors in a network, $1.53\sqrt{\lambda M^2}$ must be bigger than 2. So we can find that there must be a point $x_0 \in (0,1)$ at which $f_1(x_0) = f_2(x_0)$. When $p^{\frac{1}{2}} = x_0$ , $f$ is minimized. The value of $x_0$ depends on the number of sensors in a network.

2) $r$

$r$ is only associated with $E_{elec}$ and $\varepsilon_{fs}$ in equation (8). The total energy consumption is minimized when $\frac{2E_{elec}}{r} = \varepsilon_{fs}r$ , namely $r = \sqrt{\frac{2E_{elec}}{\varepsilon_{fs}}}$ .

### 5.2    Numerical Results

We assume that there are $N$ sensor nodes distributed randomly in a square $M*M$ region with intensity $\lambda = 0.01$. We set the same communication energy parameters as in LEACH: $E_{elec} = 50nJ/bit$ and $\varepsilon_{fs} = 10pJ/bit/m^2$. The energy for data aggregation is set to $E_{DA} = 5nJ/bit$. Each data message is 500bytes long. Hence, the radio range of each sensor node is taken as $\sqrt{\frac{2E_{elec}}{\varepsilon_{fs}}} = \sqrt{\frac{2*50}{10*10^{-3}}} = 100m$. Unless otherwise stated, all the following investigations adopt these values as their system parameters.

Since our objective is to design highly scalable clustering algorithm, the value of $M$ here is larger than in LEACH. Fig. 3 shows the total energy spent by the

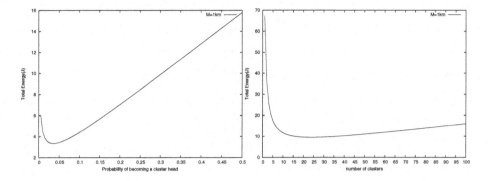

**Fig. 3.** $E[C]$ vs. $p$ in EMCA        **Fig. 4.** $E[C]$ vs. $k$ in LEACH

network against various probability of becoming a cluster head($p$) with $M = 1km$. The value of $p$ reflects the number of clusters in network, which is $\lambda M^2 p$ in our algorithm. It can be shown that when $p \approx 0.0358$, the total energy spent by the network in a frame gets the minimum of $3.334J$. For the convenience of comparison, the performance of LEACH is simulated using the same scenario. According to [9], the total energy in a frame is given by:

$$E_{total} = l(E_{elec}N + E_{DA}N + k\varepsilon_{mp}d_{toBS}^4 + E_{elec}N + \varepsilon_{fs}\frac{1}{2\pi}\frac{M^2}{k}N) \qquad (11)$$

where $k$ is the number of clusters and $d_{toBS}$ is set to $\frac{0.765M}{2}$ here. When cluster heads transmit data to the base station, the multipath model is used. Fig. 4 shows the total energy consumption as the number of clusters increases. We can see that the optimum number of clusters is between 15 and 35, and the minimum of energy spent that LEACH can achieve is about 10J.

Comparing Fig. 3 with Fig. 4, we can also find that besides the minimum energy consumption achievable when both EMCA and LEACH operate with optimal parameter values, generally, the energy consumption of EMCA is much smaller. If the time of steady-state phase lasts long enough, this energy saving can be huge.

Next, we analyze the scalability of both clustering algorithms. The total energy consumption with different values of $M$ are illustrated in Fig. 5 and Fig. 6 respectively.

We can see from Fig. 5 that, the minimal energy spent by EMCA grows almost linearly with the application area. It proves that EMCA is highly scalable and energy efficient. However, this increase of energy is much faster in Fig. 6. Although LEACH uses cluster head rotation to balance energy consumption, the energy spent in each round is uneven. Those cluster heads far away from the base station cannot afford such a big energy burden. This will greatly reduce the lifetime of the whole network, especially in large-scale area.

Another observation we can find from these two figures is that, the total energy spent in EMCA is less sensitive to the number of clusters than in LEACH. The

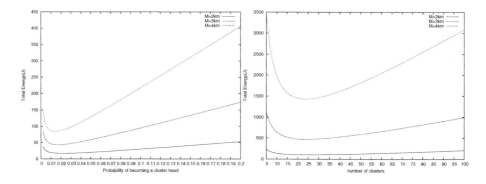

**Fig. 5.** $E[C]$ for different $M$ in EMCA     **Fig. 6.** $E[C]$ for different $M$ in LEACH

increase of 5 in the number of clusters may add several hundred joules of energy consumption in LEACH. However energy consumption almost does not change in EMCA, which can compensate for the randomicity in the cluster head election phase more or less.

## 6   Conclusion

We have proposed an Enhanced Multihop Clustering Algorithm(EMCA) for organizing sensor nodes into clusters with an objective to minimize the total energy spent in the system and improve scalability of clustering, especially for wireless sensor network of relatively large scale. We have deduced the optimal parameter values for the algorithm that minimize the energy spent in the network. It is numerically demonstrated that EMCA can reduce energy consumption significantly so as to prolong the lifetime of wireless sensor network. It is also verified to be highly scalable.

In our system model, we have supposed that the base station is at the center of the field and homogeneous sensor nodes transmit at the same power. Where the base station is located only influences the $E[C_3]$ part in total energy consumption. In heterogeneous sensor network, our algorithm can be simplified especially in the cluster heads election phase. Those sensors with less restriction on energy supply and hardware resource will take most burden of processing and communicating reasonablely. As our future work, we will research on the optimal per-hop length used in EMCA with the assumption of uniform transmitting power relaxed.

## References

1. Bharathidasan, A., Ponduru, V.A.S.: Sensor networks: an overview. (2003)
2. LanWang, Xiao, Y.: A survey of energy saving mechanisms in sensor networks. (2005)

3. Pham, T., Kim, E.J., Moh, M.: On data aggregation quality and energy efficiency of wireless sensor network protocols - extended summary. In: Proc. BROADNETS 2004. 730–732
4. Mhatre, V., Rosenberg, C.: Homogeneous vs heterogeneous clustered sensor networks: a comparative study. In: Proc. ICC 2004. Volume 6. 3646–3651
5. Ho, S., Su, X.: CuMPE: Cluster-management and power-efficient protocol for wireless sensor networks. In: Proc. ITRE 2005. 60–67
6. Selvakennedy, S., Sinnappan Sinnappan, S.: A configurable time-controlled clustering algorithm for wireless sensor networks. In: Proc. ICPADS 2005. Volume 2. 368–372
7. Bandyopadhyay, S., Coyle, E.: An energy efficient hierarchical clustering algorithm for wireless sensor networks. In: Proc. INFOCOM 2003. Volume 3. 1713–1723
8. S.G.Foss, Zuyev, S.: On a voronoi aggregative process related to a bivariate poisson process. Advances in Appl. Probability **28** (1996) 965–981
9. Heinzelman, W., Chandrakasan, A., Balakrishnan, H.: An application-specific protocol architecture for wireless microsensor networks. IEEE Trans. Wireless Comm. **1** (2002) 660–670
10. Handy, M., Haase, M., Timmermann, D.: Low energy adaptive clustering hierarchy with deterministic cluster-head selection. In: Proc. MWCN 2002. 368–372
11. Manjeshwar, A., Agrawal, D.: TEEN: a routing protocol for enhanced efficiency in wireless sensor networks. In: Proc. IPDPS 2001. 2009–2015
12. Zhao, L., Liang, X.H.Q.: Energy-efficient self-organization for wireless sensor networks: a fully distributed approach. In: GLOBECOM '04. IEEE. Volume 5. 2728–2732
13. Ye, M., Li, C., Chen, G., Wu, J.: EECS: an energy efficient clustering scheme in wireless sensor networks. In: Proc. IPCCC 2005. 535–540
14. Qin, M., Zimmermann, R.: An energy-efficient voting-based clustering algorithm for sensor networks. In: Proc. SNPD/SAWN 2005. 444–451
15. Younis, O., Fahmy, S.: Distributed clustering in ad-hoc sensor networks: a hybrid, energy-efficient approach. In: Proc. INFOCOM 2004. Volume 1.
16. Soro, S., Heinzelman, W.: Prolonging the lifetime of wireless sensor networks via unequal clustering. In: Proc. IPDPS 2005.
17. Smaragdakis, G., Matta, I., Bestavros, A.: SEP: A stable election protocol for clustered heterogeneous wireless sensor networks. In: Proc. SANPA 2004.
18. Wang, Y., Zhao, Q., Zheng, D.: Energy-driven adaptive clustering data collection protocol in wireless sensor networks. In: Proc. ICIMA 2004. 599–604
19. Gupta, G., Younis, M.: Performance evaluation of load-balanced clustering of wireless sensor networks. In: Proc. ICT 2003. Volume 2. 1577–1583
20. Mhatre, V., Rosenberg, C., Kofman, et al.: A minimum cost heterogeneous sensor network with a lifetime constraint. IEEE Trans. on Mobile Comput. **4** (2005) 4–15

# A BEB-Based Admission Control for VoIP Calls in WLAN with Coexisting Elastic TCP Flows

Boris Bellalta[1], Michela Meo[2], and Miquel Oliver[1]

[1] Universitat Pompeu Fabra, Spain
{boris.bellalta@upf.edu, miquel.oliver}@upf.edu
[2] Politecnico di Torino, Italy
michela.meo@polito.it

**Abstract.** VoIP service over WLAN networks is a promising alternative to provide mobile voice communications to compete with cellular systems. However, several performance problems appear due to *i)* heavy protocol overheads, *ii)* unfairness and asymmetry between the uplink and downlink flows and *iii)* the coexistence with other traffic flows. This paper addresses the performance of VoIP communications with simultaneous presence of bidirectional TCP traffic, showing how the presence of elastic flows drastically reduces the capacity of the system. To solve this limitation we propose a simple solution using an adaptive Admission and Rate Control algorithm which tunes the BEB (Binary Exponential Backoff) parameters as is defined in the IEEE 802.11e standard. The results show the improvement achieved on the overall system performance (in terms of number of simultaneous voice calls with QoS guarantees).

## 1 Introduction

First Access Points (AP) and wireless cards implementing the EDCA (Enhanced Distributed Channel Access) [1] are already commercially available under the WMM (Wireless MultiMedia) denomination[1]. Based on the differentiation capabilities provided by the EDCA specification, some possible solutions for guaranteeing acceptable QoS levels can be achieved, consisting in optimizing the use of the transmission resources. This process requires an entity, the call admission and rate controller, capable to decide whether new traffic flows can be accepted and what resources they can use in order to guarantee the QoS levels of already active flows.

Nowadays, voice calls over Internet (VoIP) are becoming popular, with service providers growing around the world. Most of these VoIP users already have WLAN as access network to the Internet and therefore, the bidirectional VoIP stream goes through the AP from/to a laptop or a VoIP/WLAN phone. In this situation, the VoIP flow is competing with typical TCP based flows (like P2P, FTP, e-mail, etc.) to get access to the transmission resources over the wireless channel which causes serious performance problems for the voice data.

---

[1] http://www.wi-fi.org/

Y. Koucheryavy, J. Harju, and V.B. Iversen (Eds.): NEW2AN 2006, LNCS 4003, pp. 130–141, 2006.
© Springer-Verlag Berlin Heidelberg 2006

In this paper, we investigate the performance perceived by VoIP calls in a Hot-spot scenario with the simultaneous presence of persistent TCP connections. To solve the performance limitations, we propose a combined admission control and rate shaping scheme based on the use of different values of the minimum backoff contention window ($CW_{min}$). In order to evaluate the performance of the CAC scheme and provide quantitative analytical results, we develop a model of the IEEE 802.11e specification [1]. The model captures the changes on the main protocol dynamics caused by the different settings of the MAC parameters ($AIFS$, $TXOP$ and $CW$).

## 2   A Hot-Spot Wireless Scenario

A single AP providing access to a fixed network to $n$ Mobile Nodes (MNs) is considered. The MNs and the AP use the EDCA operation mode and the DSSS PHY specifications in the 2.4 $GHz$ band [2]. We assume ideal channel conditions, i.e., no packet is lost due to channel errors or the hidden terminal phenomenon. The system parameters are reported in Table 1.

**Table 1.** System parameters of the IEEE 802.11b specification [2]

| Parameter | Value | Parameter | Value |
|---|---|---|---|
| $R_{data}$ | 2 $Mbps$ | $R_{basic}$ | 1 $Mbps$ |
| $AIFS_j = SIFS + A_j \cdot \sigma,$ | $A_j = \{7,\ 3,\ 2,\ 2\}$ | $CW_{min}$ | 32 (all AC) |
| SIFS | 10 $\mu s$ | $CW_{max}$ | 1024 (all AC) |
| TXOP | 1 (all AC, in packets) | m | 5 (all AC) |
| SLOT ($\sigma$) | 20 $\mu s$ | ACK | 112 $bits$ @ $R_{basic}$ |
| RTS | 160 $bits$ @ $R_{basic}$ | CTS | 112 $bits$ @ $R_{basic}$ |
| MAC header | 240 $bits$ @ $R_{data}$ | MAC FCS | 32 $bits$ @ $R_{data}$ |
| PLCP preamble | 144 $bits$ @ $R_{basic}$ | PLCP header | 48 $bits$ @ $R_{basic}$ |
| Retry Limit (R) | 7 | $Q$ (Queue length) | 20 $packets$ |

### 2.1   Performance of VoIP Calls with TCP Flows in a Non-QoS WLAN

The negative influence of TCP traffic over the VoIP calls is clearly shown in Figure 1. These results are obtained using the ns-2 simulator [3] with the parameters setting reported in Table 1, and the use of the $RTS/CTS$ mechanism and the TCP new-RENO version. Note the fast degradation of the VoIP throughput with TCP downlink flows and the inoperability of any VoIP call with just a single TCP uplink flow. It is also interesting to observe that, when the AP queue is saturated with VoIP traffic, the interaction with TCP traffic is reduced due to the starvation of TCP flows. Therefore, the presence of TCP traffic in both the downlink (buffer losses) and the uplink (AP starvation) leads to low performance of VoIP calls. These problems have to be solved in order to deploy a successful

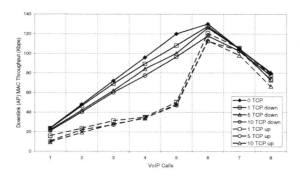

**Fig. 1.** G.279 VoIP downlink throughput in presence of *i)* downlink TCP flows and *ii)* uplink TCP flows

VoIP service over WLAN. In the downlink, a simple classification/prioritization scheme can be used (four Access Categories (AC) are defined in the EDCA [1]) where the TCP and VoIP packets can occupy separated buffers with priority to the VoIP packets. However, the main problems raise in presence of uplink TCP flows because nodes act independently from each others. The only possible solution is setting different MAC parameters (such as $AIFS$ [4], $CW_{min}$ [5] or/and $TXOP$ [6]) at each mobile node in order to reduce the interaction of these TCP flows with the VoIP calls.

## 3    A Model of the IEEE 802.11e EDCA

A user-centric model of the the IEEE 802.11e-EDCA operation mode is proposed in this section. We approximate each mobile node by a finite length queue with bulk and network-dependent service times. We assume that each mobile node carries a single traffic flow of category $AC_j$ (or equivalently, each node $i$ has only one $AC_j$ active at the same time). For the sake of simplicity, henceforth we omit the subscript $j$ as a single $AC$ is considered to be active in each MN.

### 3.1    A Mobile Node

Packets with average length $L_i$ arrive to node $i$ with rate $\beta_i$. Both the time between packet arrivals and the service time are assumed to be exponentially distributed. Each node (the AP included) is modeled by an $M/M^{[TXOP_i]}/1/Q_i$ queue with bulk services times and queue length $Q_i$ measured in packets (Figure 2). The steady state probability that $q$ packets are in the queue is denoted by $\pi_{q,i}$. The bulk service time $X_i^k$, where the super-index $k$ refers to the length of the burst, is used to approximate the TXOP (Transmission Opportunity) option of the EDCA [1]. The offered traffic load to the MAC layer, the queue utilization and the expected throughput for node $i$ are $\nu_i$, $\rho_i$, and $S_i$ respectively.

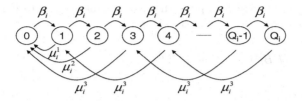

**Fig. 2.** $M/M^{[TXOP_i]}/1/Q_i$, $TXOP_i = 3$

$$\begin{cases} \nu_i = \beta_i \left( \sum_{k=0}^{TXOP_i} \pi_{k,i} X_i^k + \sum_{k=TXOP_i+1}^{Q_i} \pi_{k,i} X_i^{TXOP_i} \right), \\ \rho_i = \sum_{k=1}^{Q_i} \pi_{k,i} = 1 - \pi_{0,i}, \\ S_i = \sum_{k=1}^{TXOP_i} \pi_{k,i} \frac{k \cdot L_i}{X_i^k} + \sum_{k=TXOP_i+1}^{Q_i} \pi_{k,i} \frac{TXOP_i \cdot L_i}{X_i^{TXOP_i}} \end{cases} \quad (1)$$

Using this queuing model, we obtain simple expressions to measure the service observed by a node in terms of blocking probability, average queue length and average transmission delay (including the service time),

$$\begin{cases} P_{b,i} = \pi_{Q_i,i}, \\ EQ_i = \sum_{k=1}^{Q_i} k \cdot \pi_{k,i}, \\ ED_i = \frac{EQ_i}{\lambda_i(1-P_{b,i})}, \\ P_{L,i} = P_{b,i} + (1 - P_{b,i})P_{d,i} \end{cases} \quad (2)$$

where $P_{b,i}$ is the packet blocking probability, $EQ_i$ is the average queue occupation, $ED_i$ is the average queuing delay and $P_{L,i}$ is the probability to loose a packet, with $P_{d,i}$ as the probability that a packet is dropped at the MAC layer because the number of retransmissions have exceeded the retry limit.

## 3.2   The MAC Protocol

Letting $AEB_i$ (Arbitrary Expected Backoff) be the average number of backoff slots at each transmission attempt by node $i$, the steady state probability that the node transmits in a random slot given that a packet is ready in its transmission queue can be computed from

$$\tau_i = \frac{E[Pr(Q_i(t) > 0)]}{AEB_i + 1} = \frac{\rho_i}{AEB_i + 1} \quad (3)$$

The $AEB_i$ parameter include the $A_j$ blocked slots from the $AIFS$ interval plus the average number of backoff slots selected at each transmission attempt, $EB_i$, which can be computed from the expression presented by Tay et al. [7] as

$$EB_i = \frac{1 - p_i - p_i(2p_i)^{m_i}}{1 - 2p_i} \frac{CW_{min,i}}{2} - \frac{1}{2} \quad (4)$$

To compute the $AEB_i$ value we use an approach similar to [8]. We take into consideration that the number of transmissions observed by a node during its backoff is $p_{tr,i} \cdot EB_i$, which implies an extra number of blocked slots that the node has to wait equal to $p_{tr,i} \cdot EB_i \cdot A_i$, then

$$AEB_i = (EB_i + A_i) + p_{tr,i} \cdot EB_i \cdot A_i, \qquad p_{tr,i} = 1 - \prod_{j \neq i}(1 - \tau_j) \qquad (5)$$

where $p_{tr,i}$ is the probability that at least another node transmits in a given slot.

Being $s_i$ the instantaneous queue occupation (in packets) and $b_i$ the instantaneous packet burst length, with $b_i = s_i$ if $s_i \leq TXOP_i$ and $b_i = TXOP_i$ in any other case, the departure rates of each state for the $M/M^{[TXOP_i]}/1/Q_i$ queue depend on the $b_i$ value, $\mu_i^{b_i} = 1/X_i^{b_i}$. The service time, i.e., the time interval from the instant in which a packet enters in service until it is completely transmitted or discarded, is given by,

$$X_i^{b_i} = (M_i - 1)\left(AEB_i\alpha_i + ET_{c,i}^{ba||rts}\right) + AEB_i\alpha_i + T_{s,i}^{b_i,ba||rts} \qquad (6)$$

where $M_i$ is the average number of required transmissions, $\alpha_i$ is the average slot duration, $T_{s,i}^{b_i,ba||rts}$ is the duration of a $b_i$ packets burst transmission using the BA or the RTS/CTS access mechanism, and $ET_{c,i}^{ba||rts}$ is the average duration of a collision of node $i$. We approximate the value of $ET_{c,i}^{ba||rts}$ by,

$$\begin{cases} ET_{c,i}^{ba} \approx \dfrac{\sum_{j \neq i} \tau_j \max(T_{s,i}^{1,ba}, T_{s,j}^{1,ba})}{\sum_{j \neq i} \tau_j} \\ ET_{c,i}^{rts} = T_c^{rts} \end{cases} \qquad (7)$$

where we neglect the fact that more than two packets collide simultaneously. Note that if the RTS/CTS access scheme is used, $ET_{c,i}^{rts}$ is constant and equal for all nodes.

A node frozes its backoff counter every time the channel is sensed busy and releases it after the channel is sensed free for an $AIFS$ period. Therefore, the time between two backoff counter decrements is a random variable which depends on the behavior of the other nodes. By letting $\alpha_i$ be the average time between two backoff counter decrements, or equivalently, the average slot duration, we have

$$\alpha_i = p_{e,i}\sigma + p_{s,i}(ET_{s,i}^{ba||rts,*} + \sigma) + p_{c,i}(ET_{c,i}^{ba||rts,*} + \sigma) \qquad (8)$$

where $ET_{s,i}^{ba||rts,*}$ and $ET_{c,i}^{ba||rts,*}$ are the average durations of an observed successful transmission or a collision for node $i$ when it is performing a backoff instance. To compute $ET_{c,i}^{ba||rts,*}$ we consider that the probability that more than two stations collide can be neglected, then

$$\begin{cases} ET_{c,i}^{ba,*} \approx \dfrac{\sum_{j \neq i} \sum_{k > j, k \neq i} \max(T_{s,j}^{1,ba}, T_{s,k}^{1,ba})\left(\tau_j\tau_k \prod_{u \neq \{j,k,i\}}(1-\tau_u)\right)}{\sum_{j \neq i} \sum_{k > j, k \neq i}\left(\tau_j\tau_k \prod_{u \neq \{j,k,i\}}(1-\tau_u)\right)} \\ ET_{c,i}^{rts,*} = T_c^{rts} \end{cases} \qquad (9)$$

and

$$ET_{s,i}^{ba||rts,*} \approx \frac{\sum_{j \neq i} ET_{s,j}^{TXOP_j,ba||rts} \left( \tau_j \prod_{u \neq \{i,j\}} (1 - \tau_u) \right)}{\sum_{j \neq i} \left( \tau_j \prod_{u \neq \{i,j\}} (1 - \tau_u) \right)} \tag{10}$$

where $ET_{s,j}^{TXOP_j,ba||rts}$ is the average duration of a successful transmission from a node $j$. The probabilities $p_{e,i}$, $p_{s,i}$ and $p_{c,i}$ are related to the channel status (empty, successful transmission and collision) in a given slot when a node is in backoff.

$$p_{e,i} = \prod_{j \neq i} (1 - \tau_j) \quad p_{s,i} = \sum_{z \neq i} \tau_z \prod_{j \neq z \neq i} (1 - \tau_j) \quad p_{c,i} = 1 - p_{e,i} - p_{s,i} \tag{11}$$

In [9], the authors present a model of the IEEE 802.11 DCF, refer to it for a more detailed explanation of several model assumptions and other model details omitted in this work.

### 3.3   Model Validation

In order to validate the model, we have considered a single-hop scenario with two traffic types: elastic (maximum achievable bandwidth (saturated node) and frame length equal to 1500 Bytes) and streaming (bandwidth of 100 Kbps and frame length equal to 400 Bytes). The network comprises $n$ nodes, each node uses the BA access scheme and carries a single traffic flow. We refer to $S1$ to unsaturated (streaming) flows and we use $E1$ to refer to saturated (elastic) flows. Analytical results are compared against simulations performed using a detailed simulator of the EDCA IEEE 802.11e MAC protocol built using the COST (Component Oriented Simulation Toolkit) simulation engine [10].

In Figure 3 we plot the aggregate throughput for four $E1$ flows when the number of $S1$ flows increase. The goal could be to provide protection to $S1$ flows with respect to the $E1$ flows. Note how with a properly parameter tuning we can achieve the goal and then improve the system performance. Anyway, these results allow to validate the model as both simulation and analytical outputs match very well.

## 4   Call Admission Control

In order to solve the performance problems previously exposed about the interaction between TCP and VoIP traffic, we propose an admission and rate control scheme based on tuning the BEB parameters ($CW_{min}$) of network nodes.

### 4.1   CAC Architecture

The call admission control entity is located at the AP. When an application wants to use the cell resources it sends an $ADDTS$ (Add Traffic Specification)

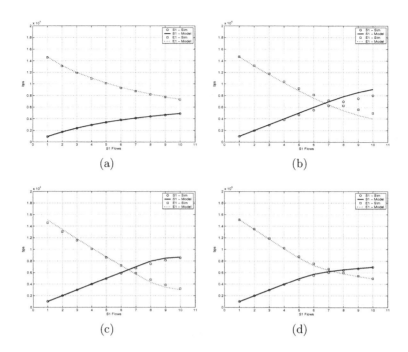

**Fig. 3.** Aggregate Throughput for $S1$ and $E1$ nodes. (a) IEEE 802.11b parameters (b) $TXOP_{S1} = 4$ (c) $AIFS_{E1} = 3$ (d) $CW_{min,E1} = 64$.

packet [1]) to the AP with the traffic profile required by the flow. Using the information provided by the application, the call admission control decides if the new state of the network is feasible. If so, it sends a positive response to the request. Otherwise, it sends a negative response and the new flow is rejected, preserving the grade of service of the active flows already in the system.

To differentiate TCP and VoIP downstream flows, the AP uses two different $AC$ queues: voice queue ($AC\_VO$) for VoIP packets and best-effort queue ($AC\_BE$) for TCP packets. The service prioritization for the voice queue is given only by a short $AIFS$ as we have considered that the $CW_{min}$ and the $TXOP$ burst parameters are equal for both queues with values 32 and 1 respectively. We refer with $\rho_{s,d}$ to the $AC\_VO$ queue utilization and with $\rho_{e,d}$ to the $AC\_BE$ queue utilization. To simplify the analysis, we assume that TCP packets are served only when the $AC\_VO$ queue is empty, then the probability to transmit a downstream TCP packet is $1 - \rho_{s,d}$. Note that, since the upstream feedback traffic is proportional to the downstream TCP traffic, we can assume a minimal impact of uplink TCP ACKs over the VoIP packets.

The VoIP and TCP upstream flows use respectively the $AC\_VO$ and $AC\_BE$ queues of mobile nodes. They use equal $TXOP = 1$ bursts, different and fixed values of $AIFS$ and different values of the $CW_{min}$ parameter (variable for TCP uplink flows and determined dynamically by the admission control entity). Let $\Phi_{CW}$ be the set of all possible $CW_{min}$ values that can be used by uplink elastic flows, with $\Phi_{CW} = \{32, 64, 128, 256, 1024\}$. When the CAC receives a new request of a VoIP call or a TCP uplink flow, it computes the suitable $CW_{min}$ value for the new and remaining active uplink elastic flows and broadcasts the new $CW_{min}$ value. If there are no VoIP flows in the system we assume that all nodes and the AP use the default value of $CW_{min}$.

### 4.2 Performance Evaluation

**Source Traffic Models**

**The G.729 flow traffic model.** The VoIP calls use the $G.729$ codec with a bandwidth of 8 $Kbps$. Each flow is modeled by a Poisson process with packet sending rate of $\beta_s = 50$ packets/second. The $RTP$, $UDP$ and $IP$ header are added to 20 Bytes of voice data, resulting in a packet length of $L = 60$ Bytes at the MAC layer entrance. Thus, a raw bandwidth of $B_s = 24$ Kbps has to be managed by the MAC layer for each VoIP flow.

**A bidirectional TCP flow model for WLANs.** The TCP traffic model used can be found in [9]. The TCP packet length considered is set to 1500 Bytes (including the TCP header).

**Model of a Cell.** Under the assumption of exponential distributions of flow arrivals and departures, the system can be described by a three-dimensional Continuous Time Markov Chain (CTMC). To solve this CTMC we suggest to break the three-dimensional CTMC in two bi-dimensional CTMCs. The first CTMC ($CTMC_A$) comprises the situation where the VoIP calls compete with uplink TCP flows and second CTMC ($CTMC_B$) the situation where downlink TCP flows compete with uplink TCP flows. The partial results of both CTMC are averaged using the approximation that with probability $\rho_{s,d}$ the system works in the situation described by $CTMC_A$ and with probability $1 - \rho_{s,d}$ the system behavior can be modeled by $CTMC_B$.

While the number of elastic flows can grow to infinity, the maximum number of streaming flows is limited by the bandwidth requirements of the voice calls to $N_{voip}^{th}$. In order to solve these infinite bi-dimensional CTMCs, we need to truncate the state space. Without loss of generality, we introduce a realistic minimum bandwidth $B_e^{min}$ required for an elastic flow which gives a maximum number of $N_{e,u}^{th}$ ($N_{e,d}^{th}$) uplink (downlink) elastic flows. The CTMCs state space is described by

$$S_A = \{(n_{e,u}, n_s) \mid S_{e,u}^{tcp}(n_{e,u}, n_s)/n_{e,u} \geq B_e^{min}, \; S_{s,d}^{voip}(n_{e,u}, n_s) > 0.97 \cdot n_s B_s\}$$
$$S_B = \{(n_{e,u}, n_{e,d}) \mid S_{e,u}^{tcp}(n_{e,u}, n_{e,d})/n_{e,u} \geq B_e^{min}, \; S_{e,d}^{tcp}(n_{e,u}, n_{e,d})/n_{e,d} \geq B_e^{min}\}$$
$$(12)$$

For both voice and elastic flows the user population is considered to be infinite with steady state arrival rates $\lambda_{e,u}$ and $\lambda_{e,d}$ for elastic flows and $\lambda_s$ for VoIP calls. The elastic flow duration is function of the bandwidth observed by the elastic flows and the flow length (amount of data to transmit) $FL_e$, with departure rate equal to $\mu_{e,x} = S_{e,x}^{tcp}(.)/(n_{e,x}FL_e)$, while VoIP calls have a fixed average duration

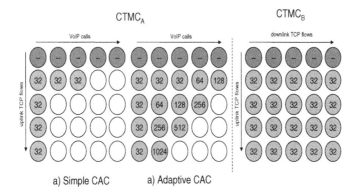

**Fig. 4.** CTMCs Example. Feasible states and $CW_{min}$ values for uplink TCP flows.

equal to $1/\mu_s$. A possible example of both CTMCs is depicted in Figure 4. We can observe how by properly tuning the $CW_{min}$ parameter the number of feasible states grows.

**Parameters.** Table 2 reports the parameters considered to test our CAC algorithm. In all cases, the flow arrival and departure rates follow a Poisson process, the flow length has an exponential distribution and, for the sake of simplicity, $B_e^{min}$ is computed to allow a maximum number $N_{e,u}^{th} = N_{e,d}^{th} = 10$ of active elastic flows in the system in all situations.

**Table 2.** Parameters for CAC evaluation

|  | Scenario 1 | Scenario 2 | Scenario 3 |
|---|---|---|---|
| $\lambda_s$ (flows/second) | *variable* | 0.0083 | 0.0083 |
| $1/\mu_s$ (seconds/flow) | 240 | 240 | 240 |
| $\lambda_{e,d}$ (flows/second) | 0.5 | 0.5 | 0.5 |
| $\lambda_{e,u}$ (flows/second) | 0.5 | *variable* | 0.5 |
| $FL_e$ (Mbits) | 1, 2 | 1, 2 | *variable* |

### 4.3   Numerical Results

Results are shown as the comparison of the performance achieved by the WLAN using a non adaptive CAC scheme called *simple CAC*, according to which all flows always use a fixed value of $CW_{min}$ equal to 32, and using the proposed CAC scheme, called *adaptive CAC*.

The results obtained using the *adaptive CAC* can be better understood if we take into consideration the following aspects: *i)* the number of feasible states of the CTMC grows (i.e., the number of coexisting uplink TCP flows and VoIP calls grows), *ii)* the *adaptive CAC* reduces the instantaneous throughput of

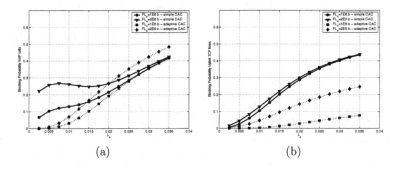

(a)                                    (b)

**Fig. 5.** a) $BP_s$ - Blocking probability VoIP calls, b) $BP_{e,u}$ - Blocking probability for elastic uplink flows

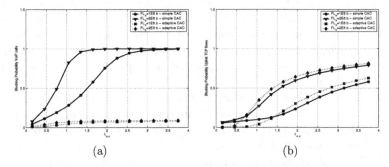

(a)                                    (b)

**Fig. 6.** a) $BP_s$ - Blocking probability VoIP calls, b) $BP_{e,u}$ - Blocking probability for elastic uplink flows

TCP uplink flows, increasing their latency, that is the time they are active in the system, and *iii)* the *adaptive CAC* is not preemptive and does not block already existing flows, it only reduces the rate of uplink TCP flows until the $CW_{min} = 1024$ value is reached.

In Figures 5 we investigate the effect of increasing the VoIP call arrival rate, in terms of blocking probability for voice calls, uplink and downlink TCP flows. Using the *adaptive CAC* scheme the blocking probability for VoIP calls is significantly lower for a reasonable range values of $\lambda_s$. The reason for the higher blocking probability of the *adaptive CAC* for greater values of $\lambda_s$ is motivated by the no-preemption characteristic as the system tends to be always occupied with uplink TCP flows when a new call request arrives. For the uplink TCP flows, the proposed scheme always show better results. The gain in both VoIP calls and uplink TCP flows is compensated by a higher blocking probability of TCP downlink flows.

In Figure 6 the arrival rate of VoIP calls is kept constant to test the sensibility of the novel scheme to variations of the traffic intensity of the TCP uplink flows.

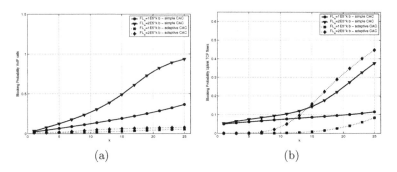

**Fig. 7.** a) $BP_s$ - Blocking probability VoIP calls, b) $BP_{e,u}$ - Blocking probability for elastic uplink flows

For both CAC schemes the blocking probability increases as the traffic intensity of TCP uplink flows increases. However, under the *adaptive CAC* the blocking probability exhibits substantially lower values, which tend to a constant value equal to 0.1, which shows the desirable insensibility property. For the uplink TCP flows, when $\lambda_{e,u} > 0.75$ flows/s, the blocking probability for uplink TCP flows is larger than for the *simple CAC*.

Finally, in Figure 7, we evaluate the impact of the elastic flow length $FL_e$. As expected, as the elastic flow length increases, the blocking probability of VoIP calls also increases. The blocking probability for uplink TCP flows is more sensitive if we use the *adaptive CAC*, showing higher values than the *simple CAC* for large $FL_e$. This result is also motivated by the rate reduction (reduction of the departure rates of the $CTMC_A$ for uplink TCP flows) which increases the normalized traffic load of TCP flows.

In all situations, the higher number of accepted VoIP calls using the *adaptive CAC* results in an increment of the VoIP throughput and then, a higher utilization ($\rho_{s,d}$) of the downlink $AC\_VO$ queue, decreasing the transmission opportunities of the TCP downlink traffic.

## 5    Conclusions

The performance of the IEEE 802.11e MAC protocol (EDCA) has been investigated by means of a novel user-centric model based on a $M/M^{TXOP}/1$ queue that was validated against simulation results. This model is used to evaluate an admission and rate control algorithm based on tuning the $CW_{min}$ parameter. The results show how using the proposed CAC algorithm the performance of VoIP calls is significanlty improved.

## Acknowledgements

This work was partially supported by Catalonian Government under the *i2CAT* (Advanced Internet in Catalonia) project, Spanish Government under project

*TIC2003-09279-C02-01*, and by the European Commission under *NEWCOM* network of excellence.

# References

1. IEEE Std 802.11e. Wireless LAN Medium Access Control (MAC) and Physical Layer (PHY) Specifications; Amendment: Medium Access Control (MAC) Quality of Service Enhancements. *IEEE Std 802.11e*, 2005.
2. IEEE Std 802.11. Wireless LAN Medium Access Control (MAC) and Physical Layer (PHY) Specifications. *ANSI/IEEE Std 802.11*, 1999 Edition (Revised 2003).
3. NS2. Network Simulator. *http://www.isi.edu/nsnam/ns/*, February 2005 (release 2.28).
4. Chun-Ting Chou, Kang G. Shin., and Sai N. Shankar. Inter-frame space (IFS) based service differentiation for IEEE 802.11 wireless LANs. In *IEEE 58th Vehicular Technology Conference, VTC 2003-Fall*, October 2003.
5. Albert Banchs, Xavier Pérez-Costa, and Daji Qiao. Providing Throughput Guarantees in IEEE 802.11e Wireless LANs. *ITC Specialist on Providing QoS in Heterogeneous Environments Seminar, Berlin, Germany*, September 2003.
6. A. Ksentini, A. Guéroui, and M. Naimi. Adaptive Transmission Opportunity with Admission Control for IEEE 802.11e Networks. *ACM/IEEE MSWIM 2005, Montreal, Quebec, Canada*, October 2005.
7. Y.C. Tay and K.C. Chua. A Capacity Analysis for the IEEE 802.11 MAC Protocol. *Wireless Networks 7, 159-171*, March 2001.
8. Paal E. Engelstad and Olav N. Østerbø. Non-Saturation and Saturation Analysis of IEEE 802.11e EDCF with Starvation Prediction. In *ACM/IEEE MSWIM 2005, Montreal, Quebec, Canada*, November 2005.
9. Boris Bellalta, Michela Meo, and Miquel Oliver. Comprehensive Analytical Models to Evaluate the TCP Performance in 802.11 WLANs. *4th Wired/Wireless Internet Communications, Bern, Switzerland*, May 2006.
10. Gilbert (Gang) Chen. Component Oriented Simulation Toolkit. *http://www.cs.rpi.edu/~cheng3/*, 2004.

# Energy Efficient Sleep Interval Decision Method Considering Delay Constraint for VoIP Traffic

Jung-Ryun Lee

R&D Center, LG-Nortel Inc., South Korea
jylee11@lg-nortel.com

**Abstract.** Power saving mode (PSM) in data networks has usually been used for non real-time traffic, such as web-browsing, to minimize the energy consumption of mobile stations (MSs). However, in IEEE 802.16e there is an attempt to adopt PSM for real-time traffic. In this paper, a new criterion is suggested for the delay constraint when PSM is applied to Voice over IP (VoIP) traffic. Based on this criterion, a method is provided for choosing the longest sleep interval within a given end-to-end delay threshold. In addition, it is shown how the trade-off between energy efficiency and sleep interval length of MS operates.

**Keywords:** Power saving mode, delay constraint, sleep interval, energy efficiency.

## 1 Introduction

PSM based on the *'Sleep-Awake-Verity'* procedure has been widely used in most wireless data networks [1]. MSs conserve their energy by powering down during sleep intervals. As a trade-off, PSM inevitably causes larger transmission delay in the base station (BS) compared to the normal mode (always-active). Thus, typical power saving mechanisms are generally applied to non real-time traffic such as web-access, which allows a longer delay than does real-time traffic.

However, in IEEE 802.16e [3], there is an attempt to use power saving mechanism with real-time traffic. Power saving class (PSC) II has been designed in [3] to support PSM for real-time service classes, such as Unsolicited Grant Service (UGS) and Real-Time Variable Rate (RT-VR) service. The UGS service offers fixed-size grants on a real-time periodic basis, in order to eliminate the overhead and latency of MS requests. Since UGS service guarantees a fixed bandwidth to MSs, it is suitable for real-time services such as VoIP without silence suppression. The RT-VR service offers real-time, periodic, and unicast request opportunities, in order to allow the MS to specify the desired bandwidth. This service is designed to support variable rate real-time traffic, such as VoIP with silence suppression or moving pictures experts group (MPEG) video [3].

PSC II has three characteristics over PSC I, the typical power saving method. (1) There is no indication of pending packets to MSs. No MS listens for traffic indications from the BS, but all are expected to listen all frames while in awake mode to eliminate signaling overhead caused by frequent downlink packet transmissions.

Y. Koucheryavy, J. Harju, and V.B. Iversen (Eds.): NEW2AN 2006, LNCS 4003, pp. 142–151, 2006.

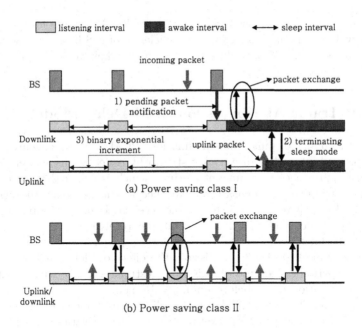

**Fig. 1.** Power Saving Class I and II in 802.16e

(2) MSs transmit uplink packets only when they are awake. In a typical PSM, such as PSC I, MS should send a message to the BS in order to send uplink packets, thus terminating sleep mode operation and entering awake mode. However, in PSC II, MSs transmit their packets only while they are awake, thus avoiding interruption of their periodic sleep intervals. This operation reduces the signaling cost caused by frequent notification of uplink traffic. (3) There is a *fixed* sleep interval. Although PSC I adopts a truncated binary exponential increment of sleep interval, PSC II adopts a fixed sleep interval, since real-time traffic is usually generated periodically. Fig. 1 illustrates the three characteristics of PSC II over PSC I.

There have been researches to develop a more energy-efficient sleep interval decision algorithm. A method that minimizes energy consumption for wireless web access using the bounded slowdown (BSD) protocol is suggested by R. Krashinsky and H. Balakrishnan [4]. The BSD protocol minimizes energy consumption while providing a guaranteed bound on RTT slowdown for request/response network traffic. The BSD protocol makes the STA stay awake during some time interval after request is sent. Thus, this algorithm is suitable for traffic with a short round-trip-time (RTT). [2] suggested an adaptive beacon listening protocol based on the numerical RTT estimation, which is suitable for long RTT traffic. However, to our knowledge, there has been no study of power saving mode for real-time traffic.

This paper is organized as follows. In Section 2, VoIP attributes and Land-to-Mobile (L-M) delay model are provided. The delay constraint in terms of VoIP traffic is suggested in Section 3. Section 4 provides the results for the longest sleep interval and the normalized energy efficiency of MS based on the method described in Section 3. Concluding remark and further work are presented in Section 5.

## 2  VoIP Traffic Attributes and L-M Delay Model

For VoIP, the analog or pulse-code modulation (PCM) voice signals are encoded and compressed into a low-rate packet stream by codec at the sender [5]. This operation requires a certain amount of time, which is called the packetization delay. In addition, to compensate for variance in the delay for the arrival of different packets, there is a de-jitter at the receiver. Usually, the packetization delay at the sender and de-jitter delay at the receiver for VoIP devices are assumed to be fixed.

In view of uplink packet transmission, the effect on delay in PSC II which is caused by periodic sleep intervals is easily estimated, since there exists only a *fixed* packetization delay before uplink VoIP packet transmissions of a MS. By contrast, the effect of PSC II on downlink packet reception is not easily calculated, since the time for a packet to arrive at the BS varies according to the variable packet processing delay in each router. Thus, in this work, L-M VoIP traffic model is considered to focus on the effect of PSC II for downlink packet transmission in the BS. As for the effect of PSC II in BS, it is expected that there will be a delay caused by PSC II in the BS, since the packets arriving at the BS should be pended during the sleep interval. In what follows, this delay is called *on-off* delay. (Refer to Fig. 2).

In general, the source-destination delay in a wired environment is assumed to be composed of two parts (fixed and variable) and have a heavy-tailed asymmetric distribution with a minimum delay [6][7]. The fixed delay (minimum delay) consists of the propagation delay and transmission delay in each router. The transmission delay varies according to the packet size and link bandwidth, but it is very small in general relative to the propagation delay. So, the sum of the transmission delay and propagation delay is regarded as to be fixed. The variable part of the delay consists of the service time in each server and is mainly

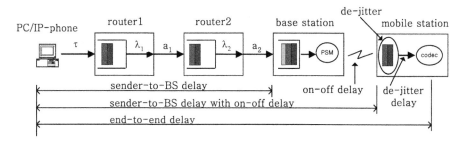

**Fig. 2.** M-L VoIP delay model

dominated by the queuing delay of each router. It may include additional delays caused by such factors as database lookup, a decryption procedure, and header modification. For the service time in each server, the evaluation begins with an exponentially distributed queuing delay with rate $\lambda_i$ at the $i$-th server by adopting the method used in [8] and [13]. Although this model may be regarded as non-realistic, (3) shows that the sender-to-BS delay model that is our target to use in this work follows a hypoexponential distribution and conforms to the property of real end-to-end delay in wired environments (heavy-tailed asymmetric distribution with a minimum delay) discussed in [6] and [8].

Suppose that sender (PC or IP phone)-to-BS VoIP traffic is transferred in a $N$ hop fashion. Let the sum of propagation delay and transmission delay from the $i$-th server to the $(i + 1)$-th server be $a_i$. The delay at the $i$-th server is expressed as

$$Y_i = X_i + a_i \tag{1}$$

where $X_i$ follows an exponential distribution with rate $\lambda_i$. Then the sender-to-BS delay with $n$ hops is expressed by

$$Y = \sum_{i=1}^{N} Y_i + \tau = \sum_{i=1}^{N} (X_i + a_i) + \tau = \sum_{i=1}^{N} X_i + a + \tau \tag{2}$$

where $a = \sum_{i=1}^{N} a_i$ and $\tau$ is the sum of the packetization delay and transmission/propagation delay in the sender[1]. Since the sum of exponential random variables follows a hypoexponential distribution [9], the pdf of $Y$ conforms to the delay property mentioned above and is given by

$$f_Y(t) = \sum_{i=1}^{N} C_{i,n} \lambda_i e^{-\lambda_i(t-(a+\tau))} \tag{3}$$

for $t \geq a + \tau$ where $C_{i,n} = \prod_{i \neq j} \frac{\lambda_j}{\lambda_j - \lambda_i}$. Notice that $f_Y(t) = 0$ for $t$ smaller than minimum delay, $a + \tau$. The cdf of $Y$ is given by

$$F_Y(t) = 1 - \sum_{i=1}^{N} C_{i,n} e^{-\lambda_i(t-(a+\tau))}. \tag{4}$$

## 3   Energy-Efficient Sleep Interval Decision Method

VoIP packets are generated with a period which is different according to the used codec. Let this frequency of packet generation be $T$. $\alpha$ and $\beta$ are defined as

---

[1] De-jitter delay is not calculated in this expression, because it is located in the receiver.

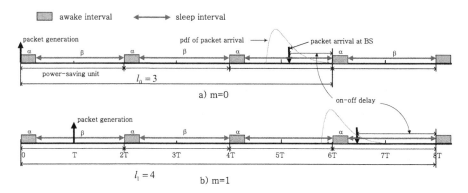

**Fig. 3.** Schematic representation of the distribution of sender-to-BS delay and on-off delay when $\alpha + \beta = 2T$

the length of awake and sleep intervals, respectively, and the sum of these two intervals $(\alpha + \beta)$ is termed the *power-saving unit* in this work. For convenience, the value of $T$ is assumed to be an integer fraction of the length of the power-saving unit; namely,

$$\alpha + \beta = kT \tag{5}$$

for some positive integer $k$.

Consider VoIP packets generated at $0, T, 2T, \ldots, (k-1)T$ when $\alpha + \beta = kT$. Fig. 3 shows a schematic representation of the distribution of sender-to-BS delay and on-off delay of each packet generated at $0$ and $T$ when $k = 2$. From (3), the cdf of the sender-to-BS delay of packet generated at the $m$-th slot becomes

$$f_{Y_m}(t) = \sum_{i=1}^{n} C_{i,n} \lambda_i e^{-\lambda_i(t-(a+\tau+mT))}. \tag{6}$$

Generally, the delay constraint is given by

$$Pr\{W > D_{thr}\} \le \delta \tag{7}$$

where $W$ is the end-to-end delay of a packet, and $D_{thr}$ and $\delta$ are the delay threshold and the maximum probability of exceeding $D_{thr}$, respectively [10]. Consider the sender-to-BS delay with on-off delay as shown in Fig. 2 and Fig. 3. By substituting the end-to-end delay $W$ with the sum of sender-to-BS delay and on-off delay $\overline{Y}$, (7) is rewritten by

$$Pr\{\overline{Y} > D'_{thr}\} \le \delta \tag{8}$$

where $D'_{thr}$ is the end-to-end delay threshold except de-jitter delay. In the following Proposition, a new criterion for a delay constraint is suggested when PSC II is applied to VoIP traffic.

**Proposition 1.** *For VoIP users under PSC II, the delay constraint (8) with $\delta < 1/k$ is satisfied if and only if, for each packet generated at m-th slot ($0 \leq m \leq k - 1$) in each power-saving unit, there exists some positive integer $l$ that satisfies the following two inequalities simultaneously:*

$$Pr\{\overline{Y}_m > (lk - m)T\} < \delta \tag{9}$$

*and*

$$(lk - m)T \leq D'_{thr} \tag{10}$$

*where $\overline{Y}_m$ is the sum of sender-to-BS delay and on-off delay for packets generated at the m-th slot of each power-saving unit.*                                    *None*

*Proof.* Suppose that a packet generated at the $m$-th slot of the $p$-th power-saving unit arrives at the BS in $(n + p)$-th power-saving unit. Then, the sender-to-BS delay with on-off delay is expressed as $n(\alpha + \beta) - mT = (nk - m)T$. Thus, $\overline{Y}_m(u) = (nk - m)T$ if $((n-1)k - m)T \leq Y_m(u) < (nk - m)T$. A sufficient condition is easily derived from the facts that $Pr\{\overline{Y}_m \geq u\}$ is a decreasing function of $u$ and $Pr\{\overline{Y} > D'_{thr}\} = \frac{1}{k} \sum_{i=0}^{k-1} Pr\{\overline{Y}_i > D'_{thr}\}$.

Let's prove the necessary condition. Since the cdf of $\overline{Y}_m$, $F_{\overline{Y}_m}$, is an increasing function, we can find the smallest integer $l_m$ for each $m$ satisfying the following inequality:

$$1 - F_{Y_m}((l_m k - m)T) < \delta. \tag{11}$$

Let $l$ be the maximum value among $l_m$. By substituting $l_m$ in (11) for $l$, (9) is satisfied. Suppose that there is no $l$ satisfying (10) for some $m$. That is to say, for all $l$, $(lk - m)T > D'_{thr}$. This means that $\overline{Y}_m$ is always larger than $D'_{thr}$, so, $Pr\{\overline{Y}_m > D'_{thr}\} = 1$. Then,

$$Pr\{\overline{Y} > D'_{thr}\} = \frac{1}{k} \sum_{i=0}^{k-1} Pr\{\overline{Y}_i > D'_{thr}\} \tag{12}$$

$$= \frac{1}{k} + \frac{1}{k} \sum_{i=0, i \neq m}^{k-1} Pr\{\overline{Y}_i > D'_{thr}\} > \frac{1}{k} > \delta, \tag{13}$$

which contradicts the delay constraint (8). Thus, there should be some integer $l$ satisfying (10) for all $m$, which ends the proof.                                    *None*

Notice that $\delta$ usually has a very small value, such as 0.03, so the assumption $\delta < 1/k$ is reasonable. $l_m$ indicates the smallest value of $n$ such that the packet generated at the $m$-th slot in the $p$-th power-saving unit satisfies (9) and (10) simultaneously in the $(n+p)$-th power-saving unit. In Fig. 3, when $D'_{thr}$ is larger than $8T$, the smallest value of $n$ satisfying (9) becomes 3 for $m = 0$. The smallest value of $n$ satisfying (9) is 4 for $m = 1$, because the probability that a packet generated at $T$ will arrive at the receiver beyond $3 * 2T$ is larger than $\delta$, which is assumed to be sufficiently small. In this case, our desired value of $l$ becomes 4.

The efficiency of PSC II is determined by the length of $\beta$ in the sense that energy efficiency improves as $\beta$ is getting larger. To maximize the energy efficiency of PSM, the value of $k$ should be as large as possible within the end-to-end delay threshold. Proposition 1 provides us with a method to search for the most energy-efficient length for the sleep interval: for all non-negative integers $m$ less than $k$, find the *largest* $k$ satisfying (9) and (10) simultaneously for some integer $l$.

## 4   Results and Discussions

### 4.1   Energy-Efficient Sleep Interval Decision

The parameters described in Proposition 1 were determined as follows. For the VoIP model, $20ms$ codec delay, $20ms$ packet generation frequency and $40ms$ dejitter were used. For the delay constraint parameters, $D_{thr}$ and $\delta$ were assumed to be $300ms$ and $0.03$, respectively [11]. The fixed delay of each routers was assumed to be $10ms$ across all routers ($a_i = a_j = 10ms$ for all $i, j$). The framing interval in IEEE 802.16e is $5ms$ and the length of the awake interval is fixed by one frame[2].

**Table 1.** List of $l$ as a function of $k$ when $N = 2$ and $\lambda_i = \frac{i}{20}$

| k | l |
|---|---|
| 1 | 7,8,9,10,11,12,13 |
| 2 | 4,5,6 |
| 3 | 3,4 |
| 4 | 3 |
| 5 | - |

Table 1 shows an example of the available values of $l$ with increasing $k$ when $N = 2$ and $\lambda_i = i/10$ for the $i$-th router. In this case, the largest value of $k$ having available values of $l$ is 4. Thus, the longest power-saving unit satisfying the delay constraint becomes $4 * 20 = 80(ms)$. By subtracting the awake interval length $5ms$, the longest sleep interval in this case becomes $75ms$.

Fig. 4 shows the longest sleep interval as a function of $N$ and $\lambda_i$. As $N$ is increasing, the fixed delay in the end-to-end delay is increasing, so, the length of the longest sleep interval is decreasing to satisfy the criterion for the delay constraint. The average delay of the variable part in the end-to-end delay is increasing as $\lambda_i$ is decreasing and this results in shorter sleep intervals. From the result, it can also be determined whether or not PSC II is applicable. For example, PSC II is not applicable when $N > 2$ and $\lambda_i = \frac{i}{50}$ or $N > 4$ and $\lambda_i = \frac{i}{40}$, respectively.

---

[2] Here, it is assumed that all pending packets for user destination are to be transmitted during one frame interval.

## 4.2   Energy Efficiency

Let the amount of the consumed energy of MS when it is in awake mode and sleep mode be $E_{awake}$ and $E_{sleep}$, respectively. $1.5W$ and $0.045W$ are used as values of $E_{awake}$ and $E_{sleep}$, respectively [12]. In this work, energy consumption

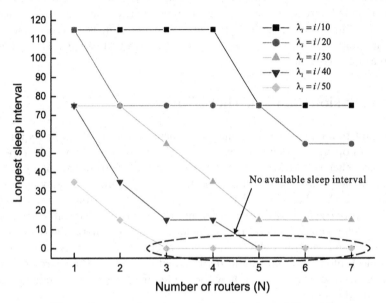

**Fig. 4.** Longest sleep interval as a function of $N$ and $\lambda_i$

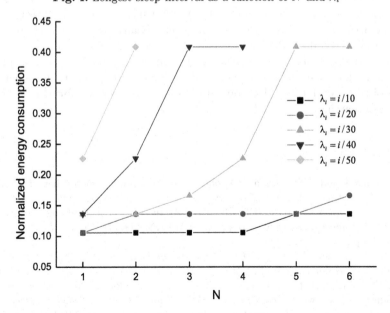

**Fig. 5.** Normalized energy consumption of MS as a function of $N$ and $\lambda_i$

of MS under PSC II is evaluated by normalized energy consumption, which is expressed by

$$\frac{\alpha E_{awake} + \beta E_{sleep}}{kT}. \tag{14}$$

Fig. 5 shows the energy consumption of MS with the same parameters used in Fig. 4. This figure shows a trade-off between energy efficiency and the length of the sleep interval. As expected, energy consumption is increasing as $N$ is increasing and $\lambda_i$ is decreasing, because the length of the longest sleep interval is decreasing accordingly.

## 5    Conclusions and Further Work

An attempt to adopt a power saving mechanism for real-time traffic requires the choice of an optimized sleep interval within the delay constraint with respect to the energy efficiency of MS. The goal of this paper is to suggest a solution for the longest sleep interval length, still satisfying requirement of delay constraint. Combining the VoIP traffic model with the periodic generation of source packets, this paper proposed a new delay constraint criterion for PSC II and it was shown that this delay constraint criterion coincides with the general delay constraint criterion. Based on this delay constraint criterion, it was suggested how the longest sleep mode interval should be designed to enhance the energy efficiency of MS. It was also demonstrated how the trade-off between energy efficiency and sleep interval operates.

As discussed in Section 2, the delay between each VoIP packet can vary instead of remaining constant. To compensate for this variation in delay (jitter), most VoIP devices use a de-jitter to smoothly (periodically) transmit their received VoIP packets to codec. Although this de-jitter size is assumed to be fixed in this work, the de-jitter size is related to the length of longest sleep interval in the sense that a longer sleep interval can be allowed as the de-jitter size is increasing. Study of the sleep interval decision with variable de-jitter size forms part of our ongoing research.

## References

1. R. Kravets and P. Krishnan., "Application-driven power management for mobile communication", *Wireless Networks*, vol. 6, no. 4, pp.263-277, 2000.
2. J. R. Lee, S. W. Kwon and D. H. Cho, "Adaptive beacon listening protocol for a TCP connection in slow-start phase in WLAN ", *IEEE Comm. Letters*, vol. 9, no. 9, pp.853-855, Sep. 2005.
3. IEEE P802.16e/D10, *"Draft IEEE Standard for Local and Area Networks-Part 16: Air Interface for Fixed and Mobile Broadband Wireless Access Systems,* Sep. 2005.
4. R. Krashinsky, H. Balakrishnan, "Minimizing energy for wireless web access with bounded slowdown", *Proceedings of the 8th annual international conference on Mobile computing and networking*, Atlanta, Georgia, USA, pp.119-130, Sep. 2002.

5. W. Wang, S. C. Liew and Li, V.O.K, "Solutions to performance problems in VoIP over a 802.11 wireless LAN", *IEEE Trans. on Vehicular Technology*, vol. 54, no. 1, pp.366-384, Jan. 2005.

6. Kao, B., Garcia-Molina, H., and Barbara, D., "Aggressive transmissions of short messages over redundant paths", *Parallel and Distributed Systems, IEEE Transactions*, vol. 5, pp. 102-109, Jan. 1994.

7. F. Cristian, "Probabilistic clock synchronization", *Distributed Computing*, vol. 3:3, 1999.

8. A. Fasbender, D. Kesdogan and O. Kubitz, "Analysis of Security and Privacy in Mobile IP", *4th International Conference on Telecommunication Systems, Modelling and Analysis*, Nashville, TN, USA, Mar. 1996.

9. M. Ross, *Introduction to Probability Models*. 7th ed. UK:Harcourt, 2000.

10. M. Andrews, K. Kumaran, K. Ramanan, A. Stolyar, P. Whiting, and R. Vijayakumar, "Providing quality of service over a shared wireless link", *IEEE Commun. Mag.*, vol. 39, pp. 150-154, Feb. 2001.

11. *Recommendation ITU-T G.114, One-Way Transmission Time*, Int'l Telecommunication Union, Geneva, 1996

12. E. S. Jung, and N. H. Vaidya, "An energy efficient MAC protocol for wireless LANs", *INFOCOM 2002, Proceedings, IEEE*, vol. 3, pp.1756-1764, Jun. 2002.

13. Korpeoglu, I., Tripathi, S.K. and Xiaoqiang Chen, "Estimating end-to-end cell delay variation in ATM networks", *Communication Technology Proceedings, ICCT '98. 1998 International Conference on*, vol. 1, pp. 472-483, Oct. 1998.

# Enhanced Fast Handovers Using a Multihomed Mobile IPv6 Node

Albert Cabellos-Aparicio* and Jordi Domingo-Pascual

Universitat Politècnica de Catalunya,
Departament d'Arquitectura de Computadors, Spain
{acabello, jordid}@ac.upc.edu

**Abstract.** Wireless technologies have rapidly evolved in recent years. IEEE 802.11 jointly with Mobile IP provides mobility to the internet. The most critical part of this technology (IEEE 802.11 and Mobile IP) is the handover. During this time the connection may be interrupted and packets can be lost or delayed. These issues makes difficult to run VoIP applications in a mobile environment. In this paper we propose an extension to the Mobile IPv6 protocol using a multihomed mobile node and a cross-layer design. We compare our proposal with Fast Handovers for Mobile IPv6 and Hierarchical Mobile IPv6.

## 1 Introduction

Wireless technologies have evolved in recent years. IEEE 802.11 [1] is one of the most used wireless technologies. In current Internet status, a user can be connected through a wireless link but he cannot move without breaking the IP communications. That's why IETF designed Mobile IP which jointly with IEEE 802.11 provides mobility to the Internet. The most critical part of this technology (IEEE 802.11 and Mobile IP) is the handover. The handover is the time spent when changing from one point of attachment (i.e. an access router) to another. During this time the mobile node is not able to send or receive data and thus, the connection may be interrupted, packets may be lost or delayed due to intermediate buffers. This issue makes difficult to run VoIP applications in such environment.

The IETF has designed two versions of Mobile IP, one for IPv4 and another one for IPv6. Mobile IPv6 [2] is very similar to Mobile IPv4 [3]. Although Mobile IPv6 is more efficient than Mobile IPv4 it still suffers of some problems, slow handover latency and signaling overhead. To address these issues the IETF has designed two protocols that extend Mobile IPv6 and improve its performance: Fast Handovers for Mobile IPv6 [4] and Hierarchical Mobile IPv6 [5]. Both solutions effectively reduce the handover latency and the signaling overhead but they are complex solutions.

---

* This work was partially funded by the Spanish Ministry under contract TSI2005-07250-C03-02 and the Generalitat de Catalunya under contract 2005-SGR-00481.

Y. Koucheryavy, J. Harju, and V.B. Iversen (Eds.): NEW2AN 2006, LNCS 4003, pp. 152–163, 2006.

This paper presents a novel Mobile IPv6 extension to improve the handover latency using a multihomed-host and a cross-layer design. Our solution reduces the handover latency to zero, packets are not lost or delayed, and does not require complex network support. Even more, it is able to run VoIP applications in mobile environments. Our solution is intended for mobile vehicles such as cars or trains where movement is restricted to a well-known predefined area.

## 2  Mobility Protocols Evaluation

This section presents an overview of the IEEE 802.11, Mobile IPv6, Fast Handovers for Mobile IPv6 and Hierarchical Mobile IPv6 protocols. A performance evaluation focused on the handover latency is also presented. We define the handover latency as the amount of time spent while packets sent or received by the MN are being lost or delayed (buffered) due to the handover operations.

### 2.1  IEEE 802.11

The Wireless LAN protocol [1] is based on a cellular architecture managed by an Access Point or AP. If a mobile node (MN) desires to join a cell, it can use passive scanning, where it waits to receive AP's Beacon Frames messages or active scanning, where it sends Probe Request frames and receives a Probe Response frame from all available APs. Scanning is followed by the Authentication Process, and if that is successful, the Association Process. Only after this phase the MN is capable of sending and receiving data frames.

Several studies have measured the wireless handover ([6] and [7]). In average, the MN spends 257ms to change from one AP to another, the most time consuming phase is the Scan Phase which takes 95% of the total handover latency time.

### 2.2  Mobile IPv6

The main goal of the Mobile IPv6 (MIPv6) protocol is to allow mobile nodes to change its point of attachment to the Internet while maintaining its network connections. This is accomplished by keeping a fixed IP address on the mobile node (Home Address or HAd). This address is unique and when the mobile node is connected to a foreign network (not its usual network) it uses a temporal address (Care-of Address or CoA) to communicate however, it is still reachable through its Had (using tunnels or with special options in the IPv6 header). MIPv6 has three functional entities, the Mobile Node (MN), a mobile device with a wireless interface, the Home Agent (HA), a router of the home network that manages localization of the MN and finally, the Correspondent Node (CN), a fixed or mobile node that communicates with the MN.

The handover latency for Mobile IPv6 has several components. When a MN is changing from one point of attachment to another it first needs to associate to the new AP, which involves Scanning, Authentication and Association ($D_{L2}$), this is the wireless part of the handover. Then, in the Agent Discovery phase the MN needs to check that the old AR is unreachable using the Neighbor Unreachability Detection ($D_{NUD}$) [24] algorithm and obtain a new temporal IP address (Care-of-Address). It also needs to ensure that the recently obtained CoA is unique on the new link using the Duplicate Address Detection algorithm ($D_{DAD}$) [25]. Finally, the MN must send three Binding Updates in the Registration phase, one to its Home Agent and two to the Correspondent Node (we suppose that the MN is only communicating with one CN, this parameter does not affect our solution). One of the CN's BU messages must be sent through the HA and the other one directly. The equation (1) models the handover latency for the Mobile IPv6 protocol.

$$
\begin{aligned}
HL_{MIPv6} = D_{L2} + D_{DAD} + D_{NUD} + RTT(MN, HA) \\
+max[(RTT(MN, HA) + RTT(HA, CN)), RTT(MN, CN)]
\end{aligned}
\tag{1}
$$

Experimental measurements show that the handover latency for the Mobile IPv6 protocol takes in average 2107ms [7], during this time packets are lost making difficult to run VoIP applications. Most of the time (87%) is spent by the $D_{DAD}$ and the $D_{NUD}$ algorithms. Those algorithms are required for IPv6 auto-configuration. Note that several improvements can be done to reduce $D_{L2}$, $D_{DAD}$ or $D_{NUD}$. However the time spent by the Binding Update and Binding Acknowledgment messages depend exclusively on the Round Trip Time between the MN and the HA or the CN.

### 2.3   Fast Handovers for Mobile IPv6

FMIPv6 is a Mobile IPv6 extension that reduces the handover latency and, during the handover it buffers packets delaying them instead of losing them. This is accomplished by allowing the MN to send packets as soon as it detects a new subnet link (IEEE 802.11 in our case) and delivering packets to the MN as soon as its attachment is detected by the new access router. In FMIPv6 the MN discovers nearby APs and then requests all the important information related to the corresponding new access router. When attachment to an AP takes place, the MN knows the corresponding new router's coordinates. Through special Fast Binding Update and Fast Binding Acknowledgment messages the MN is able to formulate a prospective new CoA. As soon as it is attached, the MN sends a Fast Neighbor Advertisement message announcing its presence. Moreover, the previous access router will tunnel and forward packets to the new CoA until the MN sends a Binding Update registering its new CoA to the HA and to the CNs hence, any packet is lost.

FMIPv6 reduces the handover latency almost to the layer 2 handover latency. Once the MN has received the Fast Binding Acknowledgement message the communications between the MN and the CN will be interrupted. Packets destined to the MN will be buffered at the new access router (and thus delayed) until the MN regains connectivity. When the layer 2 handover is finished, and the MN is connected on the new link it will announce its presence through a Fast Neighbor Advertisement (FNA) message and the new access router will forward the buffered packets. We define $D_{FNA}$ as the time between the MN regains connectivity and the FNA message has been received by the new access router. Equation (2) models the handover latency for the FMIPv6 protocol operating in Predictive mode, which is the fastest one.

$$HL_{FMIPv6} = D_{L2} + D_{FNA} \tag{2}$$

Experimental measurements show that the handover delay for FMIPv6 using the Predictive mode is 319ms in average [8]. Note that the handover latency is very close to the layer 2 handover latency because the $D_{FNA}$ is very small, moreover it does not depend on the distance to the HA or to the CN. During this time, packets are being buffered at the new access router making difficult to run VoIP applications.

## 2.4 Hierarchical Mobile IPv6

Hierarchical Mobile IPv6 (HMIPv6) is another extension of Mobile IPv6. Its main goal is to reduce the signaling overhead and to improve the handover latency. HMIPv6 introduces a new entity called Mobile Anchor Point (MAP) placed at any point of the hierarchy of a network. The MAP is essentially a Home Agent which will limit the amount of Mobile IPv6 signaling outside the local domain. The HMIPv6 handover latency is very similar to the Mobile IPv6 handover latency. If a HMIPv6-aware MN moves from one access router to another one outside the MAP boundaries it will have the same handover latency than Mobile IPv6. However, when the MN moves to another AR which is inside the MAP domain it will not need to send Binding Updates. The MN will just send one Binding Update to the MAP. Equation (3) models this handover latency.

$$HL_{HMIPv6} = D_{L2} + D_{NUD} + D_{DAD} + RTT(MN, MAP) \tag{3}$$

# 3 Proposed Solution

The proposed solution is an enhancement of Mobile IPv6 and it uses two IEEE 802.11 interfaces with a cross-layer design in the MN, figure 1 presents how it operates.

On the first phase, the MN is connected to the first AR using the first wireless interface. This interface has a configured CoA1 and it is communicating with the CN. The second interface is scanning for better connectivity. When the MN moves, the first wireless interface using layer 2 triggers will inform to the upper layers that the signal quality is becoming low. Meanwhile the second interface has detected a new access point with better signal quality. The IPv6 layer will start to obtain a new CoA for the second interface (CoA2) in the new access router. Next, in the third phase it will start to send the corresponding Binding Update messages to the CN and HA announcing that it has a new CoA while the first interface is still sending and receiving packets. Finally, in the fourth

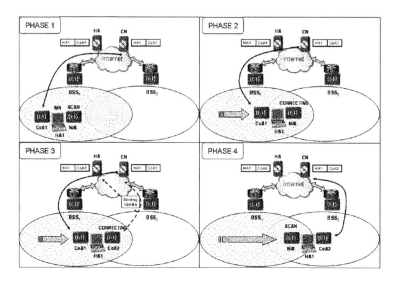

**Fig. 1.** An overview of the proposed solution

phase, when the Binding Update messages have arrived to the CN and to the HA they will stop sending packets to the CoA1 and start to send them to the CoA2. The handover is now complete and the first wireless interface will enter the scan mode, looking for better signal quality.

### 3.1    System Architecture

This section presents the different modifications required to run our proposed solution, note that all the modifications need to be done on the MN. Our solution does not require any modification on the network, the CN or the HA.

### 3.2    The IEEE 802.11 Layer

To provide seamless handovers our solution needs link-layer triggering i.e., events will be fired at the link layer module and communicated to the network layer module. Moreover, the network layer module will be able to force the layer 2 to connect or disconnect to a given Access Point. Different organizations, such as IETF and IEEE are defining a generic interface to provide link-layer triggering [9],[10] and [11].

The link-layer triggers used by our solution are based on [9] and they are divided into three categories. First, the *Events* are fired at layer 2 providing important information for the network layer. The *LinkUp* event indicates that the network layer can start sending packets because the layer 2 link has established a link whereas the *LinkDown* event indicates that the link-layer is down. Finally the *Link Quality Crosses Threshold* indicates that the signaly quality has remained low for a certain period of time and that the network may start preparing for a handover. Secondly, the *Information* is provided by the link-layer and accessible by the network layer. The *Link Quality* information is used by the network layer to check the signal quality received on a given wireless interface. Finally, the *Services* are actions that can be requested by the network layer to the link-layer. The network can request to the link-layer to *Connect* or *Disconnect* from a given AP. Moreover it can *Scan* for available APs.

The required link-layer triggers are divided into three different types. The Events are fired from the layer 2 to inform about different aspects of the link-layer connectivity. The Information can be accessed by the network layer to obtain connectivity information. Finally, the Services are actions that can be requested by the network layer to change different aspects of the link-layer. Our solution is intended for IEEE 802.11, however this list is a generic framework that can fit on almost any link-layer technology based on radio access such as GPRS or UMTS.

### 3.3   IPv6 and the Mobile IPv6 Layer

The network layer, IPv6 in our case, is responsible of maintaining at least one interface connected to an access router while the other one is scanning for better signal quality. While one interface is connected to an access router, the IPv6 layer using the Scan service defined in the previous section will scan looking for available APs. When an AP is found the network layer will connect the second interface to the new AP. If more than one AP is detected, the wireless interface will connect to the best one in terms of signal quality. Once the second wireless interface is connected to a new AP it will start to perform all the operations related with auto-configuration in order to obtain a new CoA. This CoA will not be used and any packet will be sent or received using it.

The second interface will wait until the first one raises the Link Quality Crosses Threshold event. This event indicates that the handover is imminent and that the second wireless interface needs to register its new CoA with the HA and with the CNs. In order to prevent that the second interface losses connectivity with the AP before the handover, the interface will monitor the signal quality received using the Link Quality information. If the Link Quality Crosses Thresholds is raised on the second interface it will start to Scan for other available APs with better connectivity and the whole process will start again.

If the signal quality on the second interface is high and the first one fires the Link Quality Crosses Thresholds event, the MIPv6 layer will send Binding Update messages to its HA and CN using the new CoA (CoA2). Once the corresponding Binding Acknowledgment messages have been received, and the HA and CN have changed its bindings with the new CoA the network layer will disconnect the first wireless interface. Next, the communication with the CN will continue without interruption, and the first wireless interface will start to Scan to look for better signal quality.

Figure 2 presents a schema of the proposed solution and shows that the handover latency is zero and that packets are not lost. The handover latency, as shown in figure 4 is zero. While the first wireless interface is receiving and sending

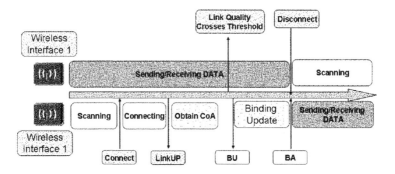

**Fig. 2.** Schema of the proposed solution

packets, the second interface will be performing all the related operations with the handover. When the Binding Acknowledgment message is received, the HA and all the CN have switched from the old CoA to the new one. All the packets in transit have been received and thus, the network layer can disconnect the first interface and start sending and receiving packets through the second one.

The time spent by the link layer to connect to the new AP is $D_{L2}$. The time spent by the IPv6 layer to obtain a new CoA is improved in comparison with Mobile IPv6. In the Mobile IPv6 case this time was modeled as $D_{NUD} + D_{DAD}$. Our solution does not require executing the Neighbor Unreachability Detection algorithm. This algorithm is required because the IPv6 layer must switch from the old access router (considered as primary) to the new one (considered as secondary) [24]. In our proposal the IPv6 layer of the second wireless interface does not have any default router configured and thus, does not need to check that the primary one is no longer reachable. Moreover, our solution incorporates Optimistic Duplicate Address Detection [12]. This new algorithm, which has not been standardized yet, aims to minimize address configuration delays in the successful case and interoperates with unmodified hosts and routers.

The handover latency for our proposal is zero. However the second wireless interface spends a certain amount of time performing the handover related operations. Due to the improvements of our solution, avoiding the $D_{NUD}$ term and reducing the $D_{DAD}$ term, this time is always lower than the Mobile IPv6 handover latency ($HL_{MIPv6}$).

### 3.4   Applicability of the Solution

The proposed solution uses two IEEE 802.11 wireless interfaces to achieve zero handover latency and no losses. It requires that within the domain where the MN will move the different APs coverage areas have an overlapped region. In other words, our system requires of this overlapped region to connect to the new access router while it is sending and receiving packets through the old one. In this section we will focus on these requirements.

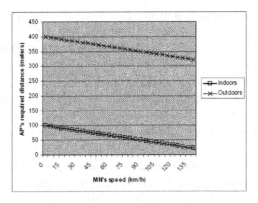

**Fig. 3.** Inter-APs required distance

According to [13] and other tech specs, APs usually have an outdoor range of 400 meters and an indoor range of 100 meters. Our system requires that during at least $HL_{MIPv6}$ (1) seconds, which is the maximum amount of time spent by one of the interface to perform all the handover related operations, the MN is able to communicate with two APs at the same time. Depending of the MN's speed we will need that the APs are close enough to provide this overlapped region. For example, if the MN is moving at 50km/h and the $HL_{MIPv6}$ is 2 seconds [7] (worst case) our system requires an overlapped region of 28 meters and thus, in an outdoors environment, two consecutives APs must be set 372 meters away one from the other.

Figure 3 shows the distance required between two consecutives AP to provide the necessary overlapped region depending on the MN's speed. If the system is deployed indoors the MN's speed will be up to 10km/h (a person walking), the required overlapped region is about 5 meters and the APs must be set around 95 meters away. If the system is deployed outdoors the MN's speed can be up to 120km/h (a MN traveling in a car) and the APs must be set 333 meters away. These results demonstrate the applicability of our proposal. Note that our solution is intended for mobile vehicles where movement is within a well known area. For instance the proposed solution can be deployed along a railway where MNs are inside the train.

### 3.5 Proposed Solution Evaluation and Comparison

In this section we evaluate different important parameters of our proposal and compare it with Fast Handovers for Mobile IPv6 and Hierarchical Mobile IPv6.

*Handover Latency:* Our proposal has zero handover latency because one of the wireless interfaces sends and receives the data packets while the other one is performing the handover operations. FMIPv6 has very low handover latency, close to the layer 2 handover latency and HMIPv6 has higher handover latency due to the IPv6 auto-configuration delays.

*Packet Loss:* Our proposal does not lose any packet. FMIPv6 does not lose any packet neither, however packets are stored in a buffer and delayed. This delay makes difficult to use FMIPv6 with real-time traffic such as VoIP. HMIPv6 suffers of great packet losses during the handover latency.

*Inter Packet Delay Variation:* During the handover, HMIPv6 losses packets. FMIPv6 does not lose packets, but packets received during the handover are buffered introducing IPDV. This also makes difficult to use it with VoIP applications. Our proposal does not introduce any kind of delay during the handover.

*Network Support:* FMIPv6 requires that access routers and MNs are FMIPv6-aware, even more it needs that each access router knows the prefix announced by its neighbors. Probably this information will be obtained through a discovery protocol, but it has not been yet defined. HMIPv6 does not require any kind of support by the access routers but it introduces a new entity called Mobility Anchor Point. The MAP must be deployed in each domain to support hierarchical handovers. Even more, all the packets destined to MNs inside the domain will be routed through the MAP becoming a Single Failure Point. The network support required by both protocols makes them more expensive to deploy and to maintain. Our solution does not require any kind of network support, just modifications on the MN. Our proposal is also interoperable with MNs and CNs that do not implement it.

*Buffering:* FMIPv6 requires that each access router incorporates a buffer where packets will be stored during the MN's handover. It requires a buffer for each MN that it is changing its point of attachment. This buffer may introduce some security issues and increments the deployment cost. HMIPv6 and our solution do not use buffers.

*State-based Protocol:* Both FMIPv6 and HMIPv6 are state-based protocols. FMIPv6 needs to track MN's localization information. When a MN changes its point of attachment, the new access router must know it in advance, store its IP address and the packets destined to it. The HMIPv6's MAP is essentially a Home Agent and it needs to store the bindings between Regional CoA and oCoAs. State-based protocols introduce security issues and have a greater cost to deploy and to maintain. Our proposal is a stateless protocol.

*Signaling Overhead:* HMIPv6's main goal is to reduce the signaling outside the MAP boundaries and it has the lower signaling one. Our solution and FMIPv6 have the same signaling overhead than MIPv6.

*Overlapping Requirements:* Our solution has hard overlapping requirements compared with the other MIPv6 extensions. FMIPv6 and HMIPv6 can run in non-overlapped networks but their performance will be severely affected because the wireless handover will increase ($D_{L2}$) producing a greater handover latency. Our proposal will perform at the same level than Mobile IPv6 in non-overlapped networks.

Table 1 summarizes the comparison and demonstrates that our proposal performs better than the other extensions to Mobile IPv6. It has better handover latency, packet losses and IPDV. This makes our proposal suitable to use it with real-time traffic applications such as VoIP. Moreover, FMIPv6 and HMIPv6 have a high deployment and maintenance cost due to the network support requirements while our proposal does not require modifying access routers and does not introduce any new entity. Finally, HMIPv6 and FMIPv6 are state-based protocols introducing security issues and Single Failure Points (the MAP) whereas our solution is a stateless protocol and pushes the complexity to the MNs, keeping the network simple.

The main issues with our proposal are that requires a hard overlapped network and two wireless interfaces for each MN. Both wireless cards can increase the energy drain, which can be a problem in battery operated devices. Regarding the wireless interfaces, nowadays the cost of an IEEE 802.11 card is very low and decreases rapidly, even more in [14] authors present a smart device driver for a Prism-based WLAN card that operates as two wireless cards. Regarding the overlapping requirements all the mobility extensions (including ours) require an overlapped network to provide seamless handovers. Our proposal allows VoIP applications to run in a mobile environment and keeps the network simple, cheap to maintain and to manage, pushing the complexity to the MN.

**Table 1.** Summary of the comparison

|  | FMIPv6 | HMIPv6 | Proposed Solution |
|---|---|---|---|
| *Handover Latency* | Low | High | 0 |
| *Packet Loss* | 0 | High | 0 |
| *IPDV* | Yes | No | No |
| *Network Support* | Yes | Yes | No |
| *Buffering* | Yes | No | No |
| *State-Protocol* | Yes | Yes | No |
| *Wireless Interfaces* | 1 | 1 | 2 |
| *Signaling Overhead* | High | Low | High |
| *Overlapping-Requirements* | Low | Low | Hard |

## 4   Related Work

New wireless standards are rapidly appearing and at the same time terminals which have radio and protocol support for two, three or even more standards are appearing. The idea of having a multihomed node to improve Mobile IPv6's performance is not new. The IETF has recently created the MONAMI6 [15] working group which will take advantage of this and will standardize how the Mobile IPv6 RFC must be modified to support multiple interfaces and thus, multiple Care-of-Addresses registrations. Finally, in [16],[20] the authors discuss the benefits of having multihomed MNs and how Mobile IPv6 can take advantage. Those proposals require modifications on the CNs, MNs and HAs and their main goal is not to improve the handover but to allow using more than one Care-of-Addresses trough multiple bindings at the same time. Our proposal does not require multiple CoA registrations.

The IETF has also recently created the Detect Network Attachment [21] working group. Its main goal is to reduce the Mobile IPv6 handover latency, specifically the IPv6 part of the handover (modeled as $D_{NUD}$ in our paper) by using link-layer triggers. They have yet published an RFC [17] discussing the goals of the working group and an internet draft [18]. We also avoid such delay in our proposal, not by using link-layer triggering but because our MN does not need to perform this operation. Our proposal use an extended version of link-layer triggers, we use a cross-layer design.

Several papers aim to enhance the handover, [22] improves the wireless handover by using neighbor graphs, a data structure which dynamically captures the mobile topology of a wireless networks. [14] also improves the IEEE 802.11 handover by virtualizing a single wireless card that is able to be connected to multiple networks simultaneously. Our proposal improves the whole Mobile IPv6 handover, not just the wireless part. Finally [19] and [23] enhances the whole Mobile IP handover by using multicast while our proposal does not require of complex network support.

## 5   Conclusions

This paper presents a novel extension to the Mobile IPv6 protocol that improves the handover performance. The handover is the most critical part of the mobile technologies because packets can be lost or delayed making difficult to run VoIP. Our solution uses a multihomed mobile node with two IEEE 802.11 cards and a cross-layer design allowing interaction between the link-layer and the network layer. While one interface is sending and receiving data packets from CNs the other is performing all the operations related with the handover.

The IETF has designed two extensions to Mobile IPv6: FMIPv6 and HMIPv6. These solutions add complexity to the network, they require of special network support and they introduce new functional entities. We claim that we must keep the network simple, cheap to maintain and to manage pushing the complexity to the edges, the MN in our case. Our solution, intended for mobile vehicles, takes advantage of this philosophy providing a simple protocol. Moreover our solution performs better than FMIPv6 and HMIPv6 and it has zero handover latency and no losses. We have not validated our proposal through simulations because this would not provide any kind of significant information. Our solution can be considered a generic framework to provide seamless handovers between any different link layer radio technologies such as GPRS or UMTS. In other words, we can apply our proposal to provide seamless handovers between IEEE 802.11 and GPRS, or UMTS using Mobile IPv6.

# References

1. IEEE 802.11: Wireless LAN Medium Access Control and Physical Layer. (1997)
2. D. Johnson, C. Perkins and J. Arkko: IP Mobility Support for IPv6 RFC 3775
3. C. Perkins: IP Mobility Support for IPv4 RFC 3344 (2002)
4. R. Kookl: Fast Handovers for Mobile IPv6 RFC 4068 (2005)
5. H. Solimant et al: Hierarchical Mobile IPv6 Mobility Management RFC 4140(2005)
6. A. Mishra, M. Shin and W. Arbaugh: An Empirical Analysis of the IEEE 802.11 MAC Layer Handoff Process Vol.33 ACM SIGCOMM Comp. Comm. Review
7. A. Cabellos-Aparicio et al: Measurement based analysis of the handover in a WLAN MIPv6 scenario Passive and Active Measurements 2005, LNCS 3431, pp 203-214
8. A. Cabellos-Aparicio et al: Evaluation of the Fast Handover Implementation for Mobile IPv6 in a Real Testbed 5th IEEE International Workshop on IP Operations and Management, Springer, LNCS 3751, pp181-190.
9. V. Gupta and D Johnston: IEEE 802.21 a Generalized Model for Link Layer Trigger IEEE 802.21 Media Independent Handoff Working Group, Mar 2004 mtg. mins
10. IEEE 802.21: A proposal for MIH function and Information Services (2004)
11. A.E Yegin et al: Link-layer triggers protocol draft-yegin-l2-triggers-00.txt (2002)
12. Nick Moore: Optimistic Duplicate Address Detection for IPv6 draft-ietf-ipv6-optimistic-dad-06.txt (September 2005)
13. D-LINK: Tech Specs for the D-Link DWL-G800AP product http://www.dlink.com/products/resource.asp?pid=267&rid=790&sec=0
14. Ranveer Chandra, Paramvir Bahl, Pradeep Bahl: Multinet: Connecting to Multiple IEEE 802.11 Networks Using a Single Wireless Card INFOCOM 2004
15. T. Ernst, N. Montavont: Mobile Nodes and Multiple Interfaces in IPv6 IETF MON-AMI6 Working Group
16. C. Ng et al: Analysis of Multihoming in Mobile IPv6 draft-ietf-nemo-multihoming-issues-04.txt (October 2005)
17. JH. Choi et al: Goals of Detecting Network Attachment in IPv6 RFC 4135 (2005)
18. A. Yegin et al: Link-Layer event notifications for Detecting Network Attachements draft-ietf-dna-link-information-03.txt (October 2005)
19. Mark Stemm et al: Vertical handoffs in wireless overlay networks Mobile Networks and Applications, Vol.3, Issue 4 (1999)
20. Ryuji Wakikawa: Multiple Care-of-Address registration draft-wakikawa-mobileip-multiplecoa-04.txt (June 2005)
21. Greg Daley, Suresh Krishnan: Detecting Network Attachment IETF WG
22. Arunesh Mishra et al: Context Caching using Neighbor Graphs for Fast Handoffs in Wireless Network INFOCOM 2004
23. Ganesha Bhaskara et al: Efficient micro-mobility using itra-domain multicast-based mechanisms (M&M) Globecom 2005
24. T. Narten et al: Neighbor Discovery in IP Version 6 (IPv6) RFC 2461 (1998)
25. Thomson S and T. Narten: IPv6 Address Autoconfiguration RFC 2462 (1998)

# The Structure of the Reactive Performance Control System for Wireless Channels

Dmitri Moltchanov

Institute of Communication Engineering,
Tampere University of Technology,
P.O. Box 553, Tampere, Finland
moltchan@cs.tut.fi

**Abstract.** To optimize performance of applications running over wireless channels, novel state-of-the-art wireless access technologies provide a number of advanced features including dynamic adaptation of protocol parameters at different layers. To exchange control information among non-adjacent layers a cross-layer signalling protocol is needed. Then, the control information should be used by a certain performance control entity to determine the set of protocol parameters resulting in best possible performance of an application in a given wireless channel and traffic conditions. For wireless access technologies with dynamic adaptation of protocol parameters design of the cross-layer performance control system is proposed. Functionalities of components of the system are isolated and described in details.

## 1   Introduction

To optimize performance of applications in fixed networks it is sufficient to control performance degradation caused by packet forwarding procedures. Even though this is not a trivial task, dealing with wireless networks we also have to take into account performance degradation caused by incorrect reception of channel symbols. It has completely different nature compared to what we dealt in fixed networks and often contributes a lot to end-to-end performance degradation. Thus, the air interface could be a 'weak point' in any quality of service (QoS) assurance model would ever be proposed for IP-based wireless networks.

It is often pointed out that although layered system design has proven itself to be efficient in wired network, it is no longer appropriate for wireless networks. In order to timely and appropriately react to changing wireless channel conditions and requirements of applications, state-of-the-art wireless access technologies should include a number of advanced capabilities. Nowadays, many of them allow to dynamically change protocol parameters at different layers of the protocol stack in response to changes in wireless channel conditions. These parameters may include the transmission power, the modulation and channel coding scheme at the physical layer, forward error correction (FEC) and automatic repeat request (ARQ) techniques at the data-link layer and some higher

Y. Koucheryavy, J. Harju, and V.B. Iversen (Eds.): NEW2AN 2006, LNCS 4003, pp. 164–176, 2006.

layer protocol parameters. In order for layers to make decisions on changes of protocols parameters, they should be aware of the current state of protocols at other layers. To exchange this control information among different non-adjacent layers of the protocol stack a signaling scheme is needed. Depending on the state of the wireless channel and application, it is then possible to dynamically change protocol parameters at different layers to obtain best possible performance for a given application at any given instant of time. To achieve this aim a performance control system is needed.

In this paper we propose the design of the performance control system that is responsible for dynamic adaptation of protocol parameters at different layers to provide the best possible performance for a given application. In Section 2 performance degradation of applications in wireless networks is considered. The need for cross-layer interactions at the air interface is explained in Section 3. The structure of the cross-layer performance control system for wireless channels is proposed in Section 4. Conclusions are given in the last section.

## 2   Performance Guarantees in Wireless Networks

Due to inherent properties of wireless channels neither deterministic nor statistic guarantees can be provided for applications in wireless networks. Indeed, there are a number of factors that affect the service process of wireless channels. The most important are stochastic nature of traffic characteristics, unstable behavior of wireless channels due to both movement of a user and environmental characteristics of landscapes, and protocols with a set of their parameters, as shown in Fig. 1. It is known that each application is characterized by its own traffic characteristics. Environmental characteristics of a landscape and movement of a user are stochastic factors determining propagation characteristics of a wireless channel. Protocols and their parameters determine how a given traffic is treated on a wireless channel. In general, performance that a given application receives running over a wireless channel is determined by properties and interactions between these components. Performance control of applications in wireless networks is then a sophisticated task involving a number of interdependent stochastic and deterministic factors. Even if the wireless channel is exclusively assigned to a user during the whole duration of the session, a certain level of performance can never be guaranteed. However, there is a possibility to optimize performance

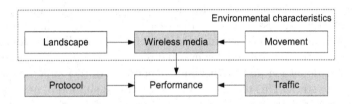

**Fig. 1.** Components that affects performance of wireless channels

of a given application for a given wireless channel conditions by varying values of protocols' parameters.

## 3   The Need for Cross-Layer Interactions

### 3.1   Cross-Layer Interactions in the Protocol Stack

Both ITU-T X.400 OSI abstract protocol model and TCP/IP protocol model separate and isolate functionalities of each layer. In these models each layer is responsible for a certain set of functions, communicates directly with the same layer of a peer communication entity and is usually unaware of specific functions of other layers. Both architectures do not allow direct communication of any kind between non-adjacent layers. Communications within the protocol stack are only allowed between adjacent layers using the so-called request-response primitives defined for the service access points (SAP). Higher layers just use the functions provided by the adjacent lower layers. The major advantage of the layered architecture is that it is possible to implement protocols at different layers independently.

To optimize performance of applications running over wireless channels new wireless access technologies call for novel design of the protocol stack at the air interface. Although interfaces between adjacent layers are preferable, there is the need for efficient, fast and direct interactions between non-adjacent layers. To exchange control information beyond the standard layered architecture, new interfaces between different protocols across the layers are required. In fact, the network layer and above layers often need direct interfaces to the data-link layer for handover support. Another example concerns transmission parameters including transmission mode, channel coding and data-link layer retransmissions which must be related to application characteristics (e.g. type of information, source coding, etc.), network characteristics, user preferences and context of use. In order to make decisions on traffic management, data-link layer protocols should be aware of higher layer including network and transport layer parameters and vice versa. In novel wireless access technologies we can refer to the air interface protocol architecture with interactions among different layers.

In this paper, we define the cross-layer design of the protocol stack as a certain design that somehow violates the respective layered structure of communication protocols [1]. Up to date there were a number of proposals for cross-layers interactions in the protocol stack for wireless channels. We usually distinguish between the following types of cross-layer interactions: creation of new interfaces, merging of adjacent layers, design coupling without new interfaces, and vertical calibration across layers [1]. In order to directly exchange information between non-adjacent layers new interfaces can be created. The information can be exchanged in the upward and downward directions. Merging of adjacent layers refers to joint definition and further implementation of two adjacent protocols in the protocol stack. This technique allows to avoid definition of new interfaces at the expense of more complicated implementation. Note that this approach is not inherently cross-layer but it still violates the layered architecture of the protocol stack. According to the design coupling no information is exchanged between non-adjacent layers at the runtime. Instead, two protocols are made aware of specific peculiarities of each other at the design phase. Finally, vertical calibration of all layers refers to the case when parameters of protocols at different layers are adjacent at the runtime such that a certain performance metric is controlled and optimized.

The goal of all abovementioned cross-layer approaches is to explicitly or implicitly exchange information between different layers of the protocol stack whether at the runtime or at the design phase for further optimization purposes. Detailed review of these approaches can be found in [1] while several examples of the cross-layer design methodologies are discussed in [2].

## 3.2   The Cross-Layer Signalling Protocols

In order for non-adjacent layers to communicate between each other the cross-layer signalling protocol is needed. Up to date there were a number of proposals for cross-layer signaling in the protocol stack. In [3] in order to communicate between TCP and data-link layer protocol in wireless IP-enabled networks authors proposed to use the wireless extension header (WEH) of IPv6 protocol. The advantage of this method is that it makes use of IP data packets as in-band signalling for information exchange between the transport and the data-link layer. Another method was proposed in [4]. In order to provide communication between different layers of the protocol stack authors proposed to use ICMP messages. According to the proposal a new message is generated whenever a certain parameter changes. A different 'network' approach is proposed in [5]. According to the proposal a special network service that gathers, stores, manages and distributes information about current parameters used at mobile hosts is introduced. Those protocols that are interested in a certain parameter can access this network service. Even though the approach is not 'cross-layer' at the glance, it still provides cross-layer capability via third party service. Usage of local profiles instead of remote network service was proposed in [6]. The concept of the method is similar to that one proposed in [5]. The difference is that information is stored locally and there is no need to access it via the network. The result is the low overhead and low delay associated with this approach. In [7] authors proposed a separate cross-layer signalling protocol for communication between layers in the protocol stack. The major advantage of this protocol is that non-neighboring layers can exchange control information directly without processing at intermediate layers. This approach, however, requires additional complexity to be introduced to the protocol stack. Authors in [7] also provided a comparison between abovementioned signalling protocols.

Those approaches, cited above, do not explicitly take into account the application of the cross-layer signalling. Indeed, the signalling protocol itself does not provide any advantages for a communicating entity. One of the most promising applications of the cross-layer signalling is the optimization of parameters of protocols at different layers in real-time (cross-layer parameters tuning) to provide the best possible performance at any instant of time for any channel and traffic conditions. Information, carried out by cross-layer signalling protocol can be (and should be) efficiently used to optimize operational parameters of protocols at different layers.

Considering the cross-layer signalling and the optimization of application performance jointly we have to choose whether to use the distributed or centralized control of wireless channel performance. Some approaches propose to use in-band signalling for exchange of control information among layers [3, 4, 7]. Other proposals suggest to implement a separated cross-layer communication system [5, 6]. The in-band cross-layer signalling proposals imply the distributed performance control strategy. According to the distributed control strategy and protocols that use in-band cross-layer signalling, layers exchange their information and this information should then be used by the performance control entities. These control entities must be implemented at each layer that participates in information exchange and allows its parameters to be dynamically controlled. Such approach is referred to as distributed performance control, requires significant modifications to be introduced *to each layer* and significantly affects the complexity of the whole system. It was pointed out in [8] that there are a number of problems associated with distributed control. Indeed, when the decision regarding the change of parameters is taken independently at each layer the resulting effect may not be straightforward. According to the centralized control and out-of-band signalling, layers export their current operational parameters to a certain external performance control entity via a predefined set of interfaces. This external entity may not only save the information of all layers but may implement the performance control features and then distribute information regarding what kind of protocol parameters should be further used. Thus, it makes an external intelligent cross-layer performance optimization system that preserves all the features of the out-of-band cross-layer signalling system. This method is centralized in nature and is generally in accordance with the signalling protocol proposed in [6].

In what follows, we propose the performance control system that is completely separated from the layered architecture of the protocol stack. The only modification to the current protocols needed to implement the system is the external interface that each controllable protocol should provide to the performance control system. The cross-layer communications between different layers is implemented using the so-called active local profiles [7], also known as shared database across layers [1]. The performance control system monitors the wireless channel performance, keep track of parameters of protocols that are controllable, use this information to estimate the current performance of a given application, decides whether it is possible to improve the performance and if so, determines and sets new parameters of controllable protocols. If the performance cannot be improved the system takes no action allowing to use old protocol parameters. For the proposed performance control system, the interface between a controllable protocol at a certain layer must have informational outputs and controlling inputs. All the logic of the performance control system is implemented by the external module.

# 4 The Structure of the System

## 4.1 Cross-Layer Performance Optimization

A number of approaches have been proposed so far for analysis of information transmission over wireless channels with error correction capabilities of the data-link layer. A review of these approaches can be found in [9]. However, only a few studies addressed the cross-layer tuning of parameters of various protocols to achieve the best possible performance for a wireless channel. Basic ideas concerning the parameters tuning across multiple layers were formulated in [10, 11]. In both studies the same concept of the so-called black-box approach for performance evaluation and optimization was simultaneously proposed. Authors in [11], however, noticed that this approach is time-consuming and suggested to look for analytical solutions for cross-layer performance optimization. Recently, such studies started to appear. Among others, one should mention studies carried out by Liu *et al.* in [12, 13]. Authors considered the wireless channel with adaptive modulation and coding schemes (AMC) and model it using the Markov model. Then, the performance characteristics at the data-link layer in terms of the mean delay of an arbitrary frame have been derived using the analytical approach. The next step was taken by Tang and Zhang in [14], where authors used the Markov model of the AMC-based wireless channel with multiple-in and multiple-out (MIMO) antenna diversity and employed the effective capacity approach to estimate the delay bound at the data-link layer.

Although all abovementioned studies explicitly take into account adaptive protocols at the physical and data-link layers they still cannot be considered as performance optimization studies. The reason is that wireless channel processes in [12, 13, 14] were modeled by homogenous ergodic Markov model that is covariance stationary in nature. Recently, important observations of the bit error process have been published in [15]. Authors found their GSM bit error traces to be non-stationary and proposed an algorithm to extract covariance stationary parts. They further used doubly-stochastic Markov processes to model these parts separately. The modeled trace is finally obtained by concatenation. Among other conclusions, authors suggest that a given bit error trace can be divided into a number of concatenated covariance stationary traces. Similar observations have been found in [16]. Note that the bit error probability is a function of the SNR value, and frame error probability is a function of bit error rate. As a result, we can expect the same properties for SNR and frame error observations. Moreover, it is important to note that even if observations are indeed stationary, covariance stationary models may not be suitable for performance control purposes. Such models assume perfect control of parameters of various protocols and allow to determine the performance characteristics experienced by applications as seen by time-averages. It does not provide any information about when, how and which protocol parameters should be changed.

Another approach was introduced by Moltchanov and Koucheryavy in [17]. Authors divided the functionality of the performance control system into four different modules. These modules include the channel prediction, channel modeling, cross-layer mapping and performance evaluation modules. In [18] authors used a simple Markov model of the Gilbert type to capture the bit error rate and lag-1 autocorrelation of the covariance stationary bit error segments and then used a special case of non-preemptive queuing system to obtain performance characteristics of the frame arrival process in terms of the mean number of lost frames. In [9] the performance evaluation model at the data-link layer was extended to obtain the probability function of the number of lost frames when ARQ and FEC procedures are simultaneously used. Subsequently, in [16] the change point statistical test for detecting changes in non-stationary wireless channel observations is proposed. Supplemented with previously developed performance evaluation model [9], the framework proposed in [17] can be used to determine parameters of protocols at different layers resulting in best possible performance at the data-link layer in presence of changes in wireless channel statistics. The lack of the clear and understandable component-based architecture is the main shortcoming of the approach. In this paper we try to fill this gap proposing abstract structure of the performance control system.

## 4.2   The Cross-Layer Performance Control System

The generic structure of the cross-layer performance control system is presented in Fig. 2, where CPOS stands for cross-layer performance optimization system. The protocol stack is logically divided into three groups of protocols. The first group consists of an application itself that falls into a certain traffic class. We assume that a network is intended to deliver four traffic classes. These are conversational, streaming, interactive and background classes. These classes are defined in 3GPP specifications and are generally in accordance with ITU-T and ETSI specifications. Applications that fall in conversational class require to preserve time relation between information entities of the traffic stream and require stringent guarantees of end-to-end delivery of the traffic entities. Examples of these applications include real-time two-way voice communications, audio multicasting, etc. Applications that are classified to the streaming class require the network to only preserve time relation between information entities of the traffic stream and do not require strict guarantees of end-to-end delivery. The most common example of these applications is the streaming video. Both interactive and background classes expect a network to guarantee the reliable delivery of information units. The difference between these two classes is that the interactive class includes applications operating in request-response mode, thus, posing additional requirements on end-to-end delay. Applications that fall in background class are usually characterized by the so-called bulk transfers and do not require bounded delay in the network.

Considering the defined traffic classes one may observe that there is strict correspondence between the traffic class and protocols at the transport and network layers. Indeed, those applications that require strict delay requirements (conversational and streaming classes) usually use (RTP/)UDP/IP as the combination of the transport and network layer protocols. UDP provides a datagram connectionless service. The main function of UDP is to add a port number to the IP address providing a socket for the application. Those applications that require a network to preserve the content of the transmission (interactive and background classes) use TCP and IP at the transport and network layers, respectively. TCP provides a connection-oriented service in the network and includes rules for establishing and terminating connections, congestion control, flow control and reliable in-order delivery of data.

Note that TCP, UDP and IP are nowadays well standardized. For the sake of interoperability with existing implementations, no modifications should be made for these protocols. Indeed, modifications to any of these protocols may require network wide modifications. For this reason, we can reasonably require that protocols of the transport and network layers and their parameters should not be controlled by the performance control system.

Wireless access technology determines how the traffic is treated at the wireless channel. It defines protocols of the data-link and physical layers and their parameters. Protocols used at these layers are specific for a given technology and may incorporate advanced features such as dynamic choice of the parameters to achieve the best possible performance for a wireless channel conditions. Since the main performance degradation in wireless networks stems from the unstable nature of wireless channel characteristics, this feature of state-of-the-art wireless access technologies may provide a feasible solution for performance control.

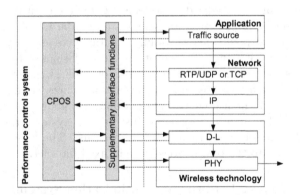

**Fig. 2.** The cross-layer performance optimization system

According to the performance control system presented in Fig. 2, the application determines the network protocol suit (TCP/IP or (RTP/)UDP/IP) to be used during the active session. This decision is made independently of the performance control system and the mapping is strict for a given application. The application implicitly notifies the CPOS regarding which protocols are used at the transport and network layers providing the traffic class on informational input. It should also provide information regarding which performance parameters should be provided at the local wireless channel. During the whole duration of the session CPOS monitors the state of the wireless channel and the state of the application. The current state of the application (traffic model) and the current state of the wireless channel (wireless channel model) and protocols parameters at the data-link and physical layers are used to determine IP layer performance parameters that should include IP packet loss rate (IPLR), IP packet transfer delay (IPTD) and IP packet delay variation (IPDV) specified in ITU-T Y.1540. Then, COPS should determine which actions should be taken to provide the best possible performance for a given application at the current instant of time, that is, which current protocol parameters should be changed and how. The list of actions should include changes of protocol parameters at the data-link and physical layers. This capability is available in most state-of-the-art wireless access technologies including EDGE, UMTS, IEEE 802.16 and HIPERLAN/2. Additionally, it may also include change of the applications' parameters (e.g. rate of the codec for video and audio applications) and change of the buffer space at the IP layer. The former capability is available for real-time applications such as streaming video. The latter one should be implemented additionally. Note that when a certain applications does not allow to change the rate at which the traffic is fed to the network, the feedback regarding the current rate should still exist. When this capability is available, controlling inputs should be provided to the source.

At the beginning of the session, the CPOS should be made aware of controllable protocols in the protocol stack. It can also be made statically at the software development phase. It is important to note that at the beginning of the session protocols can be initialized with default parameters. In this case, these parameters should be immediately communicated to the CPOS. The another approach is to setup a predefined set of initial parameters for each application. During the active session the CPOS controls the performance by setting protocol parameters in response to changing traffic or wireless channel conditions. Therefore, during the active session informational interfaces are not used.

The proposed performance control system does not guarantee that the application receives the quality it requests. Instead, the system tries to achieve the best possible performance at any instant of time. The service it tries to implement is similar to the best effort service in fixed Internet.

### 4.3   The Cross-Layer Performance Optimization System

The core of the proposed performance control system is the CPOS. The proposed structure of the CPOS is shown in Fig. 3. Three major components of this system are the real-time channel estimation module (CEM), the real-time traffic estimation module (TEM) and the performance evaluation and optimization module (PEOM). The CEM is responsible for detecting changes in wireless channel statistics and estimation of the channel state in terms of the SNR, bit or frame error models. The TEM performs the same functions for traffic observations. To achieve that, wireless channel and traffic statistics should be observed in real-time and immediately fed to the input of the respective change point analyzer blocks. Note that the usage of TEM is only mandatory for real-time applications that may have unexpectable traffic patterns.

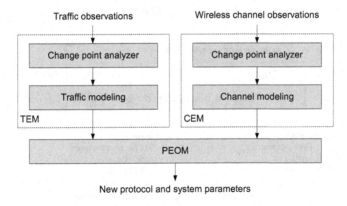

**Fig. 3.** The structure of the performance optimization subsystem

Change point analyzers test incoming observations for changes in parameters that mostly affect performance of applications running over wireless channels. Recently, emerged methods of measurement-based traffic modeling have allowed to recognize major statistical characteristics of the traffic affecting its service performance in a network [19, 20]. According to Li and Hwang [19] the major impact on performance parameters of the service process is produced by the marginal distribution of the arrival process and the structure of its autocorrelation function (ACF). Hayek and He [20] highlighted the importance of marginal distributions of the number of arrivals showing that the queuing response may vary for inputs with the same mean and ACF. It was also shown [19] that accurate approximation of empirical data can be achieved when both marginal distribution and ACF of the model match their empirical counterparts well. Recently, it was also shown that the wireless channel statistics including the mean frame error rate and the lag-1 autocorrelation of the frame error process significantly affect performance parameters of applications running over wireless channels at the data-link layer with hybrid ARQ/FEC [9]. In [18] authors considered the effect of cross-layer propagation of bit error to the IP layer with FEC procedures implemented at the data-link layer. It was found that the mean bit error rate and the lag-1 autocorrelation of the bit error process affect the performance response at the IP layers in terms of the mean number of lost IP packets. Similar conclusions have been made in [21], where authors considered the effect of bit errors on performance of applications at the IP layer.

As an indicator of the change in wireless channel characteristics, SNR process, bit error process, or frame error process can be used. The reason to use the bit error process is twofold. Firstly, it allows to abstract functionality of the physical layer of different wireless access technologies. As a result, a single cross-layer performance control system can be potentially applied for different wireless channels. Secondly, the bit error process is binary in nature. It allows to significantly decrease the complexity of the modeling algorithm as shown

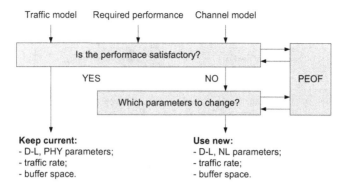

**Fig. 4.** The structure of the performance control system

in [22]. It is also allowed to use the SNR process instead of the bit error process if the relationship between the bit error probability and the SNR value is known. Finally, the frame error process can also be used. Note, that frame error statistics can be directly obtained observing operation of ARQ protocols at the data-link layer. However, monitoring of the frame error process introduces significant delay in detection of the channel state. In this case the system may not timely react to changing wireless channel conditions.

The current models of the wireless channel and the traffic source are parameterized in the respective modeling blocks and then immediately fed to the input of PEOM. Note that the current traffic and channel models are stored in the respective modeling blocks for further usage. The structure of PEOM is shown in Fig. 4. According to the system design the current traffic and channel models are directly fed to the input of the decision module. Taking the reference performance of a given application at appropriate layer (e.g. data-link or IP) as another input, this module decides whether the current performance is satisfactory. In order to take the decision the module containing the performance evaluation framework (PEOF) is activated. If the performance is satisfactory, no changes are required and current protocol parameters are further used at the data-link layer. Otherwise, the wireless channel and traffic models are used to decide whether performance can be improved and, if so, which parameters have to be changed and how. Depending on particular protocols of the protocol stack and type of the application, new protocol parameters resulting in best possible performance for a given wireless channel and traffic statistics are computed in the PEOF and then fed back to the decision module. These parameters are then used till the next change in input wireless channel and traffic statistics. The PEOF may implement the performance evaluation framework or just contain a set of pre-computed performance curves corresponding to a wide range of wireless channel and traffic statistics and different configurations of the protocol stack. Due to the real-time nature of the system the latter approach is preferable.

The change point analyzers must signal only the point when a change in either traffic or wireless channel statistics is detected. Otherwise, no actions are performed by the system except for continuous real-time monitoring of the channel and traffic statistics. Note that it is allowed for the system to be activated in response to the change in either channel or traffic statistics. In this case there is no need to parameterize the model associated with statistics that did not change. This model is stored in the respective modeling blocked and must be used as the current model. The model associated with statistics that experienced a change in a certain parameter must be re-parameterized and then used.

To synthesize the system the following tools have to be developed:

- statistical test for detecting changes in channel and traffic statistics;
- model for covariance stationary observations;
- cross-layer extension for wireless channel model to the layer of interest;
- performance evaluation model at the data-link and IP layers.

Note that the increase in FEC correction capabilities results in decrease of the throughput obtained at higher layers that may not always be tolerable for some bandwidth-greedy applications. There can also be situations when even the most powerful FEC code cannot provide satisfactory performance due to either worst channel conditions of insufficient throughput. In this case ARQ error correction can be introduced at the wireless link. Alternatively, power control system can be initiated and the latter case is seen as joint usage of the proposed performance control system and the power control capabilities of WCDMA systems.

## 5  Conclusions

For wireless access technologies with dynamic adaptation of the protocol parameters to dynamically changing wireless channel conditions the design of the cross-layer performance control system has been proposed. According to the proposed design of the performance control system it is still possible to implement protocols at different layers independently. The only requirement we have to impose is that certain protocols should be actually controllable and should export information about their current parameters using appropriate interfaces.

## References

1. V. Srivastava and M. Motani. Cross-layer design: a survey and the road ahead. *IEEE Comp. Comm.*, 43(12):112–119, Dec. 2005.
2. V. Raisinghani and S. Lyer. Cross-layer design optimizations in wireless protocol stacks. *IEEE Comp. Comm.*, 27:720–724, 2004.
3. G. Wu, Y. Bai, J. Lai, and A. Ogielski. Interactions between TCP and RLP in wireless Internet. In *Proc. IEEE GLOBECOM'99*, pages 661–666, Rio de Janeiro, Dec. 1999.
4. P. Sudame and B. Badrinath. On providing support for protocol adaptation in mobile wireless networks. *Mob. Netw. and Appl.*, 6(1):43–55, Jan./Feb. 2001.
5. B.-J. Kim. A network service providing wireless channel information for adaptive mobile applications. In *Proc. IEEE ICC'01*, pages 1345–1251, Helsinki, June 2001.

6. K. Chen, S. Shan, and K. Nahrstedt. Cross-layer design for accessibility in mobile ad hoc networks. *Wireless Pers. Comm.*, 21(1):49–76, Apr. 2002.

7. Q. Wang and M. Abu-Rgheff. Cross-layer signalling for next-generation wireless systems. In *Proc. WCNC'03*, pages 1084–1089, March 2003.

8. V. Kawadia and P.R. Kumar. A cautionary perspective on cross-layer design. *IEEE Pers. Comm.*, 12(1):3–11, Feb. 2005.

9. D. Moltchanov, Y. Koucheryavy, and J. Harju. Loss performance model for wireless channels with autocorrelated arrivals and losses. *Special Issue of Comp. Comm.*, In Press, 2005.

10. M. van Der Schaar and N. Shankar. Cross-layer wireless multimedia transmission: challenges, principles, and new paradigms. *IEEE Wir. Comm.*, 12(4):50–58, Aug. 2005.

11. Y. Koucheryavy, D. Moltchanov, J. Harju, and G. Giambene. Cross-layer blackbox approach to performance evaluation of next generation mobile networks. In *NEW2AN*, pages 266–272, St.-Petersburg, Russia, Feb. 2004.

12. S. Liu, Q. Zhou and G. Giannakis. Queuing with adaptive modulation and coding over wireless links: cross-layer analysis and design. *IEEE Trans. Wir. Comm.*, 4(3):1142–1153, May 2005.

13. S. Liu, Q. Zhou and G. Giannakis. Cross-layer modeling of adaptive wireless link for QoS support in multimedia networks. *ACM/Kluwer J. on Wir. Netw.*, In Press, 2006.

14. X. Zhang, J. Tang, H.-H. Chen, S. Ci, and M. Guizani. Cross-layer-based modeling for quality of service guarantees in mobile wireless networks. *IEEE Comm. Mag.*, 44(1):100–106, Jan. 2006.

15. A. Konrad, B. Zhao, A. Joseph, and R. Ludwig. Markov-based channel model algorithm for wireless networks. *Wireless Networks*, 9(3):189–199, 2003.

16. D. Moltchanov. State description of wireless channels using change-point statistical tests. In *Proc. WWIC'2006, To appear*, Bern, Switzerland, May 2006.

17. D. Moltchanov and Y. Koucheryavy. Some elements of the performance control system for wireless channels. In *Proc. First International Workshop on Convergence of Heterogeneous Wireless Networks*, pages Published on CD–ROM, Budapest, Hungary, July 2005.

18. D. Moltchanov, Y. Koucheryavy, and J. Harju. Cross-layer modeling of wireless channels for IP layer performance evaluation of delay-sensitive applications. *Special Issue of Comp. Comm.*, In Press, 2005.

19. S.-Q. Li and C.-L. Hwang. Queue response to input correlation functions: discrete spectral analysis. *IEEE Trans. Netw.*, 1:522–533, Oct. 1997.

20. B. Hajek and L. He. On variations of queue response for inputs with the same mean and autocorrelation function. *IEEE Trans. Netw.*, 6(5):588–598, October 1998.

21. Y.-Y. Kim and S.-Q. Li. Capturing important statistics of a fading/shadowing channel for network performance analysis. *IEEE JSAC*, 17(5):888–901, May 1999.

22. D. Moltchanov, Y. Koucheryavy, and J Harju. Simple, accurate and computationally efficient wireless channel modeling algorithm. In *Proc. WWIC*, Xanthi, Greece, May 2005.

# Erlang Capacity of a CDMA Link with Transmission Rate Control

Ioannis Koukoutsidis[1] and Eitan Altman[2,*]

[1] FORTH-ICS, P.O. Box 1385, 71110, Heraklion, Crete, Greece
jkoukou@ics.forth.gr
[2] INRIA, 2004 Route des Lucioles, B.P. 93, 06902 Sophia Antipolis Cedex, France
Eitan.Altman@sophia.inria.fr

**Abstract.** Regulating the transmission rate of non-real-time applications offers enhanced flexibility for QoS management and capacity control on a CDMA link. For a system working at saturation conditions, controlling data rates can further enhance its capacity. A more thorough investigation of this possibility and the trade-offs involved is presented in this paper. We analytically study the Erlang capacity of CDMA when scaling the physical transmission rates on the link. We consider single and multiple service classes, extending our analysis to an asymptotic regime. It is shown that the most efficient way of boosting this capacity is by lowering transmission rates, which nevertheless comes at the expense of increased transfer delays and energy consumption.

## 1 Introduction

The advent of data applications in wireless third generation networks presents new perspectives and challenges to network engineers. Data traffic (e.g. file transfers, e-mail, web-browsing) is inherently of non-real-time nature and has no guaranteed bit rate. A data application requesting transmission on a wireless link is assumed to have a fixed amount of information to send. We are primarily interested in transferring all parts of the information reliably (i.e., with a given BER), or even in a best-effort manner, and less in the duration of the transfer. The latter can be "stretched" —under appropriate rate assignment in a multiplexing scenario— in order to satisfy an obvious objective: increase the capacity of the link.

The most straightforward definition of capacity in a wireless link is the so-called *integer capacity*. This refers to the most prompt question, how many ongoing transmissions of a given service class the system can handle at a certain instant in time. In CDMA, the limit of the integer capacity, achieved (theoretically) when the transmit powers approach infinity, is called *pole capacity* [4]. Another, more probabilistic definition extends the well-known notion of *Erlang capacity* [11]. This is the arrival rate of traffic, or the amount of offered traffic the system can allow so that the probability that the QoS of any service class is not attained is sufficiently small.

---

* The work of E. Altman was supported by the EuroNGI Network of Excellence.

Y. Koucheryavy, J. Harju, and V.B. Iversen (Eds.): NEW2AN 2006, LNCS 4003, pp. 177–188, 2006.

Since in CDMA there is no inherent blocking (e.g., by a fixed number of channels), but just a gradual degradation of quality, the study of Erlang capacity is closely tied with the application of an admission control scheme, which is necessary to preserve link efficiency. The most appropriate definition of a blocking condition is when the integer capacity is exceeded, defined to occur when the interference power exceeds the background noise by a certain amount [11].

Using the blocking condition arising from the integer capacity and based on an initial work in [1], in this paper we study more thoroughly the Erlang capacity in the multiservice context of a CDMA link. Considering a system which controls the rates of transmission, we attempt an analytical modeling and characterization of the Erlang capacity behavior when scaling up and down these rates, extending to an asymptotic regime. We shall see that the flexibility of assigning elastic flows with a specific data rate has important ramifications on the system performance: the Erlang capacity of the link can be significantly increased by lowering physical transmission rates of elastic flows. However, apart from the increase in transfer time, it is revealed that rate reduction also involves an increase in the energy consumed for this transfer by a mobile device. The observations coming from this research provide useful insights regarding rate assignment in a CDMA cell, in an offline or online control scenario.

## 2   The Capacity Model

The capacity of a CDMA cell is primarily limited by interference in the uplink and the total base station output power in the downlink [4, 12]. Either side can be the bottleneck at one time or another, depending on traffic and transmission (including mobility) conditions. Here we concentrate on the uplink, for which there exists a simple and accurate modeling. A similar modeling approach exists for the downlink (see [3], [4] and the references therein), however only a rough estimate of capacity can be done and it seems hazardous to extend the model in the context of admission and rate control. Nevertheless, the principle of operation is the same (transmissions limited by a power-related constraint), and we expect the essence of our results and conclusions to carry over to this case.

Transmission rates of mobiles will be considered *fixed* or *pre-assigned*, mapping to a set of active service classes $\mathcal{K} = \{1, \ldots, K\}$. We are interested in studying the Erlang capacity of the uplink when scaling the different rates of transmission. Before getting into this, it is necessary to recap the analytical calculation of integer capacity (consult e.g., [4]).

Consider a transmission of service class $s$ at rate $R_s$ (bits/sec). For a certain bit error rate (BER), an energy per bit to interference density requirement, $E_b^{(s)}/I_0$, is prescribed at the receiver RF side. Accordingly, the minimum required *carrier to interference ratio* (CIR), denoted as $\tilde{\Delta}_s$, is:

$$\tilde{\Delta}_s = \frac{E_b^{(s)}}{I_0} \frac{R_s}{W} \ , \tag{1}$$

where $W$ is the spread spectrum bandwidth. The CIR target condition for users of class $s$ $(s = 1, \ldots, K)$ writes:

$$\frac{P_s}{N + I_{intra} + I_{inter} - P_s} = \tilde{\Delta}_s \ , \tag{2}$$

where $P_s$ is the (minimum) required received power, $N$ is the background (including thermal) noise and, considering $m_s$ users from class $s$,

$$I_{intra} = \sum_{s=1}^{K} m_s P_s \ , \quad I_{inter} = f \cdot I_{intra}$$

are the intracell and intercell interferences, respectively. Here we make the standard assumption that the intercell interference is a fraction of the intracell one [11, 4].

We then write (2) as follows:

$$\frac{P_s}{N + I_{intra} + I_{inter}} = \Delta_s \ , \quad \text{where} \quad \Delta_s := \frac{\tilde{\Delta}_s}{1 + \tilde{\Delta}_s} \ .$$

Solving for the total interference $I_{tot} = I_{intra} + I_{inter} + N$, we have:

$$I_{tot} = \frac{N}{1 - (1 + f) \sum_{s=1}^{K} m_s \Delta_s} \ . \tag{3}$$

By saying that the uplink capacity in CDMA is interference-limited, we mean that the amount of interference presented at the base station demodulator must be limited. The engineering metric considered is the *uplink noise rise*, $\Lambda := I^{tot}/N$, which must not exceed a certain value [4]. In connection with this, the *uplink load factor* is defined as $L := (1 + f) \sum_{s=1}^{K} m_s \Delta_s$.

The maximum noise rise $\Lambda_{max}$ is called the *interference margin* of the system (measured in dB). For a given maximum value of the uplink load factor $L_{max}$, based on $\Lambda_{max}$, the integer capacity is described through a linear system of equations. The following definition applies:

**Definition 1.** *Let $\mathcal{M}$ be the finite subset of vectors $\boldsymbol{m} \in \mathbb{N}^K$ for which*

$$(1 + f) \sum_{s=1}^{K} m_s \Delta_s \leq L_{max} \ . \tag{4}$$

*We define the integer capacity $\mathcal{M}_B$ of the system as the set of boundary integer points $\boldsymbol{m}$ for which*

$$\boldsymbol{m} \in \mathcal{M} \ \text{and} \ \boldsymbol{m} + e_s \notin \mathcal{M} \ , \qquad s = 1, \ldots, K$$

*where $e_s$ is the unit vector in direction $s$ in the space $\mathbb{R}^K$.*

The (theoretical) pole capacity of the system can likewise be defined as the boundary points of the set of vectors $\boldsymbol{m} \in \mathbb{N}^K$ for which $(1 + f) \sum_{s=1}^{K} m_s \Delta_s < 1$.

## 3   The Stochastic Knapsack

The condition that a feasible set of vectors should satisfy, $(1 + f) \sum_{s=1}^{K} m_s \Delta_s \leq L_{max}$, put in the context of a stochastic environment with transmissions randomly entering and leaving the network, meets the definition of a *stochastic knapscack* model [9].

We assume that data flows (a data flow is used commonly to denote a string of packets) of each class $s$ arrive for transmission on the link according to independent Poisson processes of rate $\lambda_s$. The durations of transmissions of each flow in the Poisson stream are independent, identically distributed random variables with mean $1/\mu_s$. In addition, flow durations are also independent between different streams. Similarly to Definition 1, a blocking set $\mathcal{M}_B^s$ is defined for each class $s$ (such that the integer capacity is the intersection of all blocking sets $\mathcal{M}_B^s$, $s = 1, \ldots, K$).

We designate

$$\eta := L_{max}/(1 + f) \ .$$

The system can be seen as one in which there is a total given number of resources $\eta$, and in which each class $s \in \mathcal{K}$ flow requires an amount $\Delta_s$ of resource. In this sense, $\Delta_s$, $s = 1, \ldots, K$ will also be referred to as the "knapsack values," and $\eta$ as the "knapsack capacity."

Define the load of each class as $\rho_s := \lambda_s/\mu_s$. Then from [5] we can phrase the following theorem:

**Theorem 1.** *The stationary distribution of the number of flows in the system is given by*

$$\pi_\rho(\mathbf{m}) = \frac{1}{G_\rho} \prod_{s=1}^{K} \frac{\rho_s^{m_s}}{m_s!} \ , \quad \mathbf{m} \in \mathcal{M}, \quad \text{where} \quad G_\rho = \sum_{\mathbf{m} \in \mathcal{M}} \prod_{s=1}^{K} \frac{\rho_s^{m_s}}{m_s!} \ . \quad (5)$$

*The probability $P_B^s(\boldsymbol{\rho})$ that an arriving call of class $s$ is blocked and the average global blocking probability $P_B(\boldsymbol{\rho})$ are given by $(\lambda = \sum_s \lambda_s)$:*

$$P_B^s(\boldsymbol{\rho}) = \sum_{\mathbf{m} \in \mathcal{M}_B^s} \pi_\rho(\mathbf{m}) \ , \quad P_B(\boldsymbol{\rho}) = \sum_{s=1}^{K} \frac{\lambda_s}{\lambda} P_B^s(\boldsymbol{\rho}) \ .$$

*Remark 1.* One can even further relax the statistical assumptions under which Theorem 1 holds by allowing successive flow durations to be *dependent* (however, the lifetimes between different service classes are still assumed to be independent). This follows from the more general approach in [2], provided the process of successive service times is stationary ergodic. Relaxing partially the independence assumptions has a great significance; for example, we may define correlation dependencies between the sizes of flows following the sending of an initial flow, since that initial flow usually specifies the purpose of sending data over the link.

In the original Erlang loss system with a single service class, the Erlang capacity is defined as the maximum load such that the blocking probability is smaller than a given $\epsilon$. In the multiservice case, a dissimilar definition of Erlang capacity is in order:

**Definition 2.** *The multiservice Erlang capacity is defined as the boundary of the admissible set*

$$\mathcal{A} = \{\boldsymbol{\rho} = (\rho_1, \ldots, \rho_K) | P_B^s(\boldsymbol{\rho}) \leq \epsilon_s; s = 1, \ldots, K\}$$

*of loads from each class such that the blocking probability of each class $s$ is bounded by a given value $\epsilon_s$.*

## 4   Transmission Rate Control

We consider a certain initial state of the system described by a vector of pre-assigned transmission rates for each class, $\mathbf{R} = (R_1, \ldots, R_K)$. Let the size of a data transfer be a generally distributed random variable with mean $1/\zeta_s$ for class $s$, in which case the expected transfer time becomes $(\zeta_s R_s)^{-1}$. We define an analogous load here (in terms of size) as $\nu_s := \lambda_s/\zeta_s$, and the corresponding vector $\nu_s = (\nu_1, \ldots, \nu_K)$. For notational convenience, we also write $\delta_s = \frac{E_b^{(s)}}{I_0 W}$ so that $\tilde{\Delta}_s = R_s \delta_s$ and $\Delta_s = \frac{R_s \delta_s}{1 + R_s \delta_s}$, and let $\boldsymbol{\delta} = (\delta_1, \ldots, \delta_K)$.

Our purpose is to study the behavior of the system when scaling the initial rate vector. Scaling transmission rates is here seen as maintaining the "proportional fairness" that originates from different channel conditions of each user. To this end, we use a control parameter $\alpha$ as a division factor on the vector of transmission rates. In general, it is implied that $\alpha \in \mathbb{R}^+$; however, in some cases it will be necessary to consider $\alpha \in \mathbb{N}$ and this will be explicitly mentioned. A new system setting – indexed by $\alpha$– is defined for each choice of control parameter. The rate vector is thus $\mathbf{R}(\alpha) = \mathbf{R}/\alpha$, and call durations are multiplied by $\alpha$.

We shall assume that, as the transmission rate varies, the same energy to noise ratio is required for a certain BER. This generally holds when the modulation scheme does not change. Then as a consequence $\tilde{\Delta}_s$, $s = 1, \ldots, K$ are also divided by $\alpha$. Additionally, we consider the interference margin $\Lambda_{max}$ (and consequently $\eta$) to remain the same at each instance of the controlled system (practically, the link budget remains unchanged, see e.g. [4, Tables 8.3–8.5] for different services).

The best illustration of behavior of this controlled system is obtained by studying the simplest case of homogeneous service for all mobiles.

### 4.1   Homogeneous Service

Here all classes of mobiles have an initial common value of $\Delta \equiv \Delta_s$, $s = 1, \ldots, K$, indexed by $\Delta(\alpha)$ at each setting (with a corresponding value $\tilde{\Delta}(\alpha)$). The integer capacity in the initial system is:

$$\mathcal{M}_B = max\{m \in \mathbb{N} : m\Delta \leq \eta\} = \left\lfloor \frac{\eta}{\Delta} \right\rfloor. \tag{6}$$

The new integer capacity $\mathcal{M}_{B,\alpha}$ after all rates are divided by $\alpha$ becomes

$$\mathcal{M}_{B,\alpha} = \left\lfloor \frac{\eta}{\Delta(\alpha)} \right\rfloor = \left\lfloor \eta \cdot \frac{\alpha + \tilde{\Delta}}{\tilde{\Delta}} \right\rfloor = \left\lfloor \eta \cdot \left( \frac{\alpha}{\tilde{\Delta}} - \alpha + 1 \right) \right\rfloor . \tag{7}$$

An immediate conclusion derived by comparing the integer capacity with that of the initial system is: if $\alpha < 1$, $\mathcal{M}_{B,\alpha} \leq \mathcal{M}_B$, and if $\alpha > 1$, $\mathcal{M}_{B,\alpha} \geq \mathcal{M}_B$.

By the modified transmission rates, we get a new load $\rho_\alpha = \rho\alpha$. Then the steady-state probabilities for the new system are

$$\pi_{\rho,\alpha}(m) = \frac{(\rho\alpha)^m/m!}{\sum_{j=0}^{\mathcal{M}_{B,\alpha}} (\rho\alpha)^j/j!} , \qquad m = 0, \ldots, \mathcal{M}_{B,\alpha} . \tag{8}$$

The blocking probability $P_B(\rho_\alpha)$ is immediately obtained for $m = \mathcal{M}_{B,\alpha}$. For a given small value of $\epsilon$, the Erlang capacity of the system is derived after finding the maximum supported load $\rho_\alpha$ and dividing by the parameter $\alpha$.

We focus on what happens to the Erlang capacity of the system when lowering the transmission rates, i.e., for increasing values of $\alpha$. A basic result is the following:

**Proposition 1.** *For any initial transmission rate, there exists an infinite set of increasing values of $\alpha$ for which the corresponding Erlang capacity increases.*

*Proof.* We find a condition for which the blocking probability decreases. Let $X$ be the state (number of flows) of the initial system and $Y$ the state of the one derived by dividing transmission rates by $\alpha$. Define

$$Z = \max\{0, Y - \mathcal{M}_{B,\alpha} + \mathcal{M}_B\} .$$

Note that the random variables $X, Z$ have the same support set $\mathcal{I} = \{0, 1, \ldots, \mathcal{M}_B\}$.

In order to compare blocking probabilities given the distributions at hand, we use the *likelihood ratio* ordering (see e.g., [9]). For a discrete random variable $V$ with support $\mathcal{I}$ define the *ratio function*

$$r_V(m) := \begin{cases} 0, & m = 0 \\ \dfrac{P(V = m - 1)}{V = m}, & m = 1, \ldots, \mathcal{M}_B . \end{cases}$$

A well known result for the likelihood ratio ordering is that $r_Z(m) \geq r_X(m)$ $\forall m \in I$ implies $Z \leq_{st.} X$. Then equivalently $E[f(Z)] \leq E[f(X)]$ for any non-decreasing function $f$. By taking $f(x)$ to be the indicator random variable $f(x) = 1_{\{x = \mathcal{M}_B\}}$, we can then conclude that

$$P(X = \mathcal{M}_B) \geq P(Z = \mathcal{M}_B) = P(Y = \mathcal{M}_{B,\alpha}) .$$

Thus we have a means to show the desired result by simply examining the values of $\alpha$ for which $r_Z(m) \geq r_X(m)$ for every $m = 1, \ldots, \mathcal{M}_B$. We have for these values of $m$ that

$$r_X(m) = \frac{m}{\rho} , \quad r_Z(m) = \frac{m + \mathcal{M}_{B,\alpha} - \mathcal{M}_B}{\alpha\rho} .$$

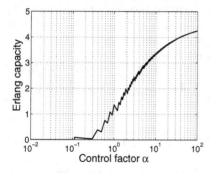

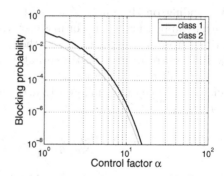

**Fig. 1.** Erlang capacity in the homogeneous service case with varying control factor $\alpha$ ($\eta = 0.5$; for $\alpha = 1$, $\Delta = 0.1$)

**Fig. 2.** Blocking probability of two classes of flows with varying decrease factor $\alpha$ ($\eta = 0.5$; for $\alpha = 1$, $\Delta_1 = 0.12$, $\Delta_2 = 0.05$, and $\rho_1 = 1$, $\rho_2 = 2$)

After a simple calculation, we arrive at the following condition:

$$\alpha \geq \frac{\frac{\eta}{\Delta} - m + c - \eta}{\frac{\eta}{\Delta(\alpha)} - m - \eta} \ . \tag{9}$$

The variable $c$, where $|c| < 1$, comes from the difference of the fractional parts of the two floor functions in $\mathcal{M}_{B,\alpha}, \mathcal{M}_B$ (its value depends on $\alpha$, without this being a problem since we know that it is always bounded). Since $\eta, \Delta, \Delta(\alpha) < 1$ and $|c| < 1$, it shows that there exists an infinite set of increasing values of $\alpha$ such that the above condition holds true $\forall m \in \{1, \ldots, \mathcal{M}_B\}$.[1]

The proof that the Erlang capacity will increase derives then as a simple consequence of the fact that in the Erlang model, the blocking probability increases with the input rate ([8]); since blocking is decreased in the new system, a larger value of the input rate can be sustained. □

*Remark 2.* When considering the *pole* capacity of the system, we refer to Theorem 3 of [1] that showed that for a system with no intercell interference and with the additional condition that $\Delta^{-1} \in \mathbb{N}$, the Erlang capacity of the system always increases for increasing $\alpha \in \mathbb{N}$.

Nevertheless, in the general case inequality (9) insinuates that the Erlang capacity may not be monotonically increasing. This can easily be verified numerically. The typical behavior of Erlang capacity in the controlled system is a saw-tooth curve, as shown in Fig. 1, resulting from the antagonism between the increase of integer capacity and load. We see that the sustainable load has a general trend to increase when transmission rates are decreased, as well as that the behavior is not monotonic.

---

[1] The calculations also lead to a $\leq$-condition; however, there exists then no single value of $\alpha > 0$ such that $r_Z(m) \geq r_X(m) \ \forall m$.

Depending on the initial state and system conditions, the boost in Erlang capacity can be significant. For instance, the Erlang capacity is increased almost 2.5 times from $\alpha = 1$ to $\alpha = 10$ in the graph. The steeper increase for smaller values of $\alpha$ also suggests that high initial rates are preferable.

In the limit $\alpha \to \infty$ there exists a finite bound on the Erlang capacity almost surely. We deduce this as follows. Take the limit of the load to the integer capacity of the system,

$$\rho^* := \lim_{\alpha \to \infty} \frac{\rho \alpha}{\left[ \eta \left( \cdot \frac{\alpha}{\Delta} - a + 1 \right) \right]} = \frac{\rho \Delta}{\eta (1 - \Delta)} .$$

Then a known result for Erlang loss systems (also cited in [9]) is that the blocking probability becomes:

$$\lim_{\alpha \to \infty} P_B(\alpha) = \begin{cases} 0 , & \text{if } \rho^* \leq 1 \\ 1 - 1/\rho^* , & \text{if } \rho^* > 1 . \end{cases}$$

The positive probability case yields $P_B = \lim_{\alpha \to \infty} P_B(\alpha) = 1 - \frac{\eta(1-\Delta)}{\rho \Delta}$. Thus if we were to let the load $\rho \to \infty$ the blocking probability would become 1 and the Erlang capacity has a finite bound almost surely. In the next subsection, a similar result for the blocking probabilities will be derived for the multiservice system.

The integer and Erlang capacity of the link become zero when bit rates increase inordinately. The maximum allowed bit rate on the link corresponds to the greatest lower bound of values of $\alpha$, for which any smaller value would lead to the integer capacity being zero.[2] Note here that even in the best case scenario of a mobile transmitting alone in a cell the rate cannot be further increased. Since in CDMA a signal is not distinguished from noise and (intercell) interference *a priori*, condition (4) arising from the interference-limited system must still be satisfied. This bounds the feasible bit rate and drives the Erlang capacity to become zero.

Overall, results in this section show that it is not possible to obtain any notable increase of Erlang capacity by increasing bit rates; the slight gain that occurs can always be outweighted by lowering these rates.

## 4.2    Multiple Classes of Service

Let us now return to the case of multiple classes of service, each with a different rate of transmission. The analysis with stochastic comparisons is much more difficult in this case (for instance, it is not possible to follow the same approach as in Proposition 1). Although monotonicity properties have been shown for the stochastic knapsack (see [9] for some main results, but also [8]), we are dealing with the *dual scaling problem* here: the knapsack capacity remains the same and the individual knapsack values decrease. What's more, these generally assume

---

[2] In reality, the maximum transmission rate can be more restrained, depending on the type of modulation used and the chip rate.

non-integral values. Due to the above difficulties we only present some analytical results for the asymptotic case, for which the two problems become, as we shall see, equivalent.

Nevertheless, a similar behavior should be anticipated, something that is confirmed by a numerical investigation. Fig. 2 shows an example of the blocking probability of two classes of flows as the transmission rate decreases with $\alpha$. The blocking probability of each class flow decreases with $\alpha$ in a saw-tooth fashion, which gradually diminishes. Of course, the fact that the blocking probability of a class-$s$ flow decreases means that the load of *some, or all* sources can be increased. Therefore according to Definition 2 the multiservice Erlang capacity of the system will be increased. The rapid fall of the curve also suggests that the increase in Erlang capacity can be significant.

**Zero and Non-zero Blocking Conditions.** We consider the asymptotic analysis of the multiservice system as $\alpha \to \infty$, $\alpha \in \mathbb{N}$, and study a condition for zero and non-zero blocking probability of any class flows.

The condition is derived by observing the equivalence of the asymptotic analysis in our case with that of the original stochastic knapsack problem [9]. The limiting regime considered therein is one with a sequence of "system instances" indexed by $N$, whereupon the load of each class and knapsack capacity are increased in line with one another, such that the following limit exists:

$$\rho_s^* = \lim_{N \to \infty} \frac{\rho_s(N)}{\eta(N)} \ , \quad s \in \mathcal{K} \ .$$

In our setting, we have that as $\alpha$ increases, the load of a class-$s$ flow in the system increases, $\rho_s(\alpha) = \alpha \rho_s$. Further, since $\tilde{\Delta}_s = \frac{E_b^{(s)}}{I_0 W} R_s$, we have $\tilde{\Delta}_s(\alpha) = \frac{\tilde{\Delta}_s}{\alpha}$, the resource demands of all users decrease. It is also $\Delta_s(\alpha) = \frac{\tilde{\Delta}_s(\alpha)}{1 + \tilde{\Delta}_s(\alpha)} = \frac{\tilde{\Delta}_s}{\alpha + \tilde{\Delta}_s}$ so that in the limit $\alpha \to \infty$, $\Delta_s(\alpha) \to \frac{\tilde{\Delta}_s}{\alpha}$. Hence, the condition $\sum_{s=1}^{K} m_s \Delta_s(\alpha) \leq \eta$ becomes, as $\alpha \to \infty$: $\sum_{s=1}^{K} m_s \tilde{\Delta}_s \leq \alpha \eta$.

Thus, in the limit the problem is equivalent to the one with increasing the knapsack capacity by a factor $\alpha$. Then we can apply known results for the blocking probability in the limiting regime. Let $\rho_s^* := \lim_{\alpha \to \infty} \frac{\alpha \rho_s}{\alpha \eta} = \frac{\rho_s}{\eta}$, and $\rho^* := \sum_{s=1}^{K} \tilde{\Delta}_s \rho_s^*$.

Then

$$\sum_{s=1}^{K} \tilde{\Delta}_s \rho_s(\alpha) - \eta = \beta \sqrt{\alpha} + o(\sqrt{\alpha}) \ , \quad -\infty < \beta < \infty \ .$$

Then from [9] (see [6] for the proof) in the limit it is

$$\lim_{\alpha \to \infty} P_B^s(\alpha) = \begin{cases} 0 \ , & \rho^* \leq 1 \\ 1 - (1 - \zeta^*)^{\tilde{\Delta}_s} \ , & \rho^* > 1 \ , \end{cases} \tag{10}$$

where $\zeta^*$ is the unique solution to $\sum_{s=1}^{K} \tilde{\Delta}_s \rho_s^* (1 - \zeta^*)^{\tilde{\Delta}_s} = 1$.

Since $\tilde{\Delta}_s = R_s\delta_s$ and $\nu_s = R_s\rho_s$, the condition $\rho^* \le 1$ is equivalent to $\sum_{s=1}^{K} \delta_s\nu_s \le \eta$.

Further, if $\rho^* \to \infty$ for *any* $s \in \mathcal{K}$, a unique solution exists only when $\zeta^* \to 1$, in which case the blocking probability becomes 1 for any class flow. Hence the load of the system cannot be increased indefinitely and as in the special case of homogeneous service there exists a finite bound almost surely.

Finally, for the case of a *critically loaded* regime, $\rho^* = 1$, where the load in the system very nearly matches capacity, the boundary of the admissible region satisfies [7]:

$$\sum_{s=1}^{K} \tilde{\Delta}_s\rho_s^* = 1 + O(\frac{1}{\sqrt{\alpha}}) \; , \tag{11}$$

which in the limit $\alpha \to \infty$ lies in a hyperplane. This result derives straightforwardly, since the analysis in [7] for the critically loaded regime holds also for non-integer knapsack values. For overloaded or underloaded regimes ($\rho^* > 1$, $\rho^* < 1$ respectively) it does not, and analogous results can be subject of a future investigation. Most importantly, further analysis is required to check what happens in the case where $\alpha \gg 1$, but not infinite. An open problem then is to investigate on the convexity of the admissible region, since convexity results are particularly important from a practical dimensioning viewpoint.

## 5    Energy Consumption

Rate control actions affect, apart from the Erlang capacity of the system, transfer delay of data and the energy consumption of a mobile device. It is understood that the transfer time may grow inordinately when the transmission rate decreases, and upper bounds should be set according to QoS constraints for each different type of data traffic.

In this section we focus on the effect on energy consumption which, especially in the uplink and for a mobile device with small battery, cannot be neglected. Sending packets consumes much more energy than receiving and can be significant when compared to the cost of being idle, particularly when a mobile is sending large amounts of data.

It can easily be shown that in saturation conditions, decreasing the rate of transmission increases the amount of energy required to maintain the same $E_b/I_0$. Consider two system settings, an initial system and the one that occurs by decreasing the rates of transmission by a factor $\alpha$. Following similar steps as in the analysis of Section 2, for a certain power to interference target at the system demodulator, the minimum received power from a mobile of class $s$ in the two settings must satisfy, in saturation conditions:

$$P_s = \frac{N\Delta_s}{1 - L_{max}} = \frac{N\dfrac{R_s\delta_s}{1 + R_s\delta_s}}{1 - L_{max}} \; , \qquad P_s(\alpha) = \frac{N\Delta_s(\alpha)}{1 - L_{max}} = \frac{N\dfrac{\frac{R_s}{\alpha}\delta_s}{1 + \frac{R_s}{\alpha}\delta_s}}{1 - L_{max}} \; .$$

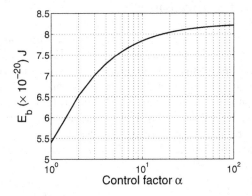

**Fig. 3.** Required received energy per bit as the transmission rate decreases

This is the average received signal power. Multiplying by the bit duration, $T_{b,s} = 1/R_s$, and taking into account that $T_{b,s}(\alpha) = \alpha T_{b,s}$ we get for the required received *energy per bit* for a mobile of class $s$ in the two systems:

$$E_b^{(s)} = \frac{N\delta_s}{(1 - L_{max})(1 + R_s\delta_s)} \quad , \quad E_b^{(s)}(\alpha) = \frac{N\delta_s}{(1 - L_{max})(1 + \frac{R_s}{\alpha}\delta_s)} \quad . \quad (12)$$

Thus the two last equations show that a higher energy consumption is required for slower transmission rates. Although slower rates yield relatively less power, the increased bit duration increases the overall $E_b$.

An example in Fig. 3 depicts the variability of the received energy per bit with the rate of transmission. The setting is as follows: $L_{max} = 0.5$ (corresponding to an interference margin of 3 dB), $N = -100$ dBm, $W = 3.84$ Mcps, $R_s = 1280$ kbps and $E_b^{(s)}/I_0 = 2$ dB. We have assumed here that the knapsack is filled exactly, i.e., $(1 + f)\sum_{s=1}^{K} m_s\Delta_s = L_{max}$.[3] Then the energy per bit is an increasing, concave function of $\alpha$. It also reaches a limit as $\alpha$ approaches infinity, $\lim_{\alpha\to\infty} E_b^{(s)}(\alpha) = \frac{N\delta_s}{1 - L_{max}}$ .

*Remark 3.* This increased energy demand may be compensated by changes in the coding and transmission scheme. Specifically, in a communication system it is known that lowering the number of bits per transmitted symbol reduces the $E_b$ for a certain BER (see e.g., [10]). Such an effect can be had in CDMA by changing the modulation scheme (e.g., from QPSK to BPSK). For a pseudonoise signal this will decrease the required energy over a chip period, and hence the required energy per bit.

---

[3] This is approximated when the granularity of the system increases, meaning there being a large number of transmissions with small resource demands. In a different case, the $E_b$ increases in a saw-tooth fashion.

## 6    Summary and Conclusions

In this work we have investigated the potential increase in Erlang capacity of a CDMA link by controlling the transmission rates of data traffic. While it might be expected that both increasing and decreasing the rates can return a benefit, a closer look shows that a considerable increase is only feasible in the latter case. The achieved gain then depends on the initial state and system conditions, and is greater for higher initial rates. But while from a system point of view the increase in Erlang capacity is desirable, undesirable effects will occur when the transmission rates are decreased inordinately, in terms of transfer delay and energy consumption.

The results in the paper reveal this multi-objective optimization problem involving capacity, delay and energy, and provide useful information for the design of an offline or online (dynamic) control scheme. Since a delay and energy cost improvement, as well as a steeper increase in capacity is observed for higher rates, for a system that can assign rates dynamically it would be suggested to offer high initial rates, and in case of congestion conditions decrease these rates to increase the Erlang capacity, to the extent that this is permitted by upper bounds on transfer delay, and while taking into account the increase in energy consumption.

## References

1. Altman, E.: Capacity of multi-service cellular networks with transmission-rate control: A queueing analysis. In: Proc. ACM Mobicom (2002) 205–214
2. Franken, P., König, D., Arndt, U., Schmidt, V.: Queues and Point Processes. John Wiley & Sons (1982)
3. Hiltunen, K., De Bernardi, R.: WCDMA downlink capacity estimation. In: Proc. IEEE VTC Spring (2000) 992–996
4. Holma, H., Toskala, A. (eds.): WCDMA for UMTS: Radio access for third generation mobile communications. 3rd edn., John Wiley & Sons (2004)
5. Kaufman, J.S.: Blocking in a shared resource environment. IEEE Trans. Commun. **29** (1981) 1474–1481
6. Kelly, F.P.: Blocking probabilities in large circuit-switched networks. Adv. Appl. Prob. **18** (1986) 473–505
7. Morrison, J.A., Mitra, D.: Asymptotic shape of the Erlang capacity region of a multiservice shared resource. SIAM J. Appl. Math. **64** (2003) 127–151
8. Nain, P.: Qualitative properties of the Erlang blocking model with heterogeneous user requirements. Queueing Systems **6** (1990) 189–206
9. Ross, K.W.: Multiservice Loss Models for Broadband Telecommunications Networks. Springer-Verlag (1995)
10. Uysal-Biyikoglou, E., Prabhakar, B., El Gamal, A.: Energy-efficient packet transmission over a wireless link. IEEE/ACM Trans. Networking **10** (2002) 487–499
11. Viterbi, A.M., Viterbi, A.J.: Erlang capacity of a power controlled CDMA system. IEEE J. Selected Areas in Commun. **11** (1993) 892–900
12. Viterbi, A.J.: CDMA: Principles of Spread Spectrum Communication. Addison-Wesley (1995)

# Dynamic Time-Threshold Based Scheme for Voice Calls in Cellular Networks

Idil Candan and Muhammed Salamah

Computer Engineering Department, Eastern Mediterranean University,
Gazimagosa, TRNC, Mersin 10 Turkey
idil.candan@emu.edu.tr,
muhammed.salamah@emu.edu.tr

**Abstract.** This paper presents a dynamic bandwidth allocation scheme for new and handoff voice calls in cellular networks. The main idea of the new scheme is based on monitoring the elapsed real time of handoff calls and according to both a time threshold ($t_e$) and a dynamically changing bandwidth threshold ($B_t$) parameters, a handoff call is either prioritized or treated as a new call. Also in this paper, we introduce a crucial general performance metric Z that can be used to measure the performance of different bandwidth allocation schemes and compare them. Z, which is a performance/cost ratio, is a function of the new call blocking probability, handoff call dropping probability and system utilization all together. The results indicate that our scheme outperforms other traditional schemes in terms of performance/cost ratio, and maintains its superiority under different network circumstances.

## 1 Introduction

With the increasing popularity of wireless communication systems, a satisfactory level of quality of service (QoS) should be guaranteed in order to manage the incoming calls more efficiently. However, establishment and management of connections are crucial issues in QoS-sensitive cellular networks due to user mobility. A wireless cellular network consists of a large number of cells, mobile users with various movement patterns and a number of applications. A cell involves a base station (BS) and a number of mobile stations (MSs). Handoff is the mechanism that transfers an ongoing call from one cell to another as a user moves through the coverage area of a cellular system. In literature, several approaches, like in [1-6] have been proposed to give priorities to handoff calls. In [1], resource reservation is made according to the staying time and switching time of an MS in order to increase bandwidth utilization. Static guard channel scheme (SGCS) in [2] gives priority to handoff calls by exclusively reserving fixed number of channels for handoff calls. Although SGCS decreases handoff dropping probability ($P_d$), it increases new call blocking probability ($P_b$) and may not utilize the system efficiently. The dynamic guard channel scheme (DGCS) in [2] is a variant of SGCS scheme where the number of guard channels is changed dynamically in order to improve the performance of the system. On the other hand,

Y. Koucheryavy, J. Harju, and V.B. Iversen (Eds.): NEW2AN 2006, LNCS 4003, pp. 189–199, 2006.

according to the fully shared scheme (FSS), discussed in [2], all available channels in the cell are shared by handoff and new calls. Thus, FSS scheme minimizes the new call blocking probability and maximizes system utilization. However, it is difficult to guarantee the required dropping probability of handoff calls. Usually, SGCS and DGCS schemes are preferred by users since they decrease $P_d$, and FSS scheme is preferred by service providers since it maximizes system utilization. As mentioned before, most of the schemes like in [3-15] prioritize handoff calls at the expense of blocking new calls. Their claim is "forced termination of ongoing calls is more annoying than blocking of newly originating calls". We believe that this is true to some extent, as the annoyance is a fuzzy term which depends on the elapsed time of the ongoing call. For example, dropping an ongoing voice call is very annoying if it does not last for a moderate duration, whereas it is not that much annoying if it is approaching to its end. Motivated with these arguments, we introduce a novel dynamic bandwidth allocation scheme for voice calls which is based on fairness among calls and outperforms the SGCS, DGCS and FSS schemes. The main idea of our Dynamic-Time-Threshold-based Scheme (DTTS) is based on monitoring the elapsed real time of voice handoff calls and according to both a time threshold ($t_e$) and a dynamically changing bandwidth threshold ($B_t$) parameters, a handoff call is either prioritized or treated as a new call.

## 2   Dynamic Time-Threshold-Based Scheme (DTTS)

In our scheme, we focus on a single cell as a reference cell in a cellular wireless network. We assume that the arrival traffic at the BS is of 'voice' type, the cell has a total capacity $B$ of 100 bandwidth units (BUs), and each call requires one unit. Fig. 1 below shows the bandwidth allocation of the DTTS scheme for new and handoff voice calls. The bandwidth threshold ($B_t$) dynamically changes between a fixed bandwidth threshold ($B_{tfixed}$) and the total capacity of the system ($B$) according to the time threshold $t_e$ values (i.e. $B_{tfixed} < B_t < B$). A new voice call or a non-prioritized handoff voice call is served if the amount of occupied bandwidth is less than $B_t$ BUs upon its arrival. However, a prioritized handoff voice call is served as long as the amount of occupied bandwidth is less than the total capacity which is $B$ BUs.

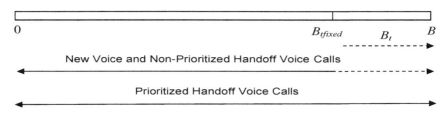

**Fig. 1.** Bandwidth Allocation of the DTTS Scheme

Fig.2 below shows the flowchart of processing a voice call in the DTTS scheme.

According to DTTS, handoff calls that have elapsed real time smaller than time threshold $t_e$ are prioritized. That is, such handoff calls are accepted as long as the amount of occupied bandwidth in the cell is smaller than the total capacity, $B$. On the other hand, handoff calls that have elapsed real time greater than or equal to time threshold $t_e$, and new calls are treated according to the guard policy. That is, such calls are accepted as long as the amount of occupied bandwidth in the cell is less than a bandwidth threshold $B_t$, where $B_t$ dynamically changes depending on the $t_e$ values. If the time threshold $t_e$ is large, the number of prioritized calls increases and the number of non-prioritized calls decreases. Therefore, the bandwidth threshold $B_t$ may be decreased. On the contrary, if the time threshold $t_e$ is small, the number of prioritized calls decreases and the number of non-prioritized calls increases. Therefore, the bandwidth threshold $B_t$ may be increased in order to increase the performance of the system. Since $B_t$ is directly proportional to $t_e$, it can be defined as

$$B_t = \frac{B_{t\,fixed} - B}{2\mu_n} \cdot t_e + B \qquad (1)$$

where $B_{tfixed}$ is taken as 90 units [16] and $\mu_n$ is the average service rate of a call. $B_t$ is dynamically changing between $B_{tfixed}$ and $B$ units.

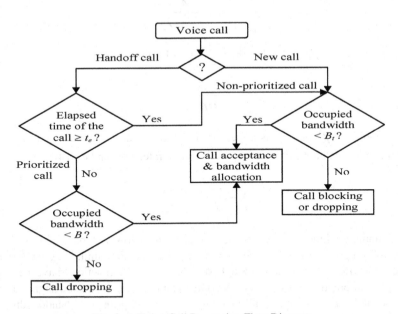

**Fig. 2.** A Voice Call Processing Flow Diagram

In our previous work [17], we proposed a static time-threshold-based scheme (STTS) for voice calls where handoff voice calls are prioritized or treated as new calls according to a time threshold parameter $t_e$. In the STTS scheme, prioritized handoff calls are accepted as long as the amount of occupied bandwidth is smaller than total capacity, $B$. However, new calls and non-prioritized handoff calls are accepted as long as the amount of occupied bandwidth is smaller than a fixed bandwidth threshold $B_t$. In the STTS scheme, the bandwidth threshold $B_t$ is *not changed dynamically* depending on the time threshold $t_e$. That is, bandwidth threshold $B_t$ is fixed ($B_t = 90$ units) in the STTS scheme. However, in the DTTS scheme presented in this paper, $B_t$ is dynamically varying between $B_{tfixed}$ and $B$ bandwidth units as illustrated in both Fig. 1 and equation (1).

## 3   Simulation Parameters and Performance Metrics

The simulation has been performed using the Ptolemy simulation tool, developed by the University of California at Berkley [18]. During simulation, more than 30 runs are taken for each point in order to reach 95% confidence level. As we mentioned before, we concentrate on a single reference cell. The interarrival times of new voice, handoff voice calls are assumed to follow Poisson processes with means $1/\lambda_n$ and $1/\lambda_h$ respectively. The call holding times also follow exponential distribution with means $1/\mu_n$ for new calls, and $1/\mu_h$ for handoff calls.

The average service time is defined as

$$\frac{1}{\mu} = \frac{\lambda_h}{\lambda_h + \lambda_n} \cdot \frac{1}{\mu_h} + \frac{\lambda_n}{\lambda_h + \lambda_n} \cdot \frac{1}{\mu_n} \tag{2}$$

The normalized offered load of the system (in Erlang) is defined as

$$\rho = \frac{\lambda_n + \lambda_h}{B\mu} \tag{3}$$

The mobility ($\gamma$) of calls is a measure of terminal mobility and is defined as the ratio of handoff call arrival rate to new call arrival rate, and can be written as

$$\gamma = \frac{\lambda_h}{\lambda_n} \tag{4}$$

The simulation input parameters used are given in following table (Table 1).

The design goals of handoff schemes should also include minimizing the Grade of Service (GoS) cost function. Although sophisticated cost functions have been proposed [19], in practice, a simple weighted average is useful for most design purposes. The weighted sum of the new call blocking probability, prioritized handoff dropping probability ($P_{d1}$) and non-prioritized handoff dropping probability ($P_{d2}$) is introduced as a measure of grade of service (GoS) and can be defined as

$$GoS = P_b + k \cdot P_{d1} + P_{d2} \tag{5}$$

**Table 1.** Simulation Parameters

| Mobility($\gamma$) | 3 |
|---|---|
| Load ($\rho$) | 0.9 Erlangs |
| Time threshold values ($t_e$) | 30, 60, 90, 120, 150 and 180sec. |
| Total bandwidth ($B$) | 100 units |
| Bandwidth threshold ($B_t$) | 91 – 99 units |
| Fixed bandwidth threshold ($B_{tfixed}$) | 90 units |
| Arrival rate of handoff calls ($\lambda_h$) | 0.6 calls/sec |
| Arrival rate of new calls ($\lambda_n$) | 0.2 calls/sec |
| New call average service time ($1/\mu_n$) | 180 sec. (Exp. Dist.) |
| Handoff call average service time ($1/\mu_h$) | 180 sec. (Exp. Dist.) |
| Average elapsed time of a handoff call | 90 sec. (Unif. Dist.) |

where $k$ is the penalty factor used to reflect the effect of the handoff dropping over the new call blocking in the *GoS* cost function. A penalty of 5 to 20 times is commonly recommended [20]. In accordance with our proposed scheme, we used the penalty for the prioritized handoff calls, whereas the non-prioritized handoff calls have the same weight as the new calls. Of course, from the mobile user's point of view, the objective is to minimize the *GoS* cost function in order to improve the performance of the system. Therefore, the performance of a system can be defined as

$$Performance = \frac{1}{GoS} \tag{6}$$

Another objective, from the service provider's perspective is to decrease the cost by increasing utilization of the system. Therefore, the cost of a system can be defined as

$$Cost = \frac{1}{Utilization} \tag{7}$$

In order to make a fair balance between both user satisfaction and service provider satisfaction, a crucial performance metric $Z$ is introduced to measure the performance of different bandwidth allocation schemes and compare them. $Z$ can be defined as

$$Z = \frac{Performance}{Cost} \qquad (8)$$

It is clear that, $Z$ is a function of new call blocking probability, handoff dropping probability and system utilization all together. Of course, the design goals of a handoff scheme are increasing the performance and decreasing the cost, which means maximizing $Z$.

## 4 Performance Results

The proposed scheme is evaluated for different bandwidth threshold ($B_t$) values. The performance measures obtained through the simulation are the blocking probability of new voice calls ($P_b$), the total dropping probability of handoff voice calls ($P_d$) which includes both prioritized and non-prioritized handoff voice call dropping probabilities, grade of service (GoS) and $Z$.

Fig. 3 below shows the relationship between $B_t$ and $t_e$ values that are calculated according to equation (1). It is clearly seen that $B_t$ decreases as $t_e$ increases. This is because, as $t_e$ increases, the number of prioritized handoff calls increases and therefore the guard bandwidth units should be enlarged to reserve enough bandwidth for these calls which can be achieved by decreasing $B_t$ .

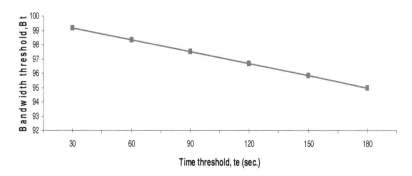

**Fig. 3.** Bandwidth threshold ($B_t$) versus time threshold $t_e$

Fig. 4 below shows the new call blocking probability ($P_b$) versus bandwidth threshold $B_t$ for different time threshold ($t_e$) values. It is seen that as $B_t$ increases, $P_b$ decreases for all $t_e$ values. Since the number of guard bandwidth units reserved for prioritized handoff calls decreases as $B_t$ increases, and therefore the chance of accepting a new call increases. The highest and lowest $P_b$ is obtained for $t_e$=90 sec and $t_e$=30 sec. The reason is that, when $t_e$=90 sec. the number of prioritized handoff calls is high compared to the number of prioritized handoff calls when $t_e$=30 sec. That is, when $t_e$=90 sec, the number of guard bandwidth units reserved for prioritized handoff calls is higher than those reserved when $t_e$=30 sec. It is worth to mention that there is no serious degradation in terms of $P_b$ for all $t_e$ values.

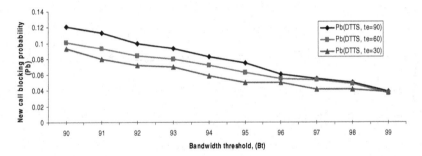

**Fig. 4.** New call blocking probability ($P_b$) versus bandwidth threshold ($B_t$) for different time threshold ($t_e$) values

Fig. 5 below shows the handoff call blocking probability ($P_d$) versus bandwidth threshold $B_t$ for different time threshold ($t_e$) values. It is seen that as $B_t$ increases, $P_d$ decreases for all $t_e$ values. This result is interesting since we know that static or dynamic guard channel schemes(SGCS and DGCS) improves $P_d$ at the expense of increasing $P_b$ and degrading system utilization. On the other hand, fully shared scheme (FSS) improves $P_b$ and system utilization at the expense of increasing $P_d$. However, our dynamic time-threshold scheme shows improvements in terms of both handoff dropping ($P_d$) and new call blocking ($P_b$) probabilities. From Fig.4 and Fig.5, it is also seen that the DTTS scheme shows good performance in terms of $P_b$ and $P_d$ for $t_e \leq 90$ sec. Therefore, it is recommended to set $t_e$ to a value less than or equal to the average service time of a handoff call, $(\frac{1}{2})\mu_h$.

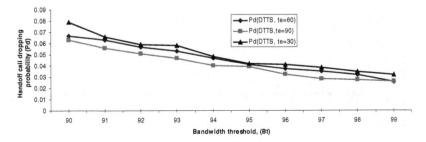

**Fig. 5.** Handoff call dropping probability ($P_d$) versus bandwidth threshold ($B_t$)  for different time threshold ($t_e$) values

Fig. 6 below shows the grade of service (GoS) cost function versus bandwidth threshold $B_t$ for DGCS, DTTS and FSS schemes for k=10. The FSS gives the worst performance and it is almost constant as expected. It is very clear that our DTTS scheme gives the best GoS compared with the others. The DTTS scheme shows improvement over the DGCS and FSS schemes for all $B_t$ values. For example, DTTS improvements over DGCS and FSS schemes reach 135% and 230% at $B_t = 97$ units respectively. It is worth to mention that similar trends of GoS as in Fig. 5 are observed for other values of $k$ and better results are obtained for $t_e$=30 and $t_e$=60 sec.

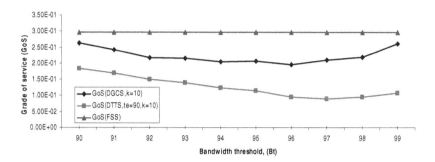

**Fig. 6.** Grade of service (GoS) versus bandwidth threshold ($B_t$) for different schemes

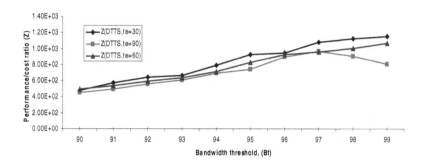

**Fig. 7.** $Z$ versus bandwidth threshold ($B_t$) for dynamic TTS scheme (DTTS)

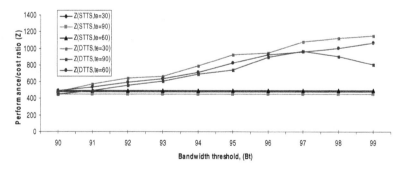

**Fig. 8.** $Z$ versus bandwidth threshold ($B_t$) for dynamic TTS scheme (DTTS) and fixed TTS scheme (STTS)

Fig. 7 below shows the performance/cost ratio ($Z$) versus bandwidth threshold $B_t$ for DTTS scheme under different $t_e$ values. It is clearly observed that, $Z$ has the highest value when $t_e = 30$ sec. It is worth to mention that $Z$ does not show a serious deterioration for other $t_e$ values.

Fig. 8 below shows the performance/cost ratio ($Z$) versus bandwidth threshold $B_t$ for static time-threshold (STTS) and dynamic time-threshold (DTTS) schemes for different $t_e$ values. It is clear that the $Z$ value of DTTS scheme is higher than $Z$ value of STTS scheme for all $t_e$ values.

Fig. 9 shows the performance/cost ratio ($Z$) versus bandwidth threshold $B_t$ for SGCS, DGCS, FSS and DTTS schemes. It is clear that, our DTTS scheme outperforms the other schemes, since it has the highest $Z$ for all $B_t$ values. As expected the $Z$ value of the FSS is constant with respect to $B_t$, because all channels are shared between new and handoff calls. The $Z$ value of the SGCS is also constant since it has a fixed $B_t$ value which is 90 units. The DTTS scheme maintains its superiority over SGCS, DGCS and FSS scheme for all $B_t$ values. For example, DTTS improvements over SGCS, DGCS and FSS schemes reaches 200%, 134% and 228% at $B_t$ =97 units respectively. It is worth to mention that the DTTS scheme maintains its superiority for other values of $t_e$.

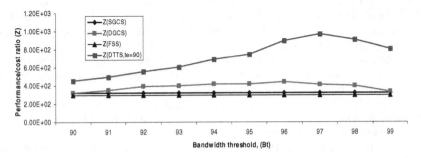

**Fig. 9.** $Z$ versus bandwidth threshold ($B_t$) for dynamic TTS (DTTS), dynamic GCS (DGCS), static GCS scheme (SGCS) and fully shared (FSS) schemes

# 5  Conclusion

In this paper, we have proposed and analyzed the performance of a new dynamic time-threshold based bandwidth allocation scheme (DTTS) for voice calls in cellular networks. The proposed DTTS scheme relies on fairness among new calls and handoff calls that may tolerate uncritical dropping. It is well known that the SGCS and DGCS aim to increase user satisfaction by decreasing the dropping probability; and the FSS aims to increase service provider satisfaction by decreasing the blocking probability and increasing bandwidth utilization. It is concluded that the proposed scheme outperforms these two extreme schemes. Hence, both users and service providers will be satisfied in the DTTS scheme. We also introduced a general performance metric $Z$ as a performance/cost ratio that can be used to compare different bandwidth allocation schemes. The results show that the DTTS scheme maintains its superiority over STTS, SGCS, DGCS and FSS scheme for all bandwidth threshold values. It is worthy to mention that DTTS can be implemented easily since it uses the elapsed real time of a call which is already recorded in all systems for billing and other purposes.

There are a number of issues that will be addressed in our future research. We are currently working on the analytical modeling of the proposed DTTS scheme. What is more, since the proposed scheme is applicable for FDMA systems, we are going to improve it in order to be applicable for CDMA systems. In addition, we are going to apply a reverse strategy of the proposed DTTS scheme for data handoff calls, such that the data handoff calls that have elapsed time greater than time threshold $t_e$ are going to be prioritized. This is because; dropping an ongoing data call is very annoying if it is approaching to its end, whereas it is not that much annoying if it has just started. We believe that our scheme can be considered as a tool for service-balance in multimedia wireless networks.

# References

1. G. S. Kuo and P.C. Ko: A Probabilistic Resource Estimation and Semi-Reservation Scheme for Flow-Oriented Multimedia Wireless Networks. IEEE Communications Magazine 39 (2001) 135–141
2. I. Katzela and M. Naghshineh: Channel Assignment Schemes for Cellular Mobile Telecommunications Systems: A Comprehensive Survey. IEEE Personal Communications Magazine, 3(3) (1996) 10-31
3. C. Oliveria, J. B. Kim and T. Suda: An Adaptive Bandwidth Reservation Scheme for High-Speed Multimedia Wireless Networks. IEEE J. Select. Areas Communications, 16(6) (1998) 858-874.
4. J. Y. Lee, J. G. Choi, K. Park and S. Bahk: Realistic Cell-Oriented Adaptive Admission Control for QoS Support in Wireless Multimedia Networks. IEEE Transactions on Vehicular Technology, 52(3) (2003) 512-525
5. Y. Kim, D. Lee and B. Lee: Dynamic Channel Reservation based on Mobility in Wireless ATM networks. IEEE Communications Magazine, 37(11) (1999) 47-51
6. Z. Xu, Z. Ye, S. V. Krishnamurthy, S. K. Tripathi and M. Molle: A New Adaptive Channel Reservation Scheme for Handoff Calls in Wireless Cellular Networks. Proc. of NETWORKING 2002,Technical Committee "Communications Systems" of International Federation for Information Processing (IFIP-TC6),Pisa-Italy, (2002) 672-684.
7. C. Chou and K.G. Shin: Analysis of Adaptive Bandwidth Allocation in Wireless Networks with Multilevel Degradable Quality of Service. IEEE Trans. Mobile Computing 3(1) (2004).
8. C. Chou and K.G.Shin: Analysis of Combined Adaptive Bandwidth Allocation and Admission Control in Wireless Networks. Proc. IEEE Infocom ,NewYork, (2002).
9. J. Wang, Q. Zeng and D. P. Agrawala: Performance Analysis of a Preemptive and Priority Reservation Handoff Scheme for Integrated Service-Based Wireless Mobile Networks. IEEE Transactions on Mobile Computing, 2(1) (2003) 65-75
10. K. Lee and S. Kim: Optimization for Adaptive Bandwidth Reservation in Wireless Multimedia Networks. The International Journal of Computer and Telecommunications Networking 38, (2002) 631-643.
11. W. K. Lai, Y. Jin, H. W. Chen and C. Y. Pan: Channel Assignment for Initial and Handoff Calls to Improve the Call-Completion Probability. IEEE Trans. On Vehicular Tech, 52 (4), (2003) 876-890.
12. D. Lee and T. Hsueh: Bandwidth-Reservation Scheme Based on Road Information for Next-Generation Cellular Networks. IEEE Trans. On Vehicular Technology 53 (1) (2004) 243-252.
13. J. Hou and Y. Fang: Mobility-Based Call Admission Control Schemes for Wireless Mobile Networks. Wireless Communications and Mobile Computing 2001,1(3), (2001) 269-282.

14. F. Hu and N. Sharma: Priority-Determined Multiclass Handoff Scheme with Guaranteed Mobile QoS in Wireless Multimedia Networks. IEEE Trans. On Vehicular Tech., **53**(1) 2004 (118-135).

15. A. Olivre: Call Admission Control and Dynamic Pricing in a GSM/GPRS Cellular Network.a dissertation submitted to University of Dublin, 2004.

16. L. Huang, S. Kumar and C. C. Kuo: Adaptive Resource Allocation for Multimedia QoS Management in Wireless Networks. IEEE Transactions on Vehicular Technology, **53**(2) (2004) 547-558

17. M. Salamah, and Idil Candan: A Novel Bandwidth Allocation Strategy for Voice Handoff Calls in Cellular Networks. Proc. of the 9th CDMA International Conference, (2004) 381

18. PTOLEMY simulation package: http://ptolemy.eecs.berkeley.edu/

19. ETSI Standard ETR 310,, Radio Equipment and Systems; Digital Enhanced Cordless Telecommunications; Traffic Capacity and Spectrum Requirements, (1996)

20. F. Barcelo: Performance Analysis of Handoff Resource Allocation Strategies through the State-Dependent Rejection Scheme. IEEE Transactions on Wireless Communications, **3**(3) (2004) 900-909

# Queue Size in a BMAP Queue with Finite Buffer

Andrzej Chydzinski

Silesian University of Technology,
Institute of Computer Sciences,
Akademicka 16, 44-100 Gliwice, Poland
Andrzej.Chydzinski@polsl.pl
http://pp.org.pl/andych/

**Abstract.** The paper presents an analysis of the queue size distribution in a finite-buffer queue fed by a BMAP (Batch Markovian Arrival Process). In particular, the time-dependent and steady-state solutions are given as closed-form formulas. In addition, computational issues are discussed and a numerical example based on IP traffic is presented[1].

**Keywords:** teletraffic modeling, finite-buffer queue, BMAP, performance evaluation.

## 1 Introduction

The batch Markovian arrival process (BMAP) is a tool of choice for traffic modeling and predictability, performance evaluation of buffering processes, congestion and admission control mechanisms etc.

It is worth recommending for several reasons. Firstly, it is able to mimic the self-similar and bursty nature ([1, 2]) of network traces, remaining analytically tractable due to its Markovian structure.

Secondly, the BMAP generalizes a wide set of processes used in teletraffic modeling. For instance, by setting the proper parameterization of the BMAP we may obtain a Poisson process, a batch Poisson process, a Markov-modulated Poisson Process (MMPP), a Markovian arrival process (MAP) or a phase-type renewal process. Some of them are classic, others, like MMPP, gained great attention in applications connected with multimedia and ATM [3]–[7]. All results obtained for the BMAP can be automatically used for the processes mentioned above.

Thirdly, the BMAP is successfully used for the modeling of aggregated IP traffic [8, 9]. In this approach, different lengths of IP packets are represented by BMAP batch sizes. What is important is that algorithms for the fitting of the BMAP parameters to recorded IP traces are available [8, 9]. Some examples of particular performance issues connected with IP networks, analyzed by means of the BMAP, can be found by the reader in [10, 11].

---

[1] This material is based upon work supported by the Polish Ministry of Scientific Research and Information Technology under Grant No. 3 T11C 014 26.

Y. Koucheryavy, J. Harju, and V.B. Iversen (Eds.): NEW2AN 2006, LNCS 4003, pp. 200–210, 2006.

In this paper we deal with the basic characteristic of a single-server queueing system, namely the queue size distribution. The queue size distribution and its parameters (average, variance) play an important role in the performance evaluation of buffering mechanisms in network devices by providing fundamental insight into the system's behaviour.

The main contribution of this paper is the Laplace transform of the transient queue size distribution in the BMAP queue with finite buffer (Theorem 1). To the best of the author's knowledge, there have been no reported results of this type yet. The classic papers by Ramaswami and Lucantoni [12]–[15] are devoted to BMAP queues with infinite buffers. Articles in which the finite-buffer model is investigated are rare and deal with stationary characteristics [16] or special cases of the BMAP, like MMPP [17], only.

The original approach applied in this paper gives the results in a closed form which permits one to easily compute the stationary as well as the time-dependent queue size distribution. All the analytical results obtained herein were checked and confirmed by means of a discrete-event simulator written in OMNET++ [18].

The paper is organized as follows. Firstly, the model of the queue and the arrival process are presented and the notation is listed (section 2). Then, in section 3, the formula for the transform of the transient queue size is proven. Furthermore, some remarks on how it can be used in practice for obtaining time-dependent and stationary queue size distributions are given. In section 4, a numerical illustration based on IP traffic is shown. In addition, computational aspects connected with coefficient matrices occurring in the main formula are discussed. Finally, conclusions are gathered in section 5.

## 2    Queueing Model and Notation

In the paper we investigate a single server queueing system whose arrival process is given by a BMAP. The service time is distributed according to a distribution function $F(\cdot)$, the buffer size (queueing capacity) is finite and equal to $b$ (including service position). In Kendall's notation, the system described is denoted by $BMAP/G/1/b$.

As regards the BMAP, it is constructed by considering a 2-dimensional Markov process $(N(t), J(t))$ on the state space $\{(i,j) : i \geq 0, 1 \leq j \leq m\}$ with an infinitesimal generator $Q$ in the form:

$$
Q = \begin{bmatrix} D_0 & D_1 & D_2 & D_3 & \cdot & \cdot \\ & D_0 & D_1 & D_2 & \cdot & \cdot \\ & & D_0 & D_1 & \cdot & \cdot \\ & & & \cdot & \cdot & \cdot \end{bmatrix},
$$

where $D_k$, $k \geq 0$ are $m \times m$ matrices. $D_k$, $k \geq 1$ are nonnegative, $D_0$ has nonnegative off-diagonal elements and negative diagonal elements and $D = \sum_{k=0}^{\infty} D_k$ is an irreducible infinitesimal generator (see [13]). It is assumed that $D \neq D_0$. Variate $N(t)$ represents the total number of arrivals in $(0, t)$, while variate $J(t)$ represents the auxiliary state (phase) of the modulating Markov process.

An alternative, constructive definition of a BMAP is the following. Assume the modulating Markov process is in some state $i$, $1 \le i \le m$. The sojourn time in that state has exponential distribution with parameter $\lambda_i$. At the end of that time there occurs a transition to another state and/or the arrival of a batch. Namely, with probability $p_i(j, k)$, $1 \le k \le m$, $j \ge 0$, there will be a transition to state $k$ with a batch arrival of size $j$. It is assumed that:

$$p_i(0, i) = 0, \qquad \sum_{j=0}^{\infty} \sum_{k=1}^{m} p_i(j, k) = 1, \quad 0 \le i \le m,$$

and the relations between parameters $D_k$ and $\lambda_i$, $p_i(j, k)$ are:

$$\lambda_i = -(D_0)_{ii}, \qquad 1 \le i \le m,$$

$$p_i(0, k) = \frac{1}{\lambda_i}(D_0)_{ik}, \qquad 1 \le i, k \le m, \quad k \ne i,$$

$$p_i(j, k) = \frac{1}{\lambda_i}(D_j)_{ik}, \qquad 1 \le i, k \le m, \quad j \ge 1.$$

In the sequel, the following notation will be of use:

$\mathbf{P}(\cdot)$ – the probability

$X(t)$ – the queue size at the moment $t$ (including service position if not empty)

$P_{i,j}(n, t) = \mathbf{P}(N(t){=}n, J(t){=}j \mid N(0){=}0, J(0){=}i)$ – the counting function for the BMAP. $N(t)$ denotes the total number of arrivals in $(0, t)$

$a_{k,i,j}(s){=}\int_0^{\infty} e^{-st} P_{i,j}(k, t) dF(t),$

$f(s) = \int_0^{\infty} e^{-st} dF(t)$ – the transform of the service time distribution

$\delta_{ij}$ – the Kronecker symbol ($\delta_{ij} = 1$ if $i = j$ and 0 otherwise)

In addition, we will be using the following $m \times m$ matrices:

$$I = m \times m \text{ identity matrix,}$$

$$\mathbf{0} = m \times m \text{ matrix of zeroes,}$$

$$A_k(s) = [a_{k,i,j}(s)]_{i,j},$$

$$Y_k(s) = \left[\frac{\lambda_i p_i(k, j)}{s + \lambda_i}\right]_{i,j},$$

$$\overline{D}_k(s) = \left[\int_0^{\infty} e^{-st} P_{i,j}(k, t)(1 - F(t)) dt\right]_{i,j},$$

$$\overline{A}_k(s) = \sum_{i=k}^{\infty} A_i(s),$$

$$B_k(s) = A_{k+1}(s) - \overline{A}_{k+1}(s)(\overline{A}_0(s))^{-1},$$

$$R_0(s) = \mathbf{0}, \quad R_1(s) = A_0^{-1}(s),$$

$$R_k(s) = R_1(s)(R_{k-1}(s) - \sum_{i=0}^{k-1} A_{i+1}(s)R_{k-i}(s)), \quad k \geq 2.$$

$$M_b(s) = R_{b+1}(s)A_0(s) + \sum_{k=0}^{b} R_{b-k}(s)B_k(s) - \sum_{k=b+1}^{\infty} Y_k(s)$$

$$- \sum_{k=0}^{b} Y_{b-k}(s)[R_{k+1}(s)A_0(s) + \sum_{i=0}^{k} R_{k-i}(s)B_i(s)].$$

and column vectors of size $m$:

$$\mathbf{1} = \text{the column vector of 1's,}$$

$$z(s) = ((s + \lambda_1)^{-1}, \dots, (s + \lambda_m)^{-1})^T.$$

## 3   Queue Size Distribution

In a BMAP queue, all the time-dependent characteristics depend on the initial queue size, $X(0)$, and the initial state of the modulating process, $J(0)$. This dependence will be represented by indices $n$, $i$ in the queue size distribution:

$$\Phi_{n,i}(t, l) = \mathbf{P}(X(t) = l | X(0) = n, J(0) = i).$$

Naturally, $l$ and $n$ vary from 0 to $b$ while $i$ varies from 1 to $m$. In the stationary case the dependence on $X(0)$ and $J(0)$ vanishes and we may simply denote the stationary queue size distribution by $p_l$ where

$$p_l = \lim_{t \to \infty} \mathbf{P}(X(t) = l) = \lim_{t \to \infty} \mathbf{P}(X(t) = l | X(0) = n, J(0) = i)$$

and $n$, $i$ can be arbitrary.

The main result of this paper is expressed in terms of the Laplace transform:

$$\phi_{n,i}(s, l) = \int_0^\infty e^{-st} \Phi_{n,i}(t, l) dt,$$

and the column vector representing different initial states of the modulating process:

$$\phi_n(s, l) = (\phi_{n,1}(s, l), \dots, \phi_{n,m}(s, l))^T.$$

**Theorem 1.** *The Laplace transform of the queue size distribution in the BMAP/G/1/b queue has the form:*

$$\phi_n(s, l) = \sum_{k=0}^{b-n} R_{b-n-k}(s)g_k(s, l)$$

$$+ [R_{b-n+1}(s)A_0(s) + \sum_{k=0}^{b-n} R_{b-n-k}(s)B_k(s)]M_b^{-1}(s)m_b(s, l), \qquad (1)$$

*where*

$$g_k(s,l) = \overline{A}_{k+1}(s)(\overline{A}_0(s))^{-1}r_b(s,l) - r_{b-k}(s,l),$$

$$r_n(s,l) = \begin{cases} \mathbf{0} \cdot \mathbf{1}, & \text{if} & l < n, \\ \overline{D}_{l-n}(s) \cdot \mathbf{1}, & \text{if} & n \le l < b, \\ \frac{1-f(s)}{s} \cdot \mathbf{1} - \sum_{k=0}^{b-n-1} \overline{D}_k(s) \cdot \mathbf{1}, & \text{if} & l = b. \end{cases}$$

$$m_b(s,l) = \sum_{k=0}^{b} Y_{b-k}(s)\sum_{i=0}^{k} R_{k-i}(s)g_i(s,l) - \sum_{k=0}^{b} R_{b-k}(s)g_k(s,l) + \delta_{0l}z(s).$$

Proof. Conditioning on the first departure moment we may write for $0 < n \le b$, $1 \le i \le m$:

$$\Phi_{n,i}(t,l) = \sum_{j=1}^{m} \sum_{k=0}^{b-n-1} \int_0^t \Phi_{n+k-1,j}(t-u,l)P_{i,j}(k,u)dF(u)$$

$$+ \sum_{j=1}^{m} \sum_{k=b-n}^{\infty} \int_0^t \Phi_{b-1,j}(t-u,l)P_{i,j}(k,u)dF(u) + \rho_{n,i}(t,l), \qquad (2)$$

where

$$\rho_{n,i}(t,l) = (1-F(t)) \cdot \begin{cases} 0, & \text{if} \quad l < n, \\ \sum_{j=1}^{m} P_{i,j}(l-n,t), & \text{if} \quad n \le l < b, \\ \sum_{j=1}^{m} \sum_{k=b-n}^{\infty} P_{i,j}(k,t), & \text{if} \quad l = b. \end{cases}$$

Similarly, if $n = 0$ then for $1 \le i \le m$ we have:

$$\Phi_{0,i}(t,l) = \sum_{j=1}^{m} \sum_{k=0}^{b} \int_0^t \Phi_{k,j}(t-u,l)p_i(k,j)\lambda_i e^{-\lambda_i u}du$$

$$+ \sum_{j=1}^{m} \sum_{k=b+1}^{\infty} \int_0^t \Phi_{b,j}(t-u,l)p_i(k,j)\lambda_i e^{-\lambda_i u}du + \delta_{0l}e^{-\lambda_i t}. \qquad (3)$$

Applying transforms to (2) and (3) yields

$$\phi_{n,i}(s,l) = \sum_{j=1}^{m} \sum_{k=0}^{b-n-1} a_{k,i,j}(s)\phi_{n+k-1,j}(s,l)$$

$$+ \sum_{j=1}^{m} \sum_{k=b-n}^{\infty} a_{k,i,j}(s)\phi_{b-1,j}(s,l) + \int_0^{\infty} e^{-st}\rho_{n,i}(t,l)dt,$$

and

$$\phi_{0,i}(s,l) = \sum_{j=1}^{m} \sum_{k=0}^{b} p_i(k,j)\phi_{k,j}(s,l)\frac{\lambda_i}{s+\lambda_i}$$

$$+ \sum_{j=1}^{m} \sum_{k=b+1}^{\infty} p_i(k,j)\phi_{b,j}(s,l)\frac{\lambda_i}{s+\lambda_i} + \delta_{0l}\frac{1}{s+\lambda_i},$$

respectively. Next, applying matrix notation we get:

$$\phi_n(s,l) = \sum_{k=0}^{b-n-1} A_k(s)\phi_{n+k-1}(s,l) + \sum_{k=b-n}^{\infty} A_k(s)\phi_{b-1}(s,l) + r_n(s,l), \quad 0 < n \le b, \quad (4)$$

$$\phi_0(s,l) = \sum_{k=0}^{b} Y_k(s)\phi_k(s,l) + \sum_{k=b+1}^{\infty} Y_k(s)\phi_b(s,l) + \delta_{0l}z(s). \quad (5)$$

Denoting $\varphi_n(s,l) = \phi_{b-n}(s,l)$ we may rewrite (4) and (5) as follows:

$$\sum_{k=-1}^{n} A_{k+1}(s)\varphi_{n-k}(s,l) - \varphi_n(s,l) = \psi_n(s,l), \quad 0 \le n < b, \quad (6)$$

$$\varphi_b(s,l) = \sum_{k=0}^{b} Y_{b-k}(s)\varphi_k(s,l) + \sum_{k=b+1}^{\infty} Y_k(s)\varphi_0(s,l) + \delta_{0l}z(s), \quad (7)$$

where $\psi_n(s,l) = A_{n+1}(s)\varphi_0(s,l) - \sum_{k=n+1}^{\infty} A_k(s)\varphi_1(s,l) - r_{b-n}(s,l)$. Now, the system of equations (6) has the following solution:

$$\varphi_n(s,l) = R_{n+1}(s)c(s,l) + \sum_{k=0}^{n} R_{n-k}(s)\psi_k(s,l), \quad n \ge 0, \quad (8)$$

where $c(s,l)$ is a function that does not depend on $n$ (see, for comparison, [19], page 343). Therefore we are left with the task of finding $\varphi_0(s,l)$, $\varphi_1(s,l)$, which is necessary for calculating $\psi_k(s,l)$, and the function $c(s,l)$. Substituting $n = 0$ in (8) we can easily obtain

$$c(s,l) = A_0(s)\varphi_0(s,l), \quad (9)$$

Substituting $n = 0$ in (6) we have

$$\varphi_1(s,l) = (\overline{A_0}(s))^{-1}(\varphi_0(s,l) - r_b(s,l)), \quad (10)$$

which reduces the problem to finding $\varphi_0(s,l)$. Using the boundary condition (7) we get

$$\varphi_0(s,l) = M_b^{-1}(s)m_b(s,l),$$

which finishes the proof.  □

Formula (1) may be used in practice in several ways. First, using the well-known limiting behaviour of the Laplace transform we can easily obtain the stationary queue size distribution:

$$p_l = \lim_{t\to\infty} \mathbf{P}(X(t) = l) = \lim_{s\to 0+} s\phi_b(s,l).$$

Instead of $b$ in $\phi_b(s,l)$, any other initial queue size can be chosen. However, using $b$ is recommended as in this case the formula (1) reduces to its simplest form, namely $\phi_b(s,l) = M_b^{-1}(s)m_b(s,l)$.

Next, we may obtain the average stationary queue size $L = \sum_{k=0}^{b} lp_l$ and all moments, for instance variance $Var = \sum_{k=0}^{b} (l - L)^2 p_l$. Furthermore, we can obtain also the time-dependent queue size distribution, average, variance etc. To accomplish that, we have to invert the Laplace transform presented in (1). For this purpose we may use, for instance, the algorithm based on the Euler summation formula [20].

## 4  Numerical Illustration

In this example we demonstrate the queue size distribution using a BMAP parameterization based on measurements of aggregated IP traffic. For this purpose, a trace file recorded at the Front Range GigaPOP (FRG) aggregation point, which is run by PMA (Passive Measurement and Analysis Project, see http://pma.nlanr.net) has been utilized[2]. The average rate of the traffic is 72 MBytes/s, with mean packet size of 869Bytes.

As IP traces are often dominated by several most frequent packet sizes [8], it is not necessary to take all possible packet sizes into account. In the sample used herein, seven packet sizes (40, 52, 552, 1300, 1420, 1488, 1500) account for 97 percent of the traffic and only these sizes were used. Using the expectation-maximization (EM) method [8] the following BMAP parameters were estimated:

$$D_0 = \begin{bmatrix} -90020.6 & 5300.2 & 11454.4 \\ 9132.5 & -126814.5 & 14807.8 \\ 2923.1 & 198.6 & -94942.6 \end{bmatrix}, D_{40} = \begin{bmatrix} 2898.0 & 3415.6 & 1365.3 \\ 3649.6 & 1510.9 & 1044.6 \\ 1943.3 & 1954.9 & 7696.6 \end{bmatrix},$$

$$D_{52} = \begin{bmatrix} 10980.1 & 8180.1 & 3284.0 \\ 5875.6 & 19512.8 & 19443.1 \\ 4064.1 & 9956.0 & 2744.3 \end{bmatrix}, \quad D_{552} = \begin{bmatrix} 1259.3 & 1143.0 & 960.4 \\ 1343.6 & 1541.2 & 3903.5 \\ 2470.5 & 1607.7 & 665.2 \end{bmatrix},$$

$$D_{1300} = \begin{bmatrix} 115.3 & 174.5 & 188.1 \\ 333.8 & 28.0 & 405.9 \\ 337.3 & 124.4 & 552.4 \end{bmatrix}, \quad D_{1420} = \begin{bmatrix} 878.9 & 65.6 & 506.4 \\ 916.4 & 31.1 & 999.1 \\ 205.0 & 138.1 & 1603.7 \end{bmatrix},$$

$$D_{1488} = \begin{bmatrix} 95.1 & 93.2 & 61.3 \\ 199.5 & 175.3 & 129.7 \\ 192.2 & 34.6 & 107.5 \end{bmatrix}, \quad D_{1500} = \begin{bmatrix} 13997.0 & 14089.9 & 9514.9 \\ 31145.3 & 9632.8 & 1052.4 \\ 17681.9 & 14814.8 & 22926.4 \end{bmatrix}.$$

The basic characteristics of the original traffic sample and its BMAP model are shown in Table 1.

In practice, packets are usually segmented into fixed size units prior to storage, which is connected with memory architecture for fast packet buffering in routers and switches [22]. Therefore, for computation of the buffer occupancy distribution we will rather use the following BMAP parameters:

$$D_1' = D_{40} + D_{52}, \quad D_9' = D_{552}, \quad D_{21}' = D_{1300}, \quad D_{23}' = D_{1420}, \quad D_{24}' = D_{1488} + D_{1500}.$$

---

[2] Precisely, one million packet headers from the trace file FRG-1137458803-1.tsh, recorded on Jan 17th, 2006, were used.

**Table 1.** Parameters of the original traffic sample and its BMAP model

|  | mean packet interarr. time [$\mu$s] | standard deviation of the interarr. time [$\mu$s] | mean packet size [Bytes] | total arrival rate [MBytes/s] |
|---|---|---|---|---|
| traffic sample | 11.467 | 11.599 | 869.18 | 72.286 |
| BMAP | 11.467 | 11.594 | 869.38 | 72.301 |

where the new indices denote numbers of occupied units, assuming typical unit size of 64Bytes [22].

We assume that the queue is served at the constant rate of 80MB/s (1310720 units/s) and the buffer size is 100kBytes. The initial state of the modulating process, $J(0)$, is distributed according to $\pi = (0.39517, 0.24563, 0.35920)$.

The matrices $A_k(s)$ and $\overline{D}_k(s)$, which appear in Theorem 1, can be calculated effectively by means of the uniformization technique (see [13] for more details). Applying this technique we get:

$$A_n(s) = \sum_{j=0}^{\infty} \gamma_j(s) K_{n,j}, \qquad \overline{D}_n(s) = \sum_{j=0}^{\infty} \delta_j(s) K_{n,j}, \tag{11}$$

where

$$K_{0,0} = I, \qquad K_{n,0} = \mathbf{0}, \quad n \geq 1,$$
$$K_{0,j+1} = K_{0,j}(I + \theta^{-1}D_0), \qquad \theta = \max_i\{(-D_0)_{ii}\},$$

$$K_{n,j+1} = \theta^{-1} \sum_{i=0}^{n-1} K_{i,j} D_{n-i} + K_{n,j}(I + \theta^{-1}D_0),$$

$$\gamma_j(s) = \frac{e^{-(\theta+s)d}(\theta d)^j}{j!}, \qquad d = \text{service time},$$

$$\delta_j(s) = \theta^j \frac{\Gamma(j+1, 0) - \Gamma(j+1, d(s+\theta))}{j!(s+\theta)^{j+1}},$$

and $\Gamma(j, x)$ denotes the incomplete gamma function.

The remaining matrices and vectors in Theorem 1 are either trivial (like $Y_k(s)$, $z(s)$) or simple functions of $A_k(s)$ and $\overline{D}_k(s)$.

Now we are in a position to present the numerical results. In Figure 1 the stationary queue size distribution for the system considered is presented. In particular, in part (I) the whole distribution is depicted while in part (II) a close-up of range 0-5kB is shown. The average queue size is:

$$L = 9.584 \text{ kBytes},$$

while its standard deviation:

$$\sqrt{Var} = 9.998 \text{ kBytes},$$

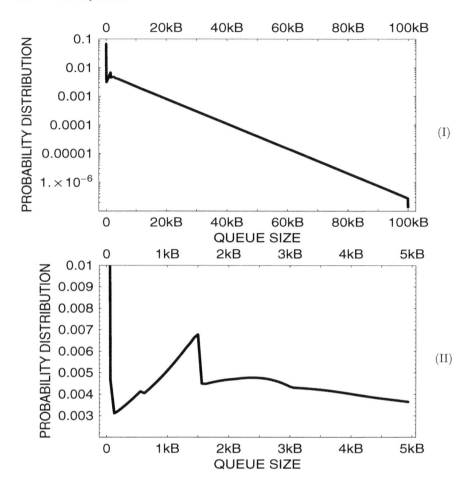

**Fig. 1.** Stationary queue size distribution. Part (I) - the whole distribution in logarithmic scale, part (II) - a close-up of the range 0-5kB.

The probability that the buffer is empty equals to

$$p_0 = 6.769 \times 10^{-2},$$

while the probability that the buffer is full is

$$p_b = 1.367 \times 10^{-7}.$$

It is striking that in the lower range, the distribution has jumps and its shape is rather complicated. This effect is typical for the batch arrival queue and the irregularities in the shape are connected with batch sizes and their combinations.

On the other hand, for larger queue sizes the shape becomes simple and the function is approximately linear (on a log-scaled plot).

It should be emphasized that the statistical structure of the BMAP, which reflects the structure of the original traffic, has a deep impact on the queue size distribution. For instance, if we replace the BMAP by the Poisson process with exactly the same arrival rate, we will obtain [3].

$$L = 470.6 \text{ Bytes},$$
$$\sqrt{Var} = 464.0 \text{ Bytes},$$
$$p_b = 3.792 \times 10^{-98}.$$

This set of numbers, which is in dramatic contrast to the previous one, illustrates how misleading it may be to neglect taking the precise statistical structure of the traffic into account.

## 5  Conclusions

In this paper, an analysis of the queue size distribution in a BMAP queue with finite buffer was conducted. It is reasonable to believe that the presented results are of practical importance due to the following reasons. Firstly, a very flexible arrival process, which among other things can model IP traffic, was considered. Secondly, the finite buffer was assumed. In a real network all elements (switches, routers etc.) have finite buffers, which causes losses and influences the network performance. Thirdly, the results were presented in a closed, easy to use form, and they permit one to obtain both, transient and stationary queue size distributions. Finally, computational remarks were given. In particular, formulas for the numerical calculation of the coefficient matrices $A_k(s)$ and $\overline{D}_k(s)$, which occur in the main theorem, were shown.

## References

1. Leland, W., Taqqu, M., Willinger, W. and Wilson, D. On the self-similar nature of ethernet traffic (extended version), IEEE/ACMTransactions on Networking 2(1): 115, (1994).
2. Crovella, M. and Bestavros, A. Self-similarity in World Wide Web traffic: Evidence and possible causes, IEEE/ACM Transactions on Networking 5(6): 835-846, (1997).
3. Shah-Heydari, S. and Le-Ngoc, T. MMPP models for multimedia traffic. Telecommunication Systems 15, No.3-4, 273-293 (2000).
4. Wong, T. C., Mark, J. W. Chua, K. C. Delay performance of voice and MMPP video traffic in a cellular wireless ATM network. IEE Proceedings Communications, Vol. 148, 302-309 (2001).
5. Wu, G. L. and Mark, J. W. Computational Methods for Performance Evaluation of a Statistical Multiplexer Supporting Bursty Traffic. IEEE/ACM Transactions on Networking, Vol. 4, No. 3, pp. 386-397, (1996).

---

[3] The formulae for the classic $M/G/1/b$ queue, which were used for obtaining this set of results, can be found by the reader in [21], page 202.

6. Kim, Y. H. and Un, C. K. Performance analysis of statistical multiplexing for heterogeneous bursty traffic in ATM network. IEEE Trans. Commun. 42(2-4) pp. 745-753, (1994).

7. Skelly, P., Schwartz, M. and Dixit, S. A histogram-based model for video traffic behavior in an ATM multiplexer. IEEE/ACM Trans. Netw. 1(4): 446-459 (1993).

8. Klemm, A., Lindemann, C. and Lohmann, M. Modeling IP traffic using the batch Markovian arrival process. Performance Evaluation, Vol. 54, Issue 2, (2003).

9. Salvador, P., Pacheco, A. and Valadas, R. Modeling IP traffic: joint characterization of packet arrivals and packet sizes using BMAPs. Computer Networks 44, pp. 335-352, (2004).

10. Landman, J. and Kritzinger, P. Delay analysis of downlink IP traffic on UMTS mobile networks. Performance Evaluation 62, pp. 68-82 (2005).

11. Klemm, A., Lindemann, C. and Lohmann, M. Traffic Modeling and Characterization for UMTS Networks, Proc. GLOBECOM 2001, pp. 1741–1746, San Antonio TX, November (2001).

12. Ramaswami, V. The $N/G/1$ queue and its detailed analysis. Adv. Appl. Prob. 12, 222-261 (1980).

13. Lucantoni, D. M. New results on the single server queue with a batch Markovian arrival process. Commun. Stat., Stochastic Models 7, No.1, 1-46 (1991).

14. Lucantoni, D. M., Choudhury, G. L. and Whitt, W. The transient $BMAP/G/1$ queue. Commun. Stat., Stochastic Models 10, No.1, 145-182 (1994).

15. Lucantoni, D. Further transient analysis of the $BMAP/G/1$ queue. Commun. Stat., Stochastic Models 14, No.1-2, 461-478 (1998).

16. Blondia, C. The finite Capacity $N/G/1$ Queue. Communications in Statistics: Stochastic Models, 5(2):273–294, (1989).

17. Baiocchi, A. and Blefari-Melazzi, N. Steady-state analysis of the MMPP/G/1/K queue. IEEE Trans. Commun. 41, No.4, 531-534 (1992).

18. http://www.omnetpp.org/

19. Chydzinski, A. The oscillating queue with finite buffer. Performance Evaluation, 57/3 pp. 341-355, (2004).

20. Abate, J. and Whitt, W. The Fourier-series method for inverting transforms of probability distributions, Queueing Systems 10: 587, (1992).

21. Takagi, H. *Queueing analysis. Vol. 2. Finite Systems.* North-Holland, Amsterdam. (1993).

22. Iyer, S., Kompella, R. R. and McKeown, N. Analysis of a memory architecture for fast packet buffers. In Proc. of IEEE High Performance Switching and Routing, Dallas, Texas, May (2001).

# A Scaling Analysis of UMTS Traffic[*]

Lucjan Janowski[1], Thomas Ziegler[2], and Eduard Hasenleithner[2]

[1] Department of Telecommunications AGH University of Science and Technology,
Al. Mickiewicza 30, 30-059 Kraków, Poland
`janowski@kt.agh.edu.pl`
[2] Telecommunications Research Center Vienna,
Donaucity Strasse 1, 1220 Vienna, Austria
`ziegler@ftw.at, hasenleithner@ftw.at`

**Abstract.** This paper reports on the results of a scaling analysis of traffic traces captured at several Gn Interfaces of an opera-tional 3G network. Using the Abry Veitch test and Variance Time Plots we find that the analyzed traffic is characterized by a non self-similar process for small time scales and strong self-similarity for larger time scales. The reasons for the time scale dependency of traffic self similarity are analysed in detail. We find that for smaller time scales the packet arrival times are independent explaining the weak correlation of the data. For larger timescales the arrival of TCP connections in the UMTS network can be considered independent. This property is intuitive because the only link where TCP connections could influence themselves is the channel originating at the mobile terminal. As a further finding we observe that the cut off point between weak and strong correlation is around the mean RTT of TCP flows.

Additionally, we propose a model which allows generating traffic similar to the measured UMTS traffic. The model contains only three parameters allowing to influence all observed changes in Log Diagram and Variance Time Plot. Comparing the output of the model with the real traffic traces we find that the model matches reality accurately.

## 1 Introduction

UMTS networks have been deployed for a few years. Especially in the Asian and European countries UMTS networks are already reasonably loaded to enable measurements and a statistical evaluation of UMTS traffic to support network planning. In this study we focus on the temporal traffic characteristics as seen at several measurement points at the Gn interfaces, between SGSNs and GGSNs, of Austria's largest UMTS/GPRS network. The traffic is monitored using DAG cards, a high performance diskarray, and a special software for message parsing as well as anonymisation. This methodology enables the storage of all packets

---

[*] This work is funded by the Austrian Kplus program and the European Union under the E-Next Project FP6-506869.

Y. Koucheryavy, J. Harju, and V.B. Iversen (Eds.): NEW2AN 2006, LNCS 4003, pp. 211–222, 2006.

traversing the edge of the UMTS core network to the public Internet, i.e. the aggregated traffic from and to the entire UMTS network.

Traffic measurements and modelling have been considered since the first telecommunication system was built. Different technologies have been described by different traffic models. Self-similarity has been shown to constitute a reasonably simple and effective analytic model of traffic dynamics in packet networks. It is simple since it contains only one parameter called the Hurst parameter [1, 2] and effective in the sense that the obtained theoretical results are similar to the measured ones [3, 4, 5]. In addition, the self-similarity phenomenon has been observed on all network layers [2, 6, 7], different kinds of network protocols [7, 8] and aggregation points [2, 8, 9]. The self-similarity helps to build models because of its natural description of long range dependence structures.

Some papers claim that a single self-similar parameter is not enough to model network traffic behaviour. Thus numerous models like multifractal or $\alpha$-stabile models have been proposed [5, 10, 11]. Publications investigating the self-similarity phenomenon often restrict themselves to estimating the Hurst parameter for the entire traffic. However, in this paper we have analysed traffic in order to understand critical properties of the traffic behaviour on different time scales.

The paper is organized as follows: Chapter 2 describes the basic properties of self-similar processes and estimation algorithms. Chapter 3 explains how measurements have been conducted. The Hurst parameter of traffic traces is analyzed in Chapter 4. In Chapter 5 an explanation of results obtained in Chapter 4 is given. In Chapter 6 a model of the observed behaviour is presented. The last chapter presents the summary of the paper.

## 2   Estimating Self-similarity

A commonly used class of self similar processes, allowing to a select the length of the observation interval such that the measurements can be considered as stationary, are self similar processes with stationary increments [1, 12]. This process is described by the Hurst Parameter $H$ and the discrete time scale $m$ of the process $z_n$. The time scale $m$ is defined as the number of values on a *basic time scale* summed up to obtain a single sample on the $m$ time scale. The basic time scale is the precision with which the traffic is measured, which corresponds to 2ms in our case. The Hurst parameter describes how strongly a time scale change influences the distribution of the process. The Hurst parameter value is within the interval $[1/2, 1)$, where $H = 1/2$ means that the process is not self-similar and $H > 1/2$ means that the process is self similar with stronger self-similarity for greater values of the Hurst parameter. The common way to evaluate whether a process belongs to the class of self similar processes with stationary increments is to estimate whether its increment process is long range dependent (LRD) taking into account the decay of its autocorrelation function.

There exist several methods for Hurst parameter estimation. As shown later on in the paper, our measurements do not contain a single self-similar parameter, but two parameters dependent on considered time scales. Thus we focus on the Abry-Veitch Test (AVT) and on Variance Time Plots (VTP), two estimators allowing to estimate the Hurst parameter in a time scale dependent manner. The Abry-Veitch test is based on a wavelet transform implying numerous advantages [13, 15, 16, 17]. First of all, if a considered process is LRD the wavelet coefficients are almost uncorrelated and certainly SRD. Therefore, well known SRD analysis can be used and a confidence interval is computable. The second important advantage is that this test allows removing all linear or even more complicated trend functions [15]. The Variance Time Plot (VTP) is a log-log plot of the process variances at different time scales $m$, allowing to estimate the Hurst parameter value. VTP estimation has some disadvantages especially in comparison to the Abry-Veitch test. A linear trend added to data traffic will dramatically change the obtained $H$ value, as described in [19, 20]. Additionally, the confidence interval cannot be computed because the confidence intervals of the variance values are not known. The biggest advantage of the VTP estimation method is that it is conceptually simple and that it follows directly from the definition of LRD. Since most of the other Hurst parameter estimation methods have the same disadvantages as VTP, we focus on VTP to illustrate results in a simple manner in addition to the Abry Veitch Test.

## 3  Traffic Collection

The result of the ftw. project METAWIN is the creation of a large scale monitoring infrastructure that is able to capture all packet headers at all layers in the protocol stack of the 3G Core Network [21]. The provider's privacy policy demands removing all private information from the traffic trace therefore there is no possibility to identify the specific users by header fields like the IMSI or the MSISDN. The measurements are done in the core network of the largest Austrian mobile provider having more than 3 millions customers. The base architecture is shown in Figure 1. Capturing has been conducted using Endace DAG cards and stored packets are associated with GPS synchronized timestamps. The measurements have been done for 100% of all UMTS and GPRS traffic from home subscribers. From the scaling analysis point of view we focus on packets including TCP/IP headers. Such traffic traces enable the evaluation of TCP specific parameters like RTT and spurious retransmission timeouts [22].

Our analysis is focused on the four the highest loaded hours. The traffic trace consists of numbers representing the amount of traffic that has been sent during 2ms time intervals; therefore a single traffic trace contains more than 8 million values. Uplink and downlink traffic traces are considered separately as different types of behaviour have been observed. In addition to the above mentioned trace, trace files which collects all TCP connections have been analyzed. The information about TCP connection length, the number of received and sent packets and the amount of data that has been downloaded and uploaded has been gathered. The traffic trace length is approximately one day. The traffic

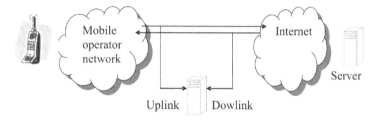

**Fig. 1.** Measurement network base schema

monitoring tool allows analyzing the duration time distribution and correlation between time duration and the amount of data sent by a particular connection.

For proprietary reasons all traffic values (e.g. the amount of data sent through an observed link during the 2ms interval) for all traffic traces are multiplied by a uniformly distributed random value before any analysis is done. From the scaling analysis point of view the operation does not change anything and has no influence on trends because a scaling analysis is focused on a correlation structure rather than on the values of traffic volume or rate.

## 4   Hurst Parameter Estimation

On the beginning we have focused on time stationarity and proved by various methods that the measurements are time stationary, details can be found in an extended version of this paper [26]. Here only a short description is given. We have started from considering mean and variance plots [23] that are used to test mean and variance stationarity. Secondly we have chacked an influence of vanishing moments on a Hurst parameter value [15]. Last three tests considered Hurst parameter time stationarity. We have used three tools: "LASS" as described in [18], a 3D-LD plot described in [8] and a 3D-VTP proposed by the author and based on the 3D-LD plot. All results have shown that the measurements are time stationary. From 3D-LD and 3D-VTP the Hurst parameter's time scale non-stationarity has been shown. Since the time scale non-stationarity of the Hurst parameter is not dependent on time, presented results are focused on the results obtained for the whole traffic trace. This allows for the analysis of a higher number of time scales.

Firstly, our data has undergone an analysis by means of the "Selfis" tool [14]. The obtained Hurst parameter values are strongly dependent on the estimation type as shown in Table 1. Such behaviour indicates time scale non-stationarity. The reason is that each estimation method is focused on a different time scale. Therefore the values obtained from a different estimation method are different if the Hurst parameter value is dependent on the time scale. The rest of the analysis is focused on finding the difference between these two types of behaviour and computing Hurst parameter values that can be associated with particular time scales. In order to find this association we have used two plots: VTP and LD plot.

**Table 1.** Hurst parameter values obtained from the "Selfis" tool

| Test name | Hurst parameter value and confidence interval if available |
|---|---|
| Variance Time Plot | 0.894 |
| Periodogram | 0.784 |
| Variance of Residuals | 0.959 |
| Whittle Estimator | 0.657 [0.653-0.660] |
| R/S Estimator | 0.823 |
| Absolute Moments Estimator | 0.782 |
| Abry-Veitch test | 0.642 [0.642-0.650] |

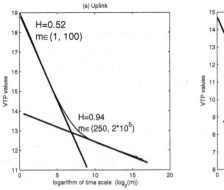
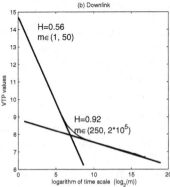

**Fig. 2.** VTP for UMTS measurement. Basic time scale is 2 ms. Hurst parameter values are computed for different time scales. Considered time scales are denoted in the Figures.

We start from VTP and compute it for all data traffic. The obtained results are presented in Figure 2.

For a strictly self-similar process like Fractional Gaussian Noise (FGN), VTP is a linear function. In Figure 2.a we can observe two linear behaviours. The first is SRD (the Hurst parameter equals 0.5) up to a time scale of 200 ms and the second is strong LRD for time scales of 0.5s and larger.

Again, we can clearly observe two different types of behaviour dependent on the time scale. Apart from that, a transition interval is observable. The only difference between Figures 2.a and 2.b is that for the downlink the value of the Hurst parameter obtained for the smaller time scales is marginally greater than for the uplink. The obtained values indicate that there is no self-similarity for smaller time scales on the uplink and for the downlink self-similarity is weak. The difference between the Hurst parameter values obtained for larger time scales is not so important because these values indicate strong self-similarity.

However, before final conclusions are drawn, the results obtained from the Abry-Veitch test are presented. Figures 3 shows the results obtained for uplink and downlink, respectively.

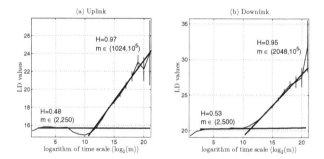

**Fig. 3.** LD plot for UMTS measurements. The basic time scale is 2 ms. Considered time scales are denoted in the Figure.

The A-V test shows the same difference in the Hurst parameter values behavior. The obtained results indicate that the analyzed traffic is characterized by a non self-similar process for small time scales (smaller than 0.5 s) and strong self-similarity for large time scales (larger than 1 s).

The detailed view in Figures 2 and 3 shows some small differences between down and uplink for particular time scales. The first observation is that VTP for the small time scales and downlink shows weak self-similarity (the Hurst parameter value is 0.56). The second observation is that uplink and downlink exhibit different types of behaviour of A-V test values around transition time scales (between 0.25s and 1s, i.e. $\log_2 m = 7$ to 9). The transition time scales are time scales for which VTP and LD values cannot be approximated by a linear function. For the uplink LD values decrease for a few average time scales and then increase rapidly. But for the downlink a decrease in LD values does not occur. Similarly to VTP the transition time scales can be clearly observed. All observations are explained in the next chapter.

## 5   Findings from the Scaling Analysis

The first finding is that the measured traffic is stationary in time in terms of the mean and variance values. The Hurst parameter value is stationary in time as well. 3D-LD and 3D-VTP plots are almost not dependent on a block number (time). This is a very important finding leading to the conclusion that we can focus only on the results presented in the Figures 2 and 3.

Considering LD plots in Figures 3 we find that packets are independent at lower timescales causing a Hurst parameter value around 0.5. For the large time scales the explanation is as follows. [24] shows that if an aggregate of TCP connections are approximated by an independent but Pareto distributed ON/OFF process, this process is self-similar with the Hurst parameter dependent on the shape parameter of Pareto distribution[1]. This leads to two conclusions. Firstly,

---

[1] According to Pareto distribution probability density function of form $f(x) = \frac{\alpha b^\alpha}{x^{\alpha+1}}$ a shape parameter is $\alpha$.

TCP connections in an UMTS network are independent. This property remains in accordance with our intuition, as the only place where TCP connections can influence themselves is the channel originating at a mobile terminal. The remaining network links are faster and so there is no congestion. On account of a small number of simultaneous connections connected to the single channel the influence among these connections is negligible. The Pareto distribution of the TCP connection length has been computed from the analyzed traffic traces and is strongly related to user behaviour [8, 25].

For the uplink LD values are decreasing at the beginning of transition time scales (the transition time scales are the time scales for which the LD plot or VTP are non linear). As a result, there is a negative correlation at the transition time scales. The reason is that there are regular periods where a source does not send any data and there are periods when data are sent. The explanation for this phenomenon is as follows: for the uplink a TCP connection sends some data, then after a burst of data which is sent, the TCP algorithm waits for ACK packets and does not send anything. It follows that for some time scales there are regular ON and OFF periods generating a negative correlation. Therefore it seems that the variation in RTT for the uplink is not higher than 0.25-0.5s because for such a time scale ($\log_2 m \approx 8$) most TCP connections behave like an ON/OFF process.

Additionally, it is important to note that transition timescales, indicating the change from short- to long range dependencies, correspond to the average RTT usually observed in UMTS networks (0.25-1sec). We have found out this correlation as result of watching traffic that has been generated by users using different technologies (GPRS, UMTS and WLAN). We have observed that for different technologies the transition timescales are different and what more the transition timescales for GPRS has superior value according to UMTS and the UMTS transition time scales has superior values according to WLAN. Since the RTT has different values for different technologies and it has motivated us to build a model where RTT would determine the transition timescales.

VTPs acknowledge the conclusion obtained from the A-V tests. The difference lies in the fact that the obtained borders of linear trends and transition time scales are not fully clear. In any of these figures it is not clear where the linear behaviour starts or finishes, therefore the values described in Figures 2 and 3 are only approximations. It is important that not all time scales are described by the A-V test. The second difference between the two considered estimation methods lies in the Hurst parameter value obtained for the downlink. The value is higher than 0.5 for VTP suggesting weak self-similarity. The difference between the A-V test and VTP is that VTP notices bursts more than the correlation structure. In the uplink there are no bursts because the core network link is much faster than the UMTS radio access network. Additionally, radio access favours the existence of some random delays between packets. Contrary to this, the downlink receives only limited numbers of packets that can be sent one by one. Therefore the bursts sent from some servers are accumulated in the queue. The small value of the Hurst parameter is a consequence of the small values of those bursts resulting from the limited space in the queues.

## 6   Modeling Traffic Dynamics

According to the findings described in the chapter 6 there are three dominating parameters to model traffic dynamics. The first is the average time of TCPs cycle (i.e. RTT). For time scales smaller than the RTT we notice independent variables. For time scales larger than the RTT, the TCP source can be approximated by a source that is always active (ON state in an ON/OFF model). Such an assumption does not take into account different source intensities. The important question is the kind of distribution to model the duration of ON-times. As proposed in [8] and proved in [24] the most accurate distribution is a heavy tailed Pareto distribution having infinite variance. The heavy tailed distribution parameter $a$ influences the length of TCP connection times. The heavy tailed parameter $a$ can be computed on the basis of the Hurst parameter according to the equation described in [24]. The LD plot and the VTP do not show a sharp change between linear behaviour of short time scales and long time scales. The explanation can be that the RTT value is not constant (that is acknowledged by previous research on the RTT in UMTS network [22]). That is why the model parameter needs to contain a parameter describing the change of the RTT.

We propose the simplest model containing all those parameters. The Mathlab algorithm is:

```
1    data=zeros(n,1);
2    for i=1:n
3      lambda=lamChange(i);
4      new=poissrnd(lambda,1,1);
5      for j=1:new
6        length=i+RTT*floor(paretornd(a,1,1));
7        if(length>n)
8          length=n;
9        end
10       index=i;
11       while(index<length)
12         data(index)=data(index)+1;
13         index=index+floor((rand(1 ,1)-0.5)*RTTmax)+RTT;
14       end
15     end
16   end
```

where paretornd$(a,1,1)$ returns a single random number from the Pareto distribution where the scale parameter is 1 and the shape parameter is $a$, rand(1,1) returns a single uniformly distributed random variable in the (0,1) range, lambda is the mean number of newly arriving connections at each time interval, and lamChange(i) is a function representing the variation in the average number of new connections.

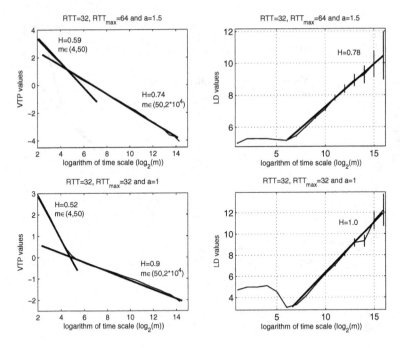

**Fig. 4.** VTP and LD plots obtained from model for RTT=32 different RTT$_{max}$ and $a$ parameters

Note that if lamChange(i)=const, the obtained results have non-stationary mean values in time. The reason is that during each time interval $i$ there are some new connections sending data in addition to the existing connections. As TCP connections have heavy tailed distributions, there is no stationary state. One possibility to keep the obtained data stationary is to change the average number of new connections in such a way that the sum of the average number of new connections at time interval $i$ and the average number of connections which start before $i$ is constant.

As we can see there is no difference in the traffic rate for this model. All TCP flows send the same amount of data after their RTT time (line 12). The RTT changes in the range RTT$\pm$RTT$_{max}/2$ with a uniform distribution. Therefore the mean RTT value is constant and if RTTmax is less than 2*RTT, the decrease in LD plot should be observed. The obtained LD and VTP for different RTT, RTT$_{max}$ and $a$ parameters are presented in the Figure 4.

The influence of RTT$_{max}$ and the change of $a$ values in the A-V test are shown in the Figure 4. The decreasing part of the LD plot is related to RTT$_{max}$. If the value is less than 2*RTT, the decreasing effect is observable. It is completely consistent with the observation of the real traffic. The real traffic RTT changes can be described by means of different distributions which need more investigation. The relation between $a$ and $H$ parameters is consistent with the theoretical results.

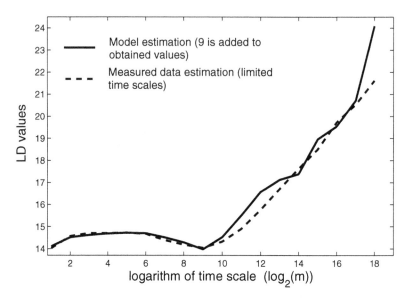

**Fig. 5.** Comparison of LD values obtained for uplink and the model with RTT=200, $RTT_{max}$=300 and $a = 1$ parameters

Figure 5 shows the comparison between LD plots obtained for the measured data at the uplink and the values obtained from the model for $a = 1$, RTT=200 and $RTT_{max}$=300. The accuracy of the curve obtained for the model is almost identical with the curve obtained for the measured data for small time scales. For larger time scales the model has more variability than the measured data. This is because the measured data consists of a lot more connections than the model. The number of new TCP connections per single time unit is limited to 40 because of computer memory limit. Nevertheless the obtained result remains very accurate.

## 7    Conclusion

Our goal has been to consider the existence of self-similarity in an operational UMTS core network. In the course of our research we have analyzed in detail the Abry-Veitch and VTP plot behaviour. One of the most crucial finding is the conclusion that TCP connections seem to be independent and therefore queue behaviour for small time scales can be computed from classic models. For small time scales, weak self-similarity (in fact burstiness) can be observed only for the downlink traffic. User behaviour influences heavy tailed TCP connection lengths and as a result strong self-similarity for long time scales can be seen. Additionally, some small differences between the downlink and the uplink traffic char-acteristic have been described in terms of RTT variation.

Additionally, we have proposed an algorithmic model which allows generating traffic similar to the measured UMTS and GPRS traffic. The model has been investigated but not presented in the paper. The model containing only three parameters makes it possible to influence all the observed LD and VTP values change.

For future work the behaviour of the traffic from/to a single base-station can be considered. Other problems are RTT and TCP flow length distributions that could be investigated more carefully. The assumption that TCP connections send the same amount of traffic for each RTT time could be investigated too. One of the most interesting points for future investigation would be to perform research on the distribution of new TCP connections that are incoming in a particular time interval. The distribution has to be dependent on the time of a day.

# References

1. Beran, J.: Statistic for Long-Memory Processes. New York: Chapman & Hall, 1994.
2. Leland, W. E., Taqqu, M. S., Willinger, W., Wilson, D. V.: On the Self-Similar Nature of Ethernet Traffic (Extended Version). IEEE Transactions on Networking, Vol. 2, 1-15, February, 1994.
3. Garrett, M., Willinger, W.: Analysis, Modeling and Generation of Self-Similar VBR Video Traffic. in Proceedings SIGCOMM94, 269-280, August 1994.
4. Mayor, G., Silvester, J.: Time Scale Analysis of an ATM Queueing System with Long-Range Dependent Traffic. INFOCOM 97
5. Gallardo, J. R., Makrakis, D., Orozco-Barbosa, L.: Use of $\alpha$-stable Self-Similar Stochastic Processes for Modeling Traffic in Broadband Networks. Performance Evaluation, v.40 n.1-3, p.71-98, March 2000
6. Lpez-Ardao, J. C., Argibay-Losada, P., Rodriguez-Rubio, R. F.: On Modeling MPEG Video at the Frame Level Using Self-similar Processes. MIPS 2004, 37-48, 2004.
7. Hong, S. H., Park, R.-H., Lee, C. B.: Hurst Parameter Estimation of Long Range Dependent VBR MPEG Video Traffic in ATM Network. Journ. Visual Commun. Image Representation, vol. 12, no. 1, pp. 44-65, Mar. 2001.
8. Uhlig, S., Bonaventure, O., Rapier, C.: 3D-LD : a Graphical Wavelet-Based Method for Analyzing Scaling Processes. In Proc. of the 15th ITC Specialist Seminar, July 2002.
9. Kalden, R., Ibrahim, S.: Searching for Self-Similarity in GPRS. The 5th annual Passive & Active Measurement Workshop, PAM 2004, France, April 2004.
10. Erramilli, A., Roughan, M., Veitch, D., Willinger, W.: Self-Similar Traffic and Network Dynamics. Proc. of the IEEE, 90(5):800-819, May 2002.
11. Riedi, R. H., Crouse, M. S., Ribeiro, V., Baraniuk, R. G.: A multifractal wavelet model with application to network traffic. IEEE Trans. Info. Theory, vol. 45, no. 3, pp. 992–1018, April 1999.
12. Stallings. W.: High-Speed Networks: TCP/IP and ATM Design Principles. Prentice Hall, 1998.
13. Rolls D. A.: Limit theorems and estimation for structural and aggregate teletraffic models, Ph.D. thesis, Queen's University at Kingston, 2003.
14. Karagiannis T., Faloutsos M.: SELFIS: A Tool For Self-Similarity and Long-Range Dependence Analysis. 1st Workshop on Fractals and Self-Similarity in Data Mining: Issues and Approaches (in KDD) Edmonton, Canada, July 23, 2002.

15. Abry P. and Veitch D.: Wavelet Analysis of Long-Range Dependence Traffic, IEEE Transactions on Information Theory, 44(1), pp. 2-15, January, 1998.
16. Roughan M. and Veitch D.: Measuring long-range dependence under changing traffic conditions, proceedings of INFOCOM'99, pp1513-1521.
17. Veitch D., Abry P.: A Statistical Test for the Time Constancy of Scaling Exponents. IEEE Transactions on Signal Processing, vol. 49, no. 10, pages 2325-2334, October 2001.
18. Stoev S., Taqqu M.: Park G., Michailidis G. and Marron J.S.: LASS: a Tool for The Local Analysis of Self-Similarity. Preprint 2004, available as SAMSI Tech. Rep. No. 2004-7.
19. Dang T. D. and Molnar S.: On The Effects of Non-Stationarity in Long Range Dependent Tests. Tech. Rep., Budapest University of Technology and Economics, 1999.
20. Molnar S. and Dang T. D.: Pitfalls in Long Range Dependence Testing and Estimation. In GLOBECOM, 2000.
21. METAWIN home page:http://www.ftw.at/ftw/research/projects/ProjekteFolder/N2.
22. Vacirca F., Ziegler T. and Hasenleithner E.: An Algorithm to Detect TCP Suprious Timeouts and its Application to Operational UMTS/GPRS Networks, Computer Networks Journal, 2006
23. NIST/SEMATECH e-Handbook of Statistical Methods, http://www.itl.nist.gov/div898/handbook/, 2002.
24. Taqqu M. S., Willinger W., and Sherman R.: Proof of a Fundamental Result in Self-Similar Traffic Modeling, Computer Communication Review, Vol. 27, pp. 5-23, 1997.
25. Barford P., Bestavro A., Bradley A. and Crovella M. E.: Changes in Web Client Access Patterns: Characteristics and Caching Implications. World Wide Web, Special Issue on Characterization and Performance Evaluation 2, 1999, 1528.
26. Janowski L., Ziegler T., Hasenleithner E.: A Scaling Analysis of UMTS Traffic, work report. www.kt.agh.edu.pl/~janowski/AScalingUMTS.pdf.

# Burst Assembly with Real IPv4 Data - Performance Assessement of Three Assembly Algorithms[*]

Nuno M. Garcia[1,2], Paulo P. Monteiro[1,3], and Mário M. Freire[2]

[1] Siemens S.A., Amadora, Portugal
[2] IT-Networks and Multimedia Group, University of Beira Interior, Covilhã, Portugal
[3] Instituto de Telecomunicações, University of Aveiro, Aveiro, Portugal
{nuno.mgarcia, paulo.monteiro}@siemens.com, mario@di.ubi.pt

**Abstract.** This paper describes the generation of data bursts using real IPv4 data as input and compares the performance of the three aggregation algorithms. The profiles of bursts are studied, in particular the mean packet delay per burst and burst inter-arrival time followed by mean burst size and mean number of packets per burst. Observations are made regarding the identification of relations between bursts and packets for the studied data traces and assessed the performance of the aggregation algorithms under study. The conclusions are generalized to IPv6 traffic.

## 1 Introduction

Data packet aggregation process was initially proposed by Amstutz in 1983 [1] as a way to benefit from the statistical multiplexing effect, or as initially described, "improved bandwidth efficiencies". This concept was later re-introduced in Optical Networks, contributing to the Optical Burst Switching Network paradigm, initially proposed by Qiao and Yoo around 1999 [2]. When referring to Optical Burst Switching (OBS), three major aggregation algorithms are used: time constrained; size constrained; both time / size constrained, also termed the hybrid algorithm.

With the perspective increasing importance of burst switching, e.g. OBS, it is of great importance to assess the burst assembly algorithms performance under real traffic conditions. The performance of aggregation algorithms has already been studied for simulated traffic [3-5], namely for research in the area of burst assembly algorithms, QoS and burst traffic statistics. To the best of our knowledge, this is the first report regarding the evaluation of burst aggregation algorithms using real traffic input data. The algorithms under study were implemented in the Aggregation Simulator package designed in Java, and the simulator was fed with real IPv4 traffic data captured by NLARN [6] in several network sites in the United States of America. These collection sites are mostly universities and research centers, and thus, traffic nature is very heterogeneous. This data is stored in a time stamped header (.tsh)

---

[*] This work has been partially funded by **Fundação para a Ciência e a Tecnologia**, Portugal, through the grant contract SFRH/BDE/15527/2004 and through CONDENSA Project contract POSI/EEA-CPS/60247/2004 and by **Siemens S.A.**, Portugal.

Y. Koucheryavy, J. Harju, V.B. Iversen (Eds.): NEW2AN 2006, LNCS 4003, pp. 223–234, 2006.
© Springer-Verlag Berlin Heidelberg 2006

format, which contains the payload stripped IPv4 datagram (or packet, as we will refer to it in this document), time stamped in microseconds at their acquisition. The standard format of the .tsh data packet header is available in [7]. Although the algorithms were applied to all files (in a batch process running in several laboratory machines) only the relevant results are shown here. The performance of each algorithm was evaluated following a set of previously defined metrics. The paper also extrapolates the conclusions on the studied IPv4 to IPv6 data conversion, following the results in [8].

The remainder of this paper is organized as follows: Section II describes the burst assembly algorithms; Section III discusses the metrics devised for assessment of the burst assembly algorithms; Section IV presents the assessment of the burst assembly algorithms and Section V presents main conclusions.

## 2   Burst Assembly Algorithms

Data packet aggregation is a process in which individual data packets are assembled together before the resulting burst is sent into the network structure. The packets in question may experience re-encapsulation (or not, depending on the supported scenario) and typically the nature and origin of the data packets under consideration is not relevant to the aggregation principle, as these may be Ethernet frames, ATM cells, IP packets, and so forth. The burst assembly process requires only the other end of the transmission link to run a complimentary burst disassembly process, retrieving the original constituent packets. In this study, IPv4 packets were used and no encapsulation of the assembled packets was performed. The disassembly mechanism should thus consider the first 20 bytes of the data burst to be an IPv4 header, and proceed to extract that packet from the aggregated data. This step is repeated until no data is left within the burst. If the network implements burst segmentation techniques, the last readable packet may be corrupt, and if so, it is discarded.

Burst assembly algorithms are constraint driven, and fall into three categories:

1) Maximum Burst Size (MBS) [9]
2) Maximum Time Delay (MTD) [10]
3) Hybrid Assembly (HA) [11, 12]

Other burst assembly algorithms, like the ones considering classes of services, build upon one of the aforementioned basic types.

In the Maximum Burst Size (MBS) assembly algorithm, the incoming data packets are aggregated consecutively into a burst, until its size exceeds the defined threshold. When this occurs, the last data packet overflowing the current burst starts a new burst, while the current burst is transmitted into the network structure.

The Maximum Time Delay (MTD) assembly algorithm was devised to prevent situations where, while using the MBS algorithm, the rate of incoming packets is so low or the arriving packets are so small, that it takes an unacceptable amount of time to fill up a single burst, resulting in excessive transmission delay for the aggregated packets. The MTD algorithm checks for the time difference between the head packet in the burst and the current local time. The burst is sent into the network as soon as that time difference exceeds the maximum delay time defined, independently of the size of the burst and of the number of packets it contains.

If the traffic flow rate is too high or the incoming packets are big, the MTD algorithm may end up creating bursts that are too big. In order to prevent such a situation, a Hybrid Assembly (HA) algorithm was devised. In this assembly scheme, both thresholds – time and size – are considered simultaneously. Incoming packets are assembled into a new burst until one of the threshold conditions is met.

# 3 Burst Assembly Simulation Evaluation Metrics

Burst assembly was performed using a simulator designed in Java. The simulator reads .tsh input files and acts on each packet trace recording, in an event driven simulation. Simulation parameters were: name of .tsh input files (in particular, the name of the folder in which the input files were stored), burst assembly type (MBS, MTD, or HA), and burst assembly thresholds. The output provided includes burst details and the metrics defined in section 3.1. Due to the large number of processed files, the simulator was designed to perform batch processing of the input files.

## 3.1  Evaluation Metrics

In recent literature, it is common to find burst sizes defined in time, which in turns allows for its measurement in bytes given a known data channel rate. Since we are using real IPv4 as tributary data for our research, the sizes are defined in bytes (B) and times are defined in microseconds (due to .tsh format limitations). For the burst itself, a suitable set of statistics was used:

a) mean packet delay per burst
b) mean number of packets per burst
c) mean burst size
d) mean burst inter-arrival time

The mean packet delay per burst is important as it provides insight on the value of average delay that a single packet suffers from due to the very nature of the burst assembly mechanism. The acceptable delay for a single data packet is one of the main criteria when selecting an adequate burst assembly algorithm. This metric was calculated by averaging the delay occurred between the arrival time of each packet and its burst departure time, assumed to be the arrival time of the last packet in the burst. The mean packet number per burst additionally shows how much processing the disassembly (complimentary to the assembly process) mechanism needs to perform to forward individual packets, but primarily it demonstrates the amount of switching effort saved in the network structure due to the statistical multiplexing effect. While the mean burst size is important to determine the occupancy of the network, the mean burst inter-arrival time will provide insight as to the ratio at which the bursts flow in the network structure. Mean burst interarrival time was calculated averaging the delay occurred between two consecutive burst departures.

One Way Delay, as defined in RFC 2679 [13] and IP Packet Delay Variation Metric, as defined in RFC 3393 [14] produced by the IETF IP Performance Metrics workgroup (IPPM) [15] were not adopted as these measure the performance of packet transmission for a given set of links (considering the transmission between source and destination), while we used particular network point packets arrival times.

## 3.2 Burst Assembly Variables

Burst Assembly was performed with several thresholds. Time thresholds used were 100μs, 500μs, 1ms, 10ms and 100ms and size thresholds used were 9KB, 64KB and 1MB. The first two constitute the maximum sizes of respectively the Ethernet Jumboframe and the IP packet. The third threshold was originally set to 4GB as consistent with the maximum size for an IPv6 Jumbogram [16], but it was soon discovered that even 1MB burst size required burst assembly times in excess of 1s (for "MEM-1111612868-1.tsh"), making the 4GB size threshold of little research interest. Hybrid aggregation was performed for the following six scenarios: size thresholds where 9KB and 64KB and time thresholds were 500μs, $10^3$μs and $10^4$μs.

# 4 Results

The results presented were obtained by simulation for the burst assembly process executed for the following three randomly chosen traces: AMP-1107250616-1.tsh, ANL-1109366607-1.tsh and MEM-1111612868-1.tsh. AMP, ANL and MEM stand for AMPATH, Miami, Florida (OC12c), Argonne National Laboratory to STARTAP (OC3c link) and University of Memphis (OC3c link), respectively [17]. Other data trace files were also subject to simulation process, producing results coherent with the ones presented in this paper. Each of the examined trace files records the activity in the selected network point for about continuous 91 seconds. Table 1 shows the activity of each of these network points.

Table 1. Network Activity for the selected trace files

|  | AMP | ANL | MEM |
|---|---|---|---|
| Packet load (in Bytes) | 876 301 540 | 246 923 708 | 159 669 572 |
| Time Span (in μs) | 89 836 840 | 90 921 971 | 91 918 790 |
| Offered Load (in MB/s) | 9.754 | 2.716 | 1.737 |

## 4.1 Results for MBS

Fig. 1 depicts the relative frequency of burst inter-arrival time for each of the studied data traces with the burst size threshold set to 9KB. As anticipated, the AMP data results in the lowest inter-arrival time plot, since 90% of the bursts arrive within 1555μs or less and this trace contains the highest traffic load. For the other traces, the same limit is observed at 616 μs and 12903μs for ANL and MEM, respectively. It is worth noting that the plots do not share the same shape – around 2500μs ANL and MEM switch tendencies – this suggests that burst inter-arrival time depends on the nature of the traffic on the network point where bursts are to be assembled.

Fig. 1 does not present the whole burst inter-arrival scale: AMP reaches its maximum at 10001, ANL at 30038, and MEM at 55702μs (for MEM, 99.95% of the burst arrive up until $35 \cdot 10^3$μs), respectively. The same behaviour is also visible for other size thresholds in all the studied traces.

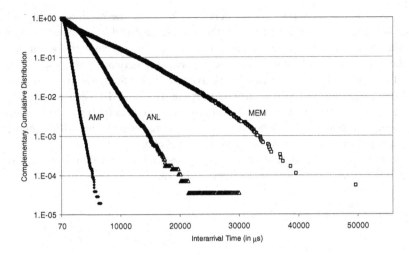

**Fig. 1.** Complementary Cumulative Distribution Function for Burst Inter-arrival Time for MBS threshold = 9KB

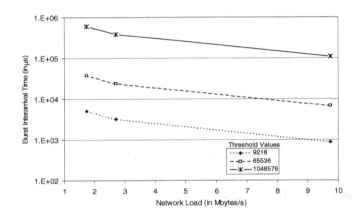

**Fig. 2.** Weighted average burst inter-arrival time vs. network load (for MBS)

The comparison between burst inter-arrival times for different size thresholds lead to the conclusion that a 2 orders of magnitude increase in the burst size (from 9KB to 1MB) results in an increase of more than two orders of magnitude for the burst inter-arrival time. This situation is depicted in Fig. 2. It also shows the influence of network offered load on burst inter-arrival time. Here, it is assumed that the nature of the traffic in the three network points is comparable, and the offered throughput was calculated using values from Table 1. When the network load increases by a factor of 5, the average burst inter-arrival time decreases by an order of magnitude. This behaviour is also verified for several burst size thresholds.

Fig. 3 depicts the average packet delay per burst for the MBS assembly scheme. When the size threshold is set to 9KB, the average packet delay per burst parameters are almost the same, for the studied data traces, diverging only when the size threshold increases. As anticipated, the higher load network point shows a lower packet average delay. The increase in the average packet delay when size threshold is increased from 9KB to 64KB is different for the data in AMP when compared to the data in ANL and MEM. ANL and MEM also here seem to follow a close behaviour, expectedly suggesting that when the link is more loaded, the MBS aggregation algorithm performs better in terms of average packet delay per burst.

Accordingly with the definition of the burst assembly algorithm, burst size is very close to burst size threshold, and no graph is shown.

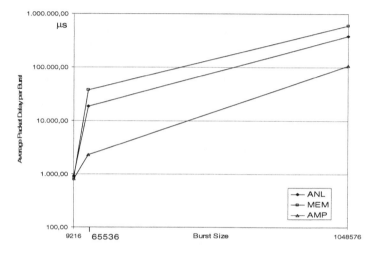

**Fig. 3.** Average packet delay per burst for MBS

## 4.2 Results for MTD

As expected, the average inter-arrival time between bursts is always higher than the time threshold defined for the burst assembly. Fig. 4 shows the relation between the weighted averages for the burst inter-arrival times for MTD plotted against burst assembly time threshold. It is clear that inter-arrival and threshold times converge when the later increase. AMP holds the lowest relation between inter-arrival and threshold times, thus performing better than ANL or MEM, which is a consequence of its higher traffic load.

There is a quasi linear relation observed in all data traces between the number of packets present in bursts and the utilised time threshold, as shown in Fig. 5. As anticipated, the higher load traffic data yields a higher number of packets per burst, and the number of packets in bursts per data traces for AMP and ANL / MEM diverge as the time threshold increases. At a higher aggregation threshold value, there is almost an order of magnitude difference in the number of packets per burst between AMP and ANL / MEM.

Burst Size depends on the MTD aggregation ratio since the faster the incoming packets arrive, the bigger the bursts are. Fig. 6 shows how data is aggregated into bursts of variable sizes when the time threshold is set to $10^4$ µs. In line with well known results [3, 18], we can see that the burst size distribution resembles a Gaussian shape when the number of bursts increases, which happens when we consider the MEM trace, following the Law of Large Numbers. The plot character for each data trace is different, even for sources with comparable traffic load such as ANL and MEM. This result suggests that the bigger the load of the aggregated traffic, the more the burst size distribution tends to the Normal distribution. The Normal distribution nature of some of the characteristics of the aggregated bursts is also visible when the packet count per burst is plotted. For burst assembly time of $10^5$ µs, bursts contain from 88 to 1887 packets following a distribution close to the one depicted in Fig. 6.

Fig. 7 shows the results when time was set to 100 µs (the lower end of the chosen burst assembly threshold scale). At this burst assembly threshold, only high intensity traffic sources achieve bursts with more than 16 packets (the maximum was one burst in AMP which contained 21 packets). This figure shows the complementary cumulative distribution function of packets in bursts for this particular burst assembly scenario. Here, bursts containing up to three packets account for 78.5% of the traffic with AMP, 95.1% with ANL and 98.3% with MEM trace files.

The average packet delay for MTD is shown in Fig. 8. Note that the average packet delay converges to the burst assembly threshold when it is higher than $10^3$ µs. Additionally, low load traces (ANL and MEM) show better performance for MTD in terms of average packet delay than AMP; in case of low time thresholds, MEM performs almost twice better than AMP, possibly because bursts contain fewer packets.

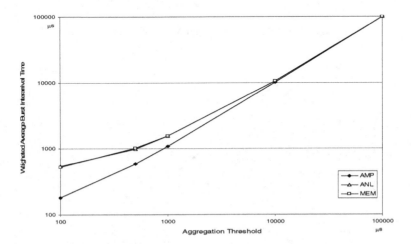

**Fig. 4.** Burst Inter-arrival Time vs. Burst Assembly (Aggregation) Time for MTD

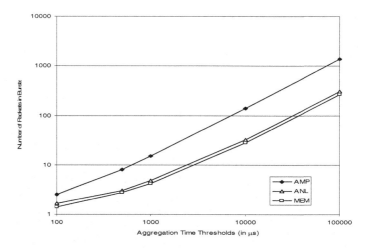

**Fig. 5.** Number of Packets in Bursts vs. Burst Assembly (Aggregation) Thresholds for MTD

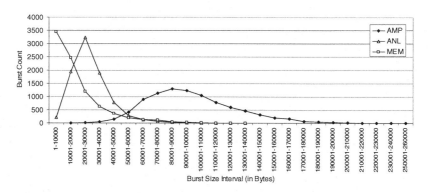

**Fig. 6.** Burst Size Histogram for MDT when time = $10^4$ $\mu s$

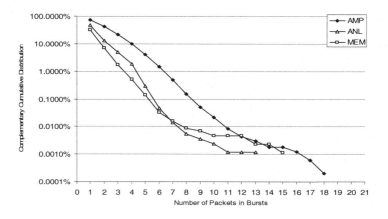

**Fig. 7.** Number of Packets in Bursts Relative Distribution for MTD = 100 $\mu s$

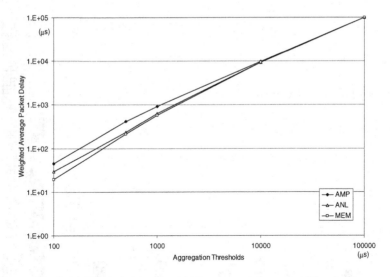

**Fig. 8.** Packets Delay (average) in Bursts vs. Aggregation Thresholds

### 4.3   Results for HA

The Hybrid Assembly algorithm combines two thresholds - time and size. Fig. 9 shows the packet inter-arrival time for three data traces, for each HA scenario. The tendency depicted in Fig. 9 remains when regarding packet delay time per burst. It is clear that ANL and MEM behave similarly, and for burst assembly times below $10^3\mu s$ inclusive, and the performance of the HA is almost constant regardless of whether the burst size threshold is set to 9KB or 64KB. AMP performs better probably because of its higher packet load and performs best with time threshold of $10^4\mu s$ and burst size 9KB than with time of $10^3\mu s$. This behaviour may be related to the change in the burst assembly threshold (from size to time or *vice-versa*). This hypothesis is confirmed by the plot of Fig. 10. Here, the nice dragon-like shape of the plot exhibits several interesting features of the HA algorithm.

The data plotted for values of bursts 7500 bytes (approximately) show the effect of the size constrained branch of the algorithm – these were the bursts that contain fast incoming packets (they were very close together in time so the time threshold was never reached). It is also visible that these bursts are responsible for a major share of the output stream (bursts whose size is bigger than 7500 bytes account for 74.70% of the total bursts count).

The lump shapes on Fig. 10 are a well know phenomena related to the size distribution of IPv4 packets [8]. 1500 bytes long IPv4 packets are very common in data traces, supposedly because of path and/or link MTU issues, and also, as a consequence of the popularity of IP over Ethernet encapsulation scheme. This graph shows local modes near the multiples of 1500 bytes, namely 1500, 3000, 4500, 6000, 7500 and finally 9000 bytes. Note that 75% of the bursts in this scenario have 7500

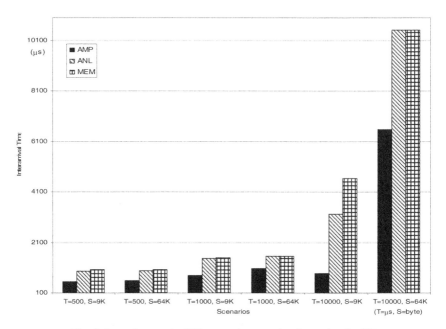

**Fig. 9.** Burst Inter-arrival Time vs. Aggregation Scenarios for HA

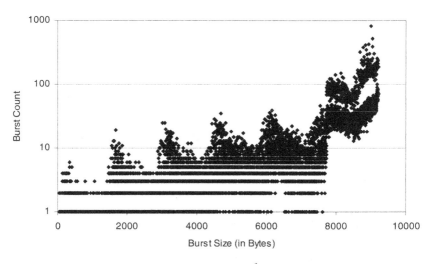

**Fig. 10.** Burst Size Histogram for HA ($T=10^3$ $\mu s$, $S= 9\ KB$) on AMP data

bytes or more, which is in line with results, in particular the ones depicted in Fig. 5 and Fig. 6 considering MTD. This behaviour is also visible in other data traces and in other aggregation scenarios.

# 5   Conclusions

The assessment of the performance of burst assembly algorithms using real IPv4 data allows for the following conclusions:

1. globally, from all the studied scenarios for the burst assembly algorithms, both MTD and HA outperform MBS in terms of delay added to the burst and/or to the packet in the burst. Therefore, as the assembly of large size bursts results in high inter-arrival times and packet delay per burst, these large bursts should only be considered for transmitting data that is not time critical, such as news server synchronization or data backup between servers. The size threshold is dependent of the traffic load on the input link, typically being bigger than 1MB for these data traces;

2. the performance of the burst assembly algorithms is also a function of the network point where the data is to be assembled, as different network sites show different performances. This feature is visible in all examined burst assembly algorithms, and several threshold values may have to be studied to reach a sustained optimum performance for a specific network point. Thus, for an optimum operation regarding the compromise between large bursts and low mean packet delay in a burst, burst assembly algorithm's thresholds or combinations of thresholds have to be dynamically set to adjust to time changing traffic conditions;

3. as IPv6 packet sizes and flow rates are expected to follow current IPv4 models, following the research presented in [8] we can expect similar results for the assembly of IPv6 packets into bursts.

## Acknowledgments

The authors would like to thank Marek Hajduczenia and Pedro Inácio for their support and many valuable discussions.

## References

1. S. R. Amstutz, "Burst Switching - An Introduction," in IEEE Communications Magazine, vol., issue 0163-6804/83/1100-0036, pp. 36-42, 1983.
2. C. Qiao and M. Yoo, "Optical burst switching (OBS) - A new paradigm for an optical Internet," *Journal of High Speed Networks*, vol. 8, issue 1, pp. 69-84, 1999.
3. S. Malik and U. Killat, "Impact of Burst Aggregation Time on Performance in Optical Burst Switching Networks," in proceedings of Optical Fibre Communications Conference, 2004, OFC 2004, Los Angeles, California, USA, 2004, vol. 2, pp. 2.
4. T. Ferrari, "End-To-End performance analysis with Traffic Aggregation," *Computer Networks*, vol. 34, issue 6, pp. 905-914, 2000.
5. A. Zapata and P. Bayvel, "Impact of burst aggregation schemes on delay in optical burst switched networks," in proceedings of LEOS 2003, Tucson, Arizona, USA, 2003.
6. National Laboratory for Applied Network Research, "National Laboratory for Applied Network Research," in http://pma.nlanr.net/, 2005, accessed at 2005-01-13.

7. National Laboratory for Applied Network Research. "tsh file format (Time Stamped Header)," in http://pma.nlanr.net/Traces/tsh.format.html, 2005, accessed at 2005-01-13.

8. N. M. Garcia, M. Hajduczenia, P. Monteiro, H. Silva, and M. Freire, "Modeling and Simulation of IPv6 Traffic," in proceedings of 7th Internet Global Congress, Global IPv6 Summit, Barcelona, 2005.

9. V. M. Vokkarane, K. Haridoss, and J. P. Jue, "Threshold-Based Burst Assembly Policies for QoS Support on Optical Burst-Switched Networks," in proceedings of Optical Networking and Communications Conference OPTICOMM 2002, Lowell, Massachussets, USA, 2002.

10. A. Ge and F. Callegati, "On Optical Burst Switching and Self-Similar Traffic," *IEEE Communications Magazine*, vol. 4, issue 3, 2000.

11. X. Yu, Y. Chen, and C. Qiao, "Performance Evaluation of Optical Burst Switching with Assembled Burst Traffic Input," in proceedings of GLOBECOM 2002, Taipei, 2002.

12. M. C. Yuang, J. Shil, and P. L. Tien, "QoS Burstification for Optical Burst Switched WDM Networks," in proceedings of Optical Fiber Communication Conference OFC 2002, Anaheim, California, USA, 2002.

13. G. Almes, S. Kalidindi, and M. Zekauskas, "One Way Delay Metrics for IP Performance Metrics (IPPM)," RFC 2679, IETF, 1999.

14. C. Demichelis and P. Chimento, "IP Packet Delay Variation Metric for IP Performance Metrics (IPPM)," RFC 3393, IETF, 2002.

15. IP Performance Metrics IETF workgroup, in http://www.ietf.org/html.charters/ippm-charter.html, accessed at June 15, 2005.

16. D. Borman, S. Deering, and R. Hinden, "IPv6 Jumbograms," RFC 2675, Internet Engineering Task Force (IETF), 1999.

17. National Laboratory for Applied Network Research, "NLANR Sites," in http://pma.nlanr.net/Sites/, 2005, accessed at 2005-02-15.

18. X. Yu, Y. Chen, and C. Qiao, "A Study of Traffic Statistics of Assembled Burst Traffic in Optical Burst Switched Networks," in proceedings of Optical Networking and Communications, Boston, Massachusetts, USA, 2002, vol. 4874, pp. 149-159.

# Multiservice IP Network QoS Parameters Estimation in Presence of Self-similar Traffic

Anatoly M. Galkin, Olga A. Simonina, and Gennady G. Yanovsky

State University of Telecommunications, Telecommunication Networks Department,
St.Petersburg, Russia
`galkinam@inbox.ru`, `simonina@bk.ru`, `yanovsky@sut.ru`
`http://www.sut.ru`

**Abstract.** This study investigates key properties of self-similar processes of multiservice traffic in IP networks. On the basis of the analytical modeling the impact of self-similarity properties on the QoS (Quality of Service) parameters (delays and losses) is shown. The results of simulation are presented.

## 1 Introduction

Traffic patterns generated by IP multimedia services are quite different from traditional Poisson models used for *circuit-switched* voice traffic. As a result, the network parameters can be underestimated if inadequate traffic models and analytical approaches are adopted. Therefore, performance analysis of network elements taking into account the self-similar nature of multiservice traffic is extremely important.

More than a decade ago W. Leland et al. have published an article on traffic self-similarity investigation [1]. This study generated a remarkable activity on traffic modeling in IP networks and its accuracy from self-similar feature standpoint. In this paper the analysis of real traffic parameters collected in Bellcore Corporation's data network during several years has been performed. On the basis of this analysis we show inappropriateness of Poisson models for some types of traffic, for example, for the case of file transfer over FTP protocol. It was shown that IP network performance evaluation based on Markov models may result in load underestimation and, as the consequence, contributes to QoS degradation. Further studies [2, 3] have shown that the data flows and their processing can be simulated successfully by processes with self-similar properties.

During the last years significant number of papers on IP networks performance analysis based on self-similar processes were published. We would like to outline the following fundamental publications [4-10, 14] devoted to IP traffic research at the network layer. Besides, a number of papers was devoted to self-similarity feature investigation of the Ethernet traffic at data-link layer [11, 12], and ATM traffic [13]. In [14-16] traffic patterns from a number of TCP applications using self-similarity models have been investigated. In S. Molnar's paper [17] it was shown that the self-similarity is observed in VoIP traffic. Rigorous mathematical description of self-similar processes was given by B. Tsibakov in [22]. Applications of self-similar processes in telecommunication networks have been described in the comprehensive book

Y. Koucheryavy, J. Harju, and V.B. Iversen (Eds.): NEW2AN 2006, LNCS 4003, pp. 235–245, 2006.

on fractal processes in telecommunications [24]. Firstly, in this study the analysis of existing models for various applications traffic is presented. Further, we propose modeling technique that describes self-similar processes quite well. We then offered an approach for QoS parameters calculation, mutual impact of elastic and real time traffic was estimated. Analytical results were verified with simulation models using Network Simulator 2.

## 2 Self-similar Processes in Multiservice IP Networks

### 2.1 Self-similarity Properties for Various Types of Traffic at Different Layers

Last years can be characterized by substantial number of results regarding an analysis of network traffic behavior at different layers. The results of studies on FTP applications traffic modeling are presented in [16, 21], in particular: [16] deals with traffic behavior description by Pareto law; [21] by Weibull law. In [17] it was proposed to describe the VoIP traffic by Pareto law. In [18, 19, 21] analysis of different applications on Application layer was performed in detail. The results of analysis confirm the presence of self-similarity properties in traffic of applications running over FTP and HTTP protocols. The conclusion that traffic generated by e-mail applications does not exhibit self-similarity feature [18] is also very important.

Based on results published in those papers we classified applications into several classes and bridged to corresponding distribution law that describes traffic on a certain layer. The results of classification are presented in the Table 1, where A and B are laws of incoming traffic and protocol data blocks size distribution, correspondingly.

**Table 1.** Distribution laws for different types of traffic in IP networks

| Traffic type | Distribution law | | Publication |
|---|---|---|---|
| | **A** | **B** | |
| **VoIP** | P | P | Molnar [17] |
| **FTP/TCP** | P | W and LN | Van Mieghem [21], Downey [16] |
| **SMTP/TCP** | M | M | Molnar [18] |
| **HTTP/TCP** | P | LN and P | Crovella [14], Van Mieghem [21] |
| **IP** | P | P | Paxson [19] |
| **Ethernet** | P | P | Taqqu [11] |
| **ATM** | D | F-ARIMA | Sadek [13] |

**Note:** M – Poisson law; W – Weibull law; P – Pareto law; LN – Lognormal law; F-ARIMA – Fractal Auto-Regressive Integrated Moving Average.

### 2.2 Insertion of Restricted Heavy-Tails Distributions

Heavy-tail distributions have an infinite variance in the range of $\alpha$, $1 < \alpha < 2$, where $\alpha$ is characteristic parameter of the self-similar process and the quadratic coefficient of variation $C^2 \to \infty$. However values of variance are limited in the real systems. For the description of self-similar process we shall enter so-called restricted distribution. The restricted distribution allows specifying the maximum value of variety that provides finite value of $C^2$, but without tail form change. In this case there is a possibility to calculate certain process parameters even the distribution's tail does not converge to zero.

As an example one may consider the model of VoIP traffic processing, which can be described in general as P/P/m system according Kendall classification. Then, in accordance with [20] we obtain:

$$C^2 = \frac{(1-\alpha)^2(L^\alpha - k^\alpha)}{\alpha(Lk^\alpha - L^\alpha k)^2}\left(\frac{L^2 k^\alpha - L^\alpha k^2}{(2-\alpha)} - \frac{\alpha(Lk^\alpha - L^\alpha k)^2}{(1-\alpha)^2(L^\alpha - k^\alpha)}\right) \qquad (1)$$

where $L$ and $k$ are maximum and minimum possible intervals between incoming protocol data blocks. The functional dependence of $C^2(\alpha)$ is shown in Fig. 1.

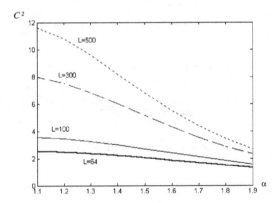

**Fig. 1.** Quadratic coefficient of variation $C^2(\alpha)$ for different $L$ values, $k = 1$, $1 < \alpha < 2$

### 2.3  Calculation of Delays

When the flow of incoming packets differs from Poisson and the process of service differs from exponential (G/G/1 systems), to calculate delays and losses one may use the results of diffusive approximation from [25]. Average value of packet number in the queue $\overline{\omega}$ and average value of delay $t_m$ in the G/G/1 system can be defined in accordance with [26]:

$$\overline{\omega} = P(\rho, m)\frac{\rho}{1-\rho} \cdot \frac{C_a^2 + C_s^2}{2}, \qquad (2)$$

$$t_m = \overline{t}_\omega + \overline{t}_s, \qquad (3)$$

$$\bar{t}_\omega = P(\rho, m)\frac{\bar{t}_s}{m(1-\rho)} \cdot \frac{C_a^2 + C_s^2}{2},\tag{4}$$

where $\bar{t}_s$ is an average time of the packet's service; $\bar{t}_\omega$ is an average time of the packet's staying in the buffer; $C_a^2$ and $C_s^2$ are quadratic coefficients of variation of incoming flow and service time distributions, correspondingly, for G/G/1.

Results of average delays' calculations are presented at Fig. 2 and Fig. 3.

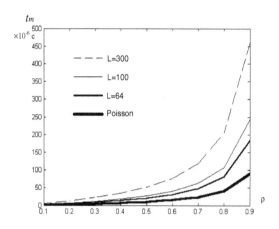

**Fig. 2.** The average delay in P/P/1 system for different $L$

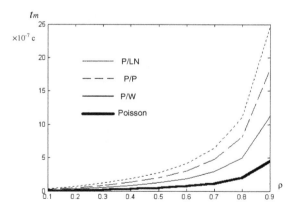

**Fig. 3.** The average delay in P/G/1 system for different distribution laws of service time

## 2.4  Calculation of Losses

Based on diffusive approximation the loss probability for G/G/1 system can be obtained as follows [26]:

$$P_{loss} = \frac{1-\rho}{1-\rho^{\frac{2}{C_a^2+C_s^2}nb+1}} \, \rho^{\frac{2}{C_a^2+C_s^2}nb} \; , \qquad (5)$$

where $C_a^2$ and $C_s^2$ are quadratic coefficients of variation of incoming flow and service time distributions, correspondingly, for G/G/1; $nb$ is a buffer size; $\rho$ is a system load. Results of loss probability calculus are presented in Fig. 4 and Fig. 5.

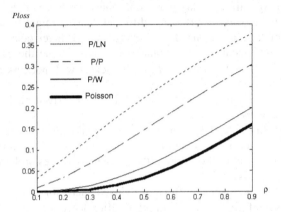

**Fig. 4.** Loss probability in P/G/1 system for different distributions of service time

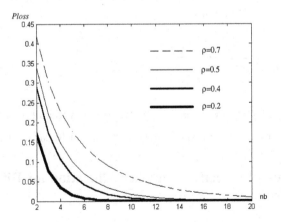

**Fig. 5.** Loss probability in P/P/1 system for different $\rho$

## 3 Real Time Traffic Losses Performance Under Jitter-Buffer Impact

Losses of real time traffic packets could affect QoS perceived by a user. It makes sense to control losses at the side of a receiver. Let us evaluate end-to-end losses as follows:

$$P_{e2e} = 1 - (1 - P_{net})(1 - P_{ter}), \tag{6}$$

where $P_{net}$ represents network losses (depends on $P_{loss}$); $P_{ter}$ represents additional losses at terminal node appearing due to acceptable delay exceeding (only for real time and streaming traffic). Delay jitter contributes end-to-end delay between network layer entities, that is why for critical end-to-end delay it might be often the case that delay jitter stimulates additional losses even network layer delay fits QoS requirements. Traffic agreement will be violated. Let us examine the case in presence of maximum acceptable delay value of $t_{max}$. all packets with network delay of $t_{net} > t_{max}$ will be considered lost.

Hence, $P_{ter} = \mathbf{P}\{t_{net} > t_{max}\}$ is the probability that arrived packet exceeded maximum acceptable delay.

The source of losses leading to maximum acceptable delay excess is a *jitter buffer*. Then let us represent a loss probability in a jitter buffer as $P_{ter}$. In this case the problem of $P_{ter}$ determination follows up to the problem of losses probability calculus in a buffer of a network node.

Packets coming to a jitter buffer are delayed there for interim needed for jitter compensation, therefore the maximum jitter buffer delay is $t_j \leq t_{max} - t_{codec} - t_{net}$, where $t_{max}$ is maximum acceptable network delay, $t_{codec}$ is a codec delay, $t_{net}$ is a network delay. Let us denote one place in a buffer by delay per time unit $t$. Then $t_{j\_max}$ is a maximum jitter buffer size. Taking into consideration (5), we obtain the following expression for $P_{ter}$ calculation:

$$P_{ter} = \frac{1-\rho}{1-\rho^{\frac{2}{C_{a\_jitt}^2 + C_{s\_jitt}^2} t_{j\_max} + 1}} \rho^{\frac{2}{C_a^2 + C_s^2} t_{j\_max}}, \tag{7}$$

where $C_{a\_jitt}^2$ and $C_{s\_jitt}^2$ are quadratic coefficients of variation of incoming flow and service time distributions correspondingly; $t_{j\_max}$ is a buffer size; $\rho$ is a system load.

## 4 Estimation of Effective Capacity Allocation for Different Types of Traffic

Formerly, the problem of capacity allocation was solved for the networks of narrow purposes in conditions of homogeneous loads [23]. The load of multiservice networks is heterogeneous and has different QoS requirements. Nowadays, the effective capacity allocation for different types of real time traffic (including VoIP) and elastic traffic is very relevant. The problem of effective capacity allocation could be stated as follows: *to determine capacity distribution for multiservice IP network for real time and elastic traffic taking into account self-similarity and QoS requirements.*

Below we provide the solution of the problem of effective capacities allocation in a point-to-point network.

For mathematical statement we use three expressions (5), (6) and (7). As a criterion we take two values Pe2e_RT and Pe2e_el, those are average packet loss probabilities for real time traffic connection and elastic one, correspondingly. For those probabilities calculation we use (5) and (7). The impact of delay is taken into account in (6), and we consider this is sufficient. For capacities allocation comparison we use the rate: $\eta = Bel/B$, where Bel is capacity dedicated for elastic traffic communication, B is an overall capacity.

Further, we set two boundary conditions:

- $\eta = 0$, only real time traffic is transferred in the network,
- $\eta = 1$, only elastic traffic is transferred in the network.

We determine the condition of QoS characteristics conformance with quality of service agreement in the following way: *Pe2e_RT < Pe2e_RT_max and Pe2e_el < Pe2e_el_max.*

Then, let us determine boundaries, *a* and *b,* forming the interval as presented in Fig. 6. Inside this interval the requirements to QoS characteristics hold true for both elastic traffic and real time traffic sharing the network.

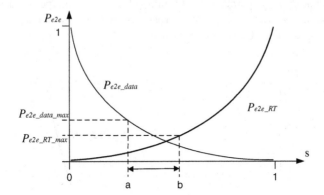

**Fig. 6.** Boundaries determining the interval of effective network capacity use for different traffic types

In what follows, the problem is solved assuming FIFO service discipline with fixed buffer size and lack of resource reservation mechanism. Incoming processes and service processes are described with Pareto distribution (P/P model). It leads us to the following equation:

$$\begin{cases} P_{loss\_el}(\eta) < P_{loss\_el\_\max} \\ P_{loss\_RT}(\eta) < P_{loss\_RT\_\max} \end{cases}, \qquad (8)$$

Substituting (5), (6), and (7) in (8) we get:

$$\begin{cases} \dfrac{1-\rho}{1-\rho^{k \cdot nb+1}}\rho^{k \cdot nb} < P_{loss\_el\_\max} \\ 1-\left(1-\dfrac{1-\rho}{1-\rho^{k \cdot nb+1}}\rho^{k \cdot nb}\right)\left(1-\dfrac{1-\rho}{1-\rho^{l \cdot nb+1}}\rho^{l \cdot nb}\right) < P_{loss\_RT\_\max} \end{cases} \qquad (9)$$

where $k = \dfrac{2}{C^2_{a\_net}+C^2_{s\_net}}$ and $l = \dfrac{2}{C^2_{a\_jitt}+C^2_{s\_jitt}}$ are determined with quadratic coefficients of variation for traffic distribution laws of network nodes and jitter buffer correspondingly.

Fig. 8 represents the example with the following parameters:

- $P_{e2e\_RT\_max} = 0.02$, maximum acceptable losses for real time traffic,
- $P_{e2e\_data\_max} = 0.03$, maximum acceptable losses for elastic traffic.
- $nb = 100$ positions, buffer size of a network node,
- $t_{j\_max} = 20$ ms, the jitter buffer size in time units,
- $B = 2$ Mbps, channel capacity,
- $\alpha = 1.5$, characteristic parameter determining burst of each type of traffic,
- The distribution laws of incoming processes and service processes in network nodes and in the jitter buffer are described with Pareto distribution.

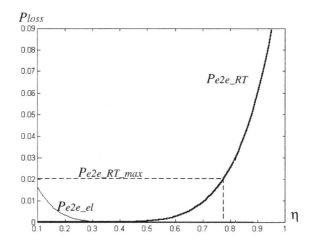

**Fig. 7.** Calculated values of elastic and real time traffic losses

As one may see in Fig. 7, elastic traffic losses appear when the real time traffic borrows more than 30% of the total bandwidth. It leads to TCP retransmissions and therefore to data traffic growth and losses increase. Results of TCP retransmissions are not considered in this paper.

## 5  Accounting for Self-similarity

Let us verify analytical results obtained in previous sections with simulation. For simulation we used Network Simulator 2 (ns2). Ns2 allows self-similar traffic genera-tors. We simulated the fragment of a network (Fig. 8) including four self-similar traf-fic sources with Pareto law (0, 1, 4, 5), one router (2) and one destination node (3).

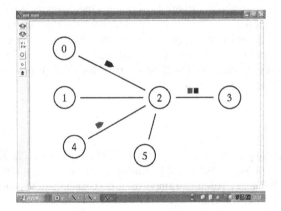

**Fig. 8.** Simulation topology

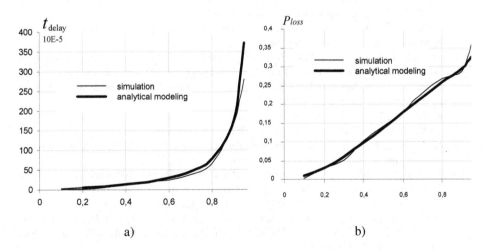

a)                                                b)

**Fig. 9.** Analytical modeling and simulation load relations: a) delay time; b) loss probability

The results of the simulation and analytic modeling for a fragment of a network in Fig. 8 are presented in Fig. 9.

In this case the simulation mistake is no more than 6%. This mistake lies within the feasible range of mistakes of engineer calculations. Thus the simulation model results confirmed analytical results with high accuracy.

## 6   Conclusions

The presence of self-similar traffic nature in IP networks leads to significant differences in QoS parameters for multiservice IP networks compared to Markov models. Calculation patterns of QoS parameters: delays and losses with consideration of traffic's self-similarity are presented in the study. It is shown that traffic of multiservice networks could be transmitted without flow control mechanisms with guaranteed QoS on condition that real time traffic borrows no more than 30% of common bandwidth.

On the basis of the diffusive approximation and simulation the estimations of delays and losses were received for G/G/1 system that can be considered as adequate model for processes in multiservice IP networks. Simulation based on ns2 package confirms the analytical results well enough.

## References

1. W. Leland, M. Taqqu, W. Willinger, D. Wilson. On the Self-Similar Nature of Ethernet Traffic (extended version). IEEE/ACM Transactions of Networking, 1994
2. R. Riedi, W. Willinger. Toward an Improved Understanding of Network Traffic Dynamics, Self-Similar Network Traffic and Performance Evaluation, 2000.
3. A. Feldman, W. Whitt. Fitting Mixtures of Exponentials to Long-Tail Distributions to Analyze Network Performance Models. Performance Evaluation, 31(8), pages 963-976, Aug. 1998.
4. M. S. Taqqu, W. Willinger and R. Sherman, Proof of a Fundamental Result in Self-Similar Traffic Modeling, Computer Communication Review 27, (1997), 5–23.
5. M. Taqqu, W. Willinger, R. Sherman. On-Off Models for Generating Long-Range Dependence, Computer Communication Review, Vol. 27, 1997.
6. M. Crovella, M. Taqqu, A. Bestavros. Heavy-Tailed Probability Distribution in World Wide Web, A Practical Guide to Heavy Tails: Statistical Techniques and Applications, 1998.
7. K. Park, T. Tuan. Multiple Time Scale Congestion Control for Self-Similar Network Traffic, Performance Evaluation, 1999.
8. K. Park, W. Willinger. Self-Similar Network Traffic: An Overview, Self-Similar Network Traffic and Performance Evaluation (ed.), Wiley-Interscience, 2000 .
9. M. Crovella, P. Barford. Measuring Web Performance in the Wide Area. ACM Performance Evaluation Review, 27(2):37—48, August 1999.
10. M. Crovella, I. Matta, L. Guo. How Does TCP Generate Pseudo-Self-Similarly? Proc. the International Workshop on Modeling, Analysis and Simulation of Computer and Telecommunications Systems (MASCOTS'01), Cincinnati, OH, August 2001.
11. W. Willinger, M. Taqqu, R. Sherman, D. Wilson. Self-Similarity through High-Variability: Statistical Analysis of Ethernet LAN Traffic at the Source Level. IEEE/ACM Transactions on Networking, 5(1), 1997.
12. A. Erramili, O. Narayan, W. Willinger Experimental Queuing Analysis with Long-Range Dependent Traffic. IEEE/ACM Transaction on Networking, Vol. 4, No. 2, pp. 209-223, 1996.

13. N. Sadek, A. Khotanzad, T. Chen. ATM Dynamic Bandwidth Allocation Using F-ARIMA Prediction Model, Department of Electrical Engineering, Southern Methodist University, Dallas, USA, 2003.
14. K. Park, G. Kim, M. Crovella. On the Relationship between File Sizes, Transport Protocols and Self-Similar network Traffic. Proc. International Conference on Network Protocols, pp. 171-180, October, 1996.
15. V. Almeida, A. de Oliveira. On the Fractal Nature of WWW and Its Application to Cache Modeling. Anais do XXIII Seminório Integrado de Software e Hardware do XVI Congresso da SBC, Recife, Agosto de 1996, Brasil 1996.
16. A. B. Downey. Lognormal and Pareto Distributions in the Internet. www.allendowney.com/research/longtail/downey03lognormal.pdf, 2003.
17. T. D. Dang, B. Sonkoly, S. Molnór, Fractal Analysis and Modelling of VoIP Traffic, NETWORKS 2004, Vienna, Austria, June 13-16, 2004
18. S. Molnór, T. D. Dang, Scaling Analysis of IP Traffic Components, ITC Specialist Seminar on IP Traffic Measurement, Modeling and Management, Monterey, CA, USA, September 18-20, 2000
19. V. Paxson, S. Floyd. Wide-Area Traffic: The Failure of Poisson Modeling // Lawrence Berkeley Laboratory and EECS Division, University of California, Berkeley, 1995
20. O.A. Simonina. Qos Parameters Estimation of Next Generation Networks // 56-th NTK PPS/SPbGUT. SPb, 2004 (in Russian)
21. Hooghiemstra, G. and P. Van Mieghem, 2001, Delay Distributions on Fixed Internet Paths, Delft University of Technology, report20011020
22. B.S Tsibakov. Teletraffic Pattern in the Terms of Self-Similar Random Process // Radiotehnika, 1999, №5 (in Russian)
23. L. Kleinrock, Communication Nets; Stochastic Message Flow and Delay, McGraw-Hill Book Company, New York, 1964.
24. O.I. Sheluhin, A.M. Tenyakshev, A.V. Osin. Fractal Processes in Telecommunications. – M.: Radiotehnika, 2003 (in Russian)
25. O.I. Aven, N.N. Gurin, Y.A. Kogan. Quality Estimation and Optimization of Computing Systems. – M. Nauka, 1982 (in Russian)
26. N.B. Zeliger, O.S. Chugreev, G.G. Yanovsky. Design of Networks and Discrete Information Communication Systems – M. Radio i svjaz, 1984 (in Russian)

# Achieving Relative Service Differentiation for TCP Traffic in Satellite IP Networks Using Congestion Pricing

Wei Koong Chai, Kin-Hon Ho, and George Pavlou

Centre for Communication Systems Research, University of Surrey, GU2 7XH, UK
{W.Chai, K.Ho, G.Pavlou}@eim.surrey.ac.uk

**Abstract.** This paper proposes the use of congestion pricing for achieving relative service differentiation in satellite IP networks. The idea is to use congestion pricing as an effective approach to control traffic rate for Transmission Control Protocol (TCP) flows, with Explicit Congestion Notification as a congestion feedback mechanism, by taking into account network users' willingness-to-pay. With multitude of competing TCP flows, congestion pricing ensures that the higher the user's willingness-to-pay, the higher the traffic throughput. We implement the congestion pricing approach on the *ns-2* simulator and evaluate its performance on both geostationary and non-geostationary satellite IP networks. Simulation results show that congestion pricing can be adopted as an effective approach for service differentiation in satellite IP networks and achieves fair relative service differentiation.

## 1 Introduction

One of the key objectives of the next generation Internet is to achieve service differentiation. The latter categorizes traffic into a set of classes to provide different treatment in Quality of Service (QoS). In recent years, two well-known architectures, Integrated Services (*IntServ*) [1] and Differentiated Services (*DiffServ*) [2], have been proposed to support service differentiation in the Internet. Meanwhile, satellite networks with vast coverage capability are vital for the future network infrastructure. As satellite networks are carrying increasing volumes of Internet traffic, the ability to provide flexible yet efficient service differentiation will be extremely important.

This paper proposes an approach to achieve *Relative Service Differentiation* (RSD) for satellite IP networks. The RSD model ensures that high-priority traffic will receive no worse QoS than low-priority traffic but it does not guarantee an absolute level of QoS to any traffic class. The key advantage of RSD is its ability to keep service differentiation consistent regardless of network load changes. To realize service differentiation, the authors in [3] claimed that pricing is an indispensable factor to be considered. They argued that proposals that define technologies in isolation of economics and implementation are fundamentally flawed. The authors in [4] also stated that the RSD model must be strongly coupled with a pricing or policy-based scheme to make higher classes more costly than lower ones. However, to the best of our knowledge, RSD that takes into account such an economic incentive for satellite networks has not been investigated. For that reason, we propose to use *congestion pricing* in order to achieve RSD for satellite IP networks.

Y. Koucheryavy, J. Harju, and V.B. Iversen (Eds.): NEW2AN 2006, LNCS 4003, pp. 246–258, 2006.
© Springer-Verlag Berlin Heidelberg 2006

Congestion pricing has been primarily investigated in terrestrial networks as an effective means for resource management and congestion control [5][6]. The fundamental concept of congestion pricing is to inform network users the congestion cost their traffic is incurring via marking and feedback mechanisms such as Random Explicit Detection (RED) [7] and Explicit Congestion Notification (ECN) [8]. The network users then react to these feedback congestion costs by adjusting the traffic sending rate according to their incentive to pay for better QoS. Such incentive is called *Willingness-To-Pay* (WTP). In theory, a user with higher WTP receives better QoS. As such, this becomes a natural mechanism for supporting RSD. Using congestion price for RSD has also an advantage of scalability as it places rate control intelligence at network edges while keeping the core network simple: each network user only reacts to feedback signals by adjusting the traffic sending rate. Although this idea is similar to that proposed for *DiffServ*, the former is much simpler because no complexity is added to the network core routers.

In this paper, we review the state-of-the art of service differentiation in satellite networks and present theoretical background on congestion pricing (Section 2 and 3). We implement congestion pricing as a window-based congestion control, taking into account network users' WTP as a weight to differentiate among their TCP flows in terms of traffic sending rate and throughputs (Section 4). We evaluate our implementation using the network simulator *ns-2* under Geostationary Earth Orbit (GEO) and non-GEO satellite IP network scenarios (Section 5). Simulation results demonstrate that congestion pricing achieves RSD in satellite IP networks through relative and fair bandwidth sharing. Finally, we conclude the paper (Section 6).

## 2   Service Differentiation in Satellite Networks

A number of proposals have been made for achieving service differentiation in satellite networks. A gateway architecture for IP satellite networks to achieve *DiffServ* via a joint resource management and marking approach is proposed in [9]. Their objectives are to minimize bandwidth wastage while satisfying QoS requirements. In [10], the authors compare several buffer management policies for satellite onboard switching to differentiate real time and non-real time traffic. The feasibility of using multiprotocol label switching for service differentiation is investigated in [11].

In the context of service differentiation for TCP flows, [12] demonstrates this possibility via joint configuration of transport-level and Medium Access Control (MAC)-level service differentiation mechanisms with extensive simulation and analysis. In [13], the authors attempt to realize *DiffServ* for TCP connections with a full-fledged ATM switch onboard with buffer management capacity. The work more related to ours is [14], where an onboard satellite waiting time priority scheduler is proposed to realize proportional differentiated services [15] in a Bandwidth-on-Demand satellite network. To the best of our knowledge, no previous work has considered an economic-incentive approach to achieve relative service differentiation in satellite IP networks, which is the focus of this paper.

## 3 Theoretical Background of Congestion Pricing

In this section, we briefly review the concept of congestion pricing. For a comprehensive introduction to congestion pricing, readers are referred to [6][16].

We follow the argument in [17] that providing differential QoS implies the use of differential pricing which in turn points to the use of congestion signals to reflect the cost of congestion. Congestion price is dynamic relative to resource loading. When a resource is not saturated, the congestion price is zero. When congestion occurs, the congestion price is an increasing function of extra load on the resource that causes undesirable effects in the network such as increase in congestion and packet loss. Users then react to the congestion price information by adjusting the offered rate according to their WTP.

A formulation of a congestion pricing model is as follows. Suppose that a network consists of a set of nodes and links. The nodes correspond to traffic sources and sinks, while router and link bandwidth are network resources. We denote by $J$ the set of resources and $C_j$ the finite capacity of each resource $j \in J$. Let a route $r$ be a non-empty subset of $J$ between a pair of source and sink, and denote by $R$ the set of all possible routes. We assume that a route is associated with a user and, thus, route and user are used interchangeably. We define a 0-1 matrix $A = (A_{jr}, j \in J, r \in R)$ and set

$$A_{jr} = 1 \qquad \text{if route } r \text{ consumes resource } j$$
$$A_{jr} = 0 \qquad \text{otherwise}$$

Consider user $r$ has a single flow with his WTP ($w_r$). Assume that $r$ ranges over the set $R$ and $j$ ranges over the set $J$, the rate adjustment algorithm [6] (also called *rate-based control*) reacted by users is given by

$$\frac{d}{dt} x_r(t) = \kappa_r \left[ w_r - x_r(t) \sum_{j \in r} A_{jr} \mu_j(t) \right]. \tag{1}$$

where

$$\mu_j(t) = p_j \left( \sum_{r \in R} A_{jr} x_r(t) \right) \tag{2}$$

$x_r$ is the offered rate and $\kappa_r$ is the feedback gain that changes how the rate adjustment behaves. As it can be seen in Eq. (1), the smaller the $\kappa_r$ the smaller the oscillation on the offered rate. The rate adjustment algorithm is to let users react to the charge of using resources by either increasing or decreasing the offered rate based on their WTP. Suppose that $\mu_j$ is the shadow price on using resource $j$, then the summation term in Eq. (1) represents the path shadow price of the user per unit flow and the term $x_r \sum_{j \in r} A_{jr} \mu_j$ represents the total charge to a user with respect to the offered rate $x_r$. We follow [5] to interpret $p_j(y_j)$ of Eq. (2) as shadow price of resource $j$ under the load $y_j$:

$$p_j(y_j) = \frac{d}{dy_j} \delta_j(y_j) \tag{3}$$

where $\delta_j(y_j)$ is the cost function under the load of $y_j$.

The subtraction of the two terms inside the square bracket in Eq. (1) gives the surplus or loss. The degree of rate adjustment depends on how large is this surplus or loss. If the WTP is higher than the total charge, the offered rate increases; otherwise it decreases. This enables the resource to send feedback signals as a shadow price at rate $\mu_j$ to each user who has traffic on it. By adjusting the offered rate, the network attempts to equalize WTP with the total charge at the equilibrium that gives an optimal offered rate. Finally, after each update interval, the offered rate becomes

$$x_r(t+1) = x_r(t) + \frac{d}{dt}x_r(t) \ . \tag{4}$$

In [16], it has been shown that the optimization objective of the above system is to maximize the difference between the total utility over all users (the first term in Eq. (5)) and the total cost incurred on all resources (the second term)

$$\text{Maximize} \quad \sum_{r \in R} w_r \log \lambda_r - \sum_{j \in J} p_j(y_j) \tag{5}$$

where $\lambda_r = \sum_{j \in r} \mu_j$.

## 4  Window-Based WTP Congestion Control in Satellite Networks

We consider a satellite access network where the bottleneck links are typically satellite up/downlinks. User nodes, associated with different WTP, connect to the satellite network through terrestrial-satellite gateway routers under the control of the satellite operator. As gateways in satellite networks are typically prone to congestion, we employ Random Early Detection (RED) queue as packet marking mechanism and ECN as feedback mechanism. We also use standard drop-tail queues for other nodes. The combined use of RED and ECN enables us to probabilistically mark packets that are causing congestion instead of dropping them.

Based on the rate-based control explained in the previous section, we implement a window-based congestion control algorithm introduced by Gibbens and Kelly [5]. Since TCP is the current dominating protocol in the Internet and its congestion control is window-based, exploring a window-based congestion control algorithm is rational as it avoids drastic changes to the entire Internet framework.

Given that users react to feedback signals by adjusting their offered rate, the TCP congestion window (*cwnd*) can be taken as the effective metric to measure the traffic rate allowed to be sent to the network. Since changing the size of *cwnd* effectively varies the number of packets that is allowed to be transmitted in a Round-Trip Time (RTT), window-based control is closely related to rate-based control [18]. The rate calculated at the source is converted to window size by multiplying it with RTT, i.e. $cwnd_r = x_r RTT$ for each user. Therefore, the rate-based and window-based control mechanisms are compatible to each other.

Initially, *cwnd* is assigned to one segment (or packet) and is adjusted for each received packet acknowledgement (ACK). The transmission of packets in *cwnd* and the receipt of their ACKs are done within a RTT. Hence, we consider a window-based congestion control algorithm [5] derived from Eq. (1) and it adjusts *cwnd* by

$$cwnd_r^+ = \overline{k}_r\left(\frac{\overline{w_r}}{cwnd_r} - f\right) \tag{6}$$

per reception of ACK, where $\overline{k}_r = k_r RTT_r$ and $\overline{w}_r = w_r RTT_r$ denotes the feedback gain and WTP per RTT and $f = \lambda_r$ is the shadow price either equals to 0 if the packet is not marked or equals to 1 if the packet is marked.

Since Eq. (6) takes into account feedback delay with queuing delay as a small component in the RTT and single resource congestion, this fits in nicely to satellite network scenarios where the propagation delay is always the dominant component in the RTT while the satellite links are the congested resources. Hence, we base our congestion control mechanism on Eq. (6). By updating *cwnd*, the congestion pricing approach differs from the conventional TCP congestion avoidance algorithm, which follows double multiplicative decrease i.e. the rate of ECN mark is proportional to the offered rate and *cwnd* is halved when congestion occurs [19].

Our implementation uses ACK packets as congestion indicators to update the *cwnd*. An ECN feedback signal is sent back through ACK to the source to react. We use the four bits ECN in [8] for this purpose. Table 1 details the four bits.

**Table 1.** ECN bit descriptions

| Bit | Description |
|---|---|
| ECN-capable (ECT) | Set when traffic flow is ECN aware |
| Congestion Experienced (CE) | Set by the router when it detects an onset of congestion |
| ECN-echo (ECNecho) | Set by sink if CE bit is set in the received packet header |
| Congestion Window Reduced (CWR) | Set by source after adjusting its *cwnd* as a response to the ECNecho set |

When there is no congestion, the source increases the *cwnd* according to the user's WTP based on Eq. (6) with $f = 0$ (i.e. when no ECN marks received, the congestion window is increased by $\overline{k}_r\overline{w}_r$ every RTT or $(\overline{k}_r\overline{w}_r)/cwnd_r$ for every ACK received since the size of *cwnd* has the equal number of ACKs that can be received within one RTT). During the congestion period, the router marks the packets that cause congestion by setting their CE bit to "1". When the receiver (sink) detects the mark, it sets the ECNecho bit in the ACK packet that it sends back to the source. Upon reception, the source, when knowing that the network resource is getting congested, adjusts its *cwnd* based on Eq. (6) with $f = 1$. The source also sets the CWR bit to indicate that it has responded to the congestion.

The total charge to a user is interpreted as the number of marked packets received per RTT. For instance, if ten packets are sent in an RTT and five of their ACKs are marked, then these marks become the charge per RTT. The condition of marking packets is based on two thresholds of RED: *minthresh* and *maxthresh*. Packets are probabilistically marked if the average queue size is between *minthresh* and *maxthresh* whereby the marking probability is a function of the average queue size. If the average queue size exceeds *maxthresh*, all arriving packets are marked.

## 5  Performance Evaluation

We implement the window-based congestion control algorithm on the *ns-2* [20]. Our simulation scenarios include GEO and non-GEO satellite networks; shown in Fig. 1 and Fig. 2 respectively. In this paper, we use Low Earth Orbit satellites for non-GEO satellite scenario. Table 2 shows the network parameters used in our simulation.

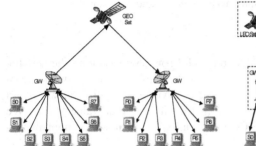

**Fig. 1.** GEO satellite topology            **Fig. 2.** Non-GEO satellite topology

**Table 2.** Network Parameters

| Link Type | Bandwidth / Delay / Buffer /Queue Type |
|-----------|----------------------------------------|
| Up/downlink | 1.5Mbyte/s / 125ms(GEO), 25ms(Non-GEO) / 60 packet / RED |
| Inter-Satellite Link | 25Mbyte/s / 1ms / 60 packet / RED |
| Terrestrial Link | 10Mbyte/s / 5μ / 60 packet / DropTail |

We select File Transfer Protocol (FTP) as the source application to provide long-duration TCP connections, which ensures that network steady state can be reached. The FTPs are configured to be transmitting packets of size 1Kbytes with packet interval set as 0.005s. For all simulations, we also set $\overline{k_r}$ as 0.05 to avoid large oscillation on the rate adjustment. We initially evaluate our implementation under the assumption that all network links are error-free and then proceed to show that the implementation also adapts to fading loss in the satellite links. As mentioned, we use RED queues for gateway routers and drop-tail queues for other nodes.

The following metrics are used to evaluate the performance of our approach:

- **TCP throughput.** we denote $t_r$ and $w_r$ as the TCP throughput and WTP of user $r$ respectively. The algorithm controls the *cwnd* (in segments) of the TCP connections, which directly influences the TCP throughput to be achieved. For this metric, we capture both the instantaneous *cwnd* and TCP throughput over time.

- **Fairness.** in theory, users with higher WTP receive higher TCP throughput. Although the RSD model does not restrict the level of QoS guarantees, the resource allocation should be done in a fair manner with respect to WTP. For fairness evaluation, we capture the *Fairness Index (FI)* [21] given by

$$FI_r = \frac{FQ_r}{IFQ_r} \tag{7}$$

where $FQ_r = t_r / \sum_{p=0}^{n} t_p$ is the fairness quotient and $IFQ_r = w_r / \sum_{p=0}^{n} w_p$ is the ideal fairness quotient. The ideal fairness result is achieved when $FI_r = 1.0$. If $FI_r < 1.0$, the sources get lower expected traffic rate with respect to their WTP, and vice versa.

### 5.1 GEO Satellite System Scenario

**TCP Throughput.** We start eight concurrent FTP sources to saturate the satellite uplink and thus create a congestion scenario in the GEO satellite network. Each source node has an FTP application. To evaluate service differentiation, we assign different WTP values to the users, starting from 1.0 with a step up value of 0.5[1]. We compare our implementation against TCP Reno, one of the major TCP implementations for the current best-effort Internet.

Fig. 3 shows the instantaneous *cwnd* of TCP Reno. As can be seen, TCP Reno does not support service differentiation. It treats all the TCP connections as equal: there is no facility for differentiating among TCP connections. This also applies to the other best-effort based TCP variants. In contrast, as shown in Fig. 4, congestion pricing manages to differentiate the sources according to their WTP. In this case, higher

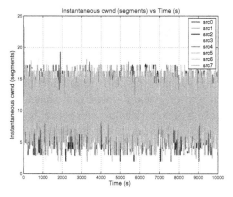
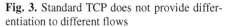
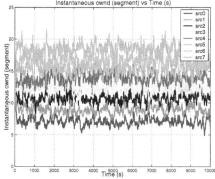

**Fig. 3.** Standard TCP does not provide differentiation to different flows

**Fig. 4.** Congestion pricing manages to differentiate TCP flows

---

[1] In simulation graphs, src0 has WTP of 1.0, src1 has 1.5, src2 has 2.0, src3 has 2.5 and so on.

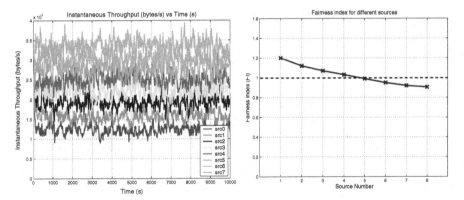

**Fig. 5.** Differentiated TCP throughput according to user's WTP

**Fig. 6.** Fairness index

WTP users consistently have higher instantaneous *cwnd*. With higher instantaneous *cwnd*, higher traffic sending rates are allowed for these sources and, therefore, higher throughput can be achieved. Fig. 5 shows different levels of instantaneous TCP throughput achieved for the FTP sources with different WTP. Fig. 4 and Fig. 5 demonstrate that congestion pricing achieves RSD.

**Fairness.** We proceed to evaluate fairness achieved by the congestion pricing approach. Fig. 6 shows that the *FI* for most of the sources is close to the ideal case. However, when comparing the results with those produced for terrestrial networks [17], we found that the degree of achieved fairness is lower for GEO satellite systems. This is attributed to the long propagation delay of satellite links, which results in slow feedback of the congestion signal to the sources. The reason that the *FI* for high WTP sources is lower than 1.0 can be explained as follows: as packets are marked continuously because of congestion, higher WTP sources would naturally receive more marked packets in proportion to the number of sent packets. As such, they would receive congestion signals in bulk and eventually lead them to further reduce their *cwnd*. In contrast, low WTP sources receive relatively less congestion feedback. Hence, the gentle slope of the performance for GEO satellite system appears. Together with the TCP throughput results presented in the previous section, congestion pricing achieves fair RSD.

**Equal WTP.** Previous evaluations are based on the scenario where each user has different WTP. In theory, users with the same WTP should receive identical TCP connection treatment. In this section, we investigate whether congestion pricing achieves equal performance for those users under a congestion scenario. We repeat the previous simulation with a small change. We assign equal WTP of 1.0 to the first two FTP sources. Fig. 7 shows that the two sources having the same WTP receive the same throughput. This ensures that congestion pricing produces reliable and consistent performance in conformance to RSD model.

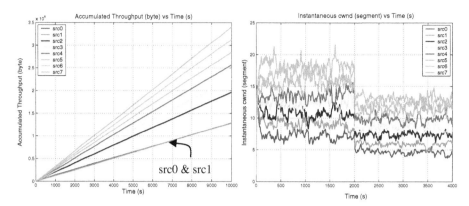

**Fig. 7.** Sources with same WTP receive same treatment in GEO satellite network scenario

**Fig. 8.** A loss in the link condition under GEO satellite network scenario

**Lossy Satellite Links.** We investigate the consistency and adaptability of our ap-proach on satellite links with fading loss. Satellite networks, particularly those using the *Ka band*, are especially susceptible to fade attenuations of signals (e.g. due to rain or bad weather conditions). Two of the more dominant fade countermeasures are *Automatic Repeat Request* (ARQ) and *Forward Error Correction* (FEC). Due to the extra delay incurred by ARQ techniques, its use is basically not recommended [22]. In this paper, we focus on the approach of counteracting fading by the use of adaptive FEC mechanisms which recover erroneous packets that is caused by channel degrada-tion by adding redundancy to packets at the physical layer. The redundancy overhead added is based on the level of signal attenuation measured whereby the redundancy of the packets is increased when C/N (*Carrier/Noise Ratio*) decreases. Therefore, the attenuation effect on a satellite link can be modelled as a decrease of bandwidth [23], as a certain amount of bandwidth has been devoted to carry redundancy overhead rather than information bit. We follow [24] to define the redundancy coefficient for source $i$ as $rc_i = IBR_{cs}/IBR_m$ where $IBR_{cs}$ is the information bit rate under clear sky condition while $IBR_m$ is the information bit rate measured at the specific in-stance. The bandwidth reduction factor is defined as $\phi_i = rc_i^{-1}$. Thus, the actual bandwidth dedicated to the transmission of information bits is given as

$$BW_i^{real} = BW_i \cdot \phi_i; \qquad \phi_i \in [0,1] \tag{8}$$

where $BW_i$ is the overall bandwidth of source $i$. We run our simulation with all sources set with initial value $\phi_i = 1.0$ (clear sky condition) and change the bandwidth reduction factor of all sources to 0.6667 at time, $t = 2000s$ to see how the congestion pricing implementation adapts to a change in the link condition. As it can be seen in Fig. 8, at $t = 2000s$, congestion pricing manages to adapt to the changes rapidly (in seconds) while sustaining a consistent and fair RSD. Note that in real world, the fading level may change rapidly, causing the fade countermeasures to continuously adapt to the new fading level. Our simulation represent the case where fading levels are segregated into *fading classes* in which the countermeasure parameter are fixed for all those fading levels within the same class. Hence, our simulation corresponds to the scenario of a transition from one *fading class* to another.

### 5.2 Non-GEO Satellite System Scenario

We repeat the previous experiments for non-GEO satellite systems. Unlike GEO satellite, the topology of non-GEO satellite is a constellation that connects satellites by inter-satellite links. Note that since our focus is on service differentiation, we ignore the issues arising from satellite mobility such as TCP connection handover.

**TCP Throughput and Fairness.** Fig. 9 and Fig. 10 show the instantaneous *cwnd* and throughput achieved by each source under a non-GEO satellite network scenario. The service differentiation achieved can be clearly seen. Compared to the GEO satellite scenario (Fig. 4 and Fig. 5), the achieved instantaneous *cwnd* and throughput in the non-GEO satellite network scenario are much lower. This is attributed to the shorter propagation delay in non-GEO satellite systems, which accelerates the congestion feedback from the network to the sources. As such, sources can rapidly react to the feedback signals, reducing *cwnd*.

On the other hand, for the evaluation of fairness, we plot the *FI* achieved by each source in Fig. 11. Compared to GEO satellite systems, it is clear that congestion pricing for non-GEO satellite systems achieves near-ideal fairness as the systems have much shorter propagation delay that enables faster reaction to congestion with more up-to-date congestion feedback signals.

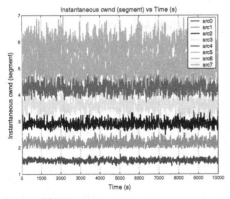

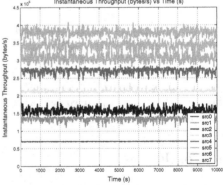

**Fig. 9.** Differentiated TCP throughput based on user's WTP in non-GEO satellite system

**Fig. 10.** Differentiated TCP throughput in the non-GEO satellite system

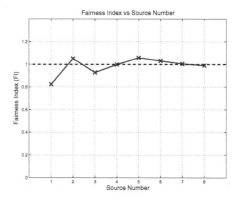

**Fig. 11.** Congestion pricing achieves fair bandwidth sharing in non-GEO satellite system

**Equal WTP and Lossy Satellite Links.** Fig. 12 and Fig. 13 show the performance of equal WTP and lossy satellite links under non-GEO satellite network respectively. We found that the results are similar to those under the GEO satellite system scenario. The sources with equal WTP consistently receive approximately equal TCP throughputs. As for lossy satellite link, the congestion pricing approach manages to achieve and maintain RSD among different sources. The plots basically follow the same trend and behaviour as those in the GEO satellite system scenario.

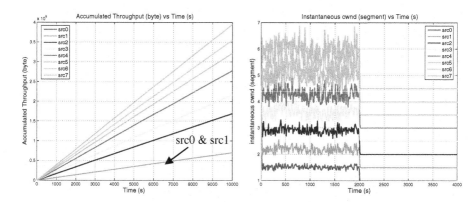

**Fig. 12.** Sources with same WTP receive same treatment in non-GEO satellite network

**Fig. 13.** A loss in the link condition under non-GEO satellite network scenario

## 6   Conclusions

An RSD scheme using congestion pricing is proposed for satellite IP networks. We have implemented the congestion pricing approach using a weighted window-based congestion control algorithm that takes into account users' WTP to react to congestion costs fed back through an ECN mechanism. Our simulation experiments include both GEO and non-GEO satellite network scenarios. Simulation results show that, for the GEO satellite network, congestion pricing achieves a fair RSD among users with different WTP and this is also the case for a network with lossy satellite links. In addition, we observe that TCP flows with the same WTP receive equal treatment in terms of throughput. Similar conclusions were also drawn for non-GEO satellite networks. However, due to shorter link propagation delay, the achieved performance such as *cwnd* and throughput are smaller than that of the GEO satellite network. In addition, we observe that in non-GEO satellite network, near-ideal fairness can be achieved. As future work, we will investigate service differentiation using Multi-level ECN which allows multiple levels of congestion to be disseminated. This will provide more accurate network status for users to react to their offered rates.

## Acknowledgement

This work has been performed within the framework of the Information Society Technologies Research Framework Program 6 (FP6) Network of Excellence SatNEx project, funded by the European Commission (EC). The financial contribution of the EC towards this project is greatly appreciated and acknowledged.

## References

1. R. Braden, D. Clark and S. Shenker, "Integrated Services in the Internet architecture: an overview," *RFC 1633*, June 1994.
2. S. Blake et al., "An architecture for Differentiated Service," *RFC 2475*, December 1998.
3. P. Key, "Service Differentiation: Congestion Pricing, Brokers and Bandwidth Futures," Proc. *ACM NOSSDAV'99*, , 1999.
4. C. Dovrolis, and P. Ramanathan, "A case for relative differentiated services and the proportional differentiation model," *IEEE Network*, September / October 1999, pp. 2-10.
5. R. J. Gibbens and F. P. Kelly, "Resource pricing and the evolution of congestion control," *Automatica*, 35, 1998.
6. F. P. Kelly et al., "Rate control for communication networks: shadow prices, proportional fairness and stability," *Journal of the Operation Research Society*, 49, 1998, pp. 237-252.
7. S. Floyd and V. Jacobson, "Early Random Detection Gateways for Congestion Avoidance," *IEEE/ACM Transaction on Networking*, vol. 1, no. 4, 1993, pp. 397-413.
8. K. K. Ramakrishnan, S. Floyd and D. Black, "The Addition of Explicit Congestion Notification (ECN) to IP," *RFC 3168*, September 2001.
9. L. S. Ronga et al., "A gateway architecture for IP satellite networks with dynamic resource management and DiffServ QoS provision," *Int'l Journal of Satellite Communications and Networking*, no. 21, 2003, pp. 351-366.
10. N. Courville, "QoS-oriented traffic management in multimedia satellite systems," *Int'l Journal of Satellite Communications and Networking*, no. 21, 2003, pp. 367-399.
11. T. Ors and C. Rosenberg, "Providing IP QoS over GEO satellite systems using MPLS," *Int'l Journal of Satellite Communications and Networking*, no. 19, 2001, pp. 443-461.
12. M. Karaliopoulos et al., "Providing differentiated service to TCP flows over bandwidth on demand geostationary satellite networks," *IEEE JSAC*, vol. 22, no. 2, 2004, pp. 333-347.
13. A. Duressi et al., "Achieving QoS for TCP traffic in satellite networks with differentiated services," *Space Communications*, vol. 17, no. 1-3, 2001, pp. 125-136.

14. W. K. Chai et al., "Scheduling for proportional differentiated service provision in geostationary bandwidth on demand satellite networks," Proc. *IEEE GLOBECOM*, 2005.
15. C. Dovrolis et al., "Proportional differentiated services: delay differentiation and packet scheduling," *IEEE/ACM Transactions on Networking,* vol. 10, no. 1, 2002, pp. 12-26.
16. F. P. Kelly, "Charging and rate control for elastic traffic," *European Transactions on Telecommunications*, 8, 1997, pp. 33-37.
17. P. B. Key et al., "Congestion pricing for congestion avoidance," Microsoft Research Technical Report MSR-TR-99-15, MSR, 1999.
18. F. P. Kelly. *Mathematical modeling of the Internet "Mathematics Unlimited – 2001 and Beyond"*. Springer-Verlag, Berlin 2001, pp. pp. 685-702.
19. V. A. Siris, C. Courcoubetis and G. Margetis, "Service differentiation and performance of weighted window-based congestion control and packet marking algorithms in ECN networks," *Computer Communications*, vol. 26, 2003, pp. 314-326.
20. Network Simulator-ns (ver. 2) [Online]. Available at http://www.isi.edu/nsnam/ns/
21. M. Shreedhar and G. Varghese, "Efficient Fair Queuing using Deficit Round Robin," Proc. *ACM SIGCOMM*, 1995.
22. M. Marchese and M. Mongelli, "On-line bandwidth control for quality of service mapping over satellite independent service access points," to appear in *Computer Networks*, 2006.
23. R. Bolla, F. Davoli and M. Marchese, "Adaptive bandwidth allocation methods in the satellite environment," Proc. *IEEE ICC*, 2001, pp. 3183-3190.
24. N. Celandroni et al., "An overview of some techniques for cross-layer bandwidth management in multi-service satellite IP networks," Proc. *IEEE GLOBECOM Workshop Advances in Satellite Communications: New Services and Systems*, pp. WO4.4:1-WO4.4:6, 2005.

# Signalling Concepts in Heterogeneous IP Multi-domains Networks

C. Chassot[1,2], A. Lozes[1,3], F. Racaru[1], and M. Diaz[1]

[1] LAAS/CNRS, 7 avenue du Colonel Roche
31077 Toulouse, France
{chassot, alozes, fracaru, diaz}@laas.fr
[2] INSA Toulouse, Dep GEI Complexe Scientifique de Rangueil
31077 Toulouse cedex, France
[3] IUT Toulouse III, Dep GEII Complexe Scientifique de Rangueil
31077 Toulouse cedex, France

**Abstract.** This paper presents a basic and global approach to design a signalling that is able to guarantee Quality of Service (QoS) in heterogeneous networks. The signalling protocol handles the requests of connections for end-to-end QoS paths between heterogeneous domains. The proposed approach is based on the use of the minimum information needed and on performance figures existing for each domain. It defines the end-to-end QoS path only as a set of inter-domain requirements between border routers; then in particular, this means that all internal domain paths have no implementation constraints and so may be implemented in the most efficient way by all domain providers and their different technology. The architecture shows how to build the sequence of the needed path equipments and of the ingress and egress routers to construct the end-to-end admission-controlled path.

## 1 Introduction

Different models to provide Quality of Service (QoS) in the Internet, IntServ [Shen97] [Wroc97] [Brad97a] [Brad97b] and DiffServ [Blak98] [Jaco99] [Hain99] have been proposed. They are used as a basis for several projects, such as in Europe MESCAL [Howa05] and EuQoS [Dugeon05], AQUILA [Enge03], TEQUILA [Myko03], CADENUS [Corte03]. It appears now that any global QoS solution is quite difficult to reach as:

- on one hand, the QoS requirements are coming from the applications and can be quite different,
- on the other hand, the multi-domains underlying network architecture must impose as weak constraints as possible on the end-to-end QoS design.

As this has been done in the Best-Effort Internet, we assert that any QoS Internet solution, to be successful, must take into account two basic commitments:

- it must be able to handle the heterogeneity of all QoS domains in all technologies,
- and it must provide all freedom needed to take all technical solutions that can be developed by all designers and implementers.

Y. Koucheryavy, J. Harju, and V.B. Iversen (Eds.): NEW2AN 2006, LNCS 4003, pp. 259–270, 2006.

Even when starting from the assumption that provisioning is adequate in a given network, it can always occur that all requests cannot be served, in particular when medium speed technologies, as wireless networks, are connected to very high speed networks, as optical networks. To solve the possible resulting overuse, connection admission control (CAC) has to be deployed. Solving the CAC problem in an end-to-end basis means to use a signalling protocol through all the domains existing in the end-to-end path. This paper deals with the design and the implementation of a simple solution to provide such an inter domain signalling.

## 1.1  The Two Approaches

Two approaches have been developed so far:

- the multi-domain provisioning defines end-to-end services having pre-defined satisfactory performances;
- when it is not possible, a signalling protocol dynamically has to check the end-to-end matching to invoke the best service with respect to the QoS requirements.

The second approach that suggests a characterization of the end-to-end performances has been proposed in [Chass05]. This paper will precise the signalling protocol that is used for defining and implementing the end-to-end CAC mechanisms.

## 1.2  NSIS Signalling Protocol

The IETF NSIS[1] working group [Hanc05a] [Fu05] considers:

- a path-coupled IntServ based approach by involving border routers to perform CAC within their domain and on their inter domain links [Sals03] [Bade05];
- a hierarchical path-decoupled management of each domain by using Bandwidth Brokers (BB) [Nich99] in charge of the intra domain CACs. Then, end-to-end CAC needs to involve all BBs of all domains in the path.

This paper presents an end-to-end basis for a path-decoupled signalling. The differences between the signalling path and the data path are given in Fig.1:

- the left part of Fig. 1 shows the *data path* given by the end-to-end routing protocol, e.g. the path given by the BGP selection of the list of the ASs[2] from the sending AS to the receiving AS, through the set of all intermediate ASs;
- the right part of Fig. 1 gives the *signalling path*, i.e. the path that has to be followed during the signalling phase, before the data transfer, by the signalling messages. This path has to be used, from one BB to the next BB, in order to reserve the resources in the different ASs that will be taken by the data path. Note that now the signalling messages have to follow in all ASs a transfer path that is different from the data path (as it has to go through all BBs in the ASs).

---

[1] NSIS: Next Step in Signalling.
[2] AS: Autonomous System.

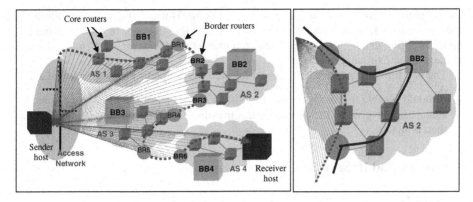

**Fig. 1.** Linking the data path and the BB signalling paths

The present NSIS proposal [Hanc05b] diverts the signalling packets towards the BBs at the entry of the domains, by the border routers. After processing, the BB sends the packets back to the router, which transmits them to the exit router.

In order to be more general, this paper discusses a direct communication between the BBs, using extended BGP tables. In particular, it follows that this approach is independent from the routing protocol as it will be described later.

The remainder of the paper is structured as follows: an end-to-end IP service characterization given in [Chass05] is recalled in section (2); the approach to address the signalling problem is described in section (3); finally, conclusions and future work are given in section (4).

### 1.3  Previous Work

[Mant04] proposes a solution based on the establishment of end-to-end pipes[3] for which bandwidth reservation has been performed. Those pipes only concern the provider's domain and are only established for a few predefined services, supported by all domains. The concatenation choice consists in the choosing of a pipe compatible with the request. Unfortunately, a strong homogeneity of the provided services is assumed and the automation of the pipe set up is an open issue. MESCAL is based on [Gode01] to quantify the performance of services supposed to be subscribed before the invocation step. The concatenation choice is based on the extended QoS class concept, recursively defined. [Fuze03] introduces mechanisms allowing the providers to exchange QoS parameters for the supported services in order to provide information on the available QoS before the SLS; this proposal is independent of the forms of the QoS request and is applicable for the domains involved in the data path. [Pereira05] considers a bandwidth management mechanism based on a mapping between Integrated Services (IntServ) and Differentiated Services (DiffServ). This mapping includes how to translate new IntServ flows into a DiffServ network. IntServ and DiffServ have the problem that the solution needs to be deployed along all the path, i.e. in all domains crossed by the path, and it appears that

---

[3] I.e. traffic aggregate of a same service class which starts from an exit border router and ends to an entry border router of another domain (not necessarily adjacent).

such a requirement is quite difficult to fulfil. [Yang05] uses service vectors allowing the choice of different successive PHB within a single domain, in order to obtain a concatenation matching the targeted requirements. However, the multi-domain context is not explicitly considered.

The two fundamental properties to be fulfilled by an open signalling protocol are:

- it should be able to use any existing routing solution, e.g. BGP or its extensions,
- and it should give to the providers the maximum freedom to implement QoS in their domains.

## 2    End-to-End IP Services Characterisation

### 2.1   Model

The usual performance characterisation of a DiffServ domain (from an entry point to an exit point) is often given in terms of maximal transit delay and/or jitter [Gode01]. For delays, in previous works, [Chas03] characterizes the performances between two edge routers by the *cumulative distribution function (CDF) of the transit delay*. Considering X, the transit delay of each packet between two edge routers as a random variable, the CDF $F_X$ is defined by $F_X(t) = P(X < t)$, where P defines, for a packet, the probability that its transit delay is lower than $t$. Such a characterization allows the required QoS to be expressed in terms of: partial or total reliability, maximal or average transit delay, bounded jitter.

### 2.2   Architecture Performance

Some measurements showed that for an IP service class different from best effort (BE), a function can be defined to be independent of: (1) the amount of QoS flows circulating in the network, and (2) the network topology, for a given load of BE traffic and a given intra domain path [Chas02] [Auri03].

In this paper, we assume that such estimation is available in the BBs, for each domain and for all couples of edge routers. Assuming that the transit delays in the different domains are independent in probability, the convolution product can be used for multiple domains. Thus, a characterization of the performances provided by each service resulting from the concatenation of these domains can be obtained and used in the end-to-end path. It follows that the domains concatenation and the CAC are performed dynamically after a QoS request, and are expressed by parameters such as maximal transit delay, maximal loss rate, etc.

## 3    Selecting Services Meeting QoS Requirements

The concatenation selection of the services is then resolved in three steps Chass05]: (1) the service classes (and their performances[4]) available on each domain of the data path(s) are discovered; (2) the end-to-end performance model is evaluated for all

---

[4] Expressed by means of the CDF of the transit delay.

service classes available on the data path by convolution; and (3) the cheapest service satisfying the QoS request is selected by an adequate algorithm.

Defining the end-to-end path, the resulting performance and later the corresponding CAC algorithm mean that:

- the data and signalling paths have to be precisely defined: this can be done in the IP world relying on the use of the BGP protocol, and on the existence of the end-to-end AS lists in the BGP tables;
- the availability check of the needed resources along the data path has to be done by all BBs in the AS path, including the verification of the conformity of the request against the subscribed contract;
- if the needed resource are available, the resources are pre-reserved if needed, then confirmed, and if not, they are released.

### 3.1 Open Signalling Protocol

**Assumptions.** In order to define a basic open solution, that does not include all details needed by a full protocol, a small set of assumptions has to be used:

- the QoS model in all domains is homogeneous, in order to allow the definition and handling of compatible or standard solutions;
- each BB knows the IP services for all its couples of border routers, or has an upper bound; note that this is consistent with the fact that a BB is the CAC/signalling equipment of its domain.

From a practical point of view, an AS is under a given sphere of management, but the ASs can have different capabilities, different services and different parameter values. As present ASs provide only accessibility, a Best-Effort service only guarantees the reachability of the destination user. For providing QoS, new QoS-Domains, or QoS-ASs have to be designed in order to guaranteed value limits to their users.

*Definition 1: QoS-Domains or Q-AS.* A QoS-domain is a domain that is either Controlled, a C-AS or Over-provisioned, an O-AS.

*Definition 2: Over-provisioned QoS-Domain.* A QoS-domain D is Over-provisioned if it is such that, for any possible communication O-in(i) entering it via an ingress border router, the domain is able to transfer this communication as an output O-out(i), to the next domain by an egress border router, by introducing a modification of its properties (Prop) that is under a well defined and accepted threshold: for all communications O-in in any O-AS: $Prop(O\text{-}out(i)) = F(Prop(O\text{-}in(i)) \leq \lambda$, where F defines the function of modification between the ingress and egress points. Of course, the domain contains enough resources to satisfying all the requests.

Prop(O-in) can have different meaning, such as bandwidth, delay or loss rate. For instance, a bandwidth guarantee through the domain will be defined as: for all i: $Bandw(O\text{-}out(i)) = Bandw(O\text{-}in(i))$.

*Definition 3: Controlled QoS-Domain.* A QoS-domain is Controlled if it not Over-provisioned and if it contains a control function to guarantee that: for all or some of the communication requests: $Prop(O\text{-}out(i)) = F(Prop(O\text{-}in(i)) \leq \lambda$.

Indeed, as the domain is not over-provisioned, all possible entering communications O-in(i) cannot be handled without modifying their properties. Then a domain is Controlled if it exits a control function able to select a subset of the entering communications, O-in-qos(i), for which it provides an outgoing O-out-qos(i), while guaranteeing the modification of their properties to be under a given threshold.

For instance, a resource manager, or Bandwidth Broker (BB), will be in charge of selecting the accepted communications, i.e. of the CAC function. Note that no resource manager is needed in an over-provisioned AS as all O-out communications are guaranteed.

**AS Configuration, Registration and CAC.** In order to deploy such architecture, three design steps have to be followed:

*1. Establishing the architecture: Configuration.* The domains have to be defined and announced, together with their BB, to form a chain of QoS-domains between two set of users. A given BB, BB(i) in AS(i) will have to receive all its adjacent AS(j) numbers and their types in terms of QoS, that can be Best Effort (BE), controlled, or overprovisonned. For BE ASs, all paths going trough them will be BE. For all controlled ASs, the BB in AS(i) will have to receive the @ of their adjacent BBs to establish paths. When one AS is overprovisonned, the set of the BBs of its adjacent controlled ASs should be given.

*2. Populating the architecture: Registration.* Upon registration the user is linked to a given controlled or over-provisioned AS. It will receive the IP @ of its BB, the BB in its domain. All QoS users will then be able to start a QoS request by asking a QoS path to their BB.

*3. Checking and reserving the QoS path: CAC.* When requesting a QoS connection, sending it request to its BB, the user will delegate the reservation of the requested parameter to the BB. The BB will find the data path, and checks the available QoS resources along the QoS path at the time of the request. If enough resources exit, it will establish the connection and start the user. If not, it will tell the user the available resources.

**Signalling Protocol.** The signalling protocol will then be handled by:

- the sender and receiver hosts, in the two different (remote access) domains,
- the registration server, that provides to the sender host the BB address,
- the BB of each domain (where the access domain BB is the local BB).

*Communication Rules.* The sequence of domains and the two data and signalling paths are given in Fig. 2. Fig. 2-a gives the simple global view of a set of domains, while Fig. 2-b shows the modification needed between the paths. It is then clear from Fig. 2 that a simple solution for signalling can be given if a solution can be found that starts from the data path and then modifies this data path when needed in order to force the signalling path to go through the BBs, while keeping all information related to the data path.

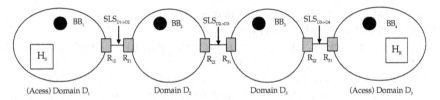

(Acess) Domain $D_1$        Domain $D_2$        Domain $D_3$        (Acess) Domain $D_3$

SLS : Service Level Specification (technical part of a DiffServ contract)

a) The BGP-based end-to-end data path

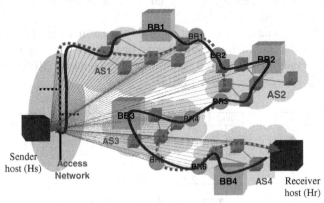

b) The BGP data path and the signalling

**Fig. 2.** Multi-domain Internet model

The simplest solution for signalling is obtained for paths for which the input and output border routers are identical for all inter-domain links. This can be derived from the fact that the global BGP will provide the inter-domain route. Modifying this inter-domain route means to modify the output of BGP, which will be quite difficult and then goes to building of the up to now unsuccessful hard path construction. Nevertheless, in all domains, all decisions are local, so any routing can be defined within one domain. Keeping for both data and signalling path the same couples of borders routers and freeing the internal routes within one domain allow any proposed solution to be based on BGP, or on any AS-to-AS external routing protocol, and to be able to modify the internal routing for the signalling and data phases. So the end-to-end inter-domain path is given, and all local parts of the end-to-end path (i.e. within each AS) can be freely selected by the AS providers, depending on local constraints.

The corresponding signalling protocol has then to be defined. In particular it should first include PDUs[5] for:

- the exchanges between the sender and the receiver hosts ($H_S$ et $H_R$ in Fig. 2-a)
- the exchanges between the sender host and its local BB ($H_S$ and $BB_1$ in Fig. 2-a)
- the creation of an end-to-end QoS path.

---

[5] PDU: Protocol Data Unit.

Let us consider Fig. 3 the case of a request coming from a client (the receiver host) which is sent to a server (the sender host).

Globally, the receiver will send the request by say a REQ_1 PDU and later the sender will answer the request by say ANS_1.

The sender sends its request to its BB (REQ_2), including the QoS request.

As the BB is in charge of the QoS discovery, and as it does not know the QoS parameters of the next ASs, some exchanges are needed for retrieving these QoS parameters: the BB requests these QoS parameters (REQ_3) and the next BB answers (ANS_3); the resource pre-reservations are also done by the next BB. This is performed along the end-to-end path.

Once all QoS parameters have been retrieved by the first BB, the adequate set of parameters has to be chosen, by example by the first BB, and the following exchanges have to confirm the reservation by a set of reservation requests (RESV_i); if needed, refresh messages can also be used (REFR_i).

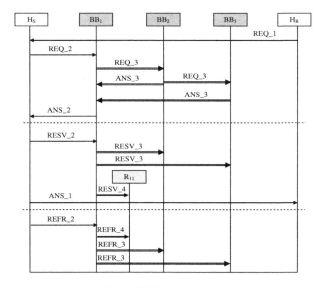

**Fig. 3.** PDUs exchanges

Two problems need to be solved for the above introduced signalling: (1) in order to propagate the request to the next BB, it must be identified, and (2) the BB must discover the ingress and egress points of its domain through which data will pass.

Our solution for the case of the interconnection of Controlled QoS-Domain is presented first; then the case of controlled QoS domains together with over-provisioned QoS domains is approached through a simple but realistic example.

*Case of an interconnection of Controlled QoS-Domain (C-AS).* Let's consider again the multi-domain environment pictured Fig. 2a).

Our solution relies on the assumption that BGP tables of border routers can be accessed by a BB. The main information contained in a BGP table[6] (see Table 1 which illustrates the BGP table of router $R_{21}$) is:

− a list of accessible destination IP network;
− for each destination: the next router's address (*next-hop*) in the adjacent domain, and the list of the domains[7] crossed (*AS path*) from the adjacent domain to the destination domain;
− for each adjacent domain: the next neighbour router's address (*neighbour*).

Table 1 provides $R_{21}$'s *next hop* view and *neighbours* view. To ensure the routing process from an ingress router (e.g. $R_{21}$) in a domain (e.g. $D_2$) to another domain (e.g. $D_4$), $R_{21}$ must know its neighbour router (here $R_{22}$), that connects it with the next adjacent domain (here $D_3$). This neighbour can be in the same domain or can be the *next hop* in the adjacent domain.

**Table 1.** BGP Table of $R_{21}$

| IP Prefix | Next hop | AS Path | | Domain | Neighbour |
|---|---|---|---|---|---|
| {IP prefix} of $D_1$ | $R_{12}$ | $D_1$ | | $D_1$ | $R_{12}$ |
| {IP prefix} of $D_2$ | $R_{31}$ | $D_3$ | | $D_3$ | $R_{22}$ |
| {IP prefix} of $D_3$ | $R_{31}$ | $D_3, D_4$ | | | |

To identify the next BB, each BB maintains a list of couples (*adjacent domain, BB address of adjacent domain*) coming from the peering agreement. When receiving a request PDU (which identifies the receiver's IP address), the BB looks at the BGP table of the ingress router; it extracts the first AS (domain) in the *AS path*, and furthermore identifies the next BB (and the ingress router's *next hop*) of this domain: the BB must have a table that contains, for all adjacent ASs, the address of the corresponding BB, and this will be part of the peering agreement. The BB is then able to propagate the request to the next BB indicating the ingress router of the next domain. For instance, receiving a REQ_3 PDU (@$IP_{dest}$ = @$IP_{HR}$, ingress router = $R_{21}$) from $BB_1$, $BB_2$ consults the BGP table of $R_{21}$ and identifies: (1) the next domain of the *AS path* leading to @$IP_{HR}$, i.e. $D_3$, and (2) the border router through which the data flow enters in $D_3$, i.e. $R_{31}$. When the $BB_3$ address is found, $BB_2$ is able to propagate the REQ_3 PDU with the $R_{31}$ address.

As explained before, the ingress router address is given in a REQ_3 PDU. In order to identify the egress router, the BB has to check the *neighbours* table of the ingress router and gets the router address corresponding to the first domain found in the *AS path* (indicated in the *next hop* table). For instance, in the previous example, $BB_2$ reads the *neighbours* table of $R_{21}$ and then identifies $R_{22}$ to be the egress router of domain $D_2$ to go to domain $D_3$.

*Case of an interconnection of Controlled QoS domains (C-AS) and over-provisioned QoS domains (O-AS)*. Let's consider the case of an interconnection of two controlled QoS access domains by over-provisioned QoS domains (O-AS):

---

[6] Distributed on each BGP device in the domain.
[7] We are assimilating domain to AS (*Autonomous System*).

typically, domains $D_2$ and $D_3$ of Fig2 b) are two O-AS and then are not provided with BB. Of course, this configuration is not the general one, but it is the one that supports the current European research networks.

To identify the next (and last) BB, the first BB has to consult the BGP table of its egress router, and then the table which contains the associations between AS and BB. From the point of view of the last BB, the ingress router discovery is trivial as it is based on the fact that access domain often have only one border router.

**Current Work.** A prototype of our protocol has been and implemented in JAVA; it is being simulated on a multi-domains emulation platform to prove its feasibility. A variant of this approach is being developed as an extended NSIS protocol in the EuQoS project.

## 4   Conclusion and Perspectives

This article addresses the problem of providing a guaranteed end-to-end QoS over multiple domains. More precisely, it proposes a simple solution to select the sequence of services that satisfies a QoS request for data through several domains. With regard to other work, it requests available service performances on the data path at the time of the request, and this can performed by choosing the service concatenation on the data path that best matches the application requirements. This leads to:

– a new approach, coupling service performance discovery and admission control, on the set of the domains on the data path;
– a signalling protocol that implements this approach under the assumption of a BB-based multi-domains Internet;
– the BB identification and ingress and egress routers discovery for each of the domains involved in the data path.

The main perspectives of this work are:

– the integration of a new Transport layer, as. DCCP, in the end-to-end service ;
– to take into account the heterogeneity of the Internet with non QoS domains;
– to finalise in EuQoS the extension and implementation of an NSIS based solution.

*Acknowledgments.* This work has been is partially conducted within the framework of the European IST EuQoS project (http://www.euqos.org).

## References

[Auri03]     Auriol G, Chassot C, Diaz M. Evaluation des performances d'une architecture de communication à gestion automatique de la QdS. $10^{\text{ème}}$ Colloque Francophone sur l'Ingénierie des Protocoles (CFIP'03), Paris, France, Sept. 2003.
[Bade05]     Bader A, Westberg L, Karagiannis G, Kappler C, Phelan T. RMD-QOSM - The Resource Management in DiffServ QoS Model. Internet draft, work in progress, October 2005
[Blak98]     Blake S, Black D, Carlson M, Davies E, Wang., Weiss W. An Architecture for Differentiated Services. RFC 2475, December 1998.

[Brad97a]    Braden R, Zhang L, Berson S, Herzog S, Jamin S. Resource ReSerVation Protocol (RSVP) - Version 1 Functional Specification. RFC 2205, September 1997.

[Brad97b]    Braden R, Zhang L. Resource ReSerVation Protocol (RSVP) - Version 1 Message Processing Rules. RFC 2207, September 1997.

[Chas02]    Chassot C, Garcia F, Auriol G, Lozes A, Lochin E, Anelli P. Performance analysis for an IP differentiated services network. International Communication Conference (ICC'02), New York, USA, May 2002.

[Chas03]    Chassot C, Auriol G, Diaz M. Automatic management of the QoS within an architecture integrating new Transport and IP services in a DiffServ Internet. 6th IFIP/IEEE International Conference on Management of Multimedia Networks and Services (MMNS'03), Belfast, Ireland, September 7-10, 2003.

[Chass05]    A user-based approach for the choice of the IP services in the multi-domains DiffServ Internet, C. Chassot, A. Lozes, F. Racaru, G. Auriol, M. Diaz, 1st IEEE Workshop on Service Oriented Architectures in Converging Networked Environments, Vienna-Austria, April 18-20, 2006

[Corte03]    Cortes G, Fiutem R, Cremonese P, D'Antonio S, Esposito M, Romano S.P, Diaconescu A. CADENUS: Creation and Deployment of End User Services in Premium IP Networks. IEEE Communications Magazine, vol. 41, n°1, January 2003.

[Dugeon05] End-to-end Quality of Service over Heterogeneous Networks (EuQoS), O. Dugeon, D. Morris, E. Monteiro, W. Burakowski, Warsaw University of Technology, M. Diaz, IFIP Network Control and Engineering for QoS, Security and Mobility, Net-Con'2005, November 14–17, 2005, Lannion, France

[Enge03]    Engel T, Granzer H, Koch B.F, Winter M, Sampatakos P, Venieris I.S, Husmann H, Ricciato F, Salsano S. AQUILA: Adapative Resource Control for QoS Using an IP-based Layered Architecture. IEEE Communications Magazine, vol. 41, n°1, January 2003.

[Fu05]    Fu X, Schulzrinne H, Bader A, Hogrefe D, Kappler C, Karagiannis G, Tschofenig H, Van den Bosch S. NSIS: A New Extensible IP Signalling Protocol Suite. IEEE Communications Magazine, Internet Technology Series, October 2005.

[Fuze03]    Füzesi P, Németh K, Borg N, Holmberg R, Cselényi I. Provisioning of QoS enabled inter-domain services. Computer Communications, vol. 26 n°10, June 2003.

[Gode01]    Goderis D, T'joens Y, Jacquenet C, Memenios G, Pavlou G, Egan R, Griffin D, Georgatsos P, Georgiadis L, Van Heuven P. Service Level Specification Semantics, Parameters and negotiation requirements. Internet draft, June 2001.

[Hanc05a]    Hancock R, Karagiannis G, Loughney J, Van den Bosch S. Next Steps in Signalling (NSIS): Framework. IETF RFC 4080, June 2005.

[Hanc05b]    Hancock R, Kappler C, Quittek J, Stiemerling M. A Problem Statement for Path-Decoupled Signalling in NSIS. Internet Draft, work in progress, May 2005.

[Hein99]    Heinanen J, Baker F, Weiss W, Wroclawski J. An Assured forwarding PHB. RFC 2597, June 1999.

[Howa05]    Howarth MP, Flegkas P, Pavlou G, Wang N, Trimintzios P, Griffin D, Griem J, Boucadair M, Morand P, Asgari A, Georgatsos P. Provisioning for inter-domain quality of service: the MESCAL approach. IEEE Com. Magazine, Vol. 43, n°6, June 2005

[Jaco99]    V. Jacobson, K. Nichols, K. Poduri. An Expedited Forwarding PHB. RFC 2598, June 1999.

[Mant04]    Mantar H, Hwang J, Okumus I, Chapin S. A scalable model for inter bandwidth broker resource reservation and provisioning. IEEE Journal on Selected Areas in Communications, Vol. 22, Issue 10, Dec. 2004.

[Myko03]    Mykoniati E, Charalampous C, Georgastos P, Damilatis T, Godersi D, Trimintzios G, Pavlou G, Griffin D. Admission Control for Providing QoS in DiffServ IP Networks: the TEQUILA Approach. IEEE Communications Magazine, vol. 41, n°1, January 2003.

[Nich99]    Nichols K, Jacobson V, Zhang L, A Two-bit Differentiated Services Architecture for the Internet. RFC 2638, July 1999.

[Pereira05] António Pereira, Edmundo Monteiro, Bandwidth Management in IntServ to DiffServ Mapping, Quality of Service in Multiservice IP Networks (QoS-IP), Catania , Italie, February 2005

[Sals03]    Salsano S, Winter M, Miettinen N. The BGRP Plus Architecture for Dynamic Inter-Domain IP QoS. First International Workshop on Inter-Domain Performance and Simulation (IPS'03). Salzburg, Austria, 20-21 February, 2003.

[Shen97]    Shenker S, Partridge C. Specification of Guaranteed Quality of Service. RFC 2212, September 1997.

[Wroc97]    Wroclawski J. Specification of the Controlled-Load Network Element Service. RFC 2211, September 1997.

[Yang05]    Yang J, Ye J, Papavassiliou S. A flexible and distributed Architecture for adaptive End-to-End QoS Provisioning in Next Generation Networks. IEEE Journal on Selected Areas in Communications, vol. 23, n°2, February 2005.

# Operator's QoS Policy in WiBro Implementation

Kyu Ouk Lee, Jin Ho Hahm, and Young Sun Kim

Electronics and Telecommunications Research Institute (ETRI)
161 Gajeong-dong, Yuseong-gu, Daejeon, 305-700, Korea
Tel.: 82-42-860-5756
kolee@etri.re.kr

**Abstract.** Four methods of mobile IP (MIP) registration, call admission control (CAC), L2/L3 mobility and DiffServ mapping are mentioned for QoS assurance according to the admission of four classes of services as unsolicited grant service, real time polling service, non-real time poling service and best effort service in WiBro implementation. At operator's first implementation stage, QoS assurance method of only L2 mobility is applied with QoS parameter and scheduling function of CAC, here CAC factors are maximum/minimum bandwidth and traffic priority. At operator's second implementation stage, QoS assurance method of L2 and L3 mobility is applied with QoS parameter of CAC and scheduling function of its CAC. Here, CAC factors are maximum/minimum bandwidth, traffic priority, maximum delay and tolerated jitter. Finally, at the third implementation stage of operators, QoS assurance for nrtPS and rtPS rtPS service class are almost same as second implementation stage, but operator will strength QoS manager functions compare to previous stage for more accurate QoS assurance.

**Keywords:** MIP, CAC, DiffServ, rtPS, PSS, RAS, ACR, ER.

## 1 Introduction

The WiBro (Wireless Broadband) is one of killing portable Internet service with fore-casted number of subscribers are 10 million by year 2010 in Korea [4], and overall network architecture of WiBro is shown in Figure 1 [6]. Main user terminals (PSS) are PDA (Personal Digital Assistant), notebook PC, smart phone and hand phone set, and main services are Internet shopping, bank and stock transactions, ticketing, material searching and downloading, on-line game and education, video telephony, VOD, E-mail, chatting and messaging, information searching of traffic, news, location, maps. Using frequency is 2.3GHz, and maximum transmission speed is 8Mbps with maximum 60km/h mobility.

The WiBro is similar to WiMax (World Interoperability for Microwave Access), W-CDMA or HSPDA (High Speed Downlink Packet Access), but there are slight technical difference between them as shown in Table 1.

The WiBro services can be classified into four categories as best effort service, non real time polling service (nrtPS), and real time polling service (rtPS), and unsolicited grant service (UGS) in IEEE 802.16 owing to the QoS requirements as Table 2 [5].

Y. Koucheryavy, J. Harju, and V.B. Iversen (Eds.): NEW2AN 2006, LNCS 4003, pp. 271 – 277, 2006.

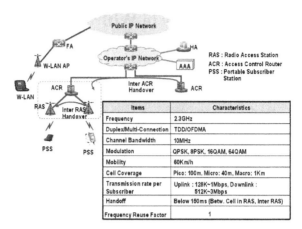

**Fig. 1.** Network Architecture of WiBro

**Table 1.** Technical Comparison among WiBro, WiMax, and W-CDMA

| Classification | WiMax | WiBro | W-CDMA (HSPDA) |
|---|---|---|---|
| Max. Trans. Speed | 70Mbps | 30Mbps | 14.4Mbps |
| Average Trans Speed per Subs. | 7Mbps | 1Mbps | 500Kbps |
| Mobility | 4Km/h | 60Km/h | 250Km/h |
| Terminals | Notebook, PDA | Notebook, PDA, Hand-phone | Hand-phone |
| Service Category | High Speed Data | High Speed Data | Voice-orient, low speed data |
| Cell Coverage | 50Km | 1Km | 3 ~ 4Km |
| Frequency Band | <11GHz | 2.3GHz | 1.9~ 2.2Ghz |
| Duplex | TDD/FDD | TDD | FDD |
| Multiple Connection | OFDM | OFDM | DS-CDMA |
| Commercial | 2006 | 2006 | 2006 |

**Table 2.** Class of Services and QoS Requirements

| Class of Services | QoS Requirements | | | Kinds of services |
|---|---|---|---|---|
| | QoS Factors | Premium | Basic | |
| UGS | Packet Delay | < 150ms | < 250ms | VoIP |
| | Delay Jitter | < 30ms | < 50ms | |
| | Packet Loss | < 0.3% | < 0.5% | |
| | Guarantee | > 99.9% | > 99.5% | |
| rtPS | Packet Delay | < 300ms | < 600ms | Video Phone, Video game, VOD (Video On Demand), AOD (Audio On Demand), Internet shopping, bank/ stock transaction |
| | Delay Jitter | < 50ms | < 100ms | |
| | Packet Loss | < 1% | < 5% | |
| | Guarantee | > 99% | > 95% | |

**Table 2.** (*continued*)

| | | | | |
|---|---|---|---|---|
| nrtPS | Packet Delay | NA | NA | High speed file trans- |
| | Delay Jitter | NA | NA | fer, Multimedia messag- |
| | Packet Loss | 0 ~ 2% | 0 ~ 5% | ing, E-commerce |
| | Guarantee | > 98% | > 82% | |
| BE | Delay/Jitter/ Loss/Guarantee | NA | NA | Web-browsing, SMS |

# 2 QoS Assurance Methods in WiBro

In this paper, 4 QoS assurance methods are applied in WiBro service, those are mobile IP (MIP) registration, call admission control (CAC), L2/L3 mobility and DiffServ mapping. The MIP registration method is the request of L3 mobility to HA (Home Agent) for WiBro service, originated at PSS and terminated at home agent (HA) through RAS (Radio Access Station), ACR (Access Control Router) and edge router (ER). HA checks the requested MIP registration and replies the MIP registration to PSS through ER, ACR and RAS. The CAC method [8] is the decisions of call accept or reject, has many parameters as bandwidth allocation, maximum/minimum traffic bandwidth, traffic priority, maximum delay and tolerated jitter. These CAC parameters are applied in WiBro network depends on service class of UGS, rtPS, nrtPS and BE. The L2 mobility is mobility offering of MAC layer in same network, in other words provision of seamless mobility from one RAS to another RAS. While, L3 mobility is mobility offering of network layer in different network, that is provision of seamless mobility from one network to another network by using mobile IP techniques.

The DiffServ mapping method is signaling transmission with hop-by-hop at each node by using DSCP code. Here, QoS marking method per service at PSS is used for handling many different services. For QoS marking and processing of L2 traffic, PSS will be marked with PRI QoS code in 802.1p field according to the subscriber's

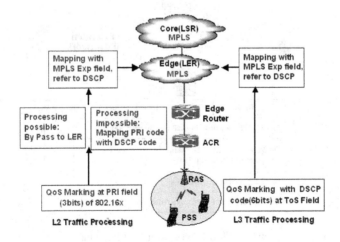

**Fig. 2.** Overall Processing Sequence of DiffServ Mapping Method

application services, and marked QoS field will be transferred to ER. ER will map the transferred QoS field with DSCP (Differentiated Service Code Point) code if ER can't process it, and pass the transferred QoS field to LER if ER can process it. LER will map the transferred DSCP code with MPLS Exp. field in core network. While,for QoS marking and processing of L3 traffic, PSS will be marked with DSCP code in ToS field according to the subscriber's application services. The marked DSCP code at PSS will be transferred to Edge router, and Edge router will pass it to LER. LER will map the transferred DSCP code with MPLS Exp field as L2 traffic processing. Traffic mapping method of L2 → L3 (Cos → DSCP), L3 → L3 (IP Prec → DSCP), L3 → L2 (DSCP → CoS), is referred to RFC 791, 2815, 1349, and overall processing sequence of DiffServ mapping method by QoS marking at PSS is shown in Figure 2 [1] [2] [3].

## 3   QoS Application Method at Each Operator's Implementation Stage

At operator's first implementation stage, QoS assurance method of only L2 mobility is applied with QoS parameter of CAC and scheduling function of its CAC. The CAC is applied differently according to class of service. For example, CAC will be accepted in case only enough resources are reserved at adjacent cell when UGS or rtPS service is requested from PSS. While, CAC will be accepted, even though self cell has enough resources when nrtPS service is requested from PSS. In case of BE service at first implementation stage, there is no reason to apply specific QoS mechanism, and specific QoS mechanisms are necessary in case of nrtPS, rtPS, and UGS services, so detail call flow procedure for these services are mentioned as follows. First of all, ①call path will be set-up through MIP registration and its replay from PSS to HA.

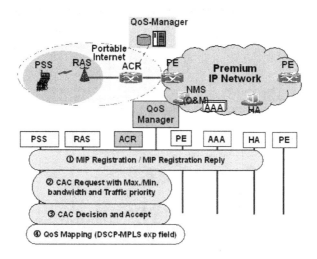

**Fig. 3.** Overall Call Flow Procedure at First Implementation Stage

Secondly ②PSS requests CAC with parameters of maximum/minimum bandwidth and traffic priority to RAS, then RAS requests its CAC to ACR that executes QoS manager. Thirdly ③ACR executes, decides and accepts CAC if cell conditions are satisfied based on requested CAC from RAS. While, ④DiffServ mapping is executed from PSS to HA for QoS assurance, and above mentioned call path set-up is also shown in Figure 3.

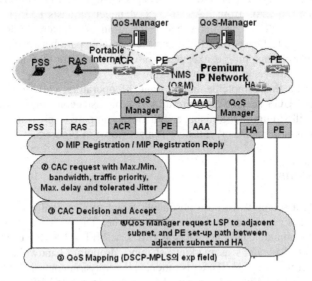

**Fig. 4.** Overall Call Flow Procedure at Second Implementation Stage

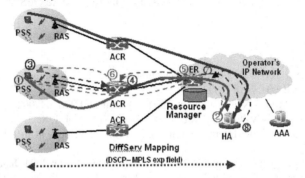

**Fig. 5.** Detail Call Flow Procedure at third Implementation Stage

At operator's second implementation stage, QoS assurance method of L2 and L3 mobility is applied with QoS parameter of CAC and scheduling function of its CAC. So, call flow procedures for nrtPS, rtPS, and UGS service class are ①call path set-up through MIP registration and its replay from PSS to HA. ②PSS requests CAC with parameters of maximum/minimum traffic bandwidth, traffic priority, maximum delay, and tolerated jitter to RAS, then RAS requests its CAC to ACR that executes QoS manager. ③ACR executes, decides and accepts CAC if cell conditions are satisfied based on requested CAC from RAS. ④QoS Manager requests LSP (Label Switched Path) to subnet of PE for available cell reservation of adjacent ACR and RAS for preparation of handoff case, then PE set up the available path between adjacent subnet and HA, ⑤DiffServ mapping is executed from PSS to HA for QoS assurance. Above-mentioned call path set-up is shown in Figure 4.

Finally, at the third implementation stage of operators, call flow procedure for nrtPS, rtPS and UGS service class are almost same as second implementation stage, but operator will strength QoS manager functions compare to previous stage for more accurate QoS assurance. Detail call flow is also shown in Figure 5.

## 4   Conclusions

Main service classes of WiBro are unsolicited grant service, real time polling service, non-real time poling service and best effort service. There are many QoS assurance methods in telecommunication networks as DiffServ, MPLS, RSVP-TE, SLA, and many congestion control methods as marking, shaping, dropping, bandwidth allocation, but four methods of mobile IP (MIP) registration, call admission control (CAC), L2/L3 mobility and DiffServ mapping are mentioned for QoS assurance in WiBro networks.

At operator's first implementation stage, QoS assurance method of only L2 mobility is applied with QoS parameter of CAC and scheduling function of its CAC. The CAC is applied differently according to class of service, and CAC factors are maximum/minimum bandwidth and traffic priority. At operator's second implementation stage, QoS assurance method of L2 and L3 mobility is applied with QoS parameter of CAC and scheduling function of its CAC. Here, CAC factors are maximum/minimum bandwidth, traffic priority, maximum delay and tolerated jitter. Finally, at the third implementation stage of operators, QoS assurance for nrtPS, rtPS, and UGS service class are almost same as second implementation stage, but operator will strength QoS manager functions compare to previous stage for more accurate QoS assurance.

## References

1. Kyu Ouk Lee, "KT-NGN Backbone Selection and Network Evolution," ETRI Document, ETRI-NGN-BB-004, Nov., 2002.
2. Kyu Ouk Lee, "Korea NGN Evolution," ETRI Document, ETRI-BT-002, June, 2002.
3. Kyu Ouk Lee, QoS Evaluation of KT-NGN", APCC2003, Vol. 1, Sept. 2003.
4. Jong Suk Ko, "Portable Internet Service & Network Development in KT," KRnet 2004, June 2004.
5. Joogho Jeong, "System Technology for Portable Internet," KRnet 2004, June 2004.
6. Kyu Ouk Lee, Hae Sook Kim, "Long term Plan of Portable Internet", ETRI Document, ETRI-0610-2004-0669, Nov. 2004.

7. Kyu Ouk Lee, Hae Sook Kim, "QoS, IPv6, Security Adaptation in KT Portable Internet Network," ETRI Document, ETRI-0610-2004-0675, Nov. 2004.

8. Y. Ran Haung et al., "Distributed Call Admission Control for a Heterogeneous PCS Network," IEEE Transactions on Computers, Vol. 51, No. 12, pp. 1400 – 1409, Dec. 2002.

9. Kyu Ouk Lee, "QoS in Portable Internet," ICACT2005, Feb. 2005.

# An Energy Efficient Tracking Method
# in Wireless Sensor Networks

Hyunsook Kim, Eunhwa Kim, and Kijun Han[*]

Department of Computer Engineering,
Kyungpook National University, Daegu, Korea
{hskim, ehkim}@netopia.knu.ac.kr, kjhan@knu.ac.kr

**Abstract.** Since it is difficult to replace batteries of the sensor nodes that are once deployed in the field, energy saving is one of the most critical issues for object tracking in wireless sensor networks. It is desirable that only the nodes surrounding the mobile target should be responsible for observing the target to save the energy consumption and extend the network lifetime as well by reducing the number of participating nodes. The number of nodes participating in object tracking can be reduced by an accurate estimation of the target location. In this paper, we propose a tracking method that can reduce the number of participating sensor nodes for target tracking. We show that our tracking method performs well in terms of energy saving regardless of mobility pattern of the mobile target.

## 1 Introduction

Tracking moving targets is emerging as applications such as wild animal habit monitoring and intruder surveillance in military regions. The sensors are used to collect information about mobile target position and to monitor their behavior pattern in sensor field. A mobile object tracking system is a complex mechanism that is accompanied with collaborative works between nodes.

Tracking of the mobile targets has lots of open problems to be solved including target detection, localization, data gathering, and prediction. In the localization problem, excessive sensors may join in detection and tracking for only a few targets. And, if all nodes have to always wake up to detect a mobile target, there are a lot of waste of resources such as battery power and channel utilization.

Actually, power conservation is one of the most critical issues in object surveillance and tracking since the sensor nodes that are once deployed in the sensor field would be difficult to replace a battery. Energy dissipation in sensors is various depending on condition of each sensor, for basic sensing operations, for powering the memory and CPU, and for communication between nodes or sink. So, if each node uses timely its energy to execute tasks, the network lifetime may be extended as a whole. Therefore, each sensor must minimize its battery power usage for desired longevity of network operation, which can be accomplished by properly managing sensor's operation. When a target moves around far away from the sensing range of a

---

[*] Correspondent author.

Y. Koucheryavy, J. Harju, V.B. Iversen (Eds.): NEW2AN 2006, LNCS 4003, pp. 278–286, 2006.

certain node, the nodes do not need to keep wake up for participating in tracking of the mobile target. This raises the necessity for prediction of the moving path of the mobile target to maintain the number of participating nodes in tracking as small as possible.

Many tracking protocols in large-scale sensor networks have been proposed to solve the problems concerned with tracking of the mobile targets from various angles [1, 2].

Krishnamurthy *et al.* [3, 4] proposed an energy efficient technique for using a sleep schedule where the nodes go to the sleep state when there is no need to take part in sensing. In [5], they explored a localized prediction approach to power efficient object tracking by putting unnecessary sensors in sleep mode. They proposed a convey tree for object tracking using data aggregation to reduce energy consumption. In [2], they attempt to solve the problem of energy savings based on the estimating the location of a mobile target. And they studied the frequency of tracking. We apply above sleep scheduling mechanisms to our study basically.

In general, the target localization is estimated successively based on the predicting of the next location, which is a result of the current measurement at a sensor and the past history at other sensors. The ideas of utilizing predictions to reduce overheads in mobile computing systems have appeared in the literature. Localization based on prediction is applied to cellular network to reduce the paging overhead by limiting search space to a set of cells that mobile users may enter [2]. The predictor-corrector methods are used in sensor network object tracking to minimize the error both in the measurements and in the prediction regardless of mobility model. Brooks *et. al.* present self-organized distributed target tracking techniques with prediction based on Pheromones, Bayesian, and Extended Kalman Filter techniques [6, 7]. In [8], Minimum Square Estimation is applied to localize the tracking of a mobile target. In [9], the future reading at a sensor is predicted, given the past reading history and the spatial and temporal knowledge of readings from surrounding sensors. But its localization accuracy is poor, because it is not a feasible solution for prediction of linear movement. Some algorithms such as [14] predict the future target position based on the assumption of a linear target trajectory. Such prediction-based approaches, however, do not produce a good result when the prediction is once wrong.

The goal of this paper is to propose an efficient tracking method that can minimize the number of participating nodes in mobile target tracking to extend the network lifetime. In this paper, we present a tracking method by predicting the location of the mobile target in 2-dimensional wireless sensor network, based on linear estimation.

The rest of this paper is organized as follows. Our proposed tracking method is presented in Section 2. Next, in Section 3, we present some simulation results. Finally, Section 4 concludes the paper.

## 2 Proposed Tracking Method

With rapid advances in sensor fabrications, recent sensors are designed to be power aware, changing their condition (e.g., shut down sensing processor or radio) when they do not need to run the components to perform a given task in a sensor field. Most sensors can operate under the three different conditions: Active, Idle and Sleep. It is important to completely shut down the radio rather than put it in the idle mode when it needs not sensing. Power management of sensor components is very important because energy consumption depends on their duties.

To save energy resource and thus extend the network lifetime, it is desirable that only the nodes that surround the mobile target are responsible for observing the target. For example, when the target passes through the $t_1$ point as shown in Fig. 1, all nodes do not need to join in the task for tracking a mobile target. Instead, it is more energy efficient for only the nodes $S_1$ around the mobile object to join in collecting information of the target and performing collaborative work among them. Other nodes located far from the target do not need to waste their powers to monitor the target. If we can predict the next location of the mobile object in advance, we can organize the group membership dynamically which should join in tracking mission. As shown in Fig. 1, for example, the number of participating nodes may be minimized, which allows us to further extend the whole network lifetime if we predict future location of the mobile target accurately.

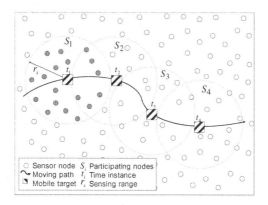

**Fig. 1.** A concept of tracking of a mobile object

As the mobile object moves, the tracking nodes may migrate to the moving direction of the target to keep on monitoring as shown in Fig. 1, where a thick line indicates the moving path of the mobile target and the blacked circles inside the dotted circle are tracking nodes at time $t_1$. Thus, sensor nodes need to control their states by themselves based on prediction of target's movement.

We assume a sensor network where $N$ sensors with the same communication and sensing range are distributed randomly in the environment that is being monitored. We also assume that each node knows its own location by using GPS or other location awareness techniques.

And we utilize triangulation for localization of a mobile target. Consequently, at least 3 sensors join the target detection and tracking with surveillance. Also each node keeps information about its neighbors such as location through the periodically message change. And each individual sensor node is equipped with appropriate sensory devices to be able to recognize the target as well as to estimate its distance based on the sensed data. Further, we assume that we predict the location of the mobile targets every one second (or minute), and each sensor records the movement pattern of the mobile object. Basically, we use a *moving average estimator* to predict the future location of the mobile target based on the measurement of direction and the velocity of the mobile target.

## 2.1 Approximate Prediction Step

First, since we assume that the mobile target does not change its direction or speed so abruptly in the sensing field, the location of the mobile object at the time instant of $(n+1)$ is approximately predicted by estimating the velocity when the mobile target will move during the time interval [n, n+1]. Fig. 2 shows the concept of tracking of the mobile target.

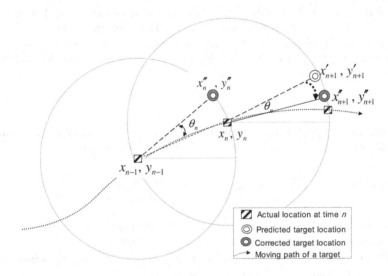

**Fig. 2.** Tracking of the mobile target in wireless sensor network

Given the current location $(x_n, y_n)$, the first predicted location of the mobile object at the time instant of (n+1), denoted by $(x'_{n+1}, y'_{n+1})$ is

$$x'_{n+1} = x_n + \tilde{v}_{x(n+1)} \tag{1a}$$

$$y'_{n+1} = y_n + \tilde{v}_{y(n+1)} \tag{1b}$$

where $\tilde{v}_{x(n+1)}$ and $\tilde{v}_{y(n+1)}$ represent the future speed estimates of the mobile object in the direction of $x$ and $y$, respectively, during the time period [n, n+1].

These speed estimates based on the previous history are given by

$$\tilde{v}_{x(n+1)} = \frac{\sum_{i=n-h+1}^{n} \tilde{v}_x(i)}{h} \tag{2a}$$

$$\tilde{v}_{y(n+1)} = \frac{\sum_{i=n-h+1}^{n} \tilde{v}_y(i)}{h} \tag{2b}$$

where $h$ is a predefined number of the past history based on which we predict the next moving factor. Hence, the future speed is a moving average of acceleration of an object. That is, the present movement of a mobile object means a reflection of the patterns of the moving history.

The estimate obtained in this step makes it possible to exactly predict the future location of the mobile object that moves linearly. However, the estimate is no longer effective when the mobile object moves in non-linear fashion since only velocity information is used to predict the future location. So, we need a correction mechanism to get a more exact estimate.

### 2.2 Correction Step

Let us express the prediction error by the angle between the actual location and the previously predicted, denoted by $\theta_n$ as shown in Fig. 2. Then we have

$$\cos \theta_n = \frac{x_n - x_{n-1}}{\sqrt{(x_n - x_{n-1})^2 + (y_n - y_{n-1})^2}} - \frac{x_n'' - x_{n-1}}{\sqrt{(x_n'' - x_{n-1})^2 + (y_n'' - y_{n-1})^2}} \tag{3}$$

The estimate obtained in step 2.1 will be corrected based on the prediction error encountered in the previous step to make an accurate estimation.

Finally, we can predict the next location ( $x_{n+1}''$ , $y_{n+1}''$) of the mobile target by correcting an angle from the first predicted location.

The new location of a mobile target can be induced by rotational displacement as theta $n(\theta_n)$, which is shifted by centering the previous actual target location. Eq.(4) shows the formula of rotational displacement.

$$\begin{pmatrix} x_{n+1}'' \\ y_{n+1}'' \end{pmatrix} = \begin{pmatrix} \cos \theta_n & -\sin \theta_n \\ \sin \theta_n & \cos \theta_n \end{pmatrix} \begin{pmatrix} x_{n+1}' \\ y_{n+1}' \end{pmatrix} \tag{4}$$

Tracking in our system is performed by the following procedure.

1. Discovery: When a sensor node around the mobile object detects the target and initializes tracking, it becomes 'estimation node' which acts as a master node temporarily.
2. Localization: A set of nodes those become aware the appearances of the mobile target compute the target's current position. The coordinates of the mobile target may be accomplished by the triangulation and their collaborative works.

3. Estimation: An estimation node predicts the future movement path of the mobile target, and transmits message about the approaching location to its neighbor nodes. The prediction is carried out by two steps: approximate a prediction and correction step that is explained above. The moving factors of a mobile target, such as direction and velocity, can be obtained by sensor nodes through collecting moving patterns of the tracked target.

4. Communication: As the mobile target moves, each node may hand off an initial estimate of the target location to the next node in turn. At that time, each node changes its duty cycle along the movement of the target [10].

For energy saving, each node operates the state scheduling by itself. For example, detectable nodes within sensing range of a node 'a' in Fig. 3, which is near the target are activated, and they participate in tracking including localization, monitoring, and prediction.

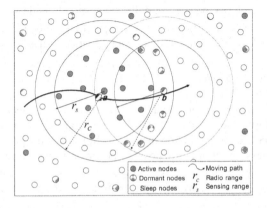

**Fig. 3.** State change at each sensor node as the mobile target moves

So, they consume an amount of energy i.e., processor and sensor activating, transmitting and receiving. And the dormant node 'b' can hear message about appearances of the mobile target in neighborhood, but cannot detect the target since the sensing range differs from its communication range. Obviously, all dormant nodes do not need to be activated. Energy consumed by the dormant nodes is small enough to be negligible comparing with that by the activated nodes. Some nodes that are located around the next position of the target wait for sensing. The others shut down the radio after hearing the message from the master node and they wake up from sleep when their duty cycle is over. When it wakes up, it first senses its region and hears message. If a sensor node cannot detect the target and does not receive any message for target appearance, it turns off its radio and goes to sleep for saving energy. And the rest of the sensor nodes run independently according to their duty cycles. Thus, the prediction of the mobile target in the tracking system reduces energy consumption in each node and extends network lifetime.

## 3 Simulation

We evaluate the performance of our method through simulation results. We carry out the experiments to measure missing rate and wasted energy. The network dimension for our experiments is [200, 200] and 500 nodes are randomly deployed within the region.

To model the movement behavior of the mobile target, we use the Random Waypoint Mobility model (RWP) and Gauss-Markov Mobility (GMM) model. RWP is a simple mobility model based on random directions and speeds. The mobile object begins by staying in one location for a certain period of time. Once this time expires, the mobile object chooses a random destination in the field and a speed. And then travels toward the newly chosen destination at the selected speed. Upon arrival, the mobile object pauses for a specified time period before starting the process again. GMM model is a model that uses one tuning ($\alpha$) parameter to vary the degree of randomness in the mobility pattern. Initially, each mobile node is assigned a current speed and direction. Specifically, the value of speed and direction at the $n^{th}$ instance is calculated based upon the value of speed and direction at the $(n-1)^{th}$ instance and a random variable [11]. Energy consumption used for simulation is based on some numeric parameters obtained in [2]. Table 1 shows the parameters and the default values of the mobile target.

Our prediction method is compared with the *least squares minimization*(LSQ) to evaluate the performance of accuracy. LSQ is a common method used for error reduction in estimation and prediction methods. LSQ solves the problem of estimating by minimizing the sum of the squares of the error terms corresponding to each distance sample. In other words, LSQ tries to get the estimate by minimizing $\sum_{i=1}^{n}(\|\hat{\phi}_i - p_i\| - d_i)^2$ , where $\hat{\phi}_n$ is estimate and $\|\hat{\phi}_i - p_i\|$ is the Euclidean distance between the estimated coordinate of the mobile device and the beacon or receiver at position $p_i$ [12].

**Table 1.** Parameters for simulation

| Description | Value | Description | Value |
|-------------|-------|-------------|-------|
| Processor activate | 360 mW | Radio receiving | 369mW |
| Processor sleep | 0.9 mW | Sensing range | 20 |
| Sensor activate | 23 mW | Radio range | 35 |
| Radio transmission | 720 mW | Target speed / unit | 10m |

### 3.1 Missing Rate

To evaluate how accurately our tracking method works, we observe the missing rate, which is defined as the percentage that sensor nodes fail to detect the mobile target.

As shown in Fig. 4, our scheme offers a smaller missing rate than LSQ regardless of mobility model used in simulation. In RWP, since the moving pattern of the mobile

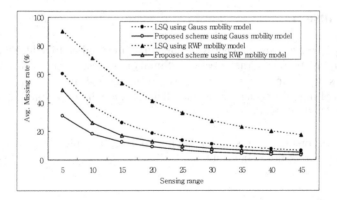

**Fig. 4.** Missing rate

target is random, the prediction error increases. As the sensing range becomes larger, the missing rate decreases as well. Due to the inaccurate prediction of location, some nodes miss the target because the real location of the target is out of the sensing range.

Fig. 5 shows the wasted energy that is defined as the amount of consumed power due to incorrect information of prediction over all nodes. From Fig. 5, we can see the effect of prediction. Obviously energy consumption is greatly influenced by an accuracy of prediction. If the sensor nodes stay awake to track the mobile target while the target is moving out of the sensing range, they consume unnecessary energy. As described earlier, we can extend the network lifetime by avoiding such unnecessary energy consumption at nodes that do not need to join in tracking. This figure indicates that our scheme can decrease the number of participating nodes and thus reduce the energy consumption too.

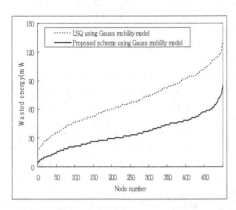
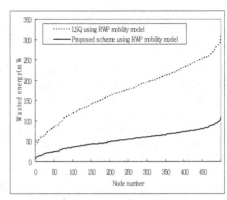

(a) Wasted energy in GMM model          (b) Wasted energy in RWP model

**Fig. 5.** Wasted energy

# 4   Conclusion

In this paper, we propose an energy efficient tracking method to reduce the number of nodes participating in target tracking. Simulations results show that our method contributes to saving energy and thus extending the network lifetime as well regardless of mobility pattern of the mobile target including Random Waypoint model and Gauss Markov model.

**Acknowledgement.** This research is supported by Program for the Training of Graduate Students for Regional Innovation.

# References

1. C. Gui and P. Mohapatra, "Power conservation and quality of surveillance in target tracking sensor networks", *Proceedings of the ACM MobiCom,* Philadelphia, PA, September, 2004
2. Y. Xu, J. Winter, and W.-C. Lee, "Prediction-based strategies for energy saving in object tracking sensor networks", *Proceedings of the Fifth IEEE International Conference on Mobile Data Management (MDM'04)*, USA, January, 2004
3. T.Yan, T,He, and J, Stankovic, "Differentiated surveillance for sensor networks", *Proceedings of ACM SenSys'03,* Los Angeles, CA, November, 2003
4. T. He, S. Krishnamurthy, J. Stankovic, T. Abdelzaher, L. Luo, R. Storelu, T. Yan, L. Gu, J. Hui, and B.Krogh, "Energy-efficient surveillance system using wireless sensor networks", *Proceedings of the 2$^{nd}$ International Conference on Mobile Systems(MobiSYS),* Boston, USA, 2004
5. Y.Xu and W.-C.Lee, "On Localized Prediction for Power Efficient Object Tracking in Sensor Networks", *Proceedings of the 1'st International Workshop on Mobile Distributed Computing(MDC)*, May, 2003
6. R. Brooks and C. Griffin, "Traffic model evaluation of ad hoc target tracking algorithms", *International Journal of High Performance Computing Applications*, 2002
7. R. Brooks and C. Griffin and D. S. Friedlander, "Self-organized distributed sensor networks entity tracking", *International Journal of High Performance Computer Applications*, 2002
8. D. Li, K. Wong, Y.Hu and A. Sayeed, "Detection, Classification, Tracking of Targets in Micro-sensor Networks", *IEEE Signal Processing Magazine,* pp. 17-29, March, 2002
9. S. Goel and T. Imielinski, "Prediction-based monitoring in sensor networks: Taking Lessons from MPEG", *ACM Computer Communication Review*, vol. 31, no. 5, 2001
10. F. Zhao, J. Liu, J. J. Liu, L. Guibas, and J. Reich, "Collaborative signal and information processing: An information directed approach", *Proceedings of the IEEE,* 2003
11. T. Camp, J. Boleng and V. Davies, "A survey of mobility models for ad hoc network research", *In Wireless Communication & Mobile Computing (WCMC): Special issue on Mobile Ad Hoc Networking*, vol. 2, no. 5, 2002
12. A. Smith, H. Balakrishnan, M. Goraczko and N. Priyantha, "Tracking Moving Devices with the Cricket Location Systems", *Proceedings 2$^{nd}$ ACM MobiSys,* pp. 190-202, 2004
13. H. Yang and B. Sikdar, "A Protocol for Tracking Mobile Targets using Sensor Networks", *Proceedings IEEE Workshop Sensor Network Protocols and Applications,* May, 2003

# On Incoming Call Connection Service in a Ubiquitous Consumer Wireless World

Máirtín O'Droma, Ivan Ganchev, and Ning Wang

Telecommunications Research Centre, ECE Dept., University of Limerick, Ireland
{Mairtin.ODroma, Ivan.Ganchev, Ning.Wang}@ul.ie

**Abstract.** This paper[*] proposes a new consumer-oriented incoming call connection service (ICCS) provision. The architecture and implementation design of this ICCS is conceived to underpin the evolution of the proposed future ubiquitous consumer wireless world (UCWW) environment, founded on the consumer-based business model (CBM). ICCS design issues, architecture, core signalling flow elements for ICCS setup and operation, and standardisation issues are presented and discussed.

## 1 Introduction

The novel ICCS proposal is presented in this paper as an integral element of the UCWW environment, which in turn is seen as emanating from the adoption of the CBM, [1-10]. It is well understood that the realisation of this new CBM-UCWW environment will require new technological design and development on a number of fronts, and the agreeing of associated standardisation. An important one of these technological challenges is the realisation of a new consumer-oriented ICCS provision. This is distinct from the present subscriber-oriented ICCS and techno-logical infrastructure, which is really the underpinning of the subscriber-based business model (SBM).

**Consumer-oriented ICCS defined.** In legacy SBM for fixed 'wired' communications, the subscriber and the local loop (LL) connection to the subscriber's terminal (the LL medium and end points are inherently integrated, permanent and persistent) has a unique network identity (e.g. a telephone number) assigned by the network provider. In wireless networks the matter is different in two important ways. Firstly the LL only comes into existence for a call, and for the duration of the call (including calls which are for user-network maintenance communications). The LL medium and time-slot/frequency band resources are shared with others, giving it the characteristics of a virtual LL. Secondly the medium used by this virtual LL is not owned by the wireless access network provider (ANP). Thus a MU with a unique identity should be able to open a LL connection with any AN and through a mutual identification and authentication process purchase and obtain services from the AN. This is part of the

---

[*] This publication has emanated from a research conducted with the financial support of Science Foundation Ireland under the Basic Research Grant Ref. No. 04/BR/E0082.

Y. Koucheryavy, J. Harju, V.B. Iversen (Eds.): NEW2AN 2006, LNCS 4003, pp. 287–297, 2006.
© Springer-Verlag Berlin Heidelberg 2006

essence of the CBM. And the technology and service to enable such a 'consumer' MU to be able to receive incoming calls is what is to be understood as a consumer-oriented ICCS. For this a 3P-AAA SP infrastructure would need to be in place, [e.g. 9 &10].

This kind off ICCS would best be managed and controlled by extra-network ICC SPs which, through standardised protocol interfaces, will work with ANPs providing ICCS support and which will work directly with the consumer MUs. Because growing numbers of teleservices do not require ICCS, in the future only some ANPs will include ICCS support in the profile of their service offerings. If the MU desires such a service it will purchase it from the ANP it chooses under some consumer-based contract and notify its ICCS SP of the arrangement.

The infrastructural role of these ICCS SPs seems open to a wide range of business opportunities, e.g. they may specialise in meeting a whole range of different MUs or groups of MUs (e.g. a company or enterprise) ICCS needs, which can be quite sophisticated and variable, with a variability as dynamic as the lives of the MUs themselves.

**Enhanced roaming.** Besides availing of existing roaming infrastructures, in keeping with the inherent CBM philosophy, a key design goal in the proposed ICCS architecture is to enable the mobile user (MU) to roam horizontally and vertically under the MU autonomous control, effecting thus mobility through user-created integrated heterogeneous networking. This kind of mobility is potentially global where appropriate standards are adopted and implemented.

**ICCS and 3P-AAA.** Operation of this consumer-oriented ICCS, and especially with this kind of mobility and roaming capabilities, relies on the existence of other core elements of the CBM-UCWW proposal, in particular the realisation and creation of Third-Party Authentication, Authorization, and Accounting (AAA) service providers (3P-AAA SP), [ c.f. 9, 10 and elsewhere], which provide autonomous AAA services.

There has been considerable related study at a conceptual level about system architecture and services integration in ubiquitous next generation networks (UNGN). For instance Akyildiz et al., [11], introduce a novel architecture for 'ubiquitous mobile communication' using a third-party interoperating agent. However, the constraints inherent in the subscriber relationship in SBM are not considered there. Alonistioti et al. in [3] introduce a generic architecture for integrated systems and services where the potential and option for transition to CBM is set out, and Chaouchi et al. in [4] present a policy-based framework to support the integration of 4G networks and services where the potential within an application CBM regime is being addressed. These papers do not address the UCWW paradigm and its consequence infrastructural innovations such as consumer-oriented ICCS treated herein.

This paper sets out a proposal for this new ICCS infrastructure including elements of the core architecture and signalling protocol required to realise it. It is organized as follows: Basic features and attributes of ICCS are present in section 2; design issues and the proposed ICCS architecture - section 3, signaling messages and flows - section 4, testbed implementation issues - section 5, and conclusions - section 6.

## 2  Incoming Call Connection Service (ICCS)

External ICC SPs and collaborating ANPs as portrayed above are integral to the consumer-oriented ICCS realisation. The division of functionality between the ICC SP, the ANP and the MU is set out below. The existence of 3P-AAA infrastructure is presumed throughout.

**ICCS operation – an outline.** Considering the matter first at a conceptual level, the following is how ICCS might operate for the traditional incoming call. An ANP in accepting the custom of a MU will provide a unique, globally significant, temporary contact address (CA), which will be associated with the unique personal address of the MU / the mobile terminal (MT). The MU/MT will send this CA to its own ICC SP, which will then be empowered to connect any incoming calls to the MU, and in accordance with the MU's instructions. This could be a re-direction functionality, i.e. re-directing incoming calls to the unique network address either directly or indirectly by informing the caller MT of the present unique MU's CA. Naturally in regard to this unique CA the ANPs can provide (e.g. offer as part of their service) all existing mobility and roaming services to the user as in present 2G and 3G networks, and the MU may choose to accept some but not others. What is also clear is that there is now a new autonomous mobility and roaming control in the power of the MU. This is on top of existing mobility and roaming encountered for instance in cellular networks. The latter will be applied under a slightly different arrangement, i.e. applied to the CA which the MU has taken out (temporarily) with the ANP. In this sense the new ICCS is backward compatible with existing wireless networking mobile and roaming infrastructure.

**ABC&S and ICCS.** The consumer MU has the possible, inherent in CBM, of choosing other options on other networks in relation to ICCS as well as other services which may better meet their subjective always best connected and served (ABC&S) requirements. In fact even in regard to ICCS support the MU may contract with more than one ANP, e.g. one may be used for personal or family incoming calls satisfactorily matching economic and QoS profile for these calls, and another for business callers matching a profile requiring the best QoS available – a typical likely scenario in future CBM-UCWW. The MU will be able to purchase flexible, simple or sophisticated ICCS from its ICC SP, (e.g. inform its ICC SP how to direct calls from different callers and caller types; time/day/week/location configurations). The business possibilities for ICC SP in this field are manifold.

**Addressing.** Each consumer MU will have at least one permanent, personal, global, network-independent, and geography-independent identifier through which, for the purpose of ICCS, the MU can always be addressable. Through this a MU will be enabled to make or receive calls anywhere, not as today's global roamer but as a 'local' to the network from which such services are being purchased by the MU as a consumer. IPv6 addresses could be used as this identifier. A large tranche of IPv6 addresses could be set aside as IP MU identifiers (IP-MUI). The network-independent attribute is an important difference in these from present IP addresses.

IP addresses once purchased by MUs must be locked, i.e. it must be possible only for the owner of that number to use it. Thus a certain level of control of sale would be required, e.g. to be sold only through 3P-AAA SPs, together with a standardised protocol to safeguard ownership through use for instance of a public-key certification system based on X.509.

**MU personal address identifier (PAI).** In meeting the need of linking the IPv6 address to a more user-friendly naming convention the MU may also acquire and use an address identifier similar to a network address identifier (NAI) [12]. However unlike the NAI, this address identifier will not point to a network but to a user only. Thus it might be more appropriately called a MU personal address identifier (PAI). There may be some advantage in associating PAI with a specific ICC SP. The discovery of IP-MUI and PAI can be achieved in similar ways to finding a person's phone number or email address today. These ways will continue to improve and benefit from technological developments.

# 3   ICCS Design Goals

Consumer oriented ICCS challenges raised in the ABC&S UCWW paradigm may be translated into ICCS design goals.

- Roaming should be supported in secure manner across heterogeneous network domains with their different access technologies, network protocols, signalling formats, authentication and authorization schemes.
- MU driven vertical handover [2, 13] should be supported without conflict with the maintenance of ICCS, thus enabling a MU to move between different network technologies (both wired – accessed through WLAN/PAN interfaces –, and wireless – e.g. different generations of cellular networks) according to the MU ABC&S decision processes.
- There should be no conflict with a MU gaining access to different services from different SP/ANPs and their ICCS arrangements with their chosen ANP(s) for ICCS support.
- Besides realising 'enhanced' roaming (outlined above), backward compatibility with existing wireless, mobile and roaming infrastructure services and facilities should be fully maintained.
- The ICCS architecture should provide an end-to-end solution for converged networks, including integrated heterogeneous networking, and employ global roaming with QoS support in accordance with the IETF standards.

## 3.1   ICCS Architectural Framework

A new consumer-oriented ICCS architectural framework is set out here, Fig.1. It allows the convergence of multi-access network, including UMTS, WLAN, WMAN, DSL etc. For call control this architecture employs a hybrid signalling protocol including elements from MIP, HMIP, Diameter, COPS, and NSIS so that seamless ICCS is offered when user moves from one access network to another.

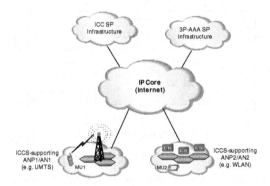

**Fig. 1.** ICCS architectural framework

The key entities are the ICC SPs, the ANPs that support ICCS, 3P-AAA SPs and of course the MU/MTs. In delivering ICCS to MUs the ANPs and the SPs liase and communicate with each other over an IP core network (e.g. the Internet).

Naturally the protocol interactions between entities here will require standardisation, especially the triangle of relationships between MU/MTs seeking ICCS, the ICC SPs, and the ICCS-supporting ANPs.

*ICC SP* performs all the ICCS activities primarily to enable an MU to receive an incoming call, and to do that in accordance with the MU desires. In effect it performs a pointer or re-director function for callers seeking to connect to the called MU. For 'call re-directing' to the called MU a global, unique, temporary forwarding address or contact address (CA) is required. MUs will be responsible for supplying their ICC SPs with this CA and any other associated information. Standardization should allow a MU to use two or more ANPs for ICCS support, each of course with its own CA. The ICC SP must also provide the MUs with the ability to manage and control their incoming calls. The sophistication and user-friendly nature of the incoming call management services that an ICCS may offer a MU is likely to evolve as users learn to manage their communications, and will become a market-differentiation factor among ICC SPs.

*3P-AAA SP* provides authentication, authorization and accounting for all ICCS selling and buying transactions. These services are supplied following the normal manner of delivering AAA services in the UCWW environment, e.g. in [9] wherean extension of the IETF Diameter protocol has also been proposed as a candidate for these AAA interactions.

*ICCS-supporting ANPs* operating different types of ANs (i.e. UMTS, WLAN, and xDSL) and providing ICCS support for MUs wishing to purchase this.

### 3.2  ICC SP Infrastructure

Figure 2 depicts the components of the ICC SP infrastructure and their lines of interaction; the components are briefly described below.

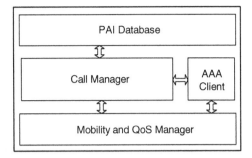

**Fig. 2.** Functional components of the ICC SP infrastructure

*Call Manager* is mainly responsible for handling the signalling and interacting with the other components, including communications with ANP(s) providing ICCS support.

*AAA Client* interacts with the 3P-AAA SP infrastructure in order to authenticate the MU, authorize the use of ICCS, and provide accounting and invoice information about the ICCS usage to 3P-AAA SP for billing purposes.

*PAI Database* is mainly used to store MUs profiles and preferences, including IP-MUI/PAI and current MU's CA. A policy repository could be integrated with the PAI database to simplify the overall service management.

*Mobility and QoS Manager* is an optional component, which provides general mobility management (e.g. based on HMIPv6) and QoS management (e.g. based on DiffServ and IntServ/RSVP).

## 4   ICCS Signalling Messages and Flows

### 4.1   ICCS Signalling Messages

The ICCS signalling is based on a combination of existing standard protocols (such as NSIS, Diameter, SIP, and COPS), more precisely on their extensions. For the purposes of ICCS new signalling messages must be defined (and perhaps introduced in some of these protocols). A skeleton set is presented in Table 1, with the focus of ICCS operation for a MU (MU2) seeking to call another MU (MU1); the signalling for this is addressed in the following section.

### 4.2   Signalling Flows-Receive Incoming Call

A sample signalling flow for the MU successfully receiving ICC is illustrated in Fig. 3. In this case MU2/MT2 wants to call MU1/MT1. We assume that MU2 (the caller) already knows the IP-MUI/PAI of MU1 (the callee), and both MU1 and MU2 have already purchased ICCS, assumed for our purposes here, from the same ICC SP. An end-to-end interaction of signalling flow may be separated into three phases: *advertisement, discovery and association (ADA), ICCS session,* and *payment.*

**Table 1.** ICCS signalling messages

| | |
|---|---|
| *ICCS-ADV* | Periodically sent by WBC SP over the WBC to **advertise** the configuration and services of ICC SP. |
| *ICCS-REQ* | Sent by MT/MU to ICCS-supporting AN/ANP to **request** CA for this MT/MU for ICCS purposes. |
| *ICCS-REQ-ACK* | Sent by ICCS-supporting AN/ANP to MT/MU to **acknowledge** *ICCS-REQ* (includes allocated ICCS-CA, – if approved). |
| *ICCS-UPD* | Sent by MT/MU to ICC SP to **update** the MT's/MU's current ICCS-CA (includes allocated ICCS-CA). |
| *ICCS-UPD-ACK* | Sent by ICC SP to MT/MU to **acknowledge** the update of ICCS-CA. |
| *ICCS-QRY* | Sent by calling MT/MU to ICC SP to **query** the current ICCS-CA of the called MT/MU (includes called MU's PAI). |
| *ICCS-QRY-ACK* | Sent by ICC SP to calling MT/MU to **acknowledge** ICCS-QRY (includes current ICCS-CA of the called MT/MU, if direct connection to the called MT/MU is approved – a MT/MU profile issue). |
| *ICCS-STP* | Sent by calling MT/MU (via ANs/ANPs and core IP) towards called MT/MU to **setup** new ICC. |
| *ICCS-STP-ACK* | Sent by called MT/MU back to calling MT/MU to **acknowledge** ICC setup. |
| *ICCS-RLS* | Sent by one MT/MU to the other MT/MU to **release** existing ICC. |
| *ICCS-RLS-ACK* | Sent back by the other ICC party (MT/MU) to **acknowledge** ICC release. |

*ADA procedure.* When the consumer MT1/MU1 moves into a new domain – e.g. roaming - it may seek a suitable ANP, e.g. through review of WBC advertisements, [6]. For ICCS support it will select an ANP offering this service, say ANP1. MT1 sends a request for association with AN1 in which includes the MU1's 3P-AAA certificate, which AN1/ANP1 needs to check to ensure that MU1 is registered with a 3P-AAA SP and will pay for the ANP communications services via the 3P-AAA SP. If association is granted, ANP1 also sends its own 3P-AAA certificate to MU1 for an inspection. After this mutual authentication MU1 can start using the communications service of AN1/ANP1.

On the other side MT2/MU2, assuming it is also a roaming consumer MU, executes similar ADA procedure and associates itself with AN2/ANP2, which is not necessarily supporting ICCS.

When a MT/MU finishes using the communications services of AN/ANP, this AN/ANP produces accounting and invoice information for services rendered. Bills are paid through the 3P-AAA SP following standard procedures.

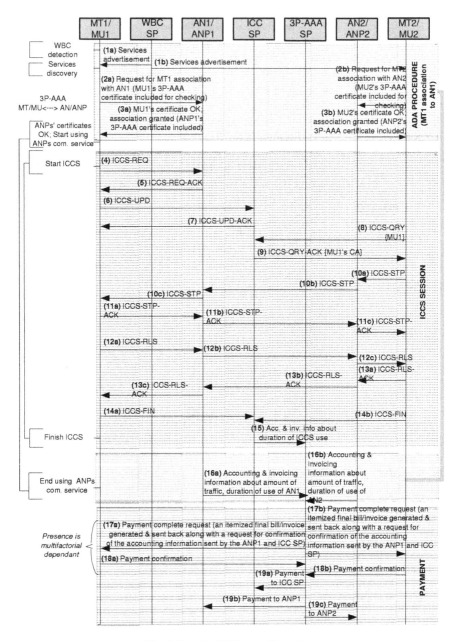

**Fig. 3.** Sample ICCS signalling flows

**ICCS session.** Following the association procedure with the ANP1, the MU1 will seek an ICCS contact address (ICCS-CA) by sending *ICCS-REQ* to AN1/ANP1 and, after receiving it (included in *ICCS-REQ-ACK*), will communicate it to its ICC SP by means of *ICCS-UPD* message. After that the MT1/MU1 may start receiving and accepting incoming calls, e.g. from MT2/MU2. To set up an ICC with MT1/MU1, MT2/MU2 first needs to obtain the current MU1's ICCS-CA from the ICC SP (message *ICCS-QRY*). After obtaining it (*ICCS-QRY-ACK*), MT2/MU2 can start setting up an ICC with MT1/MU1 by means of *ICCS-STP* message. This message propagates towards MT1/MU1 via AN2/ANP2-core_IP-AN1/ANP1 and, if MT1/MU1 acknowledges it, an ICC is established between both MTs/MUs. At the end of ICC one of the communication parties will initiate an ICC release (*ICCS-RLS*) and the other party will confirm the release by means of *ICCS-RLS-ACK*.

***Payment.*** At the end of ICCS and ANP's communications service, the 3P-AAA SP sends payment complete requests/confirmations to the paying entities, e.g. calling MU, called MU, or both MUs. There are many possible options in procedures here. The relevant shares of the payment are sent by 3P-AAA SP to the ICC SP and the ANPs for the provision of their services.

## 5  Testbed Implementation

Besides the theoretical proposal, a suitable testbed is required to get real experiments on how ICCS architecture and signalling protocol would function and perform.

The testbed presents numerous challenges, the requirement of creating a hetero-geneous networks environment, with their varying access technologies, signaling formats, and QoS mechanisms, being a particularly demanding one. Developing relationships between these and MU preferences, and the criteria to regulate these are also challenging. Secondly, another challenge is the scenarios design and realisation, i.e. roaming mobility -how the MUs would move during an experiment, disconnecting from some ANPs and connecting to others, which services would be required. A few speculative MU profile criteria may be set down e.g. the manner by which MUs would like to receive ICCS according to different roles, e.g. professional, parent/family member, or student. Numerous test scenarios are being conceived. Thirdly, protocol interactions between the ICC SPs' servers and the ANP ICCS support entities, which are network technology independent, are needed. Different elements of protocols for call management, QoS, AAA, and policy management will require detailed design. Extensive work on the development of the ICCS testbed is already underway. The MTs used in the testbed are laptops and PCs equipped with Bluetooth and WLAN adapters, which are carried around according to a scenario experiment. Non-server PC and laptops will install the MU protocols and will act as hosts for incoming and outgoing calls operated through ICCS servers in real time. The protocol and application implementation is being based on Linux.

## 6  Conclusions

This paper has presented aspects of the core elements of a novel consumer-oriented Incoming Call Connection Service (ICCS), which will be supplied by SPs external to access network providers (ANPs). Key functions and capabilities of those ANPs supplying ICCS support services on request from a MU and in conjunction with ICC SPs were set out. The process of a consumer, roaming, wireless MU's ICC SP redirecting incoming caller's request to a unique temporary address is described. The MU may have more that one such consumer contract (i.e. be handling more than one unique ANP temporary addresses for ICCSs) with different overlapping ANPs depending on the MU's Always Best Connected and Served (ABC&S) requirements at the time in question. A key MU attribute for this, and in general for realising a consumer-based business model (CBM), is MUs having personal, network-independent and geography-independent addresses. This implies using a new global numbering system; here, as in [2], we propose a large swath of IPv6 addresses be setaside for this purpose.

Aspects of design issues and an approach towards possible signalling protocol solutions to realise this novel consumer-oriented ICCS have been set out. The ICCS architectural framework proposed meet a the design goals of realising operation over integrated heterogeneous networks and global roaming according to CBM-UCWW principles. The design satisfying the goal of backward compatibility with existing mobility and roaming facilities available today under the subscriber-based business model; i.e. they will continue to be possible under this architecture, and work in tandem with some significant consumer oriented roaming enhancements. The ICCS signalling solutions suggested seek to maintain harmony with IETF standards including those addressing provision of end-to-end QoS reservation and policy-based management.

This ICCS proposal needs as yet extensive testing. A CBM testbed network design is well advanced over which initial conceptual proofing and serviceperformance, robustness, scalability and efficiency tests may be performed.

# References

1. O'Droma, M., Ganchev, I., Morabito, G., Narcisi, R., Passas, N., Paskalis, S., Friderikos, V., S.Jahan, A., Tsontsis, E., Bader, C., Rotrou, J., Chaouchi. H.: Always best connected enabled 4G wireless world. Proc. of the 12th European Union IST Mobile and Wireless Communications Summit, Aveiro, Portugal (2003) 710-716
2. O'Droma, M., Ganchev, I.: Techno-Business Models for 4G. Proc. of the Int. Forum on 4G Mobile Communications, King's College London, U.K. (2004) 53
3. Alonistioti, N., Passas, N., Kaloxylos, A., Chaouchi, H., Siebert, M., O'Droma, M., Ganchev, I., Faouzi, B.: Business Model and Generic Architecture for Integrated Systems and Services: The ANWIRE Approach. CD Proc WWRF 8bis meeting, 8 pages, Beijing, China. (2004)
4. Chaouchi, H., Pujolle, G., Armuelles, I., Siebert, M., Carlos, B., Ganchev, I., O'Droma, M., Houssos, N.: Policy based networking in the integration effort of 4G networks and services. IEEE 59th VTC (Spring). Milan, Italy. (2004) 2977-2981
5. O'Droma, M., Ganchev, I.: Enabling an Always Best-Connected Defined 4G Wireless World. Annual Review of Communications, Vol. 57. Int. Engineering Consortium, Chicago, Ill, U.S.A. (2004) 1157-1170
6. Flynn, P., Ganchev, I., O'Droma. M.: Wireless Billboard Channels-Vehicle and Infrastructural Support for Advertisement, Discovery and Association of UCEE Services. Annual Review of Communications, Vol 57. Int. Engineering Consortium, Chicago, Ill. U.S.A. (2004) 1157-1170

7. O'Droma, M., Ganchev, I., Siebert, M., Bader, F., Chaouchi, H., Armuelles, I., Demeure, I., McEvoy. F.: A 4G Generic ANWIRE System and Service Integration Architecture. ACM Mobile Computing and Communications Review, Vol.10, No.1. (2006) 13-30.
8. Chaouchi, H., Armuelles, I., Ganchev, I., O'Droma, M., Kubinidze, N.: Signalling Analysis in Integrated 4G Networks. Wiley Int. Journal of Network Management, Volume 16, Issue 1. (2006), 59-78
9. McEvoy. F., Ganchev, I., O'Droma, M.: New Third-Party AAA Architecture and Diameter Application for 4GWW. IEEE PIMRC, Berlin (2005) Proc CD 11-14.
10. Ganchev, I., McEvoy. F. O'Droma, M.: New 3P-AAA Architectural Framework and Supporting Diameter Application. WSEAS Transactions on Communications. (2005) 176-185
11. Akyildiz, I.F., Mohanty, S., Jiang, X.: A ubiquitous mobile communication architecture for next-generation heterogeneous wireless systems. IEEE Communications Magazine, Vol. 43, Issue 6. (2005). 29-36
12. Aboba, B., Beadles. M.: The Network Access Identifier. RFC 2486, IETF. (1999)
13. Jung-Ho Lee, Sang-Hee Lee, et al.: Fast End-to-End Mobility Support using SIP for Vertical Handoffs. Lecture Notes in Computer Science, Vol. 3042. (2005) 1390-1394

# Evaluating the Effect of Ad Hoc Routing on TCP Performance in IEEE 802.11 Based MANETs

Evgeny Osipov[1,*] and Christian Tschudin[2]

[1] RWTH Aachen University, Department of Wireless Networks, Kackertstrasse 9,
D–52072 Aachen, Germany
[2] University of Basel, Computer Science Department, Bernoullistrasse 16, CH–4056
Basel, Switzerland

**Abstract.** In this paper we analyze the impact of different ad hoc routing schemes on TCP traffic in multihop ad hoc wireless networks. Our hypothesis is that beyond a certain network size the uncontrolled broadcast transmission of routing messages will seriously degrade the performance of the TCP protocol. We experimentally determine the admissible operation ranges of MANETs where the level of such degradation is still acceptable for end users.

## 1 Introduction

The motivation behind studying the effect of routing traffic on TCP performance in multihop wireless ad hoc networks is twofold. On one hand, the research on benchmarking different ad hoc routing protocols is mainly conducted using UDP based CBR traffic wherefore, there is no established methodology for the analysis of TCP+routing interactions in MANETs. On the other hand, the TCP protocol in IEEE 802.11 based MANETs performs poorly even without routing and it is interesting to explore to which extend the routing traffic exacerbates the situation. Generally speaking the second problem is a direct reason for the first one. It is the unstable and unpredictable performance of TCP in MANETs which motivated the choice of CBR traffic in pioneering (e.g. [1]) and subsequent studies of ad hoc routing protocols.

In [2] the authors show that multiple TCP sessions suffer from a severe unfairness in multihop wireless networks with static routing and no mobility; At the end some sessions gain full access to the network capacity completely shutting down the unlucky competitors. The problem is linked to the inability of TCP's congestion control to differentiate the packets losses due to radio interferences from those induced by network congestion [3, 4, 5]. Obviously, the traffic generated by a routing protocol increases the contention at the physical layer and will cause packet losses in addition to those caused by radio collisions between the TCP segments themselves.

---

* A significant part of this work has been performed in the University of Basel while the first author was working there.

Y. Koucheryavy, J. Harju, and V.B. Iversen (Eds.): NEW2AN 2006, LNCS 4003, pp. 298–312, 2006.

## 1.1   Contribution of the Paper

In this paper we consider reactive ad hoc routing schemes operating in ad hoc networks with potentially perfect connectivity and no mobility. With such settings we are able to isolate the traffic generated by the routing protocol during the path maintenance phase. Based on the AODV [6] and LUNAR [7] on-demand ad hoc routing protocols we study the effect of *four* distinct traffic patterns on the quality of ongoing TCP sessions.

For each considered routing scheme we determine the network configurations in terms of the number of sustained TCP sessions and the number of nodes in the network where the ad hoc routing does not seriously affect the quality of TCP communications. We call the set of such network configurations the "admissible operation range" (AOR) for the particular routing protocol. There is a limit of AOR, which we call the "routing ad hoc horizon", beyond which the uncontrolled broadcast transmission of routing messages results in a serious degradation of TCP performance. We estimate the "Equivalent routing load" which indicates the level of this degradation.

## 1.2   Outline

We develop the topic as follows. We first review the existing experimental studies of interactions between ad hoc routing and TCP in Sect. 2. After that in Sect. 3 we introduce the problem of evaluating the effect of routing on TCP performance and outline our methodology. Section 4 is the main section where we develop our approach to analyze TCP+routing interactions in IEEE 802.11 based ad hoc networks. We summarize the material and conclude the paper in Sect. 5.

## 2   Related Work

We found very few papers which investigate the interactions between TCP *and* ad hoc routing protocols. An initial study of the effect of routing on the quality of TCP communications is presented by Dyer and Boppana in [8]. Their study concentrates on TCP throughput that is achievable using different ad hoc routing schemes. Doing such measurements the authors mainly evaluate the quality of the particular routing scheme with respect to the path recovery times. The authors do not explicitly consider particular TCP-over-MANET problems.

The work of Perkins and Hughes [9] presents an evaluation of the effect of DSR and AODV on TCP performance. The authors find that when DSR is used as a routing protocol TCP achieves higher throughput than in the case of AODV.

In [10] Nahm et al. describe problems of interoperation between TCP and on-demand ad hoc routing protocols. They observe that TCP causes overreaction of the routing protocol, which degrades the quality of the end-to-end connection. The authors show cases where TCP, occupying the available bandwidth, prevents propagation of the control routing messages. This leads to large re-connect intervals, which in turn reduce the TCP throughput. The authors propose to use the known mechanisms for reducing TCP traffic load including an adaptive window increment schemes and delayed acknowledgments techniques.

The work in [11] from Uppsala University studies the interactions between TCP, UDP and routing protocols in MANETs. The authors show that the broadcast traffic generated by dynamic routing adds instability to ad hoc networking. In particular hidden terminal and channel capture effects cause instability of routes due to losses of control routing messages caused by the competing data traffic. This results in a very high packet loss rate for UDP traffic and long repeated timeouts of TCP.

## 3    Problem Statement

The performance of ad hoc routing protocols can be evaluated either with respect to the efficiency of establishing and maintaining the routing path between the communicating peers or with respect to the effect which routing traffic places on the quality characteristics of data flows. In both cases the process of the experimental performance evaluation is a complex task. In this paper we do not evaluate the considered routing protocols with respect to the quality of the path discovery phase. We concentrate on the second aspect of the routing performance and evaluate the degree of quality degradation for TCP connections due to routing activity.

### 3.1    Current Performance Metrics and Why They Are Not Informative for Evaluating TCP+routing Interactions

The major problem with the evaluation of TCP+routing interactions is the absence of TCP specific performance metrics for ad hoc routing protocols. In the literature related to the performance analysis of routing schemes [1, 8, 12, 13] we found very few *quantitative* performance metrics. The informational RFC 2051 [14] summarizes these metrics as follows:

1. Average number of control routing bits transmitted per data bit delivered, sometimes referred as to "Normalized routing load". This metric measures the overhead produced by transmission of routing messages.
2. Average number of data bits transmitted per data bit delivered, also referred as to "Packet delivery ratio". This metric is normally interpreted as a measure of the quality of data delivery within the network.

These two metrics are used mainly for two purposes. Firstly, to quantify a load produced by routing traffic and secondly, to assess the effect of routing on UDP based communications for which the packet loss rate is an illustrative performance characteristic. However, in the case of TCP communications, none of them is able to characterize the overall quality of the ongoing sessions.

As an example of why current performance metrics are not informative consider a scenario depicted in Fig. 1 and assume that the propagation of a routing message issued by Node X causes a loss of one data packet issued by Node Y.

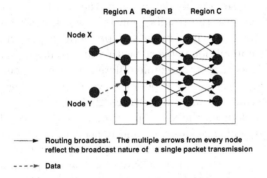

**Fig. 1.** Illustrative example of inability of the "normalized routing load" metric to capture the spatial effect of routing on data transmission

During the measurement period we observe 13 transmissions of routing messages in the entire network. The question is: How relevant is the information that 13 routing packets caused the loss of one data packet?

In fact only transmissions in Region B may cause the loss of our data packet if these transmission events coincide in time. If the transmission from Node Y happens at the same time as routing transmissions in Region A, Node Y will refrain from the transmission because it will sense the medium as busy. The nodes in Region C are located three hops away from Node Y, therefore transmissions from this region would not interfere with our data packet. Finally, only transmissions in Region B will cause the loss of the data packet because of the hidden terminal effect. Therefore the actual effect of routing traffic on the data transmission in our example is four routing packets per one lost data packet. The major conclusion from this simple observation is that the "normalized routing load" metric does not reflect the *spatial* effect from the routing activity on the quality of data sessions.

As for the metric "Packet delivery ratio", it is indeed an informative measure of the routing activity but only with respect to UDP traffic. Evaluation of TCP performance is a more challenging task. The quality of TCP communications is characterized by a *set* of metrics including the throughput and the fairness metrics. In this respect knowing the proportion of the number of delivered TCP data segments to the number of emitted data segments, which would be captured by the "Packet delivery ratio" metric, is certainly not enough to assess the degree of service degradation for the particular TCP flow.

## 3.2   Methodology Outline

We see the assessment of the routing performance based on direct observation of transmission events for routing packets as problematic due to obvious difficulties of capturing the *spatial* effect of the routing traffic described above. We adopt the methodology behind the construction of the "Packet delivery ratio" metric and benchmark different routing schemes by monitoring the performance of TCP traffic. For this we first identify a set of TCP specific performance metrics.

Our methodology is based on the definition of the "optimal" reference TCP performance. For this we use the results from [15], which describes an *ingress throttling* mechanism for achieving stable and predictable performance of the TCP protocol in multihop ad hoc networks with static routing.

In our simulation experiments we enable different routing schemes and monitor the deviation of the resulting TCP performance from the reference one. Up to a certain network scale the effect from transmissions of the control routing messages will be tolerable from the user's point of view. However, beyond that the routing traffic itself will cause a severe and unacceptable degradation of the TCP quality. We call the set of network configurations (in terms of the network size and the number of sustained TCP sessions) where routing traffic does not become a reason for poor TCP performance the *admissible operation range* (AOR) of MANETs.

Finally, at the border of the AOR, which we call the *routing ad hoc horizon*, we compute the *"Equivalent routing load"*. This metric reflects the level of quality degradation for the data traffic caused by a specific ad hoc routing scheme.

## 4    Admissible Operation Range, Routing Ad Hoc Horizon and Equivalent Routing Load

### 4.1    General Experimental Setup and TCP Performance Metrics

For our simulation experiments we use the topology in Fig. 2(a). There we have three geographical areas with sources of TCP sessions, forwarder nodes and sink nodes. The forwarding area begins and ends one wireless hop away from the area with the sources and the sinks, respectively. The size of the forwarding area ensures *three hops* communications between each source and sink. The number of nodes in the forwarding area assures potential connectivity for each source-destination pair. In this topology we are able to vary two parameters: The number of competing TCP sessions and the number of nodes in the network. By increasing the number of simultaneously active sessions we increase the intensity of the on-demand routing traffic. By increasing the number of nodes in the forwarding region we increase the duration of the broadcast bursts, since more nodes are involved in the (re-) broadcasting process.

We used the ns-2 network simulator[3] (version 2.27) and TCP Newreno as the most popular variant of the protocol. We set the value of the TCP maximum segment size (MSS) to 600 B. The data transmission rate of all devices is 2 Mb/s; RTS/CTS handshake is disabled; Other ns-2.27 parameters have default values. In all simulations we used continuous FTP traffic from all sources.

**Considered routing protocols.** For the experiments we use AODV-UU [16], the stable and the RFC compliant implementation of AODV from Uppsala University. As well as LUNAR stable implementation which is available from [17].

---

[3] [Online]. Available: http://www.isi.edu/nsnam/ns/

Two variants of AODV protocol are considered. In the first variant the HELLO mechanism is enabled to maintain the connectivity between neighbors further on we refer to this variant of AODV as to AODV-HELLO. In the second AODV variant, further referred as to AODV-LL, the route maintenance is done by means of explicit link layer feedback. In the former case the loss of connectivity between nodes forwarding traffic of a specific connection is detected by three missing HELLO messages; in this case the route maintenance procedure is invoked and the problem is signaled back to corresponding sources. In AODV-LL a packet loss during the transmission is detected by the link layer; the problem is then immediately reported to the routing engine, which in turn invokes the route maintenance operations.

The first variant of LUNAR uses standard settings as described in [7]. The second variant has a modified periodic refresh timer as we describe later on in Sect. 4.3.

Overall, the two variants of AODV and the two variants of LUNAR produce *four* distinct patterns of routing traffic in the networks. These patterns fit into *two* major types of the broadcast traffic invocation: (a) Error-driven (AODV-LL and opportunistic LUNAR) and (b) periodic (AODV-HELLO and LUNAR).

**Used performance metrics.** We assess TCP performance using the following set of metrics:

- Combined (total) TCP throughput of all existing in the network TCP flows. Denote this metric $Thr_{tot}$;
- Unfairness index $u$:

$$u = 1 - \frac{(\sum_{i=1}^{N} Thr_i)^2}{N \sum_{i=1}^{N} Thr_i^2}. \tag{1}$$

It is the opposite to the classic Jain fairness index, where $Thr_i$ is the throughput of FTP session $i$ and $N$ is the number of active sessions. A value of 0 for the unfairness index $u$ means that all TCP sessions receive the same share of the network's total capacity, while an unfairness of 1 means that only few sessions are monopolizing all bandwidth;
- The *unsmoothness* metric as defined in Appendix A for the qualitative assessment of the TCP session progress.

## 4.2   Reference "Optimal" TCP Performance

In order to obtain a stable and predictable behavior of the TCP protocol in multihop IEEE 802.11 based ad hoc network we use the *ingress rate throttling* scheme presented in [15]. There the *max-min* fairness framework from the wireline Internet is adapted to the specifics of MANETs and practically implementable mechanisms are proposed to enforce the model. The two major components of this solution are: (a) The usage of an ideal throughput achieved by a multihop TCP flow for characterizing the boundary load of a geographical region traversed by a multihop TCP session and (b) the rate throttling mechanism for reducing

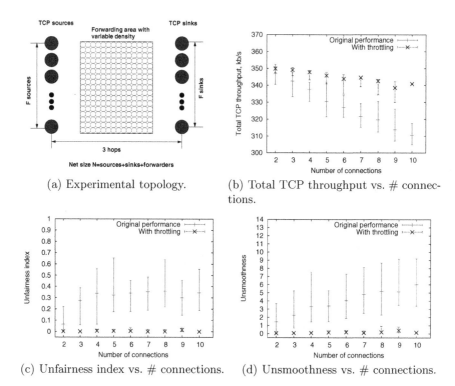

(a) Experimental topology.

(b) Total TCP throughput vs. # connections.

(c) Unfairness index vs. # connections.

(d) Unsmoothness vs. # connections.

**Fig. 2.** Experimental topology and reference "optimal" TCP performance in multihop ad hoc networks achieved with the ingress throttling scheme

the output rate at sources of TCP sessions in order to control the load in their bottleneck regions. The throttling level for each TCP session is given by (2).

$$r_i^{ingress} \leq \frac{Thr_{max}(h, MSS, TX_{802.11})}{\rho_{max}}. \tag{2}$$

In the above formula $Thr_{max}(h, MSS, TX_{802.11})$ is the maximal throughput of a TCP flow traversing $h$ hops, transmitting data segments of size $MSS$ while $TX_{802.11}$ is the used transmission rate at the physical layer along the path of the flow and $\rho_{max}$ is the maximum number of competing TCP flows along the path of a specific session (*path density*). The value of the maximal TCP throughput for the particular set of parameters can be either formally or experimentally estimated and made available at the source nodes. The above transmission rate limit is used to set the delay parameter of the scheduler for the queue with locally generated packets at layer 2.5 at each source of TCP connections. The delay parameter is computed as $\Delta = \frac{MSS_i}{r_i^{ingress}(h, MSS, TX_{802.11})}$.

The two major properties of ingress throttling [15], which we utilize for benchmarking the routing protocols, are: (a) the smooth progress of shaped TCP flows and (b) the optimal utilization of the network when all competing TCP flows are active. We illustrate these properties by an experiment as follows. We use the topology depicted in Fig. 2(a). The routes for all connections are statically assigned prior to the start of communications. We vary the number of connections in the network from 2 to 9 and activate them simultaneously; For each case we run 30 simulations with enabled and disabled ingress rate throttling.

The sub-figures 2(b) – 2(d) show the dynamics of the combined (total) TCP throughput in the network, the unfairness index and the unsmoothness metrics for different numbers of competing connections. All graphs show the mean values and the range between minimal and maximal values of the corresponding metrics over 30 simulation runs. In each run we seeded the random number generator of ns-2 randomly.

From sub-figure 2(b) we observe that performing the throttling actions, the resulting total throughput is equal to or larger than in the case without throttling. This demonstrates the increase in network utilization due to throttling. From sub-figures 2(c) and 2(d) we can see an improvement of the quality of TCP communications with respect to the unfairness and the unsmoothness metrics. When the ingress throttling is enabled all flows are smooth and free from long interruptions; overall almost perfect fairness is achieved.

It is important to point out the stability of the observed performance values. As it is visible from the graphs, the service offered by the plain combination of TCP and MAC 802.11 is highly variable and to a large extend unpredictable.

### 4.3   Admissible Operation Ranges

The stability and predictability of TCP performance demonstrated above is the needed reference behavior of the TCP protocol which we utilize for the analysis of the ad hoc routing schemes. When adding a routing protocol to the network with *enabled* ingress throttling scheme, the resulting performance will deviate from the observed "optimal" one. We predict that the uncontrolled broadcast transmissions that are present in all popular ad hoc routing schemes impose a limit on network size and number of sustained TCP sessions which we call the "admissible operation range".

**Description of AOR experiments.** In order to determine the "admissible operation range" we performed a series of simulation based experiments. In all experiments the *ingress throttling* is enabled. Each experiment is performed for the particular routing protocol with a fixed number of simultaneously active TCP sessions and a variable number of nodes in the forwarding area of the topology in Fig. 2(a).

We start with the minimum number of the forwarding nodes to ensure connectivity for every TCP connection (two nodes). We run simulations with this configuration up to 30 times and measure the worst *unsmoothness* amongst the competing flows and the *unfairness index* in each run. If after 30 runs the average of the worst unsmoothness is less than 1 and the worst unfairness index is less than 10% we increase the number of nodes in the forwarding area and repeat the experiment. We continue to increase the number of forwarding nodes until either of the metrics goes beyond the corresponding threshold. At this point we record the last network configuration (the number of TCP sessions and the network size) and the resulting combined (total) TCP throughput. After that we increase the number of simultaneously active TCP connections and repeat the sequence of experiments.

During the experiments with AODV-HELLO the best measured values of the *unsmoothness* metric is between one and two. For this variant of AODV we determine the *"weak AOR"* as it is described below.

The results for the two variants of AODV protocol and for LUNAR are shown in Fig. 3. The shaded areas in the figures show the admissible operation range where the *unsmoothness-unfairness* property is acceptable for all competing TCP flows. Outside the shaded area either the value of the *unsmoothness* or the *unfairness* metric becomes unacceptable for at least one end user.

The left slopes of the AOR show the minimal network configurations where the particular number of distinct connections is possible. For example, four distinct three-hop flows are possible when we have one source node, four destinations and two forwarding nodes, that is in total 7 nodes in the network. The right border of the corresponding AORs represents larger network configurations for the particular number of TCP sessions. For example, in the case where AODV with link layer feedback is used, four connections with the acceptable *unsmoothness* and *unfairness* metrics can exist in a network with four distinct sources, four destinations and 22 forwarding nodes, that is 30 network nodes in total.

**AODV-LL.** As we observe from Fig. 3(a) the least impact on the *ideal* TCP behavior is introduced by the routing traffic pattern of AODV with link layer feedback enabled. This is because of the error-driven invocations of broadcast transmissions. In AODV-LL the broadcast activities are initiated as a reaction to packet losses reported by the link layer to the AODV engine. In the case of a packet loss the protocol assumes that the connectivity to the corresponding neighbor is lost and initiates the route recovery procedure. When the ingress throttling scheme is enabled, the collisions between data packets are not frequent (this is indicated by the close to zero unsmoothness metric in Fig. 2(d)). In this case the reason for a higher packet loss rate, hence more frequent invocation of the path recovery phase, is the routing activity itself. As we observe in networks of up to 30 nodes, the broadcast traffic does not introduce enough overhead to force TCP flows into the routing induced slow start phase. However, beyond 30 nodes the broadcast bursts caused by every lost packet are long enough to cause the invocation of the slow start phase at a TCP sender; this leads to stammering TCP flow progress and worse fairness figures.

**AODV-HELLO.** The most unstable effect on the quality of TCP sessions is introduced by the traffic patterns produced by AODV-HELLO. In all experiments the *unsmoothness* of the competing TCP flows was between one and two at best. The analysis of the communication traces reveals that this is due to transmissions of HELLO messages. Despite of the *small* size, their frequent and independent emissions from multiple nodes does not allow any of the competing

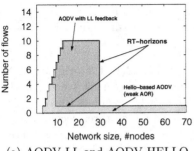

(a) AODV-LL and AODV-HELLO.

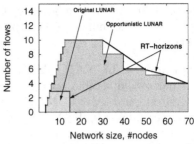

(b) Original and opportunistic LUNAR.

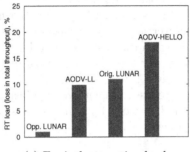

(c) Equivalent routing load.

**Fig. 3.** Admissible operation ranges and equivalent routing load for the considered routing protocols

TCP sessions to progress smoothly through the network. Even though we could not observe perfectly smooth flows as we did in the case of AODV-LL, we decided to delimit the admissible operation range where $0 < Unsmoothness < 2$ and the unfairness index (1) is less than 10%. We labeled this admissible operation range for AODV-HELLO as "weak AOR" in Fig. 3(a) .

**LUNAR.** As for the effect of the routing traffic pattern generated by LUNAR, we observe that it is very similar to that of AODV-HELLO. The smaller shaded area in Fig. 3(b) shows the admissible operation range for the original LUNAR with its forced three seconds complete route re-discoveries. The difference is that the horizon with LUNAR is stable, meaning that the *unsmoothness* metric for all flows within the AOR is equal to or less than 1 and the fairness is close to perfect. The stability of LUNAR's admissible operation range can be explained by less frequent initiations of the broadcast traffic than in the case of AODV-HELLO. Certain similarities in the right border of the AOR are explained by the fact that LUNAR's three second forced route rediscovery interval is chosen with the reference to the HELLO interval in AODV: It corresponds to two HELLO rounds which are needed by AODV to determine a change of the route. However, in the case of LUNAR we have a strict right border. This can be explained by the difference in the nature of broadcast patterns generated by the two protocols.

If in the case of AODV-HELLO we have short frequent one hop broadcasts, in the case of LUNAR we have less frequent but more massive flooding waves, which result in stammering TCP progress with increasing number of nodes in the network.

**Opportunistic LUNAR.** As can be seen in Fig. 3(b) the original LUNAR has a narrow admissible operation range. As an experiment we changed the route refresh strategy of LUNAR to one which closely resembles the error-driven pattern of AODV-LL. To do this we disabled the forced three second complete route rediscovery mechanism. Instead, we retain a route as long as there are packets arriving to the forwarding engine: With every new data packet we shift the route timeout three seconds into the future. This modification allowed us to create an *"opportunistic"* version of LUNAR.

This change allowed us to significantly extend the horizon. Moreover, we achieved even better characteristics in comparison to AODV-LL. Now the right edge of the admissible operation range stretches to bigger networks. This is because we do not interpret every loss of a packet as an indication of the route breakage as it is the case with AODV-LL. Instead, we react on invocations of slow starts in TCP which are less frequent events, given that the ingress throttling scheme is deployed in all sources. By doing so we significantly decrease the frequency of broadcast bursts.

### 4.4   Routing Ad Hoc Horizon and Equivalent Routing Load

In Sect. 3 we presented the reasoning for why the estimation of the routing load is complex and not an obvious task. Recapitulating, this is due to the difficulties to capture the spatial effect of the broadcast traffic on a particular data session. Because of this the averaged metrics for routing load do not reflect the spatially distinct effects of the routing traffic on the ongoing data sessions. In the previous subsection we described a way to quantify the admissible operation region for reactive ad hoc routing protocols and TCP communications. The *routing ad hoc horizon* represents the critical network size where TCP flows maximally use the network capacity which is not consumed by transmission of routing messages and have an acceptable end user quality. Beyond this horizon, TCP flows begin to suffer from routing induced unfairness. In this case the progress of affected flows becomes disrupted, potentially leading to long disconnection intervals and loss of network utilization. The important property of the stability of communications within the AOR is that we can indirectly estimate the routing load produced by one or another protocol.

We define an *"equivalent routing load"* metric as follows. Assuming that our ingress throttling is deployed in all sources and that we measure the total TCP throughput in the network without routing ($Thr^{tot}_{norouting}$), in the case where all competing flows are active the network is fully utilized. Now we measure the total TCP throughput on the right border of the admissible operation range ($Thr^{tot}_{withrouting}$). In this case the competing TCP flows maximally use the capacity that is left from the routing traffic. Then the quantity $(1 - \frac{Thr^{tot}_{withrouting}}{Thr^{tot}_{norouting}}) \cdot 100\%$

reveals the reduction in the throughput in per-cent[4] due to the routing activity. Figure 3(c) shows the equivalent routing load for the considered protocols.

We may observe that in addition to the very narrow admissible operation range the HELLO-based AODV consumes more useful data throughput than other protocols. As for the modified opportunistic LUNAR protocol, its "equivalent routing load" is the smallest among all routing protocols. This is because the routing activity in this case happens only during the path establishment of each flow. After that the ingress throttling scheme prevents invocation of slow starts in the stable network as we have in our experiments. Therefore no further routing traffic is involved after all sessions were successfully established until the end of simulations.

### 4.5   Summary

In this section we developed our approach for the analysis of the effect of ad hoc routing traffic on the quality of TCP communications. For each of the the two considered routing protocols and static topologies we determined the admissible operation range, which is the set of network configurations where routing and TCP traffic can peacefully co-exist.

We found that 10 - 20 nodes is a critical network size for routing schemes which utilize periodic broadcast invocations in their route maintenance phase. With such protocols only few TCP connections may exist with an acceptable quality. It is worth noting that this type of broadcast invocations is present in the specification of a new generation reactive routing protocol for MANETs. In the recent draft for *DyMo* [18] the HELLO mechanism for the connectivity maintenance between the neighbors is inherited from AODV. In contrast, our analysis shows that a "TCP-friendly" ad hoc routing protocols should use mainly an error-driven form of broadcast invocations.

## 5   Conclusions

In this paper we analyzed the effect of different routing traffic patterns on the performance of TCP during stable operations of MANETs. We found that quantifying the routing load is difficult even in static networks without mobility. We highlighted the fact that the current performance metrics for the evaluation of routing protocols are not informative enough for the evaluation of TCP+routing interactions.

We suggested a methodology for an indirect measurement of the routing load and quantification of the routing effect on TCP performance. Using this technique we were able to show that the routing traffic itself can be a reason for poor TCP performance in MANETs.

We analyzed two routing protocols with four distinct routing traffic patterns and were able to identify the scaling region of MANETs (in terms of *the number*

---

[4] We intentionally do not express the "Equivalent routing load" in bits per second since this value would be specific to the particular transmission speed picked in the experimental environment (in our case ns-2).

*of simultaneously active TCP sessions* and *the number of network nodes*) beyond which the traffic generated by the routing protocol significantly degrades the performance of TCP. We call this region the *"admissible operation range"* of MANETs.

Our major conclusion regarding the effect of routing traffic patterns on TCP communications is that periodic, non error-driven broadcasts of even short messages is harmful for data communications and leads to narrowing the operational region of MANETs.

Further studies have to add mobility to the scene, which we expect to become another limiting factor for the admissible operation range.

# References

1. J. Broch, D. Maltz, D. Johnson, Y-C.Hu, and J. Jetcheva, "A performance comparison of multi-hop wireless ad hoc network routing protocols," in *Proc. ACM MobiCom'98*, Dallas, TX, USA, 1998.
2. K. Xu, S. Bae, S. Lee, and M. Gerla, "TCP behavior across multi-hop wireless networks and the wired Internet," in *Proc. WoWMoM'02*, Atlanta, GA, USA, Sep. 2002.
3. H. Balakrishnan, V. Padmanabhan, S.Seshan, and R. H. Katz, "A comparison of mechanisms for improving TCP performance over wireless links," in *Proc. ACM SIGCOMM'05 workshops*, Philadelphia, PA, USA, Aug. 2005.
4. H. Elaarag, "Improving TCP performance over mobile networks," *ACM Computing Surveys*, vol. 34, no. 3, Sept. 2003.
5. S. Xu and T. Saadawi, "Does the IEEE 802.11 MAC protocol work well in multihop wireless ad hoc networks?" *IEEE Communications Magazine*, Jun. 2001.
6. C. Perkins, E. Belding-Royer, and S. Das, "Ad hoc on-demand distance vector (AODV) routing," IETF RFC 3561, Jul. 2003. [Online]. Available: http://www.rfc-editor.org/rfcsearch.html
7. C. Tschudin, R. Gold, O. Rensfelt, and O. Wibling, "LUNAR: a lightweight underlay network ad-hoc routing protocol and implementation," in *Proc. NEW2AN'04*, St. Petersburg, Russia, Feb. 2004.
8. T. Dyer and R. Boppana, "A comparison of TCP performance over three routing protocols for mobile ad hoc networks," in *Proc. ACM MobiHoc'01*, Long Beach, CA, USA, Oct. 2001.
9. D. Perkins and H. Hughes, "Investigating the performance of TCP in mobile ad hoc networks," *International Journal of Computer Communications*, vol. 25, no. 11-12, 2002.
10. K. Nahm, A. Helmy, and C.-C. J. Kuo, "TCP over multihop 802.11 networks: issues and performance enhancement," in *Proc. ACM MobiHoc'05*, Urbana-Campaign, Illinoise, USA, May 2005.
11. C. Rohner, E. Nordström, P. Gunningberg, and C. Tschudin, "Interactions between TCP, UDP and routing protocols in wireless multi-hop ad hoc netwrorks," in *Proc. IEEE ICPS Workshop on Multi-hop Ad hoc Networks: from theory to reality (REALMAN'05)*, Santorini, Greece, Jul. 2005.
12. S.-J. Lee, E. Belding-Royer, and C. Perkins, "Scalability study of the ad hoc on-demand distance vector routing protocol," *International Journal of Network Management*, vol. 13, no. 2, Mar./Apr. 2003.

13. S. R. Das, C. Perkins, and E. M. Royer, "Performtmance comparison of two on-demand routing protocols for ad hoc networks," in *Proc. IEEE Infocom'00*, Tel-Aviv, Israel, Mar. 2000.

14. S. Corson and J. Macker, "Mobile ad hoc networking (MANET): Routing protocol performance issues and evaluation considerations," IETF Infromational RFC 2501, 1999. [Online]. Available: http://www.rfc-editor.org/rfcsearch.html

15. E. Osipov, "On the practical feasibility of fair TCP communications in IEEE 802.11 based multihop ad hoc wireless networks," Ph.D. dissertation, University of Basel, Switzerland, 2005.

16. AODV-UU implementation, Uppsala University. [Online]. Available: http://core.it.uu.se/AdHoc/AodvUUImpl

17. Uppsala University ad hoc implementation portal. [Online]. Available: http://core.it.uu.se/AdHoc/ImplementationPortal

18. I. Chakeres, E. Belding-Royer, and C. Perkins, "Dynamic MANET on-demand (DYMO) routing," IETF draft (work in progress), 2005. [Online]. Available: http://www.ietf.org/internet-drafts/draft-ietf-manet-dymo-02.txt

19. Ch.Tschudin and E. Osipov, "Estimating the ad hoc horizon for TCP over IEEE 802.11 networks," in *Proc.MedHoc'04*, Bodrum, Turkey, Jun. 2004.

# A    Definition of the "unsmoothness" Metric

As it is shown in [19] it is important to account for interruptions in TCP flow progress since some of the flows may suffer from long no-progress intervals although the throughput and fairness metrics would report positive network performance.

In order to qualitatively assess the progress of each TCP session we construct an "unsmoothness" metric as is illustrated in Fig. 4. The construction of the metric is done during the analysis of communication traces. After the end of each test run we parse the packet trace and record the progress of the received sequence numbers in time for each flow and the corresponding start times for every session. The stop time for all sessions is assumed to be the same. With this

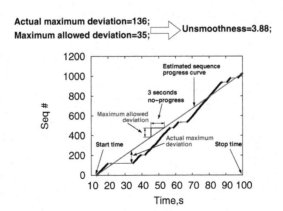

**Fig. 4.** Computation of the unsmoothness metric

information we estimate the ideal curve of the sequence number progress. It is shown by the straight line in Fig. 4. After that we analyze the recorded process of sequence number arrivals for each flow and compare each received sequence number with the estimated "smooth" value for the corresponding time. The result of the comparison is the absolute deviation of the actual sequence number from the estimated value. Obviously, in most of the measurements we will have some deviation from the estimated curve even for a perfectly smooth flow due to rounding errors. In order to allow some small deviations we compute the maximum allowed deviation. This value corresponds to the three seconds no-progress time which was chosen based on our empirical observations that longer no-progress times definitely indicate the presence of the TCP unfairness in ad hoc networks. Finally, we compute the "unsmoothness" metric as in (3).

$$Unsmoothness = \frac{max(|Deviation_{actual}|)}{Deviation_{allowed}^{max}}. \tag{3}$$

The "unsmoothness" metric is always larger than zero. We say that the quality of a TCP flow is acceptable for an end-user if $Unsmoothness \leq 1$.

# Multi-path Routing Protocols in Wireless Mobile Ad Hoc Networks: A Quantitative Comparison

Georgios Parissidis, Vincent Lenders, Martin May, and Bernhard Plattner

Swiss Federal Institute of Technology
Communications Systems Group
Gloriastrasse 35, 8092 Zurich, Switzerland
{parissid, lenders, may, plattner}@tik.ee.ethz.ch

**Abstract.** Multi-path routing represents a promising routing method for wireless mobile ad hoc networks. Multi-path routing achieves load balancing and is more resilient to route failures. Recently, numerous multi-path routing protocols have been proposed for wireless mobile ad hoc networks. Performance evaluations of these protocols showed that they achieve lower routing overhead, lower end-to-end delay and alleviate congestion in comparison with single path routing protocols. However, a quantitative comparison of multi-path routing protocols has not yet been conducted. In this work, we present the results of a detailed simulation study of three multi-path routing protocols (SMR, AOMDV and AODV_Multipath) obtained with the ns-2 simulator. The simulation study shows that the AOMDV protocol achieves best performance in high mobility scenarios, while AODV_Multipath performs better in scenarios with low mobility and higher node density. SMR performs best in networks with low node density, however as density increases, the protocol's performance is degrading.

## 1   Introduction

The standardization of wireless communication (IEEE 802.11) [1] in 1997 for Wireless Local Area Networks (WLANs) offered the opportunity for inter communication of mobile, battery equipped devices and showed the way of a revolutionary method of communication that extends the well-established wired Internet. Nowadays, mobile devices are becoming smaller, lighter, cheaper and more powerful, addressing the augmenting needs of users. While radio communication for wireless networks is standardized and many problems have been resolved, networking protocols for intercommunication are still in experimental state. The successful and wide-spread deployment of ad hoc networks strongly depends on the implementation of robust and efficient network protocols.

Single path routing protocols have been heavily discussed and examined in the past. A more recent research topic for MANETs are multi-path routing protocols. Multi-path routing protocols establish multiple disjoint paths from a source to a destination and are thereby improving resilience to network failures

Y. Koucheryavy, J. Harju, and V.B. Iversen (Eds.): NEW2AN 2006, LNCS 4003, pp. 313–326, 2006.

and allow for network load balancing. These effects are particularly interesting in networks with high node density (and the corresponding larger choice of disjoint paths) and high network load (due to the ability to load balance the traffic around congested networks). A comparison of multiple multi-path protocols is therefore particularly interesting in scenarios of highly congested and dense networks.

Up to now, no extensive simulations and quantitative comparison of multi-path routing protocols have been published. In the present paper, we fill this gap by presenting an evaluation and comparison of three wireless ad hoc multi-path routing protocols, namely SMR [2], and two modifications or extensions of AODV [3]: AOMDV [4] and AODV_Multipath[5]. With the help of the *ns-2* simulator, we examine the protocol performance under a set of network properties including mobility, node density and data load. The comparison focuses on the following metrics: data delivery ratio, routing overhead, end-to-end delay of data packets and load balancing. In addition, the AODV protocol is included as a reference single path routing protocol to compare multi-path with single path routing in general. In the context of the present work, we do not target at achieving high performance values or proposing a new protocol that outperforms existing protocols. Furthermore, our study focuses on application scenarios applied in small mobile devices with limited power and memory resources such as handhelds or pocket PCs. Therefore, in our simulations we assumed a networking interface queue size of 64 packets.

The contribution of the present paper is three-fold: i) we show in the comparison that: AODV_Multipath performs best in static networks with high node density and high load; AOMDV outperforms the other protocols in highly mobile networks; SMR offers best load balancing in low density, low load scenarios; ii) we demonstrate that multi-path routing is only advantageous in networks of high node density or high network load; and iii) we confirm that multi-path routing protocols create less overhead compared to single path routing protocols.

The remainder of this paper is organized as follows. In section 2 we present the routing protocols that are used in the performance evaluation. The methodology of the performance evaluation as well as the simulation environment are presented in section 3. The results of the quantitative comparison of multi-path routing protocols are discussed in section 4. Related work and concluding remarks are presented in sections 5 and 6, respectively.

## 2    Routing Protocols

In this work, we consider multi-path routing protocols with the following fundamental properties: (i) The routing protocol provides multiple, loop-free, and preferably node-disjoint paths to destinations, (ii) the multiple paths are used simultaneously for data transport and (iii) multiple routes need to be known at the source. Multi-path routing protocols that have been proposed for mobile ad hoc networks and satisfy the above-mentioned requirements are:

1. *SMR* (Split Multi-path Routing) [2]
   SMR is based on DSR [6]. This protocol attempts to discover maximally disjoint paths. The routes are discovered on demand in the same way as it is done with DSR. That is, the sender floods a Route REQuest (RREQ) message in the entire network. However, the main difference is that intermediate nodes do not reply even if they know a route to the destination. From the received RREQs, the destination then identifies multiple disjoint paths and sends a Route REPlay (RREP) packet back to the source for each individual route. According to the original proposal of SMR, we configure our implementation to establish at maximum two link disjoint (SMR_LINK) or at maximum two node disjoint (SMR_NODE) paths between a source and a destination.

2. *AOMDV* (Ad hoc On demand Multi-path Distance Vector routing) [4]
   AOMDV extends AODV to provide multiple paths. In AOMDV each RREQ and respectively RREP defines an alternative path to the source or destination. Multiple paths are maintained in routing entries in each node. The routing entries contain a list of next-hops along with corresponding hop counts for each destination. To ensure loop-free paths AOMDV introduces the *advertised_hop_count* value at node $i$ for destination $d$. This value represents the maximum hop-count for destination $d$ available at node $i$. Consequently, alternate paths at node $i$ for destination $d$ are accepted only with lower hop-count than the *advertised_hop_count* value. Node-disjointness is achieved by suppressing duplicate RREQ at intermediate nodes.

   In our simulations we consider four *alternative* configurations of the AOMDV protocol depending on the type (link or node disjoint) and the maximum number of multiple paths the protocol is configured to provide:
   (a) *AOMDV_LINK_2paths*: Maximum two link-disjoint paths.
   (b) *AOMDV_LINK_5paths*: Maximum five link-disjoint paths.
   (c) *AOMDV_NODE_2paths*: Maximum two node-disjoint paths.
   (d) *AOMDV_NODE_5paths*: Maximum five node-disjoint paths.
   To avoid the discovery of very long paths between each source-destination pair the hops difference between the shortest path and the alternative paths is set to five for all AOMDV protocol configurations.

3. *AODV_Multipath* (Ad hoc On-demand Distance Vector Multi-path) [5]
   AODV_Multipath is an extension of the AODV protocol designed to find multiple node-disjoint paths. Intermediate nodes are forwarding RREQ packets towards the destination. Duplicate RREQ for the same source-destination pair are not discarded and recorded in the RREQ table. The destination accordingly replies to all route requests targeting at maximizing the number of calculated multiple paths. RREP packets are forwarded to the source via the inverse route traversed by the RREQ. To ensure node-disjointness, when intermediate nodes overhear broadcasting of a RREP message from neighbor nodes, they delete the corresponding entry of the transmitting node from their RREQ table. In AODV_Multipath, node-disjoint paths are established during the forwarding of the route reply messages towards the source, while in AOMDV node-disjointness is achieved at the route request procedure.

4. *AODV* (Ad hoc On demand Distance Vector) [3]
   We use the AODV as a reference on demand single path routing protocol.
   AODV is used as a benchmark to reveal the strengths and the limitations of
   multi-path versus single path routing.

Summarizing the presentation of the routing protocols, we list the essential
properties of the multi-path protocols:

- SMR: The protocol calculates link and node disjoint paths. The maximum
  number of paths is set to two. The source is aware of the complete path
  towards the destination.
- AOMDV: The maximum number of paths can be configured, as well as
  the hop difference between the shortest path and an alternative path. The
  protocol calculates link and node disjoint paths.
- AODV_Multipath: The protocol establishes only node disjoint paths. There
  is no limitation on the maximum number of paths.

## 3   Methodology

We next describe the methodology we used to compare the different routing
protocols.

**Simulation environment:** We use a detailed simulation model based on ns-
2 [7]. The distributed coordination function (DCF) of IEEE 802.11 [1] for wireless
LANs is used at the MAC layer. The radio model uses characteristics similar to
Lucent's WaveLAN radio interface. The nominal bit-rate is set to 2 Mb/sec and
the radio range is limited to 250 meters; we also apply an error-free wireless
channel model.

**Movement model:** We use the random waypoint model [8] to model node
movements. The random waypoint movement model is widely used in simulations
in spite of its known limitations [9]. The simulation time is 900 seconds while
the pause time varies from 0 seconds (continuous motion) to 900 seconds (no
mobility) [0,30,60,120,300,600,900 seconds]. Nodes move with a speed, uniformly
distributed in the range [0,10m/s].

**Network size and communication model:** We consider 4 network sizes with
30, 50, 70, and 100 nodes in a rectangular field of size 1000m x 300m. We vary
the number of nodes to compare the protocol performance for low and high
node density. Traffic patterns are determined by 10, respectively 20 CBR/UDP
connections, with a sending rate of 4 packets per second between randomly
chosen source-destination pairs. Connections begin at random times during the
simulations. We use the identical traffic and mobility patterns for the different
routing protocols. Simulation results are averaged values of 20 scenarios with
different seed. Data packets have a fixed size of 512 bytes and the network
interface queue size for routing and data packets is set to 64 packets for all
scenarios.

**Scheduling of data packets:** A sender uses all available paths to a destination simultaneously. Data packets are sent over each individual path with equal probability. When one path breaks, the source stops using that path but does not directly initiates a new route request. Only when all available paths are broken a new route request is initiated.

**Protocol implementation:** The original source code of AOMDV [4] and AODV_Multipath [5] protocols in ns-2 is used in our performance evaluation. The implementation of SMR in ns-2 is adopted from [10]. For all protocols, we extend the implementation to use multiple paths simultaneously.

**Metrics:** We use the following five metrics to compare the performance of the multi-path routing protocols.

1. *Routing overhead.* The routing overhead is measured as the average number of control packets transmitted at each node during the simulation. Each hop is counted as one separate transmission.
2. *Average number of paths.* The average number of paths is the amount of paths that are discovered per route request.
3. *Data packet delivery ratio.* The data packet delivery ratio is the ratio of the total number of delivered data packets at the destination to the total number of data packets sent.
4. *Average end-to-end delay of data packets.* The average end-to-end delay is the transmission delay of data packets that are delivered successfully. This delay consists of propagation delays, queueing delays at interfaces, retransmission delays at the MAC layer, as well as buffering delays during route discovery. Note here, that, due to the priority queuing of routing messages, queueing delays for data traffic packets can be higher that the normal maximum queuing delay of a 64 packet queue.
5. *Load balancing.* Load balancing is the ability of a routing protocol to distribute traffic equally among the nodes. We capture this property by calculating the deviation from the optimal traffic distribution.

## 4    Simulation Results

We present in this section the simulation results for the comparison of the multi-path routing protocols. In addition, we compare the results with the results obtained with AODV to emphasize the benefits of multi-path versus single path routing. The results are presented individually per routing metric. We summarize the main findings of the comparison at the end of this section.

### 4.1    Routing Overhead

In general, SMR produces more control overhead than the AODV-based multi-path routing protocols. This is caused by the fact that SMR rebroadcasts the same RREQ packets it receives from multiple neighbors. In the following, we discuss in detail the routing overhead for each individual scenario.

*Low density, low load:* The routing overhead in networks with low node density and low traffic load is shown in figure 1(a). We clearly observe the higher overhead of SMR compared to the AODV-based routing protocols. Interestingly, the version of SMR which computes link disjoint paths (SMR_LINK) produces more overhead than the variation which determines node disjoint paths (SMR_NODE). The reason is that the source waits until all existing paths break before sending a new route request, and the probability that two paths break is lower if they are node-disjoint than otherwise.

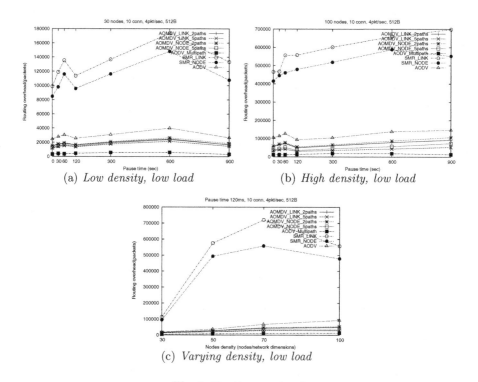

(a) *Low density, low load*

(b) *High density, low load*

(c) *Varying density, low load*

**Fig. 1.** Routing overhead

When comparing AODV, AOMDV, and AODV_Multipath, we see that all three protocols have a similar control overhead, only the overhead of AODV is slightly larger. Indeed, multi-path routing protocols require less control messages for routes to destination nodes that have been previously requested. Therefore, the saving in terms of overhead originates from connections with the same destination node.

*High density, low load:* In figure 1(b), we plot the routing overhead for a higher node density (100 nodes on a square of the same size as before). The absolute number of control packets at each node is higher than with 30 nodes. However, the trends remain the same.

*Varying density, low load:* The effect of node density on the routing overhead in a scenario with moderate mobility is illustrated in figure 1(c). The routing overhead augments slightly with increasing node density for the AODV-based protocols. However, the routing overhead of SMR starts to decrease when the number of nodes exceeds 50. The reason is that for more than 50 nodes, the network becomes congested and many control packets are dropped. We will see later that for such networks, the delivery ratio of data packets is below 10 %.

## 4.2   Average Number of Paths

We next look at the ability of the different routing protocols to find multiple paths. For this, we measured the average number of discovered routes per route request. The result is plotted for the low density and low load scenario in figure 2(a) and for the high density and high load scenario in figure 2(b). AODV_Multipath is clearly the protocol which finds the most paths. However, as we will see later when looking at the packet delivery ratio, many discovered paths are not usable when the nodes are very mobile. Note that AOMDV and SMR tend to find on average significantly less paths than their upper limit. AOMDV configured to find a maximum of 5 paths (node-disjoint or link-disjoint) finds approximately on average at most 2 paths. AOMDV and SMR when configured to find a maximum of 2 paths find approximately on average at most 1.4 paths.

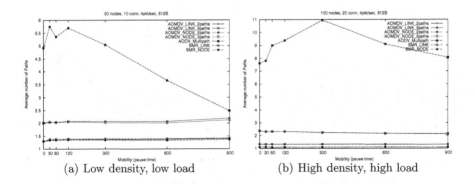

(a) Low density, low load                    (b) High density, high load

**Fig. 2.** Average number of paths

## 4.3   Data Packet Delivery Ratio

The data packet delivery ratio is now presented for the three different scenarios.

*Low density, low load:* As expected, in sparse networks with low traffic load, multi-path routing does not improve the performance compared to single path routing in terms of successful packet delivery. As we see in figure 3(a), the packet delivery ratio in this scenario is equal for AODV and all variants of AOMDV independent of the node mobility.

Surprisingly, the performance of SMR and AODV_Multipath is even worse compared to single path routing. AODV_Multipath severely suffers from packet losses when the network becomes dynamic. This is mainly because the protocol finds much more paths than the other protocols (see figure 2(a)) and a source tries to use all of them until *all* become stale. Detecting that a route is stale is time-consuming since with 802.11, a broken link is only detected by retransmitting a packet at the MAC layer multiple times without receiving an acknowledgement. SMR overloads the network with control messages and data packets are dropped at full buffers of intermediate nodes. Even in the static case (900 seconds pause time), SMR and AODV_Multipath have a packet delivery ratio which is approximately 10% below the ration of AODV.

*High density, high load:* Figure 3(b) shows the benefits of multi-path routing versus single path for dense networks with high traffic load. In this case, both AODV_Multipath and AOMDV clearly outperform AODV. Comparing AODV_Multipath and AOMDV, the performance strongly depends on the node mobility in the network. When the network is static, AODV_Multipath achieves the best performance (almost 80% delivery ratio). When the network is highly mobile (pause time less than 400 seconds), AOMDV has a higher delivery ratio.

Apparently, SMR has a very poor performance in this scenario (below 5 % delivery ratio). This is easy to understand when we consider the routing overhead of SMR (see figure 1). SMR produces an extensive amount of control packets which overloads the network dramatically. The network is so congested that only routing packets are queued and most of the data packets are dropped.

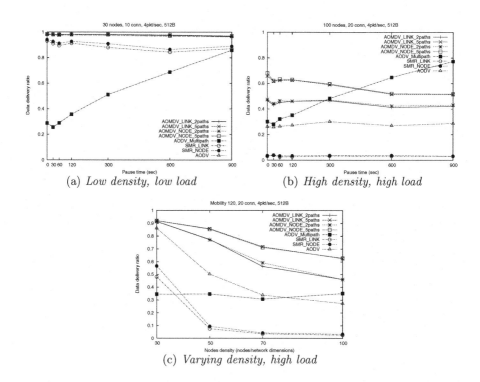

(a) *Low density, low load*    (b) *High density, high load*

(c) *Varying density, high load*

**Fig. 3.** Data packet delivery ratio

*Varying density, high load:* In figure 3(c), we also plot the packet delivery ratio versus the node density for a network with a large amount of traffic. We fixed the pause time to 120 seconds (moderate mobility) which is in favor of AOMDV as we have seen in the previous figure. By increasing the network density, we see that protocol's performance decreases except AODV_Multipath. We conclude that the performance of AOMDV_Multipath is independent of the network density when the amount of traffic is high.

## 4.4   Average End-to-End Delay of Data Packets

We have just seen that multi-path routing outperforms single path routing in terms of delivered data packets when the traffic load in the network is high. We now compare the average end-to-end delay of the multi-path routing protocols in these scenarios. We differentiate between three cases: high node mobility with a pause time of 0 sec (see figure 4(a)), moderate node mobility with a pause time of 120 sec (figure 4(b)) and no node mobility (900 sec pause time, see figure 4(c)).

End-to-end delay for the SMR protocol is higher than all protocols. In low mobility and low node density scenarios end-to-end delay is approximately 650ms, while in higher mobility and node density the end-to-end delay augments to 1200ms. The high routing overhead of SMR penalizes data packets, therefore high buffering delays contribute to high end-to-end delay.

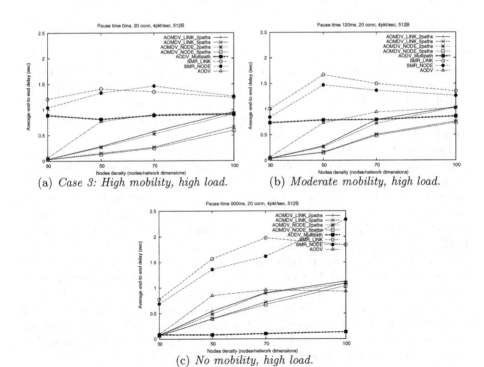

(a) *Case 3: High mobility, high load.*          (b) *Moderate mobility, high load.*

(c) *No mobility, high load.*

**Fig. 4.** End-to-end delay

AODV_Multipath is an exception as the protocol converges in low mobility scenarios (in bold in figures 4(a), 4(b), 4(c)). In high mobility, end-to-end delay is approximately 900 ms, while in moderate mobility and no mobility decreases to 750 ms and 200 ms respectively.

AOMDV achieves the smallest end-to-end delay compared to the other protocols, including the AODV, in high and moderate mobility. The average number of hops a data packet travels with multi-path routing is higher than with single path routing. Data packets are equally distributed among all available paths independent of hops difference between the shortest and an alternative path towards the same destination. However, in a congested network, multi-path routing AOMDV manages to distribute the traffic on less congested links and data packets experience smaller buffering delay on intermediate nodes.

AOMDV_LINK_5paths and AOMDV_NODE_5paths, achieve in general lower end-to-end delay than AOMDV_LINK_2paths, AOMDV_NODE_2paths, AODV, AODV_Multipath and SMR with regard to node density and mobility. Taking into consideration figure 2(b) that presents the average number of paths that are available at each sender for high node density and data load versus mobility we observe that multi-path routing is beneficial if the number of multiple paths is between 2 and 3. The high number of multiple routes that AODV_Multipath calculates is not awarded with better performance as many paths break with higher probability. Therefore multi-path routing becomes beneficial if it provides one or two additional link or node disjoint paths.

### 4.5   Load Balancing

Load balancing is the ability of a routing protocol to equally distribute the traffic among all nodes. Load balancing is for example useful to maximize the network lifetime when the networked devices are battery-powered. It is also helpful to distribute the traffic equally to avoid single bottlenecks in the network where most traffic is passing trough. We first look at load balancing in sparse networks with low load and then consider networks with increased node density.

*Low density, low load:* In figure 5(a) load balancing with low node density, low data load, and moderate mobility (120s pause time ) is presented. We sorted the nodes according to the number of data packets they forwarded on the x-axis and plot the percentage of data packets each node has forwarded on the y-axis. Since there are 30 nodes in the network, an optimal load balancing would result in 3.33 %. We see that SMR and AOMDV achieve a better distribution of the traffic between the different nodes.

We also plot the results in the low density and low data load case for different pause times in figure 5(b). Instead of plotting the percentage of forwarded packet, we plot the standard deviation from the average value. Thus, a value of 0 results in optimal load balancing. SMR and AOMDV have lower standard deviation than AODV_Multipath independent of node mobility. Furthermore, increased mobility result in better load balancing of the data traffic among the nodes. However, this result is intuitive as mobility favors data traffic dispersion.

*Varying density, low load:* To circumvent the effect of mobility, we illustrate the standard deviation of the total forwarded data packets from the optimum value in static scenarios versus node density in figure 5(c). AOMDV and SMR disseminate data across all nodes better than AODV_Multipath.

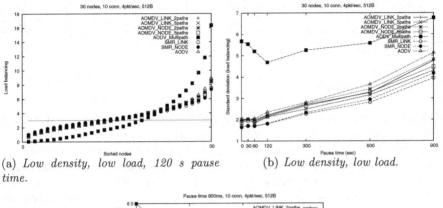

(a) *Low density, low load, 120 s pause time.*

(b) *Low density, low load.*

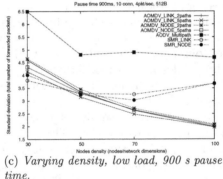

(c) *Varying density, low load, 900 s pause time.*

**Fig. 5.** Load balancing

## 4.6  Discussion of the Results

In Table 1, we summarize the performance of each multi-path routing protocol for the different metrics. We differentiate three network regimes: (a) low density and low load, (b) high density, high load and low mobility, and (c) high density, high load, and high mobility. Recapitulating the performance evaluation of the three multi-path routing protocols, we find that:

1. Multi-path routing achieves in general better performance than single path routing in dense networks and networks with high traffic load.
2. AOMDV achieves the best performance in scenarios with high node mobility.
3. AODV_Multipath performs best in relatively static scenarios.
4. The performance of SMR is poor in dense networks and networks with high traffic load because of the immense control traffic generated.

**Table 1.** Comparison of multi-path routing protocols. (++: very good, +: good, 0: neutral, -:poor). (a) Low density, low load, (b) High density, high load, low mobility (c) High density, high load, high mobility. (AODV_M corresponds to AODV_Multipath.)

| | (a) | | | (b) | | | (c) | | |
|---|---|---|---|---|---|---|---|---|---|
| | AOMDV | AODV_M | SMR | AOMDV | AODV_M | SMR | AOMDV | AODV_M | SMR |
| Routing overhead | ++ | ++ | 0 | + | ++ | - | + | + | - |
| Packet delivery ratio | ++ | 0 | + | + | + | - | ++ | 0 | - |
| Average e2e delay | + | + | + | + | ++ | 0 | + | + | + |
| Load balancing | + | 0 | ++ | + | + | + | + | 0 | 0 |

# 5  Related Work

Multi-path routing is not a new concept and has already been proposed and implemented in packet and circuit switched networks. In circuit switched telephone networks, alternate path routing was proposed in order to increase network utilization as well as to reduce the call blocking probability. In data networks, the Private Network-to-Network Interface (PNNI) signalling protocol [11] was proposed for ATM networks. With PNNI, alternate paths are used when the optimal path is over-utilized or has failed. In the Internet, multi-path routing is included in the widely used interior gateway routing protocol OSPF [12]. Multi-path routing alleviates congestion by re-routing data traffic from highly utilized to less utilized links through load balancing. The wide deployment of multi-path routing is so far prevented due to the higher complexity and the additional cost for storing extra routes in routers.

However, wireless ad hoc networks consist of many features that differentiate them from conventional wired networks. The non-employment of multi-path routing in the standardized routing protocols used in the Internet today does not imply that multi-path routing is not an appropriate and promising solution for wireless mobile ad hoc networks. The unreliability of the wireless medium and the dynamic topology due to nodes mobility or failure result to frequent path breaks, network partitioning, and high delays for path re-establishments.

The above-mentioned characteristics of mobile ad hoc networks constitute multi-path routing a very promising alternative to single path routing as it provides higher resilience to path breaks especially when paths are node disjoint [13], [14], alleviates network congestion through load balancing [15] and reduces end-to-end delay [16], [17].

In [18] the effect of the number of multiple paths on routing performance has been studied using an analytical model. The results show that multi-path routing performs better than single path if the number of alternative paths is limited to a small number of paths. Simulation results of demonstrated that with multi-path routing end-to-end delay is higher since alternate paths tend to be longer. However, a radio link layer model is not included in the simulations, thus multiple interference is not captured.

Most of the multi-path routing protocols are implemented as extensions or modifications of existing single path routing protocols like the proactive DSDV [19] and OLSR [20], or the reactive on demand protocols: AODV [3] or DSR [6]. Analysis and comparison of single path and multi-path routing protocols in ad hoc networks has been conducted in [21]. There, protocol performance is examined with regard to protocol overhead, traffic distribution, and throughput. The results reveal that multi-path routing achieves higher throughput and increases network capacity. As the dimensions of mobile ad hoc networks are spatially bounded, network congestion is inherently encountered in the center of the network since shortest paths mostly traverse the center of the network. Thus, in order to route data packets over non-congested links and maximize overall network throughput, a protocol should target at utilizing the maximum available capacity of the calculated multiple routes. The authors concluded that routing or transport protocols in ad hoc networks should provide appropriate mechanisms to push the traffic further from the center of the network to less congested links.

# 6  Conclusions

The objective of the present paper is to provide a quantitative comparison of multi-path routing protocols for mobile wireless ad hoc networks. At the same time, we examine and validate the advantages and the limitations of multi-path versus single path routing in general. Our study shows that the AOMDV protocol is more robust and performs better in most of the simulated scenarios. The AODV_Multipath protocol achieves best performance in scenarios with low mobility and higher node density. SMR performs best in networks with low node density, however the immense routing overhead generated in high node density degrades protocol's performance.

In addition, we demonstrate that the establishment and maintenance of multiple routes result in protocol performance degradation. We found that the use of two, maximum three, paths offers the best tradeoff between overhead and performance. Furthermore, protocols with high routing overhead perform badly since the routing messages fill the queues and generate data packet losses.

Compared to single path routing, our results validate the better performance of multi-path routing, especially in networks with high node density. Despite the increased routing overhead per route, the total routing overhead is lower. Furthermore, even when multiple disjoint paths are longer than the shortest path, the overall average end-to-end delay is smaller, particularly in high density scenarios. We conclude that multi-path routing in general, distributes the traffic over uncongested links and, as a consequence, the data packets experience smaller buffering delays.

# References

1. IEEE: Ieee 802.11. IEEE Standards for Information Technology (1999)
2. Lee, S., Gerla, M.: Split multipath routing with maximally disjoint paths in ad hoc networks. Proceedings of the IEEE ICC (2001) 3201–3205

3. Perkins, C.E., Belding-Royer, E.M., Das, S.: Ad hoc on-demand distance vector (aodv) routing. RFC 3561 (2003)
4. Marina, M., Das, S.: On-demand multipath distance vector routing in ad hoc networks. Proceedings of the International Conference for Network Procotols (ICNP) (2001)
5. Ye, Z., Krishnamurthy, Tripathi, S.: A framework for reliable routing in mobile ad hoc networks. IEEE INFOCOM (2003)
6. D., J., D, M.: Dynamic source routing in ad hoc wireless networks. Mobile Computing (ed. T. Imielinski and H. Korth), Kluwer Academic Publishers. Dordrecht, The Netherlands. (1996)
7. ns 2: Network Simulator: (http://www.isi.edu/nsnam/ns/.)
8. Broch, J., Maltza, D.A., Johnson, D.B., Hu, Y.C., Jetcheva, J.: A performance comparison of multi-hop wireless ad hoc network routing protocols. Proceedings of the 4th annual ACM/IEEE international conference on Mobile computing and networking. Dallas, Texas, United States. (1998) 85–97
9. Bettstetter, C., Resta, G., Santi, P.: The node distribution of the random waypoint mobility model for wireless ad hoc networks. IEEE Transactions on Mobile Computing 2(3) (2003) 257–269
10. Wei, W., Zakhor, A.: Robust multipath source routing protocol (rmpsr) for video communication over wireless ad hoc networks. ICME (2004)
11. Forum, A.: Private network-to-network interface specification version 1.0. http://www.atmforum.com/standards/approved#uni (1996)
12. version 2, O.S.P.F.O.: (Rfc 2328)
13. Tsirigos, A., Haas, Z.: Multipath routing in the presence of frequent topological changes. IEEE Communications Magazine 39 (2001)
14. Valera, A., Seah, W.K.G., Rao, S.V.: Cooperative packet caching and shortest multipath routing in mobile ad hoc networks. IEEE INFOCOM (2003)
15. Ganjali, Y., Keshavarzian, A.: Load balancing in ad hoc networks: Single-path routing vs. multi-path routing. IEEE INFOCOM (2004)
16. Wang, L., Shu, Y., Dong, M., Zhang, L., Yang, O.: Adaptive multipath source routing in ad hoc networks. IEEE ICC 3 (2001) 867–871
17. Pearlman, M.R., Haas, Z.J., Sholander, P., Tabrizi, S.S.: On the impact of alternate path routing for load balancing in mobile ad hoc networks. Proceedings of the 1st ACM international symposium on Mobile ad hoc networking & computing. MobiHoc (2000)
18. Nasipuri, A., Castaneda, R., Das, S.R.: Performance of multipath routing for on-demand protocols in mobile ad hoc networks. Mob. Netw. Appl. 6(4) (2001) 339–349
19. Perkins, C., Bhagwat, P.: Highly dynamic destination-sequenced distance-vector routing (DSDV) for mobile computers. In: ACM SIGCOMM'94 Conference on Communications Architectures, Protocols and Applications. (1994) 234–244
20. Jacquet, P., Muhlethaler, P., Qayyum, A.: Optimized link state routing protocol. Internet draft, draft-ietf-manet-olsr-00.txt (1998)
21. Pham, P., Perreau, S.: Performance analysis of reactive shortest path and multipath routing mechanism with load balance. INFOCOM, San Francisco, CA, USA (2003)

# Effect of Realistic Physical Layer on Energy Efficient Broadcast Protocols for Wireless Ad Hoc Networks

Hui Xu, Manwoo Jeon, Jinsung Cho, Niu Yu, and S.Y. Lee[*]

Department of Computer Engineering
Kyung Hee University, Korea
{xuhui, imanoos}@oslab.khu.ac.kr, chojs@khu.ac.kr,
{niuyu, sylee}@oslab.khu.ac.kr

**Abstract.** Previous work on energy efficient broadcast protocols for wireless ad hoc networks are based a commonly used physical layer model called "Path-loss model" which assume two nodes can communicate if and only if they exist within their transmission radius. In this paper, we analyze the effect of realistic physical layer on energy efficient broadcast protocols. We employ a more realistic log-normal shadowing model for physical layer and consider two link layer operating models: EER (end-to-end retransmission) and HHR (hop-by-hop retransmission). Networks with omni-antennas and directional antennas are dealt with separately. Based on above models, we analyze how to adjust actual transmission radius for transmission nodes and relay nodes to get the trade-off between maximizing probability of delivery and minimizing energy consumption. From our analysis based on shadowing model, we have derived the appropriate transmission range. The results presented in this paper are expected to improve the performance of broadcast protocols under realistic physical layer.

## 1 Introduction

Wireless ad hoc networks have emerged recently because of their potential applications in various situations such as battlefield, emergency rescue, and conference environments [1-4]. Ad hoc networks are without a fixed infrastructure; communications take place over a wireless channel, where each node has the ability to communicate with others in the neighborhood, determined by the transmission range. In such network, broadcast is a frequently required operation needed for route discovery, information dissemination, publishing services, data gathering, task distribution, alarming, time synchronization, and other operations. In a broadcasting task, a message is to be sent from one node to all the other ones in the network. Since ad hoc networks are power constrained, the most important design criterion is energy and computation conservation, broadcast is normally completed by multi-hop forwarding. We study position-based efficient broadcast protocols in which location information facilitates efficient broadcasting in terms of selecting a small forward node set and appropriate transmission radiuses while ensuring broadcast coverage. The optimization criterion is minimizing the total transmission power. There exist a lot of energy efficient

---

[*] Corresponding author.

Y. Koucheryavy, J. Harju, and V.B. Iversen (Eds.): NEW2AN 2006, LNCS 4003, pp. 327–339, 2006.
© Springer-Verlag Berlin Heidelberg 2006

broadcast protocols and their proposals are as following: first set up broadcast tree, and then at each transmission the transmission nodes will adjust their transmission radius to the distance between transmission nodes and relay nodes.

However, existing energy efficient broadcast protocols are normally based on a commonly used physical layer model called "Path-loss model" which assume that two nodes can communicate if and only if they exist within their transmission radius. In this paper, we take more realistic models into consideration. For physical layer, we employ a universal and widely-used statistic shadowing model, where nodes can only indefinitely communicate near the edge of the communication range. For link layer, we consider two operating models: EER (end-to-end retransmission without acknowledgement) and HHR (hop-by-hop retransmission with acknowledgement). In addition, energy efficient broadcast protocols in networks with omni-antennas and networks with directional antennas are dealt with separately. Based on realistic physical layer, we apply existing reception probability function and analyze how to choose the actual transmission radius between transmission nodes and relay nodes. We show how the realistic physical layer impact the selection of transmission radius in energy efficient broadcast protocols and present the trade off between maximizing probability of delivery and minimizing energy consumption in the selection of transmission radius. From our analysis, we have derived the appropriate transmission range. The results presented in this paper are expected to improve the performance of broadcast protocols under realistic physical layer.

The remainder of the paper is organized as follows: Section 2 presents related work and offers some critical comments. In Section 3, we introduce our system model, including realistic physical layer and link layer model. In Section 4 we define metrics used in our analysis, i.e. packet reception probability and expected energy consumption. In Section 5 we analyze the effect of realistic physical layer on energy efficient broadcast protocols and derived appropriate transmission radius. In Section 6, we present our conclusions and future work.

## 2  Related Work

In wireless ad hoc networks, the most important design criterion is energy and computation conservation since nodes have limited resources. Except reducing the number of needed emissions, radius adjustment is also a good way to further reduce the energy consumption. Energy efficient broadcast protocols aim to select a small forward node set and appropriate transmission radiuses while ensuring broadcast coverage. Broadcast oriented protocols achieve the objective but considers the broadcast process from a given source node. For example, the well-known centralized algorithm is a greedy heuristics called BIP [5] (Broadcast Incremental Power). It is a variant of the Prim's algorithm that takes advantage of the broadcast nature of wireless transmissions. Basically, a broadcast tree is computed from a source node by adding nodes one at a time. At each step, the less expensive action to add a node is selected, either by increasing the radius of an already transmitting node, or by creating a new emission from a passive one. As a consequence of the "wireless broadcast advantage" property of omni-antennas systems, all nodes whose distance from Node $i$ does not exceed $r_{ij}$ will be able to receive the transmission with no further energy expenditure at Node $i$.

The use of directional antennas can permit energy savings and reduce interference by concentrating transmission energy where it is needed. While using directional antenna, the advantage property will be diminished, since only the nodes located within the transmitting node's antenna beam can receive the signal. In Fig. 1, only $j$, $l$ can receive the signal, while $k$ cannot receive the signal. Applying the incremental power philosophy to network with directional antennas, the Directional Broadcast Incremental Power (DBIP) algorithm [6] has very good performance in energy saving.

**Fig. 1.** Use of directional antenna

Our work has been inspired by recent research work made in [7-10]. Mineo Takai, et al [7] focused on the effects of physical layer modeling on the performance evaluation of higher layer protocols, and have demonstrated the importance of the physical layer modeling even if the evaluated protocols do not directly interact with the physical layer. The set of relevant factors at the physical layer includes signal reception, path loss, fading, interference and noise computation, and preamble length. I. Stojmenovic, et al [8-10] presented guidelines on how to design routing and broadcasting in ad hoc networks taking physical layer impact into consideration. They apply the log normal shadow fading model to represent a realistic physical layer to derive the approximation for probability $p(d)$ of receiving a packet successfully as a function of distance $d$ between two nodes. They proposed several localized routing schemes for the case when position of destination is known, optimizing expected hop count (for hop by hop acknowledgement), or maximizing the probability of delivery (when no acknowledgements are sent). They considered localized power aware routing schemes under realistic physical layer. Finally, they mentioned broadcasting in ad hoc network with realistic physical layer and propose new concept of dominating sets to be used in broadcasting process.

In this paper, we employ the system model of previous work constructed and analyze the effect of realistic physical layer on energy efficient broadcast protocols.

# 3   System Model

## 3.1   Physical Layer Model

The most commonly used radio model for the study of wireless networks is the so-called path loss model. This model assumes that the received signal power at distance $d$ is equal to $c \cdot d^{-\beta}$, where $c$ is a constant and $\beta$ is the path loss exponent. The path loss exponent depends on the environment and terrain structure and can vary between 2 in free space to 6 in heavily built urban areas [11]. Normally the signal power at distance $d$ predicted by the path loss model is called the *area mean power*. Path loss model assumes radio signals can be received correctly when their power exceeds a minimum threshold value $\gamma$. With this assumption, the path loss model results into a perfect circular coverage area around each node with radius $R = (c/\gamma)1/\beta$. However, this is an unrealistic assumption in most practical situations. In reality the received power levels may show significant variations around the area mean power [12].

The log-normal shadowing [13] model that we use in this paper is more realistic than the path loss model because it allows for random signal power variations. Due to those variations, the coverage area will deviate from a perfect circular shape and consequently, some short links could disappear while long links could emerge (see Figure 2).

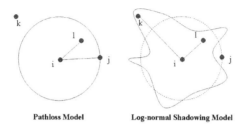

Pathloss Model        Log-normal Shadowing Model

**Fig. 2.** Coverage area comparison of Radio Models

The shadowing model consists of two parts. The first one is known as path loss model which predicts the mean received power at distance $d$, denoted by $\overline{P_r(d)}$. It uses a close-in distance $d_0$ as a reference. $\overline{P_r(d)}$ is computed relative to $P_r(d_0)$ as follows.

$$\frac{P_r(d_0)}{P_r(d)} = (\frac{d}{d_0})^{\beta} \tag{1}$$

$\beta$ is the path loss exponent and $P_r(d_0)$ can be computed from free space model. The path loss is usually measured in dB. So from Eq. (1) we have

$$\left[ \frac{\overline{P_r(d)}}{P_r(d_0)} \right]_{dB} = -10\beta \log \left( \frac{d}{d_0} \right) \tag{2}$$

The second part of the shadowing model reflects the variation of the received power at certain distance. It is a log-normal random variable, that is, it is of Gaussian distribution if measured in dB. The overall shadowing model is represented by

$$\left[ \frac{P_r(d)}{P_r(d_0)} \right]_{dB} = -10\beta \log \left( \frac{d}{d_0} \right) + X_{dB} \tag{3}$$

where $X_{dB}$ is a Gaussian random variable with zero mean and standard deviation $\sigma_{dB}$. $\sigma_{dB}$ is called the shadowing deviation, and is also obtained by measurement. Eq. (3) is also known as a log-normal shadowing model.

## 3.2 Antenna Model

We study networks with not only omni antennas but also directional antennas. The use of directional antennas can permit energy savings and reduce interference by concentrating transmission energy where it is needed. We use a directional antenna propagation model [14] as shown in Fig. 3, where the antenna orientation $\varphi$ ($0 \leqslant \varphi < 2\pi$) of node is defined as the angle measured counter-clockwise from the horizontal axis to the antenna boresight, and the antenna directionality is specified as the angle of beamwidth $\theta_f$ ($0 \leqslant \theta_f < 2\pi$). Table 1 shows the antenna classification based on above model.

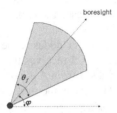

boresight

**Fig. 3.** Directional antenna propagation model

**Table 1.** Antenna classification

|  | **Omni-directional** | **Modestly directional** | **Highly directional** |
|---|---|---|---|
| Antenna Directionality | fixed beamwidth | fixed beamwidth | variable beamwidth |
| Antenna Orientation | unsteerable | steerable | steerable |

In this paper we only focus on the modestly directional antenna which has the following characteristics:

1. Beamwidth of each antenna cannot be adjusted, i.e., $\theta_f$ is fixed for any node.

2. Orientation of each antenna can be shifted to any desired direction to provide connectivity to a subset of the nodes that are within communication range.

3. A single antenna beam is provided for each session in which a node participates.

The transmission power needed by node $v$ to transmit to node $u$ in its antenna beam using beamwidth $\theta_f$ is

$$c \cdot r^{\beta} \frac{\theta_f}{2\pi} \tag{4}$$

where $\beta$ is the path loss, $r$ is the transmission radius and $c$ is constant coefficient.

### 3.3  Link Layer Model

We consider two different operating models [15]:

a) **End-to-End Retransmissions (EER):** where the individual links do not provide link-layer retransmissions and error recovery.
b) **Hop-by-Hop Retransmissions (HHR):** where each individual link provides reliable forwarding to the next hop using localized packet retransmissions.

For HHR case, we employ a communication protocol between two nodes proposed in [8-10]. After receiving any packet from sender, the receiver sends $u$ acknowledgements. If the sender does not receive any acknowledgement, it will retransmit the packet. They also derive the expected number of messages in this protocol as measure of hop count between two nodes. The count includes transmissions by sender and acknowledgments by receiver. They assume both the acknowledgement and data packets are of the same length.

Let $S$ and $A$ be the sender and receiver nodes respectively, and let $|SA| = d$ be the distance between them. Probability that $A$ receives the packet from $S$ is $p(d)$. Probability that $S$ receives one particular packet from $A$ is $p(d)$ and the probability that it does not receive the packet is $1- p(d)$. Therefore, the probability that $S$ does not receive any of the $u$ acknowledgements is $(1 - p(d))^u$. Thus, the probability that $S$ receives at least one of $u$ acknowledgements from $A$ is $1-(1 - p(d))^u$. Therefore, $p(d)(1-(1 - p(d))^u)$ is the probability that $S$ receives acknowledgement after sending a packet and therefore stops transmitting further packets. Thus, the expected number of packets at $S$ is $1/[ p(d)(1-(1-p(d))^u)]$. Each of these packets is received at $A$ with probability $p(d)$. If received correctly, it generates $u$ acknowledgements. The total expected number of acknowledgements sent by $A$ is then $up(d)/[p(d)(1-(1-p(d))^u)] = u/[(1-(1-p(d))^u)]$. The total expected hop count between two nodes at distance $d$ is then $1/[ p(d)(1-(1-p(d))^u)]+u/[(1-(1-p(d))^u)]$.

## 4  Metrics Under Realistic Physical Layer Model

The broken link problem caused by realistic physical layer is relatively easy to understand as shown in Fig. 2. Since energy efficient broadcast protocols aim to minimize energy consumption, they select a small forward node set and choose right the distance between senders and relay nodes as actual transmission radius. However, under realistic physical layer, e.g. shadowing model, they may get very poor network

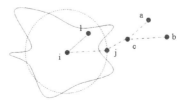

**Fig. 4.** Network coverage problem under practical model

coverage because of link availability problem. In Fig. 4, node $j$ and $c$ have been selected as forward nodes, while because the broken link between source node $i$ and $j$, node $c$, $a$ and $b$ also can't be reached.

In order to apply existing protocols without having to redesign them, the actual transmission radius for source node and forward nodes should be redesigned. To define appropriate transmission radius under practical models, we observed two different metrics: reception probability and expected energy consumption.

## 4.1 Packet Reception Probability

In shadowing model, the probability that the received power at a location $d$ exceeds a threshold value $\gamma$ can be given as:

$$P_r[P_r(d) > \gamma] = 0.5(1 - erf[\frac{\gamma - \overline{P_r(d)}}{\sqrt{2}\sigma}]) \tag{5}$$

While exact computation of packet reception probability $p(d)$, for use in routing and broadcasting decision, is a time consuming process, and is based on several measurements (e.g. signal strengths, time delays and GPS) which cause some errors. It is therefore desirable to consider a reasonably accurate approximation that will be fast for use. I. Stojmenovic, et al [8-10] derives the approximation for probability of receiving a packet successfully as a function of distance $d$ between two nodes. Having in mind an error within 4% the model is

$$p(r,d) = \begin{cases} 1 - \dfrac{(\frac{d}{r})^{2\beta}}{2} & 0 \le d < r \\ \dfrac{(\frac{2r-d}{r})^{2\beta}}{2} & r \le d \le 2r \\ 0 & others \end{cases}$$

where $\beta$ is the power attenuation factor with fixed value between 2 and 6, and $r$ is transmission radius with $p(r, d=r) = 0.5$. Fig.5 shows the packet reception probability with approximation $p(r, d)$ when $\beta$ is 2.

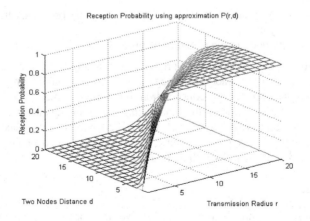

**Fig. 5.** Reception probability with approximation $p(r, d)$

## 4.2 Expected Energy Consumption

Assume now that two nodes are at distance $d$, but a packet is sent with transmission radius $r$ and with omni-antenna or fixed directional antenna $\theta_f$. The exact transmission power with omni-antenna is then $cr^\beta$ where $c$ is a constant; the exact transmission power with fixed directional-antenna $\theta_f$ is then $cr^\beta \dfrac{\theta_f}{2\pi}$. In addition, $c$ is assumed to be 1 for simplicity of analysis in next section. The packet reception probability at distance $d$ is $p(r, d)$.

In EER framework, the sender sends a packet and the receiver may or may not receive the packet, which depends on the probability of receiving. Therefore, the expected energy consumption is $r^\beta \cdot p(r,d)$ for transmission between two nodes in networks with omni-antenna; $r^\beta \dfrac{\theta_f}{2\pi} \cdot p(r,d)$ for transmission between two nodes with directional antenna of fixed beam width of $\theta_f$.

In HHR case, a message is retransmitted between two nodes until it is received and acknowledged correctly; after receiving any packet from the sender, the receiver sends acknowledgement. According to section 3.3 when $u$ equals 1, the expected number of transmitted packets is $1/p^2(r,d)$ and the expected number of acknowledgements is $1/p(r,d)$. Therefore, the expected energy consumption is $r^\beta(1/p^2(r,d)+1/p(r,d))$ for transmission between two nodes in networks with omni-antenna; $r^\beta \dfrac{\theta_f}{2\pi}(1/p^2(r,d)+1/p(r,d))$ for transmission between two nodes with directional antenna of fixed beam width of $\theta_f$. In addition, $\theta_f$ is assumed to be 1 for simplicity of energy analysis in next section.

# 5    Analysis of Realistic Physical Layer Effect

We extend broadcast oriented protocol to work in realistic physical layer environment:

1. Apply the selected broadcast oriented protocol to set up broadcast tree and determine the set of rely nodes.
2. Use $r$ for each rely node in the actual transmission.

Step 1 of the above process varies from protocol to protocol. As for the metric to decide the value of $r$, there exists a trade-off or negotiation between maximizing probability of delivery and minimizing energy consumption. We propose the following rules: for broadcasting in wireless network with omni-antennas, minimizing energy consumption is the primary metric; otherwise, for network with directional antennas, maximizing probability of delivery will be the primary metric, since transmission coverage overlapping is much fewer than that in networks with omni-antennas.

## 5.1 EER Case

In EER case, a sender sends a packet and a receiver may or may not receive the packet which depends on the reception probability. The reception probability function is

$P(r,d) = (1-(d/r)^{2\beta}/2)$ for $d < r$, $((2r-d)/r)^{2\beta}/2$ for $r \le d \le 2r$, and 0 for all the other $d$ where $d$ is the distance before sender and receiver and $r$ is actual transmission range.

For network with directional antennas, since maximizing probability of delivery is our primary metric, at least we have to guarantee the reception probability no less than 0.5; however if the reception probability is near 1, the energy consumption will be too high. Since after probability reaches around 0.9, the acceleration of $P(r, d)$ curve decreases greatly, therefore we choose [0.5 0.9] as the acceptable reception probability scope. From formula in section 4.1 we can find that if $r>d$, the scope of reception probability is [0.5, 1]; otherwise, if $r<d$, reception probability will be less than 0.5. Since we should guarantee the reception probability no less than 0.5, we will only use $P(r,d) = (1-(d/r)^{2\beta}/2)$ for $d < r$. For any value of $\beta$, $2 \le \beta \le 6$, if we want to get the relationship of $d$ and $r$ ($r>d$) for certain reception probability $\alpha$, we can set up the formula as $1-(d/r)^{2\beta}/2 = \alpha$, then we get $r = [2(1-\alpha)]^{-1/2\beta}d$. Therefore, in order for reception probability to be in the range of [0.5 0.9], the transmission radius should be in the range of $[d \;\; (1/5)^{-1/2\beta}d]$. We can verify it through Fig. 6, where $\beta=2$, $d=10$, 20 and 30. According to our proposal, we can choose the transmission radius in the scope of [10 15], [20 30] and [30 45] respectively. In Fig. 6(a), the according reception probability is in the scope of [0.5 0.9]; in Fig. 6(b), the according expected energy consumption is in the scope of [47 202], [197 811] and [447 1824] respectively.

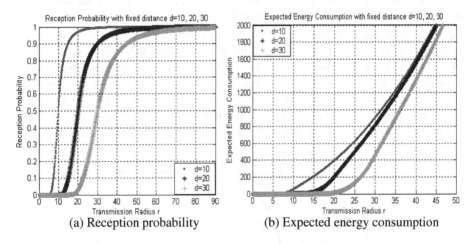

(a) Reception probability                (b) Expected energy consumption

**Fig. 6.** Reception probability and expected energy consumption with fixed distance $d$

For network with omni-antennas, minimizing expected energy consumption is primary metric. We know as transmission $r$ increases, the expected energy consumption will also increase. Therefore, we want to choose the transmission radius $r$ value as small as possible. Whereas, even minimizing energy consumption is the primary metric, we still cannot neglect the reception probability. According our proposal above, which is selecting $r$ in the scope $[d \;\; (1/5)^{-1/2\beta}d]$, and getting the reception probability scope [0.5 0.9], by guaranteeing reception probability not less than 50%, we decide to choose $d$ as the transmission radius $r$.

## 5.2 HHR Case

In HHR case, a message is retransmitted between two nodes until it is received and acknowledged correctly; after receiving any packet from sender, the receiver sends $u$ acknowledgements. Considering the characteristic of link layer in HHR case, it's better to be employed in networks with directional antennas, which represent one to one transmission model. In addition, we can find the link layer has already guaranteed successful reception, therefore our research moves to minimizing the expected energy consumption between two nodes. According to section 4.2, the total expected energy consumption is $r^\beta(1/p^2(r,d)+1/p(r,d))$, that is, the combination of consumption at sender $S$ and receiver $A$. Therefore our work is transferred to maximize the reception probability at sender $S$ and receiver $A$.

For any value of $\beta$, $2 \leq \beta \leq 6$, for receiver $A$, the relationship of $d$ and $r$ $(r>d)$ for certain reception probability $\acute{a}$ is $r = [2(1-\alpha)]^{-1/2\beta} d$, then in order for reception probability to be in the range of [0.5 0.9], the transmission radius should be in the range of $[d \ (1/5)^{-1/2\beta} d]$ ; however, for sender $S$, the relationship of $d$ and $r$ $(r>d)$ for certain reception probability $\acute{a}$ is $r = [2(1-\alpha^{1/2})]^{-1/2\beta} d$, then in order for reception probability to be in the range of [0.5 0.9], the transmission radius should be in the range of $[[2(1-(0.5)^{1/2})]^{-1/2\beta} d \ [2(1-(0.9)^{1/2})]^{-1/2\beta} d]$. Therefore considering the reception probability of both sender $S$ and receiver $A$, our proposal can be extended as the following: in HHR case, we choose $r$ from the scope of $[[2(1-(0.5)^{1/2})]^{-1/2\beta} d \ (1/5)^{-1/2\beta} d]$, where for sender $S$ the scope of reception probability is [0.5 0.9) and for receiver $A$ the scope of reception probability is within (0.5 0.9]. We can verify it through Fig. 7, where $\beta=2$, $d=10$, 20 and 30. The reception probability at sender $S$ and at receiver $A$ with fixed distance $d$ when $\beta$ is 2 is showed in Fig. 7. According to our proposal, we can choose the transmission radius in the scope of [11.4, 15], [22.9 30] and [34.3 45] respectively. In Fig. 7, for sender $S$, the scope of reception probability is [0.5 0.8] and for receiver $A$, the scope of reception probability is within [0.7 0.9].

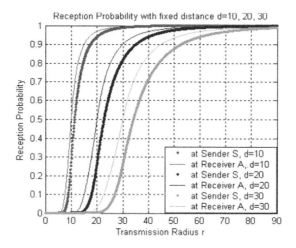

**Fig. 7.** Reception probability with fixed distance $d$=10, 20, 30

In HHR case, because of the characteristic of link layer, the number of transmission between two nodes is more than one, therefore expected hop count and expected energy consumption will be higher than that in EER case. Fig. 8 shows the total expected hop count and energy consumption including sender $S$ and receiver $A$ ($\beta$= 2).

We can verify whether our proposal of choosing $r$ from the scope of $[[2(1-(0.5)^{1/2})]^{-1/2\beta}d \quad (1/5)^{-1/2\beta}d]$ is reasonable or not. The total expected hop count and energy consumption with fixed distance $d$=10, 20, 30 when $\beta$ is 2 is showed in Figure 8. According to our proposal, we can choose the transmission radius in the scope of [11.4 15], [22.9 30] and [34.3 45] respectively. Fig. 8(a) shows that if the transmission radius $r$ is not less than the distance 10, 20 and 30 respectively, expected hop count will be less than 5 and also at last decrease to a constant number. Fig. 8(b) shows that the expected energy consumption can get minimum value when $r$ is around 11.4, 22.9 and 34.3 respectively; whereas if $r$ is larger than those values, the expected energy consumption will increase. Therefore, even if $r$ is larger than 15, 30 and 45 respectively, we can get the minimum expected hop count, but because the expected energy consumption will be larger, so we still cannot choose $r$ larger than 15, 30 and 45 respectively. In a word, our proposal for HHR case is to choose the transmission radius $r$ in the scope of $[[2(1-(0.5)^{1/2})]^{-1/2\beta}d \quad (1/5)^{-1/2\beta}d]$, which can get good performance at expected hop count and energy consumption.

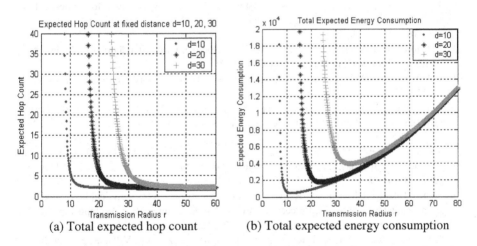

(a) Total expected hop count          (b) Total expected energy consumption

**Fig. 8.** Total expected hop count and energy consumption with fixed distance $d$

# 6   Conclusions

In this paper we investigated energy efficient broadcast protocols with and without acknowledgements and presented the trade off between maximizing probability of delivery and minimizing energy consumption for ad hoc wireless networks with realistic physical layer. In EER case, for network with omni-antennas, we decide to choose the distance $d$ between two nodes as transmission radius; in network with directional antenna, we propose to choose the transmission radius in the scope of

$[d_{(1/5)^{-1/2\beta}}d]$ to maximize the probability of delivery. In HHR case, the link layer protocol is not suitable to one-to-all communication; therefore we only consider networks with directional antennas. We propose to choose the transmission radius in the scope of $[[2(1-(0.5)^{1/2})]^{-1/2\beta}d_{(1/5)^{-1/2\beta}}d]$, which can get good performance at expected hop count and energy consumption.

Currently, we are designing new broadcast protocols based on our analysis and we will compare the performance of our new protocols with that of existing broadcast protocols under realistic physical layer.

## Acknowledgement

The research was supported by the Driving Force Project for the Next Generation of Gyeonggi Provincial Government in Republic of Korea.

## References

1. S. Basagni, M. Conti, S. Giordano, I. Stojmenovic (eds.): Mobile Ad Hoc Networking. IEEE Press/Wiley, July (2004)
2. J. M. Kahn, R. H. Katz, and K. S. J. Pister: Next century challenges: Mobile networking for "smart dust". International Conference on Mobile Computing and Networking *(MOBICOM)* (1999) 271–278
3. G. Giordano, I. Stojmenovic (ed.): Mobile Ad Hoc Networks. Handbook of Wireless Networks and Mobile Computing, Wiley (2002) 325-346
4. G.J. Pottie and W.J. Kaiser: Wireless integrated network sensors. Communications of the ACM, vol. 43, no. 5 (2000) 551–558
5. J. E. Wieselthier, G. D. Nguyen, and A. Ephremides: On the construction of energy-efficient broadcast and multicast trees in wireless networks. Proc. IEEE INFOCOM, Mar (2000) 585-594
6. J.E. Wieselthier, G.D. Nguyen, A. Ephremides: Energy-Limited Wireless Networking with Directional Antennas: The Case of Session-Based Multicasting. Proc. IEEE INFOCOM (2002) 190-199
7. Mineo Takai, Jay Martin and Rajive Bagrodia: Effects of wireless physical layer modeling in mobile ad hoc networks. Proceedings of the 2nd ACM international symposium on Mobile ad hoc networking & computing (2001) 87-94
8. I. Stojmenovic, A. Nayak, J. Kuruvila, F. Ovalle-Martinez and E. Villanueva-Pena: Physical layer impact on the design of routing and broadcasting protocols in ad hoc and sensor networks. Journal of Computer Communications (Elsevier), Special issue on Performance Issues of Wireless LANs, PANs, and Ad Hoc Networks, Dec (2004)
9. Johnson Kuruvila, Amiya Nayak and Ivan Stojmenovic: Hop count optimal position based packet routing algorithms for ad hoc wireless networks with a realistic physical layer. IEEE Journal on selected areas in communications, vol.23, no.6, June (2005)
10. J. Kuruvila, A. Nayak, I. Stojmenovic: Greedy localized routing for maximizing probability of delivery in wireless ad hoc networks with a realistic physical layer. Algorithms for Wireless And mobile Networks (A_SWAN) Personal, Sensor, Ad-hoc, Cellular Workshop, at Mobiquitous, Boston, August (2004) 22-26

11. T. Rappaport: Wireless Communications, Principles and Practice. Upper Saddle River Prentice-Hall PTR (2002)
12. R. Hekmat, Piet Van Mieghem: Interference Power Sum with Log-Normal Components in Ad-Hoc and Sensor Networks. WiOpt (2005) 174-182
13. Network Simulator - ns-2, http://www.isi.edu/nsnam/ns/.
14. Song Guo, Oliver W. W. Yang: Antenna orientation optimization for minimum-energy multicast tree construction in wireless ad hoc networks with directional antennas. MobiHoc (2004) 234-243
15. Suman Banerjee, Archan Misra: Minimum energy paths for reliable communication in multi-hop wireless networks. MobiHoc, Switzerland (2002) 146-156

# Localized Broadcast Oriented Protocols with Mobility Prediction for Mobile Ad Hoc Networks

Hui Xu, Manwoo Jeon, Shu Lei, Jinsung Cho[*], and Sungyoung Lee

Department of Computer Engineering
Kyung Hee University, Korea
{xuhui, imanoos, sl8132}@oslab.khu.ac.kr, chojs@khu.ac.kr,
sylee@oslab.khu.ac.kr

**Abstract.** Efficient broadcasting protocols aim to determine a small set of forward nodes to ensure full coverage. Position based broadcast oriented protocols, such as BIP, LBIP, DBIP and LDBIP work well in static and quasi static environment. While before they can be applied in general case where nodes move even during the broadcast process, the impact of mobility should be considered and mobility control mechanism is needed. In existing mobility management, each node periodically sends "Hello" message and based on received messages construct local view which may be updated at actual transmission time. In this paper, we proposed proactive and predictive mobility control mechanism: node will only send request to collect neighbors' info before transmission which conserves energy consumption of periodical "Hello" messages; once receiving location request command, nodes will send twice at certain interval; based on received locations, nodes will predict neighbors' position at future actual transmission time, use predicted local view to construct spanning tree and do efficient broadcast operation. We propose localized broadcast oriented protocols with mobility prediction for MANETs, and simulation result shows new protocols can achieve high coverage ratio.

## 1 Introduction

Broadcasting a packet to the entire network is a basic operation and has extensive applications in mobile ad hoc networks (MANETs). For example, broadcasting is used in the route discovery process in several routing protocols, when advising an error message to erase invalid routes from the routing table, or as an efficient mechanism for reliable multicast in a fast-moving MANET. In MANETs with the promiscuous receiving mode, the traditional blind flooding incurs significant redundancy, collision, and contention, which is known as the broadcast storm problem [1].

We study position-based efficient broadcast protocols in which location information facilitates efficient broadcasting in terms of selecting a small forward node set and appropriate transmission radiuses while ensuring broadcast coverage. The optimization criterion is minimizing the total transmission power. Broadcast oriented protocols consider the broadcast process from a given source node. For instance,

---

[*] Corresponding author.

Y. Koucheryavy, J. Harju, and V.B. Iversen (Eds.): NEW2AN 2006, LNCS 4003, pp. 340–352, 2006.
© Springer-Verlag Berlin Heidelberg 2006

Wieselthier et al. [2] proposed greedy heuristics which are based on Prim's and Dijkstra's algorithms. The more efficient heuristic, called BIP [2] for broadcasting incremental power, constructs a tree starting from the source node and adds new nodes one at a time according to a cost evaluation. BIP is "node-based" algorithm and exploits the "wireless broadcast advantage" property associated with omni-antennas, namely the capability for a node to reach several neighbors by using a transmission power level sufficient to reach the most distant one. Applying the incremental power philosophy to network with directional antennas, the Directional Broadcast Incremental Power (DBIP) algorithm [3] has very good performance since the use of directional antennas provide energy savings and interference reduction.

All the protocols that have been proposed for broadcast can be classified into two kinds of solutions: centralized and localized. Centralized solutions mean that each node should keep global network information and global topology. The problem of centralized approach is that mobility of nodes or frequent changes in the node activity status (from "active" to "passive" and vice-versa) may cause global changes in topology which must be propagated throughout the network for any centralized solution. This may results in extreme and un-acceptable communication overhead for networks. Hence, because of the limited resources of nodes, it is ideal that each node can decide on its own behavior based only on the information from nodes within a constant hop distance. Such distributed algorithms and protocols are called localized [4-8]. Of particular interest are protocols where nodes make decisions based solely on the knowledge of its 1-hop or 2-hops neighbors to them. LBIP [9] (Localized Broadcast Incremental Power) and LDBIP [10] (Localized Directional Broadcast Incremental Power) are localized algorithms which are respectively based on BIP and DBIP.

However, all above broadcast schemes assume either the underlying network topology is static or quasi-static during the broadcast process such that the neighborhood information can be updated in a timely manner. In this paper we consider a general case where nodes move even during the broadcast process, making it impractical to utilize centralized algorithms. However, experiment results show that existing localized algorithms also perform poorly in terms of delivery ratio under general case where nodes move even during the broadcast process. There are two sources that cause the failure of message delivery: collision, the message intended for a destination collides with another message which can be relieved by a very short (1ms) forward jitter delay; mobile nodes, a former neighbor moves out of the transmission range of the current node (i.e., it is no longer a neighbor).

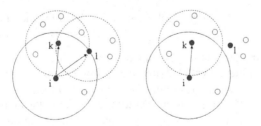

**Fig. 1.** Impact of mobility on delivery ratio

In Fig. 1, when node $l$ moves out of the transmission range of $i$, the nodes along the branch rooted at $l$ of the broadcast tree will miss the message. The majority of delivery failures are caused by mobile nodes. Therefore before localized algorithms can be applied in general mobile scenario, certain mobility control mechanism is needed which inversely may force original localized algorithms to be modified to adapt to it. In most existing mobility management, each node emits a periodic "Hello" message to advertise its presence and position at a fixed interval $\Delta$. "Hello" intervals at different nodes are asynchronous to reduce message collision. Each node extracts neighborhood information from received "Hello" messages to construct a local view of its vicinity (e.g., 2-hop topology or 1-hop location information).

There are three main problems in existing mechanisms: 1) Energy consumption problem, each node periodically emitting "Hello" message will cause a lot of energy consumption; 2) Updated local view: within "Hello" message interval, nodes moving will cause updated neighborhood information; 3) Asynchrony problem: asynchronous sample frequency at each node and asynchronous "Hello" intervals will cause asynchronous position information for each neighbor in certain local view.

In this paper we propose a unique proactive and predictive solution to address main problems in existing mechanisms: 1) Proactive neighbors' location collection to conserve energy: node will only send location request before transmission which conserves the energy consumption of periodic "Hello" messages; 2) Mobility prediction mechanism to get neighborhood information at future actual transmission time: once receiving location request command, nodes will send location twice at certain interval; based on received neighborhood location information, nodes will predict neighbors location information in future actual emitting time $T$, use this predicted information to construct spanning tree and do efficient broadcast operation in future time $T$; 3) Synchronization: since we predict all neighbors location information at the same time $T$, therefore we achieve synchronization for all neighbors. We apply above mobility management mechanism in localized broadcast oriented protocols and simulation results show that our mobility prediction mechanism greatly improved coverage ratio.

The remainder of the paper is organized as follows: In Section 3, we present our localized broadcast oriented protocols with our proposed mobility prediction model with omni and directional antennas. Section 4 shows our simulation work and results. In Section 5, we present the conclusion.

## 2 Localized Broadcast Oriented Protocols with Mobility Prediction

The application of our proposed mechanism in localized broadcast oriented protocols is as follows:

- Source node $S$ initiates its neighborhood location table and emits location request ($LR$) using omni-antenna with maximum transmission range ($MTR$).
- At the same time, $S$ stores its current location ($CL$) in location table $A$; after a certain time interval $\Delta T1$, stores its current location again in location table $B$.
- Any node which receives location request, for example $U$, at once emits its current location using omni-antenna of maximum transmission range with 1st location remark ($1LRM$). After time interval $\Delta T1$, $U$ collects its current location again and emits it with 2nd location remark ($2LRM$).

- *S* starts receiving neighbors' location information (*NL*). *S* stores neighbors' location with *1LRM* in table *A*, and neighbors' location with *2LR* in table *B*.
- After waits for time interval Δ*T2* to guarantee receiving all neighbors' location, *S* starts to predict own and all neighbors' location at future time *T*.
- *S* calculates localized broadcast spanning tree based on predicted future own and neighborhood location information.
- *S* enters into broadcast process at time *T*: broadcasting packet *P* as calculated instruction and also including relay nodes *ID*.
- Any node, for example *V*, which receives packet *P* checks whether it's in relay nodes list. If in rely list, *V* works as source node; otherwise, does nothing.

Fig.2 is the work flow of the proposed localized algorithm. In the following we will present 4 stages of our algorithm in detail.

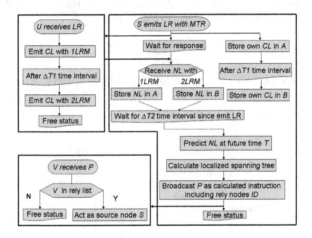

**Fig. 2.** Work flow of the proposed algorithm

## 2.1 Proactive Neighbors' Location Collection Stage

Node will only send location request before transmission which conserves the energy consumption of periodic "Hello" messages in most existing mobility mechanisms.

- Source node *S* emits location request.
- After transmission delay time interval (*TD*), any node *U* receives location request and ideally at once emits its current location with *1LRM*. While sometimes there is receiving process delay (*ED*). After a certain time interval Δ*T1*, *U* emits its current location with *2LRM*.
- After waits for time interval Δ*T2*, *S* starts to predict all neighbors' location at future actual transmission time *T*. In Fig. 3 *RT* represents the redundant time interval for guarantee receiving all neighbors' location information and *PD* is prediction delay for deciding the value of *T*.

In Fig.3 we show the time flow of our proactive mechanism.

In our mechanism $S$ needs to set timer with interval $\Delta T2$ to trigger prediction process; $U$ should set timer with interval $\Delta T1$ to trigger the sending of its 2nd current location; $S$ need to set $PD$ to decide the value of $T$. From above time flow, we can calculate that $\Delta T2 = 2*TD + \Delta T1 + ED + RT$. In ideal model, $ED$ equals zero, $TD$ depends on wireless network and $RT$ can be set depending on designer redundancy requirement. The main issue for us is how to decide $\Delta T1$ and $PD$. We propose relationships between those two parameters and average mobile speed $v$. The principle for those relationships is to predict accurate location at broadcast process time $T$. As $v$ increases, $\Delta T1$ and $PD$ should become smaller to finish prediction and broadcast before nodes change move direction. Therefore, the formula is:

$$\Delta T1 = \beta \frac{1}{v} + k \qquad (1)$$

where $k$ represents the minimum time interval and $\beta$ is the coefficient to adjust the deceleration speed according to designers requirement.

$$PD = \alpha \frac{1}{v} + \Delta \qquad (2)$$

where $\Delta$ represents the minimum space for high mobile speed scenario and $\alpha$ is the coefficient to adjust the deceleration speed according to designers requirement.

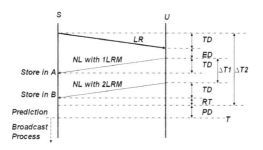

**Fig. 3.** Time flow of the proposed proactive mechanism

### 2.2 Mobility Prediction Stage

The objective of mobility prediction is to calculate neighborhood location information at future actual emitting time $T$, based on location information in table $A$ and $B$. Node will use this predicted information to construct spanning tree and start efficient broadcast process at time $T$. Another contribution of our proposed mobility prediction model is that since we predict all neighbors' location information at the same time $T$, therefore we achieve consistent future view for all neighbors.

Camp et al. [11] gave a comprehensive survey on mobility models for MANETs. They discussed seven different synthetic entity mobility models, among which Random Walk, Random Waypoint and Random Direction are simple linear mobility models. Therefore, we construct our mobility prediction model as shown in Fig. 4.

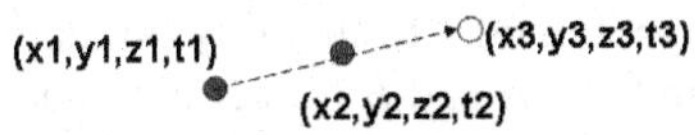

(x1,y1,z1,t1)                    (x3,y3,z3,t3)
              (x2,y2,z2,t2)

**Fig. 4.** Mobility prediction Model

If we know location of mobile node $S$ at time $t1$ and $t2$, we can predict nodes' location at time $t3$ by:

$$x3 = x1 + \frac{x2-x1}{t2-t1}*(t3-t1) \qquad x3 = x2 + \frac{x2-x1}{t2-t1}*(t3-t2)$$
$$y3 = y1 + \frac{y2-y1}{t2-t1}*(t3-t1) \qquad y3 = y2 + \frac{y2-y1}{t2-t1}*(t3-t2) \tag{3}$$
$$z3 = z1 + \frac{z2-z1}{t2-t1}*(t3-t1) \quad \text{or} \quad z3 = z2 + \frac{z2-z1}{t2-t1}*(t3-t2)$$

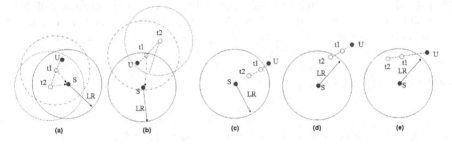

**Fig. 5.** Possible situations for node $U$

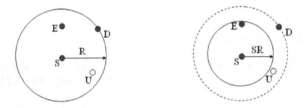

**Fig. 6.** Function of smaller neighborhood range (SR)

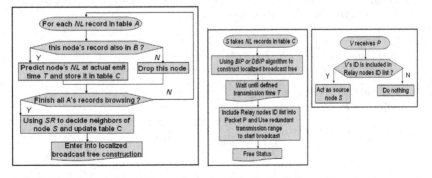

**Fig. 7.** The process of mobility prediction    **Fig. 8.** Work flow of broadcast process

To apply above mathematic model to mobility prediction process, we have to take locations respectively from table $A$ and $B$ for certain node $U$. We have stored location at time $t1$ in table $A$ and location at time $t2$ in table $B$. Fig. 5 lists the possible situations for node $U$. Fig. 5 (a) is the normal situation where node $U$ is the neighbor of $S$ at both time $t1$ and $t2$. Then we can apply our prediction model with node $U$'s record in both table $A$ and $B$. In Fig. 5 (b), node $U$ moves out of the neighborhood area of node $S$ at time $t2$. Therefore, we can just drop the node which only appears in table $A$. However, the most serious situation is that when node $S$ emits location request, node $U$ doesn't receive location request while after that at $t1$ or $t2$ time, $U$ is neighbor of node $S$. Fig. 5 (c) (d) (e) display those situations where node $U$ can appear at any position of the neighborhood of node $S$. However, the probability greatly decreases as distance between two nodes decreases. Therefore, we can only correct the problem when node $U$ appears in the outer neighborhood area of node $S$. Our solution is after prediction process to decide neighbors of node $S$ by smaller neighborhood range stand ($SR$). Then we can guarantee accurate neighborhood location information with high probability. Fig. 6(a) shows the predicted local view of node $S$, where node $U$ is not include even in fact it is the neighbor of $S$. By applying $SR$, node $S$ achieves smaller but accurate local view which is shown in Fig. 6(b).

The value of $SR$ is related to average mobile speed $v$. We propose a relationship formula which can be written as:

$$SR = R - kv ,$$ (4)

where $R$ represents original maximum transmission range and $k$ is the coefficient to adjust the slope according to designers' requirement.

According to all above analysis, we propose the process of mobility prediction which is shown in Fig. 7.

### 2.3   Broadcast Process Stage

In Fig. 8 we show the work flow of broadcast process. We absorb the redundant proposal of Wu and Dai [12 -16] to use a longer transmission range to start real broadcast process. A pseudo code of BIP is shown in Fig. 9.

---

**Input:** given an undirected weighted graph $G(N,A)$, where $N$: set of nodes, $A$: set of edges
**Initialization:** set $T:=\{S\}$ where $S$ is the source node of broadcast session. Set $P(i):= 0$ for all $1 \leq i \leq |N|$ where $P(i)$ is the transmission power of node $i$.
**Procedure:**
**while** $|T| \neq |N|$
       **do** find an edge $(i,j) \in T \times (N-T)$ such that
           incremental power $_{\Delta P_{ij}} = d_{ij}^{\alpha} - P(i)$ is minimum.
       **add** node $j$ to $T$, i.e., $T := T \cup \{j\}$.
       **set** $P(i) := P(i) + _{\Delta P_{ij}}$.

---

**Fig. 9.** Pseudo code of BIP

The incremental power philosophy, originally developed with omni antennas, can be applied to tree construction in networks with directional antennas as well. At each step of the tree-construction process, a single node is added, whereas variables involved in computing cost (and incremental cost) are not only transmitter power but beam width $\theta$ as well. In our simple system model, we use fixed beam width $\theta_f$, that means for adding a new node, we can only have two choices: set up a new directional antenna to reach a new node; raise length range of beam to check whether there is new node covered or not. A pseudo code of DBIP is shown in Fig. 10.

---

**Input:** given an undirected weighted graph $G(N,A)$, where $N$: set of nodes, $A$: set of edges
**Initialization:** set $T:=\{S\}$ where $S$ is the source node of broadcast session. Set $P(i):=0$ for all $1 \leq i \leq |N|$ where $P(i)$ is the transmission power of node $i$.
**Procedure:**
**while** $|T| \neq |N|$

    **do** find an edge $(i,j) \in T \times (N-T)$ with fixed beam width $\theta_f$ such that $\Delta P_{ij}$ is minimum;

if an edge $(i,k) \in T \times T$ raising the length range of beam can cover node $j \in (N-T)$, incremental power $\Delta P_{ij} = d_{ij}^\alpha \dfrac{\theta_f}{2\pi} - P(i)$; otherwise, $\Delta P_{ij} = d_{ij}^\alpha \dfrac{\theta_f}{2\pi}$.

    **add** node $j$ to $T$, i.e., $T := T \cup \{j\}$. **set** $P(i) := P(i) + \Delta P_{ij}$.

---

**Fig. 10.** Pseudo code of DBIP

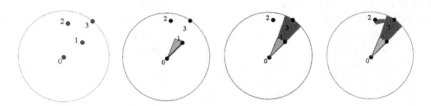

**Fig. 11.** Nodes addition in LDBIP

Fig. 11(a) shows a simple example of DBIP in which the source node has 4 local neighbor nodes 0, 1, 2, and 3. Node 1 is the closest to Node 0, so it is added first; in Fig. 11(b), an antenna with beam width of $\theta_f$ is centered between Node 0 and Node 1.

Then we must decide which node to add next (Node 2 or Node 3), and which node (that is already in the tree) should be its parent. In this example, the beam from Node 0 to Node 1 can be extended to include both Node 1 and Node 3, without setting up a new beam. Compared to other choices that setting up a new beam from Node 0 to Node 2, or from Node 1 to Node 2, this method has minimum incremental power. Therefore, Node 3 is added next by increasing the communication range of Node 0 and Node 1. In Fig. 11(c), finally, Node 1 must be added to the tree. Three possibilities are respectively to set up a new beam from Node 0, 1, 3. Here we assume that Node 3 has minimum distance. Then in Fig. 11 (d) we set up a new beam from Node 3 to Node 2.

## 3  Performance Evaluation

In this section, we present the performance evaluation of localized broadcast oriented protocols with our mobility prediction.

### 3.1  Simulation Environment

We use ns-2.28 [17] and its CMU wireless extension as our simulation tool and assume AT&T's Wave LAN PCMCIA card as wireless node model which parameters are listed in table 1.Since our purpose is to observe the effect of our mobility control mechanism, all simulations use an ideal MAC layer without contention or collision. Simulations apply ideal physical layer, that is, free space and two ray ground propagation model where if a node sends a packet, all neighbors within its transmission range will receive this packet after a short propagation delay. Table 2 displays parameters for wireless networks which are used in our simulation. In our simulation network, 100 nodes are placed in a fixed area network (900mx900m) which is relatively dense network. For each measure, 50 broadcasts are launched.

**Table 1.** Parameters for wireless node model

|  | **AT&T's Wave LAN PCMCIA** |
|---|---|
| Frequency | 2.4GHZ |
| Maximum transmission range | 250m |
| Maximum transmit power | 0.281838 W |
| Receiving power | 0.395 watts |
| Transmitting power | 0.660 watts |
| Omni-antenna receiver/transmitter gain | 1db |
| Fixed beam width of directional antennas | 30° |
| Directional-antenna receiver/transmitter gain | 58.6955db |
| MAC protocol | 802.11 |
| Propagation model | free space / two ray ground |

**Table 2.** Parameters for wireless networks

| **Parameters** | **Value** |
|---|---|
| Simulation Network Size | 900mx900m |
| Nodes number | 100 |
| Simulation time | 50m |
| Packet size | 64k |
| Transmission delay | 25us |

The mobility model used in our simulation is random waypoint model [11, 18-21] which is widely used in simulating protocols designed for mobile ad hoc networks. In this model, a mobile node begins by staying in one location for a certain period of time (i.e., a pause time). Once this time expires, the mobile node chooses a random destination in the simulation area and a speed that is uniformly distributed between *[minspeed, maxspeed]*. The mobile node then travels toward the newly chosen destination at the selected speed. Upon arrival, the mobile node pauses for a specified time period before starting the process again. In addition, the model is sometimes simplified without pause times.

## 3.2  Simulation Results

To evaluate the performance of localized broadcast protocols with mobility control management, we define some parameters: RAR, EC and SRB. The RAR (Reach Ability Ratio) is the percentage of nodes in the network that received the message. Ideally, each broadcast can guarantee 100% RAR value. While in mobile environment because of nodes mobility, RAR may be less than 100% and then RAR becomes more important in performance evaluation in mobile ad hoc networks. To investigate energy efficiency issue, we observe EC (total power consumption) over the network when a broadcast has occurred. In addition, under mobile simulation environment, the energy consumption includes not only the energy consumption for broadcasting message, but also that for propagation in mobility control process. We also observe the FNR (Forward Node Ratio) which is the percentage of nodes in the network that retransmit the message. A blind flooding has a FNR of 100%, since each node has to retransmit the message at least once.

In our simulation, nodes average moving speed varies from 0m to 160m per second. We compare protocol performance when it doesn't employ mobility prediction

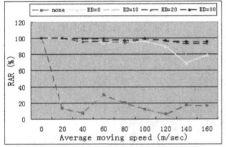

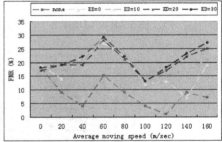

**Fig. 12.** RAR of LBIP                **Fig. 13.** FNR of LBIP

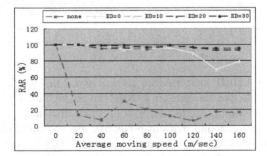

**Fig. 14.** EC of LBIP

mechanism with that when it applies prediction mechanism while redundant transmission range varies from 0 to 30. In figures, "none" presents the simulation result when protocol doesn't employ mobility management mechanisms and "ED" is the redundant transmission range.

Fig.12-14 shows LBIP performance comparison. In Fig.12, it's obvious that once employ our predictive mechanism, LBIP can get very high broadcast coverage ratio. As redundant transmission range increases, the RAR can nearly reach 100%. While Fig 13 shows that at that time forward nodes ratio will also increases, which reflects energy consumption increase shown in Fig. 14. The reason why energy consumption is very low when we didn't apply our proposal is that at that time retransmission nodes number decreases greatly and few nodes really receive message because of nodes mobility, inaccurate location information and corresponding inaccurate retransmission instruction.

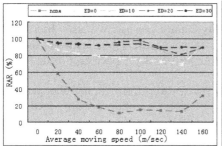

**Fig. 15.** RAR of LDBIP

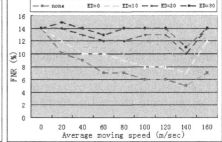

**Fig. 16.** FNR of LDBIP

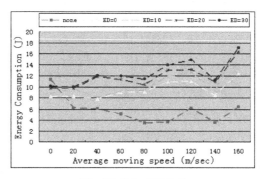

**Fig. 17.** EC of LDBIP

Fig.15-17 shows LDBIP performance comparison. In Fig.15, it's obvious that once employ our predictive mechanism, LDBIP can get high broadcast coverage ratio. As redundant transmission range increases, RAR value also increases. While compare Fig.12 and Fig 15, we can find that our predictive mechanism works better in LBIP than in LDBIP. In other words, our proposal works better in networks with omni-antennas than that with directional antennas. That's because omni-antennas have much more coverage redundancy. While directional antennas can conserve energy consumption and avoid interference, we can see from Fig. 14 and 17 that the energy consumption of LDBIP is nearly 50% of that of LBIP. In Fig 16, the FNR also correspondingly reflects LDBIP can conserver more energy since fewer nodes will do retransmission compared to LBIP. Also the energy consumption of LDBIP is very low when we didn't apply our proposal. The reason is the same with that in LBIP.

## 4 Conclusions

In this paper, we proposed a new mobility control mechanism which is proactive and predictive. The goal of our mechanism is to guarantee high broadcast coverage and energy efficient issue. Therefore, in our proposal we employ proactive Request-Response model to collect neighbors' location information to save energy which is consumed for periodic "Hello" messages in previous existing mobility control mechanisms. We propose mobility prediction mechanism to predict the neighbors' actual location at the actual transmission time by which we avoid updated information. We apply our mobility prediction mechanism into localized broadcast oriented protocols, such as LBIP and LDBIP. To apply our mechanism we modified existing protocols to adapt to our mechanism and simulation results show that our proposal greatly increased the broadcast coverage of localized broadcast protocols.

## Acknowledgement

This work was supported by grant No. R01-2005-000-10267-0 from Korea Science and Engineering Foundation in Ministry of Science and Technology.

## References

1. Y.-C. Tseng, S.-Y. Ni, and E.-Y. Shih: Adaptive Approaches to Relieving Broadcast Storms in a Wireless Multihop Mobile Ad Hoc Network. IEEE Trans. Computers, vol. 52, no. 5, May (2003)
2. J. E. Wieselthier, G. D. Nguyen, A. Ephremides: On the construction of energy-efficient broadcast and multicast trees in wireless networks. Proc. IEEE INFOCOM (2000) 585-594
3. J.E. Wieselthier, G.D. Nguyen, A. Ephremides: Energy-Limited Wireless Networking with Directional Antennas: The Case of Session-Based Multicasting. Proc. IEEE infocom (2002)
4. P. Bose, P. Morin, I. Stojmenovic, J. Urrutia: Routing with guarantee delivery in ad hoc networks. ACM/Kluwer Wireless Networks (2001) 609-616
5. T. Chu, I. Nikolaidis: Energy efficient broadcast in mobile ad hoc networks. In Proc. Ad-Hoc Networks and Wireless (ADHOC-NOW), Toronto, Canada (2002) 177-190
6. W. Peng, X. Lu: On the reduction of broadcast redundancy in mobile ad hoc networks. In Proc. Annual Workshop on Mobile and Ad Hoc Networking and Computing (Mobi-Hoc'2000), Boston, Massachusetts, USA (2000) 129-130

7.  A. Qayyum, L. Viennot, A.Laouiti: Multipoint relaying for flooding broadcast messages in mobile wireless networks. In Proc. 35th Annual Hawaii International Conference on System Sciences (HICSS-35), Hawaii, USA (2002)
8.  J. Wu, H. Li: A dominating-set-based routing scheme in ad hoc wireless networks. In Proc. 3rd Int'l Workshop Discrete Algorithms and Methods for Mobile Computing and Comm (DIALM'99), Seattle, USA (1999) 7-14
9.  F.Ingelrest, D.Simplot-Ryl: Localized Broadcast Incremental Power Protocol for Wireless Ad Hoc Networks. 10th IEEE Symposium on Computers and Communications, Cartagena, Spain (2005)
10. Hui Xu, Manwoo Jeon, Lei Shu, Wu Xiaoling, Jinsung Cho, Sungyoung Lee: Localized energy-aware broadcast protocol for wireless networks with directional antennas. International Conference on Embedded Software and Systems (2005)
11. T. Camp, J. Boleng, and V. Davies: A Survey of Mobility Models for Ad Hoc Network Research. Wireless Comm. & Mobile Computing (WCMC), special issue on mobile ad hoc networking: research, trends and applications, vol. 2, no. 5 (2002) 483-502
12. J. Wu and F. Dai: Efficient Broadcasting with Guaranteed Coverage in Mobile Ad Hoc Networks. IEEE Transactions on Mobile Computing, Vol. 4, No. 3, May/June (2005) 1-12
13. J. Wu and F. Dai: Mobility Control and Its Applications in Mobile Ad Hoc Networks. Accepted to appear in Handbook of Algorithms for Wireless Networking and Mobile Computing, A. Boukerche (ed.), Chapman & Hall/CRC (2006) 22-501-22-518
14. J. Wu and F. Dai: Mobility Management and Its Applications in Efficient Broadcasting in Mobile Ad Hoc Networks. In Proceedings of the 23rd Conference of the IEEE Communications Society (INFOCOM 2004), vol.1, Mar (2004) 339-350
15. J. Wu and F. Dai: Mobility Control and Its Applications in Mobile Ad Hoc Networks. IEEE Network, vol.18, no.4, July/Aug (2004) 30-35
16. J. Wu and F. Dai: Mobility-Sensitive Topology Control in Mobile Ad Hoc Networks. In Proceedings of the 18th International Parallel and Distributed Processing Symposium (IPDPS 2004), Apr (2004) 502-509
17. Network Simulator - ns-2, http://www.isi.edu/nsnam/ns/.
18. N. Bansal and Z. Liu: Capacity, delay and mobility in wireless ad-hoc networks. In INFOCOM, April (2003)
19. Guolong Lin, Guevara Noubir, and Rajmohan Rajamaran: Mobility models for ad-hoc network simulation. In Proceedings of Infocom (2004)
20. X. Hong, M. Gerla, G. Pei, and Ch.-Ch. Chiang: A group mobility model for ad hoc wireless networks. In ACM/IEEE MSWiM, August (1999)
21. B. Liang and Z. H. Haas: Predictive distance based mobility management for pcs networks. In INFOCOM, March (1999)

# Delay Sensitive and Energy Efficient Distributed Slot Assignment Algorithm for Sensor Networks Under Convergecast Data Traffic[*]

İlker Bekmezci and Fatih Alagöz

Computer Networks and Research Laboratory (NETLAB)
Bogazici University, Istanbul, 34342, Turkey
{ilker.bekmezci, alagoz}@boun.edu.tr

**Abstract.** The scarcest resource for most of the wireless sensor networks (WSNs) is energy and one of the major factors in energy consumption for WSNs is due to communication. Not only transmission but also reception is the source of energy consumption. The lore to decrease energy consumption is to turn off radio circuit when it is not needed. This is why TDMA has advantages over contention based methods. Time slot assignment algorithm is an essential part of TDMA based systems. Although centralized time slot assignment protocols are preferred in many WSNs, centralized approach is not scalable. In this paper, a new energy efficient and delay sensitive distributed time slot assignment algorithm (DTSM) is proposed for sensor networks under convergecast traffic pattern. DTSM which is developed as part of the military monitory (MILMON) system introduced in [16], aims to operate with low delay and low energy. Instead of collision based periods, it assigns slots by the help of tiny request slots. While traditional slot assignment algorithms do not allow assigning the same slot within two hop neighbors, because of the hidden node problem, DTSM can assign, if assignment is suitable for convergecast traffic. Simulation results have shown that delay and energy consumption performance of DTSM is superior to FPRP, DRAND, and TRAMA which are the most known distributed slot assignment protocols for WSNs or ad hoc networks. Although DTSM has somewhat long execution time, its scalability characteristic may provide application specific time durations.

## 1 Introduction

Developments in micro-electro-mechanical systems (MEMS) technology have enabled to integrate battery operated sensor, computational power and wireless communication components into a small size, low cost device [1]. These tiny sensor nodes, which consist of sensing, data processing, and communicating components, leverage the idea of sensor networks based on collaborative effort of a large number of nodes [2]. Dominant factor in energy consumption for sensor nodes is communication and the most common way to reduce energy consumption is to turn off the radio circuit when it is not needed. TDMA has a natural advantage over contention based medium

---

[*] This work is supported by the State Planning Organization of Turkey under NGSN Project and Bogazici University.

Y. Koucheryavy, J. Harju, and V.B. Iversen (Eds.): NEW2AN 2006, LNCS 4003, pp. 353–364, 2006.
© Springer-Verlag Berlin Heidelberg 2006

access methods [3]. In TDMA, nodes listen and send data in a certain schedule. One of the main problems of TDMA based networks is slot assignment. In this paper, a new distributed time slot assignment mechanism (DTSM) is proposed for TDMA based sensor networks. The most important design considerations of this new mechanism are energy efficiency, delay, and scalability.

Distributed time slot assignment is not a new topic for wireless networks. FPRP (Five-Phase Reservation Protocol for Mobile Ad Hoc Networks) [4], E-TDMA (Evolutionary-TDMA Scheduling Protocol) [5] are some examples of distributed scheduling protocols for ad hoc networks. Most of the ad hoc network algorithms are developed for peer to peer data traffic. However, data traffic in WSN is mostly convergecast. Existing slot assignment algorithms for ad hoc networks can not satisfy energy and delay requirements of WSNs under convergecast traffic. Wireless sensor networks need a delay sensitive and energy efficient slot assignment algorithm under convergecast data traffic. This is why slot assignment algorithms developed for ad hoc networks can not be directly applied to WSN.

There are some researches about distributed time slot assignment for sensor networks. Patro et.al. has proposed Neighbor Based TDMA Slot Assignment Algorithm for WSN [6]. A mobile agent visits every node and assigns a proper slot. This method reduces required number of slots and increases channel utilization. However, it is not energy efficient. Copying and running the agent consumes high amount of energy. Kanzaki et. al. has also proposed an adaptive slot assignment protocol for WSN [7]. However, the main design objective is channel bandwidth, not delay or energy efficiency. Another distributed slot assignment algorithm for sensor networks is presented in [8]. It reduces delay for broadcast, convergecast, and local gossip traffic patterns for different grid topologies. However, in many sensor network applications, sensor nodes are deployed randomly. In addition to this difficulty, it does not consider energy consumption; its design consideration is only to minimize delay. SMACS [9] uses a different distributed time scheduling algorithm. After a series of handshaking signals, neighbor nodes can agree on a frequency and time pair to construct a link. SMACS produces a scalable and reliable flat network. However, SMACS needs FDMA as well as TDMA, but sensor nodes are so tiny and limited that current sensor nodes cannot meet the requirements of SMACS. DRAND [10] is a randomized dining philosophers algorithm for TDMA scheduling of wireless sensor networks. This algorithm is the first distributed implementation of RAND [11], a commonly used, centralized channel assignment algorithm. Randomized dining philosophers approach is scalable and robust. However, in DRAND, before having a schedule, nodes communicate with each other using a contention based MAC protocol, and it increases energy consumption. μMAC has another slot assignment mechanism that includes a contention period [12]. TRAMA [13] is a TDMA-based sensor network system and it includes a distributed slot assignment mechanism. It has random access period to be able to assign proper slots, and its random access period is also contention based. In TRAMA, all the nodes have to listen to medium in random access periods. It increases energy consumption of TRAMA.

In this paper, a new delay sensitive, energy efficient distributed time slot assignment algorithm, DTSM, is proposed for wireless sensor networks. The design considerations of DTSM are delay, energy consumption, and scalability. There are a number of advantages of DTSM design over the existing designs. Firstly, unlike existing slot assignment protocols that includes 802.11 like contention sessions, nodes in DTSM contend in time slots, like FPRP [4]. In 802.11 like contention based sessions, all nodes must listen to medium and keep their radio circuits open during contention based session [14]. This strategy may consume high amount of energy. Contention in time slots results in lower energy consumption. Another important feature of DTSM is its convergecast traffic aware design. In convergecast traffic, data relays from nodes into the sink. The sink collects all the data produced by the nodes. DTSM assumes that nodes always forward data to their neighbors that are with lower hop number. In order to decrease delay, DTSM assigns the slots on the basis of the hop number of the nodes. Unlike the other slot assignments, DTSM allows to assign the same slots into the nodes within two hop region, if the assignment allows convergecast traffic.

The rest of this paper is organized as follows. In Section II, system design of the proposed algorithm is introduced. In Section III, performance results and discussion are presented. Finally, the conclusion is given in Section IV.

## 2   System Design

DTSM is a new distributed time slot assignment protocol for sensor networks under convergecast traffic. It is developed to operate with low energy and low delay. Because of its distributed nature, DTSM can be run in any network size without central node. It does not need any additional MAC layer support. It has a new mechanism to reduce delay, so it can be used in delay sensitive applications, like military monitoring. Its main design considerations are delay, energy consumption, and scalability. Before explaining the details of DTSM, assumptions are presented as follows:

- Sensor nodes will be immobile.
- Radio channel is symmetric.
- Before running DTSM, all nodes must synchronize their clock.

### 2.1  Description

Instead of random access period like TRAMA [13] or DRAND [10], DTSM uses time slots to exchange scheduling signals. The slot organization of DTSM is presented in Figure 1. DTSM assigns the slots in reservation frames. Every reservation frame begins with an advertisement slot. The nodes that receive valid slot in the last reservation frame send a special signal in this slot. All the other nodes listen to this slot and if a node receives a signal in the slot it means the reservation frame for its hop number is about to begin and it can compete for slot assignment. Every reservation frame is used for corresponding hop numbered nodes. According to this hop numbered structure, in the first reservation frame, the nodes with hop number one can get slots. After that, the nodes with hop number two get the slots and so on.

Slot assignments for a specific hop number are performed in reservation slots. Each reservation slot corresponds to a specific available data transmission slot. In this case, the number of reservation slots and the number of available data transmission slots for a particular hop number are the same. There are a certain number of reservation cycles in each reservation slot. The number of reservation cycles for each

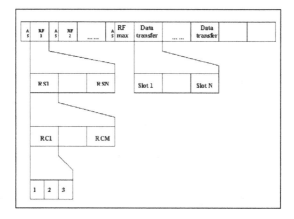

AS: Advertisement Slot, RF: Reservation Frame, RS: Reservation Slot, RC: Reservation Cycle

**Fig. 1.** Slot organization of DTSM

reservation slot is a constant and it is a parameter of DTSM algorithm. There are three tiny slots in each reservation cycle and signal exchanges are realized in these slots.

## 2.2  Signal Exchange

In traditional slot assignment, no node within two hop radius can get the same slot. However, if the only traffic in the network is convergecast, this rule can be relaxed. If data flow through the higher hop numbered nodes to lower hop numbered nodes, convergecast traffic can be realized. DTSM assumes that a node with hop number $h$ sends its data only to a node with hop number $h-1$. In such a network, the only collision that must be handled is between the nodes with hop number $h$ and the nodes with hop number $h-1$. Even if they are in two hop neighborhood, the nodes that are with the same hop number can get the same slot, because they will never communicate. Figure 2 shows a sample slot assignment for a certain topology.

In Figure 2, node A and B are with hop number $h$, node C and D are with hop number $h-1$. In this topology, C can hear A, but cannot hear B. D can hear B but cannot hear A. A and B can hear each other. In traditional slot assignment algorithms, A and B can not have the same slot. When A and B have the same slot number, they can not send data each other. However, traffic is generally convergecast in WSN and if all data will be forwarded from $h$ to $h-1$, convergecast traffic can be realized. The only requirement of a sensor network is to be able to send data to lower hop numbered nodes. In Figure 2, A must be able to send data to C and B must be able to send data to D. The only requirement is that A and C or B and D can not get the same slot. If A and B do not communicate, they can get the same slot. In Figure 2, A and B has allocated the fifth slot. A can send its data to C and B can send its data to D. Although the same slot is assigned in two node neighbor nodes, network is still collision free for convergecast traffic.

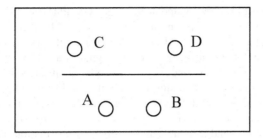

**Fig. 2.** Sample slot assignment

In Figure 2, A and B can get the same slot, because C can get the signal of A and D can get the signal of B at the same time. However, if D or C can hear A and B, the signals of A and B may collide and A and B can not get the same slot.

According to this new situation, signal exchange is designed as follows: The first slot of a reservation cycle is request slot. In this slot, every node that receives a signal in the last advertisement slot requests a slot with a certain probability. Let us assume that the hop number of the node is $h$. For a valid request, it sends a signal in request slot. Only the nodes with hop number $h-1$ may suffer from the collision of the requests. The nodes that are with hop number $h-1$ listen to request slot. If the node receives a jammed signal in this slot, it means there is a collision. The nodes that can get a valid request in the first slot send an approval signal in the second slot which is approval slot. The nodes in the hop number $h$ listen to approval slot. If a node has sent a request and if it can receive a signal at approval slot, it can get a valid slot. In the last slot which is confirmation slot, the node can get a valid slot sends a signal. The other nodes with hop number $h$ listen to confirmation slot.

At the end of this signal exchange some nodes can get valid slots. The nodes that can not have a valid slot try to get one in the next reservation cycles. Reservation cycles are repeated for a certain number of times in a reservation slot. If a node can not get a valid slot in a reservation slot, it continues to compete in the reservation cycles of the next reservation slot.

## 2.3  Updating Slot Request Probability

At the beginning of signal exchange, every node can send a request signal with a certain probability. This probability is not constant. It is calculated as $1/Nc$, where $Nc$ is the number of contender nodes in one hop neighborhood. $Nc$ is updated at the end of each reservation cycle. $Nc$ must be forecasted as realistic as possible. If $Nc$ is forecasted larger than it is, contention probability will be lower than it should be and slot assignment algorithm may take longer than it is needed. If $Nc$ is forecasted smaller than it is, contention probability becomes larger than it should be and collisions increases. $Nc$ should be updated according to the result of reservation cycle. If a neighbor node achieves to get a valid slot, number of contenders decreases. If there is a collision, $Nc$ must be increased to decrease contention probability. If nothing happens, in other words, if reservation cycle is idle, $Nc$ should be decreased to increase contention probability.

DTSM slot request probability update strategy is similar to FPRP [4]. FPRP is also adopted from Rivest's pseudo-Baynesian Broadcasting Algorithm [15], which is designed for distributed single hop ALOHA broadcast network. According to DTSM strategy, if a node can not receive or send any signal in a reservation cycle, it is idle. In idle state, $N_c$ is decreased by one. If a node sends a request in the first slot and if it can not get approval in the second slot, it is a collision. For a collision situation, $N_c$ is incremented by $(e-2)^{-1}$. If a node sends a request and receives an approval in the second slot, it is a success for itself. It gets a valid slot and it does not contend anymore. If a node that does not send a request and receive a confirmation, it means there is a success one hop away. In addition to $N_c$, a new value must be calculated to update $N_c$ for success state. This new value, $N_b$, represents the number of the nodes that has no valid slot and can not contend due to a success within one hop. The assumption is that if there is a success one hop away, a portion of its neighbors which is modeled as R can not contend. After a success one hop away, $N_b$ must be incremented R times $N_c$, and $N_c$ must be recalculated as 1-R times $N_c$. The complete structure of updating slot request probability is as follows:

```
At the beginning of each reservation slot N=N , N =0.
                                          c   b   b
for the first reservation slot N =the number of slots
                                c
that will be distributed in one reservation frame)
For each reservation cycle
Contention probability=1/ N .
                           c
If the state is
  Idle:        N  = N  -1, if N  >=1.
                c    c         c
  Collision:   N  = N  + (e-2)
                c    c         -1
  Success one hop away:
  (node does not contend in this reservation cycle)
N  = N  -1, if N  >=1.
 c    c         c
              N  = N  + R*  N .
               b    b        c
              N  = (1-R)*  N .
               c            c
```

## 2.4  Handling Delay Problems

One of the most important design issues for wireless sensor networks is delay. If application is delay sensitive, like military monitoring or surveillance as in [16], data latency can be very important. In a military monitoring system, the existence of enemy should be reported as soon as possible. Reducing delay is possible by the help of assigning time slots carefully. The rule is that smaller hop numbered nodes should get higher slot numbers. In order to realize rescheduling, time frame is divided into $u$ sub time frames. If the whole time frame has $s$ slots, a sub time frame has $s/u$ slots. The slot number assigned to a node with hop number $h$, must be in $(u-((h-1) \bmod u))^{th}$ sub time frame. In this way, the slot numbers of consecutive hop numbered nodes belong to consecutive sub time frames. Sensor node can get the number of sub time frames, $u$, from the sink's synchronization signal and calculate its sub frame number.

An example helps to understand clearer. Let us assume that the nodes in Figure 3 are one hop away from its consecutives. In this particular network, time frame has 30 time slots and there are 3 sub slots. In this case, the first sub slot is from 21 to $29^{th}$ slots, the second is from 11 to $20^{th}$ slots, and the third one is from 2 to $10^{th}$. The first

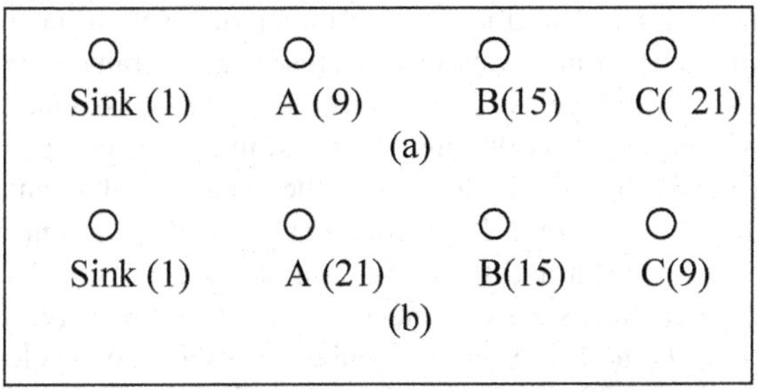

| | | | |
|---|---|---|---|
| O | O | O | O |
| Sink (1) | A (9) | B(15) | C( 21) |
| | (a) | | |
| | | | |
| O | O | O | O |
| Sink (1) | A (21) | B(15) | C(9) |
| | (b) | | |

**Fig. 3.** Example network and time slots(a) A regular slot assignment. (b) DTSM.

slot is reserved for the sink. Figure 3(a) is an example of a slot assignment. Figure 3 (b) is a slot assignment based on DTSM. The relay of an event from C to the sink takes 70 time slots for a sensor network in Figure 3(a). However, it takes only 21 time slots for the rescheduled network in Figure 3(b).

## 3  Performance Results

Performance of DTSM is discussed according to delay, energy consumption, and running time. A simulator is implemented to compare DTSM with FPRP, TRAMA and DRAND. Sensor network is assumed to be composed of Berkeley's Motes [17]. Power consumption of the radio transceiver, is 13.5mW, 24.75mW, in receiving and transmitting respectively [17]. Simulation area is assumed as a circle with 1000 m. diameter. The sink is placed at the center of the simulation area. The locations of the nodes are uniformly distributed over the simulation area. Simulator parameters are presented in Table 1.

**Table 1.** Simulation parameters

| Parameters | |
|---|---|
| Shape of the sensor network area | Circle |
| Diameter of sensor network | 40 unit. |
| Transmission Range | 1.5 unit |
| Sensing Range | 1 unit |
| Bit rate | 19200 bit/sec |
| Receive energy (one bit) | 0.7 μJoule |
| Transmit energy (one bit) | 1.29 μJoule |
| Number of bits in one signal exchange time slot (including synchronization bits) | 5 |
| Time for one data transfer time frame | 1 second |
| Number of sub frames (DTSM), u | 5, 10, 20, 40 |
| Contention Probability Parameter for DTSM (R) | 0.8 |

Another parameter for simulation is the number of reservation frames for DTSM and the number of reservation cycles in each reservation frame. Number of reservation frames and reservation cycles should be set so that slot assignment algorithm can assign valid slot with high probability and it should minimize energy consumption and run time. In order to find optimum parameters, the existence of a central coordinator is assumed for FPRP and DTSM. If a node can get the current slot, reservation cycles are continued to repeat and if there is a node that can not get valid slot in the current slot, reservation frames are continued to repeat. In this way, minimum number of slots is assigned to the nodes. Average number of reservation cycles and reservation frames are calculated after 20 runs with central coordinator and the calculated numbers are used as parameters. The most important factor that affects these parameters is node density. Number of reservation cycles and reservation frames is set for each node density. Simulation results have shown that the parameters that are calculated with this methodology can assign valid slots with more than %99.5 probability.

## 3.1 Delay

Delay sensitive applications like military monitoring, may suffer latency. DTSM has a special mechanism for handling delay problem. Delay performance of DTSM is compared with FPRP. Any other slot assignment mechanism that has no delay handling mechanism is expected to result like FPRP. In the simulation, one data transfer time frame is assumed as one second. If a packet is composed of 64 bits, and a node can send one packet in one data slot, one data slot takes 3.3 ms. One sub time frame for DTSM is composed of 9 slots, if node density is 1. In our simulation, sub time frames are 5, 10, 20 and 30. Delay is related with the distance between event and the sink. In order to investigate delay performance, 100 events are generated with different distance from the sink. Simulation is repeated for 20 times. Figure 4 shows average delay of DTSM and FPRP.

FPRP delay increases with the distance linearly. If distance between event and the sink is 20 unit, in other words in region 10th region, FPRP delay exceeds 10 seconds. If it is assumed that 1 unit is 30 m., FPRP can report a 600 m. away event within 10 seconds. If the application is delay sensitive, for example military monitoring or intruder detection system, this delay is unacceptable.

DTSM is successful to decrease delay with its sub frame structure. Especially, when the distance is long and sub frame number is high, delay difference between DTSM and FPRP may increase up to 9 times. DTSM with 20 sub time frame can report an 600 m. away intruder in only 1,1 second. Delay performance of DTSM is acceptable for most of the delay sensitive applications.

Although sub frame number affects delay performance, it is not always directly proportional with sub frame number. The delay performance of DTSM follows a step pattern related with average hop number between event and the sink. For example, average hop number for 5th region is 10, and delay of DTSM-20 and DTSM-10 is approximately the same. After the 5th region, while delay of DTSM-10 increases, delay of DTSM-20 still stays almost constant. The same structure can be found for DTSM-5. This step pattern is closely related with average hop number of the region and the number of sub frames. Average hop number of the 5th region is 10 and delay

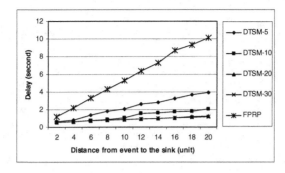

**Fig. 4.** Delay performance of DTSM and FPRP

of DTSM-10, DTSM-20, DTSM-30 and DTSM-40 is very close for region 5. If region number is higher than five, average hop number exceeds 10 and delay of DTSM-10 starts to increase. While delay of DTSM-10 increases, DTSM-20, DTSM-30 and DTSM-40 stays approximately constant. It shows that sub frame number must be chosen according to average hop number of the sensor network. If sub time frame number is lower than maximum hop number, delay increases.

### 3.2 Energy Consumption

Energy is one of the most critical resources for sensor networks. Slot assignment mechanism of a sensor network must be energy saver like any other algorithm used in sensor networks. Energy consumption to run DTSM is compared with the other time scheduling mechanism for WSN. The comparison is based on only the energy consumption of radio circuits.

TRAMA nodes can sleep %90 of the time at most. In other words, TRAMA nodes have to use its radio circuit with %10 of the time for slot assignment. It takes considerable energy. DRAND uses 802.11 like signal exchange mechanism and it needs very large amount of signaling. FPRP does not use any additional MAC layer for slot assignment. However, it is not optimized for energy consumption. DTSM is a distributed slot assignment algorithm designed for minimum energy consumption.

One of the most important parameters for energy consumption for DTSM is node density. The simulation results for different node densities are presented in Figure 5 to compare energy consumption of DTSM, FPRP, DRAND and TRAMA. Energy consumption of DTSM, FPRP and DRAND increases with the increasing node density. While increasing structure of DTSM and DRAND is polynomial, energy consumption of FPRP increases linearly. Only TRAMA is not affected by node density. It is clear that time slot assignment protocols that include contention period, like TRAMA or DRAND consume much more energy than slot assignment protocol based on tiny time slots, like FPRP or DTSM. Although FPRP is also successful when it is compared with contention period based methods, DTSM performs approximately 4 times better than FPRP.

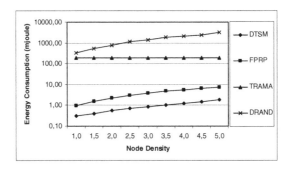

**Fig. 5.** Energy consumption comparison of different time slot assignment algorithms

### 3.3 Running Time

DTSM and FPRP are compared about their running times. FPRP is fully parallel and distributed algorithm. All the nodes can run FPRP algorithm at the same time. The reservation process for a given node only involves nodes within a two-hop radius, and is thus a local process. No coordination is necessary with more distant nodes. By keeping the reservation process localized (and running simultaneously over the entire network), the FPRP is insensitive to the network size. Its running time is constant. However, nodes run DTSM according to their hop numbers. It follows a wave pattern from the lowest hop number to maximum hop number. In the first reservation frame, the nodes which are one hop away from the sink allocate a certain set of slots. In the second reservation frame, the nodes which are two hop away from the sink allocate the slots and so on. In this case, running time of DTSM is fully dependent on the maximum hop number of the network.

Running time of FPRP is as follows:

$R_{FPRP}$= 5* time for one signal exchange slot * total number of reservation cycles in reservation frame.    (1)

Running time of DTSM for one hop is as follows:

$R_{DTSM}$= 3* time for one signal exchange slot * total number of reservation cycles in a reservation frame * maximum hop number + time for advertisement signal*maximum hop number.    (2)

Simulation model is used to compare running times. In Figure 6, average running time of DTSM with different maximum hop numbers and FPRP are compared for different network sizes, if it is assumed that every signal exchange slot has 5 bits and there are two guard bits between each signal exchange slots. Average running times are calculated with 20 simulation runs.

FPRP run time does not change with network size. Running time of DTSM is dependent on the size of the network and it is proportional to the maximum hop number of the sensor network. When node density is 5 and maximum hop number is 20, DTSM run time is 2,5 seconds. Simulation results have shown that the diameter of

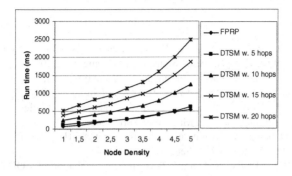

**Fig. 6.** Running time of DTSM and FPRP for different network size

such a network can be 50 unit. For a sensor network with 50 unit diameter, DTSM takes 2,5 seconds. For a typical sensor network, if 1 unit is assumed as 30 m, DTSM can assign time slots in a sensor network with 1500m diameter in 2,5 seconds. For the same network size, FPRP can assign time slots in slightly more than 500 ms. Although FPRP can run much faster than DTSM, DTSM is also acceptable even for time critical applications.

While FPRP and DTSM can run in the order of seconds, simulation implemented in [16] has shown that DRAND takes approximately 25 seconds when node density is 1. DRAND run time fits a quadratic curve with varying node densities. When node density is 5, DRAND takes approximately 240 seconds which is much longer than run time of FPRP and DTSM. There is no specific running time for TRAMA. Nodes get the slots in random access period whenever they need. It is a dynamic structure and this is why there is no specific running time for TRAMA. So TRAMA is not included in running time performance comparison.

## 4 Conclusion

In this paper, a new delay sensitive and energy efficient distributed slot assignment algorithm for wireless sensor networks, DTSM, is proposed. It assumes data traffic of sensor network is convergecast and data always flow from higher hop numbered nodes to lower hop numbered nodes. Although hidden node problem does not allow assigning the same node within two hop neighbors, DTSM can assign the same slot within two hop neighbors by the help of convergecast traffic assumption. In order to compare DTSM and common distributed slot assignment algorithms, a simulation model is developed. Extensive set of simulation results show that delay and energy consumption performance of DTSM is superior to that of FPRP, DRAND and TRAMA. Although DTSM can realize low energy consumption and delay, its running time proportionally increases with sensor network area. Fortunately, DTSM can run under in acceptable run time even for a large wireless sensor network such as a network with 1500 m diameter.

# References

1.  G. Asada, et al. Wireless Integrated Network Sensors: Low Power Systems on a Chip. In Proceedings of the European Solid State Circuits Conference, 1998.
2.  Akyildiz, I. F., Su, W., Sankarasubramaniam, Y., and Cayirci, E., "Wireless Sensor Networks: A Survey," Computer Networks (Elsevier) Journal, pp.393-422, March 2002.
3.  I. Demirkol, C. Ersoy, and F. Alagöz, MAC Protocols for Wireless Sensor Networks: a Survey, IEEE Communications Magazine, April 2006.
4.  Zhu, C. and Corson, A Five-Phase Reservation Protocol (FPRP) for Mobile Ad Hoc Networks. Wireless Networks 7, 4 (Sep. 2001), 371-384.
5.  V. Loscrì, F. De Rango, S. Marano, Performance Evaluation of AODV Protocol over E-TDMA MAC Protocol for Wireless Ad Hoc Networks, Lecture Notes in Computer Science, Volume 3124, Jan 2004, Pages 417 - 424
6.  Ranjeet Kumar Patro, Kannan Perumal Neighbor Based TDMA Slot Assignment Algorithm for WSN, IEEE Infocom 2005 poster session, Miami, March 2005.
7.  Kanzaki, A., Hara, T., and Nishio, S. 2005. An adaptive TDMA slot assignment protocol in ad hoc sensor networks. In Proceedings of the 2005 ACM Symposium on Applied Computing (Santa Fe, New Mexico, March 13 - 17, 2005).
8.  Sandeep S. Kulkarni, M.Arumugam. TDMA Service for Sensor Networks, 24th Int. Conference on Distributed Computing Systems Workshop, vol.05, no.5, pp. , 604-609, May 2004.
9.  K. Sohrabi, J. Gao, V. Ailawadhi, and G. J. Pottie, Protocols for self-organization of a wireless sensor network, IEEE Personal Communications, vol. 7, pp. 16 - 27, October 2000.
10. Injong Rhee, Ajit Warrier and Lisong, Randomized Dining Philosophers to TDMA Scheduling in Wireless Sensor Networks, Technical Report TR-2005-21, Department of Computer Science, North Carolina State University, April 2005.
11. S. Ramanathan, A unifed framework and algorithms for (T/F/C)DMA channel assignment in wireless networks, Proc. of IEEE INFOCOM, 1997, pp. 900,907.
12. Andre Barroso, Utz Roedig, and Cormac J. Sreenan. MAC: An Energy-Efficient Medium Access Control for Wireless Sensor Networks. In Proceedings of the 2nd IEEE European Workshop on Wireless Sensor Networks (EWSN2005), Istanbul, Turkey, February 2005.
13. V. Rajendran, K. Obraczka, J.J. Garcia-Luna-Aceves, Energy-Efficient, Collision-Free Medium Access Control for Wireless Sensor Networks, Proc. ACM SenSys 03, Pages: 181 - 192, Los Angeles, California, 5-7 November 2003.
14. Brian P Crow, Indra Widjaja, J G Kim,Prescott T Sakai. IEEE 802.11 Wireless Local Area Networks, IEEE Communication Magazine, September 1997, pp. 116-126.
15. R. L. Rivest. Network Control by Baynesian Broadcast (Report MIT/LCS/TM-287). MIT, Laboratory for Computer Science, Cambridge, MA, 1985.
16. I. Bekmezci, F. Alagöz, A New TDMA Based Sensor Network for Military Monitoring (MIL-MON), Proceedings of IEEE MILCOM 2005, Atlantic City, NJ, 17-20 October 2005.
17. Wei Ye, et al. An EnergyEfficient MAC Protocol for Wireless Sensor Networks. In Proceedings of the IEEE Infocom, pages 1567--1576, New York, NY, USA, June 2002.

# Real-Time Coordination and Routing in Wireless Sensor and Actor Networks

Ghalib A. Shah, Muslim Bozyiğit[1], Özgür B. Akan, and Buyurman Baykal[2]

[1] Department of Computer Engineering,
Middle East Technical University, Ankara, Turkey 06531
{e135333, bozyigit}@metu.edu.tr

[2] Department of Electrical and Electronics Engineering,
Middle East Technical University, Ankara, Turkey 06531
{akan, baykal}@eee.metu.edu.tr

**Abstract.** In Wireless Sensor Actor Networks (WSAN), sensor nodes perform the sensing task and actor nodes take action based on the sensed phenomena in the field. To ensure efficient and accurate operations of WSAN, new communication protocols are imperative to provide sensor-actor coordination in order to achieve energy-efficient and reliable communication. Moreover, the protocols must honor the application-specific real-time delay bounds for the effectiveness of the actors in WSAN.

In this paper, we propose a new real-time coordination and routing ($RCR$) framework for WSAN. It addresses the issues of coordination among sensors and actors and honors the delay bound for routing in distributed manner. $RCR$ configures sensors to form hierarchical clusters and provides delay-constrained energy aware routing (DEAR) mechanism. It uses only cluster-heads to coordinate with sink/actors in order to save the precious energy resources. The DEAR algorithm integrates the forwardtracking and backtracking routing approaches to establish paths from source nodes to sink/actors. In the presence of the sink in WSAN, it implements the centralized version of DEAR (C-DEAR) to coordinate with the actors through the sink. In the absence of sink or ignoring its presence, there is a distributed DEAR (D-DEAR) to provide coordination among sensors and actors. Cluster-heads then select the path among multiple alternative paths to deliver the packets to the actors within the given delay bound in an efficient way. Simulation experiments prove that $RCR$ achieves the goal to honor the realistic application-specific delay bound.

## 1 Introduction

Recent advances in the field of sensor networks have led to the realization of distributed wireless sensor networks (WSN). A WSN is composed of large number of sensor nodes, which are densely deployed in the sensor field in a random fashion with a sink node. The task of sensor nodes is to detect the events in the sensors field and route them to the sink node, which is responsible for the monitoring of the field.

Y. Koucheryavy, J. Harju, and V.B. Iversen (Eds.): NEW2AN 2006, LNCS 4003, pp. 365–383, 2006.
© Springer-Verlag Berlin Heidelberg 2006

Recently, the capabilities of the WSN are extended to include the actor nodes responsible with taking action against the detected events [1]. Such architecture is called a wireless sensor and actor networks (WSAN), where a small numbers of actors, as compared to sensors, are spread in the sensor field as well. Actors are mostly mobile and resource-rich devices and can be thought to form a mobile ad hoc network of their own. This paradigm of WSAN is capable of observing the physical world, processing the data, making decisions based on the sensed observation and performing appropriate actions. Typically, the architecture of a WSAN consists of sensors which sense the phenomena, a sink that collects the data from the sensors to process and actors that act upon the command sent by the sink. In the literature, such architecture is known as *semi-automated architecture*. An architecture in which sensor nodes send information to the actor nodes directly without the involvement of sink node is called an *automated architecture* [1]. It is apparent that the communication path in a *semi-automated architecture* introduces significant delay, which is not acceptable for delay-sensitive applications. For example, consider a military application where sensors in the battlefield will detect the movement of red forces and send the information to the sink which is situated in a remote command and control station. The sink then triggers an action through an actor to counter in the threat area. In this case unnecessary delay is introduced due to sensor-sink communication which could be removed if the actors can take localized actions without the involvement of the sink; depending on the data sent by the sensors. Hence, the most challenging task in WSAN is the coordination between sensors and actors to provide real-time response.

In WSAN, the effective sensor-actor coordination requires the sensors to know the right actors and the routes to reach them. Moreover, it requires delay estimates for all possible routes. In addition to energy constraints as in WSN, WSAN also imposes timing constraints in the form of end-to-end deadlines. Clearly, there is a need for real-time communication protocols for WSAN, which provide effective sensor-actor coordination while consuming less energy.

There have been considerable efforts to solve the routing problem in wireless sensor networks [3], [5], [6], [10], [11], [12], [13], [14]. However, these protocols do not consider the heterogeneity of WSAN. Moreover, none of these protocols provide sensor-actor coordination and real-time routing. A coordination framework [2] for WSAN has been proposed that is an event-based reactive model of clustering. Cluster formation is triggered by an event so that clusters are created on-the-fly to optimally react to the event itself and provide the required reliability with minimum energy expenditure. Reactive cluster formation algorithms have the disadvantage that they consume precious time on event occurrence for cluster formation. Hence for real-time coordination such an architecture is not suitable. Moreover, cluster to actor routing in [2] is done using greedy geographical approach. A packet forwarding node finds the next hop node according to the greedy approach failing to do so results into a packet loss as the packets enters into a void region. Since the work assumes that the network is dense therefore it does not propose any void region prevention or recovery mode implementation.

Consequently, there exists no unified solution which addresses the real-time co-ordination as well as routing problem for the heterogeneous WSANs.

In this study, we propose a real-time coordination and routing ($RCR$) frame-work, which addresses the sensor-actor coordination with real-time packet deliv-ery in the *semi-automated architecture* as well as *automated architecture*. $RCR$, incorporates the two components, namely DAWC and DEAR. DAWC is our heuristic clustering protocol used to dynamically configure the sensor nodes in the form of clusters to achieve energy efficiency. Whereas, RCR achieves the real-time demand $\tau$ of packet delivery through our delay-constrained energy aware routing (DEAR) protocol that is the first and foremost aim of the protocol. The DEAR protocol, described in Section 3, establishes a backbone network by inte-grating the forward tracking and backtracking mechanism that provides all the possible routes towards target nodes (sink/actors). The path selection criterion is based on the packet delay as well as the balance consumption of energy of sen-sor nodes. In the presence of the sink in WSAN, it implements the centralized version of DEAR (C-DEAR) to coordinate with the actors through the sink. On the other hand, when there is no sink or central node in WSAN, it provides the distributed version of DEAR (D-DEAR) for coordination among sensors and actors. Performance evaluation study reveals that $RCR$ addresses the real-time coordination and routing requirements of WSANs.

The remainder of the paper is organized as follows. In Section 2, we present the cluster formation procedure. We discuss the route path computation and sensor-actor coordination in Section 3. Performance evaluation and results are considered in Section 4. Finally, the paper is concluded in Section 5.

## 2   Hierarchical Configuration of Sensor Nodes

To provide a real-time coordination among sensors and actors in WSAN, $RCR$ configures the sensor nodes hierarchically. The operations of the routing pro-tocols are discussed in Section 3. The configuration of sensor nodes to achieve real-time coordination is discussed in this section. We propose so called Dy-namic Weighted Clustering Algorithm (DAWC). The operations of DAWC con-sist of cluster formation of sensor nodes, delay budget estimation for forwarding a packet from the cluster-heads and to guarantee the packet delivery within the given delay bound $\tau$.

There are many clustering algorithms [4], [7], [8], [9] proposed in the literature but unlike these studies, DAWC is neither periodical clustering procedure nor the cluster size is fixed in terms of hops. It adapts according to the dynamic topology of the sensor and actor networks. Our work is motivated from the previous work "A Weight Based Distributed Clustering" [8]. However, unlike [8], DAWC provides the cluster formation procedure to cope with the dynamic number of hops in a cluster and provides support for real-time routing. Cluster formation is based on the weighting equation formulated in the Section 2.2, which sets weight to different application parameters according to the applications need. DAWC adapts to the variation in the sensors field and can be optimized accordingly.

Once the cluster has been formed, cluster-heads get estimates of *delay budget*[1] of their member nodes. The delay budgets of member nodes help to build the delay-constrained energy efficient path.

When the sensor nodes are not uniformly deployed in the sensor field, the density of nodes could be different in different zones of the field. Choosing an optimal number of clusters $k_{opt}$, which yield high throughput but incur latency as low as possible, is an important design goal of DAWC.

## 2.1   Optimal Clustering

In this section, we evaluate the optimal number of clusters $k_{opt}$ and, hence, the optimal size of clusters. We assume the uniform deployment of sensor nodes and devise a formula to find out $k_{opt}$. Later, we see the implication of it to non-uniform deployment for optimal configuration in Section 2.1.

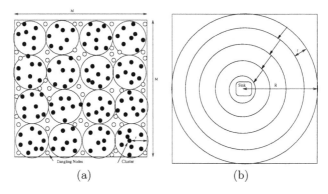

(a)                                (b)

**Fig. 1.** Network model to formulate the optimal clusters. Fig 1(a) represents the model to find the probability of DN nodes. Fig 1(b) illustrates the model of routing packets from cluster-heads to the sink node.

**Optimal Clusters in the Field.** Each member node transmits its data packet to the cluster-head. Let $r$ be the transmission radius of each node regardless of its functioning. In the clustering process, there is some probability that a number of dangling nodes[2] (DN ) may exist due to the density of nodes or coverage of the elected cluster-head. Let us first find out the probability of such nodes. To do that, we map the sensor field ($M \times M$) to non-overlapping circles of radius $r$ as shown in Fig 1(a) and assume that the nodes lying outside the boundary of the circle are DN nodes and the others are member nodes. These DN nodes require affiliating to cluster-head through the nodes insides the circle (member nodes). The squared field $M^2$ can be packed by $M^2/(2r)^2$ non-overlapping circles of radius r.

---

[1] *delay budget* is the time to deliver the data packet from the cluster-head to the member node.

[2] Nodes which have not joined any group or cluster are referred as *dangling nodes*.

Thus, the probability $\gamma$ of a multi-hop member is

$$\gamma = \frac{M^2}{(2r)^2} \times \frac{(2r)^2 - \pi r^2}{M^2} \approx 0.214$$

Let $E_{elec}$ be the energy consumed by the electronic circuitry in coding, modulation, filtration and spreading of the signal. Whereas, $\epsilon_{amp}r^2$ is the energy consumed for signal amplification over a short distance $r$. Thus, the energy consumed by each member node in transmitting a packet of size $l$ is

$$E_{Member} = l(E_{elec} + \epsilon_{amp}r^2(1 + \gamma))$$

The above equation can be simplified by taking the area as circle given in Eq. 16 of [15].

$$E_{Member} = l(E_{elec} + \epsilon_{amp}\frac{M^2(1 + \gamma)}{2\pi k})$$

Let us assume that the sensory field is covered by a circle of radius $R$, where the sink node lies at the center of this circle as shown in Fig 1(b). This assumption is made for routing packets from cluster-heads to the sink. The assumption is reasonable because it is less likely that a node lying outside the boundary of circle will be elected as cluster-head due to the low weight than the nodes inside the circle. Cluster-heads do not extend their transmission range to transmit packets directly to the sink node and, therefore, has the same radius $r$ as member nodes. We adapt the multi-hop model proposed by [17] to route packets from cluster-head to the sink.

In the model, a circle is divided into concentric rings with the thickness $r$. The energy spent to relay the packet from outside ring towards inside ring is $l(2E_{elec} + \epsilon_{amp}r^2)$. The number of hops $\Gamma$ require to route packet from cluster-head to sink node can be calculated by $\frac{R}{r}(1 - \hbar)$. Where $\hbar$ is the probability that the cluster-head is close enough to the sink to directly transmit packets. This probability can be calculated by using the nodes distribution in the rings given in [17].

$$\hbar = \frac{r}{R}\sum_{i=1}^{R/r}\frac{R^2 - (ir)^2}{M^2}$$

Since packets from the cluster-heads far from the sink node are relayed through intermediate nodes. Therefore, if $\Lambda(i)$ is the number of neighbors of a node $i$ then $\Lambda(i) \times E_{elec}$ is the energy consumed by the electronic circuitry of the neighbors of forwarding node $i$ during the propagation of packet. The number of neighbors ($\Lambda$) of any node can be found as

$$\Lambda = n\frac{\pi r^2}{M^2}$$

Hence, the energy consumed in routing data from cluster-head to sink is measured as

$$E_{CH-Sink} = l(\Lambda E_{elec} + E_{elec}$$
$$+ (2E_{elec} + \epsilon_{amp}r^2 + \Lambda E_{elec})\Gamma).$$

The total energy dissipated by the network is

$$
\begin{aligned}
E_{total} = l(&(n + n\Lambda)E_{elec} \\
&+ k(2E_{elec} + \epsilon_{amp}r^2 + \Lambda E_{elec})\Gamma \\
&+ n\epsilon_{amp}\frac{M^2(1+\gamma)}{2\pi k})
\end{aligned}
$$

For $r < R$, the optimum value of $k$ can be found by taking the derivative of above equation with respect to $k$ and equating to zero

$$
k_{opt} \approx \sqrt{\frac{n(1+\gamma)}{(2\pi(1 + \frac{2E_{elec}}{\epsilon_{amp}r^2} + \frac{\Lambda E_{elec}}{\epsilon_{amp}r^2}))\Gamma}} \times \frac{M}{r} \tag{1}
$$

It is noteworthy that the optimal value depends on the transmission range $r$ of the nodes. For long range of transmission, the value of optimal clusters $k_{opt}$ is small. This is in contrast to the optimal clustering in SEP [16] that is independent of range parameter. For example, Let us assume that $E_{elec} = 50nj/bit$ and $\epsilon_{amp} = 10pj/bit/m^2$ for experiments and $n = 100$, $M = 100$ with the sink at center of the field ($x = 50, y = 50$). Then the value of radius $R$ is obtained by drawing a circle at $x = 50, y = 50$ to cover the field. The estimated value is $R = 60$ and let set the range $r$ of individual nodes to 25. In this scenario, we obtain the value of $k_{opt} \approx 10$. By increasing the range of nodes to 40 meters, we obtain $k_{opt} \approx 7$. Whereas, the value of $k_{opt}$ in SEP [16] is 10 regardless of the transmission coverage of individual nodes.

**Optimal Cluster Size.** Besides choosing the optimal value $k_{opt}$ for number of clusters, the number of member nodes in a cluster is as important as the number of clusters. The optimal value of member nodes $M_{opt}$ helps in load balancing of clusters and ensures efficient MAC functioning. Head nodes use more energy than the member nodes. Since the sensors are energy-constrained devices and a cluster-head is selected from the homogeneous nodes, the number of member nodes in a cluster should ensure the longevity of the cluster-head as long as possible.

When the deployment is uniform then the $M_{opt}$ can be easily found by $n/k_{opt}$. However, for non-uniform deployment, the number of member nodes depends on the density in a particular zone of the sensor field. Therefore, we put the maximum and minimum limits $M_{Min}$ and $M_{Max}$ respectively on the size of cluster such that we still achieve $k_{opt}$ clusters in non-uniform deployment. Suppose $M_i$ is the number of neighboring nodes of any $i$th node and $Max(M_i)$ is the maximum number of neighboring nodes that any of the $i$th neighbor node have. We measure density of nodes in a particular zone by comparing the neighbor nodes $M_i$ with $M_{opt}$. We can conclude that the deployment is:

$$
M_i/M_{opt} > 1, dense
$$
$$
M_i/M_{opt} \approx 1, uniform
$$
$$
M_i/M_{opt} < 1, sparse
$$

We set the limits $M_{Min}$ and $M_{Max}$ as:

$$M_{Max} = Max(M_{opt}, Max(M_i))$$

That is, the maximum of $M_{opt}$ and maximum number of neighbors of any cluster-head at the time of cluster formation.

$$M_{Min} = M_{opt} \times Min(M_{opt}, Max(M_i))/M_{Max}$$

These limits allow the configuration to manage the dense as well as sparse deployment of nodes.

## 2.2 Cluster Formation

The first phase of DAWC is to form $k_{opt}$ number of clusters. During the formation of clusters, each cluster-head gets the *delay budget* of each of its member node. The delay budget is used to identify an appropriate node to send delay-constrained data packet. The cluster election procedure is based on calculating weight for each sensor node in the sensor field and it chooses the head that has the maximum weight. The weighting equation is given in cluster-head election procedure. We define weight threshold of the cluster-head to rotate the cluster-heads responsibility among all the potential nodes. A cluster is not strictly organized to 1-hop but it accepts the membership of a node that could not reach any cluster in the first phase of cluster formation. Therefore, a cluster can include n-hop members, for n ≥ 1. Although the operations of the protocol starts after the first phase of cluster formation, there may still exist some DN nodes.

We assume that the nodes are aware of their geographical locations through some localization devices like GPS. In the next section, we describe the details of computing the *delay budget* and cluster formation procedure is presented in Section 2.2.

**Delay Measurement.** When nodes are initially deployed in the field, every node $i$ broadcasts its ID, which is added in the neighbors list by all the nodes that receive this broadcast. A node that receives this broadcast, computes the delay $delay_s$ of the packet received from its neighbors along with the delay budget $delay_r$. $delay_s$ is the delay of the packet experienced and $delay_r$ is the delay that the sender estimated when some packet was received from the receiver i.e $delay_r : sender \leftarrow receiver, delay_s : sender \rightarrow receiver$.

The total delay in transmitting a packet from one node to a node in its neighbor is measured by the following factors: queue, MAC, propagation and receiving delay represented by $T_q$, $T_{Mac}$, $T_{Prop}$ and $T_{Rec}$ respectively. The wireless channel is asymmetric that does not imply any synchronization mechanism. Therefore, the delay is measured partially at both the sender and receiver. Sender measures the delay $L_s$ until the start of transmission that includes the queue delay as well as the MAC contention delay. Whereas, the receiver adds the factor $L_r$ as sum of propagation delay and receiving delay to get the total packet delay $L_h$.

$$L_s = T_q + T_{Mac}, L_r = T_{Prop} + T_{Rec}$$

The hop latency $L_h$ can be computed as sum of these factors:

$$L_h = L_s + L_r$$

Initially, the delay is measured by exchanging the *hello* beacons. Each node maintains this value in its neighborhood table that contains the fields $\{ID,$ $delay_r, delay_s, energy, weight\}$. To get more close to the accurate measurement of packet delay, the delay value is updated when the events flow from member nodes to cluster-head. It is due to the fact that the size of data packet may differ than the *hello* beacon that may experience different delay.

For the d-hop member of a cluster, packets are forwarded by the intermediate nodes to the cluster-head. Each intermediate nodes calculates the delay $L_h$ of the packet and forwards the packet to next hop by adding its $L_h$ in $L_s$. After following through some intermediate nodes, cluster-head gets the packet and adds its factor $L_r$ as the receiver of the packet. Hence the cumulative delay $delay_s$ of a member node $d$ hops away from its cluster-head is computed as:

$$delay_s = \sum_{i=1}^{d} L_{h_i} \tag{2}$$

Member nodes compute the delay estimate $delay_s$ of their cluster-head in this way through cluster-head *announcement* beacon. When member nodes broadcasts *hello* beacons, they put the $delay_s$ of cluster-head as $delay_r$ into the beacon. Cluster-head gets the delay budget $delay_r$ for its members and use this value in routing.

**Cluster-Head Election Procedure.** DAWC effectively combines the required system parameters with certain weighting factors to elect cluster-heads. Values of these factors can be chosen according to the application needs. For example power control is very important in CDMA-based networks. Thus, weight of the power factor can be made larger. In order to achieve the goal of energy saving, $RCR$ minimizes the frequency of clusters reformations. It is achieved by encouraging the current cluster-heads to remain cluster-heads as long as possible. That is why we have included the time of being cluster-head in computing weight. Similarly, if resource-rich devices are deployed to work as cluster-heads then the weighting factors of distance can be made large and time of being head can be kept small. The operation of the cluster-head election procedure is outlined as follows:

Each node $i$ maintains a list of its neighbors. Each entry of the neighbor list contains node ID and its weight $W_i$ computed on the basis of the selected parameters. Once the neighbor list is ready, the cluster-head election procedure is initiated for the first time. Each node $i$ does the following:

 – Calculates $D_i$ as the average distance to its neighbors, $M_i$ as the total number of its neighbors, $E_i$ as its energy and $T_i$ as the time being head in the past.
 – Computes weight $W_i = (c_1 D_i + c_2 E_i + c_3 M_i ) / c_4 T_i$, where the coefficients $c_1, c_2, c_3, c_4$ are the weighting factors for the corresponding parameters.

- Elects the node $i$ as the cluster-head if it has $M_i$ in the range of its minimum $M_{Min}$ and maximum $M_{Max}$ threshold of nodes and has the highest weight among its neighbors.
- Sets its threshold $W_{Th} = cW_i$, where c is the reduction factor to readjust the threshold.
- The cluster-head keeps computing its weight and when the weight goes down to its threshold $W_{Th}$, it triggers the cluster-head election procedure.

To save energy, we do not periodically reform clusters. In each round, cluster-head recomputes its weight and compares with its threshold value. If $W_i$ of cluster-head $i$ is higher than its $W_{Th}$ value then it keeps functioning as head. if $W_i < W_{Th}$ then it checks whether its $W_i$ is also lower than any of its member node weight. If so, it withdraws itself to function as cluster-head and cluster election procedure is initiated.

The pseudo-code of the operations executed by a sensor node in each round of cluster formation is reported in Algorithm 1.

---

**Algorithm 1.** Elect Cluster-head

---

1: Pseudo-code executed by each node $N$ in each round
2: $W_{max} = 0$
3: **for all** i in $\Lambda(n)$ **do**
4:　　**if** $W_{max} < W_i$ **then**
5:　　　　$W_{max} = W_i$
6:　　**end if**
7: **end for**
8: $W_i =$ my-weight()
9: **if** $status = NONE$ **then**
10:　　**if** $W_i > W_{max}$ **then**
11:　　　　announce-head()
12:　　　　$W_{th} = W_i \times c$
13:　　　　where $c$ is the threshold factor
14:　　**else if** $status = HEAD$ **then**
15:　　　　**if** $W_i < W_{th}$ **then**
16:　　　　　　**if** $W_i < W_{max}$ **then**
17:　　　　　　　　withdraw-head()
18:　　　　　　**else**
19:　　　　　　　　$W_{th} = W_i \times c$
20:　　　　　　**end if**
21:　　　　**end if**
22:　　**end if**
23: **end if**

---

If nodes could not join any cluster during the first phase then DAWC accommodates these DN nodes as follows. When a high weight node in a group of dangling nodes have the number of neighbor nodes smaller than the lower limit $M_{Min}$, it decreases this value locally by one and then retires three times. Each try is made during the periodic *hello* beacon. It continues until its $M_i$ becomes equal to $M_{Min}$ or it joins any cluster. If $M_i$ reaches to $M_{Min}$ then it announces itself as cluster-head and the other dangling nodes have chance to join this head. In this way, $M_{Min}$ is reduced to manage the sparse zone of sensors. While the $M_{Max}$ is set to disallow the nodes to make the cluster-heads overloaded in dense zone.

### 2.3   Neighboring Cluster Discovery

The sink or actors can be multi-hop away from the source clusters, packets are then forwarded through intermediate clusters. Clusters are linked with each other to provide multi-hop cluster routing. Some member nodes within a cluster can hear the members of neighboring clusters or heads, such nodes act as *gateways*. It is also possible that there will be multiple *gateways* between two clusters. Cluster-heads keep record of all of these *gateways*.

We build a set of forwarding gateway nodes $GS$, for each cluster-head, for routing packets to neighboring clusters. Let $SM_i$ be the set of members of cluster-head $H_i$ and $SM_j$ be the set of members of neighboring cluster-head $H_j$. $H_i$ maintains a set of gateway nodes $GS_i$ such that

$$GSi(Hi) = \{x \in SM_i / H_j \in \Lambda(x) \quad \forall y \in SM_j \wedge y \in \Lambda(x) \quad \forall i \neq j\}$$

Where $\Lambda(x)$ is the set of neighbors of node $x$. A member node $x$ of cluster-head $H_i$ belongs to the gateway set $GS_i$ of head $H_i$ if either $H_j$ or some member $y$ of $H_j$ exists in the neighbors set of $x$. The attributes of the elements of $GS_i$ are $\{AdjacentHead, Energy, Delay, Hops\}$. These attributes help the cluster-heads in selecting a particular item from the set $GS$. We will describe the selection criteria in detail in Section 3.3.

Once the cluster formation is complete, each cluster gets the neighbor clusters list along with the gateways to reach them. The route computation is discussed in the next section.

## 3   Delay-Constrained Energy Aware Routing (DEAR)

The main aim of $RCR$ framework is to provide real-time routing in WSAN with least energy consumption. $RCR$ achieves this by clustering sensors hierarchically and then selecting the path on the basis of end-to-end (E2E) deadline ($\tau$) as well as balanced energy consumption. We propose a delay-constrained energy aware routing (DEAR) algorithm to deliver packets from the source clusters to the target nodes (Sink/Actors) in WSAN. A similar idea of delay-constrained least cost routing has been proposed in [18], [19]. Unlike these protocols, we have combined the forward-tracking and back-tracking approach to reduce the cost of path establishment. We establish a distributed single path, in which cluster-head selects the outgoing link such that the packet deadline is meet with efficient energy consumption. An energy efficient link does not merely mean the low cost link but a link that can satisfy the delay constraint and it balances the energy consumption on all the outgoing links.

## 3.1   Network Model

Before going into the details of the algorithm, we model the network as a connected directed graph $G = (V, E)$. The set of vertices $V$ represents the sensor nodes, where $|V| = n$. $E$ is the set of directed edges such that an edge $e(u \rightarrow v) \in E$ if $(u, v) \in V$. Two non-negative real value functions R(e), the available energy resource of node $v \in V$ on the outgoing link $e(u \rightarrow v) \in E$, and $\Delta(e)$, the delay experienced by the data packet on the corresponding link, are associated with the edges. These real values are used to compute the weight W(u,v) of the link $e(u \rightarrow v) \in E \vee (u, v) \in V$. The weight of an edge $e(u \rightarrow v) \in E$ can be defined as follows:

$$W(u, v) = R(e)/\Delta(e), \quad where \quad u, v \in V$$

Links are presumably asymmetrical because the R(e) and $\Delta(e)$ for the link $e(u \rightarrow v)$ may not be same while going in the opposite direction of this link $e(v \rightarrow v)$. The existence of alternative paths between a pair of vertices $u, v \in V$ provides the possibility of some paths being shorter than others in terms of their associated cost. We need to find out a minimum spanning acyclic subgraph of G having high total weight.

Let $s$ be a source node and $d$ be a destination node, a set of links $e_1 = (s, v_2), e_2 = (v_2, v_3), ..., e_j = (v_j, d)$ constitutes a directed path P(s,d) from $s \rightarrow d$. The weight of this path is given as follows:

$$W[P(s, d)] = \sum_{e \in P(s,d)} W(e)$$

Likewise, the E2E delay experienced by following the path P(s,d) is measured as:

$$\Delta[P(s, d)] = \sum_{e \in P(s,d)} \Delta(e)$$

After the formation of clusters, we can have a vertices subset $H$ of the set $V$ such that the elements in $H$ are only the cluster-heads and has an associated integral function $hops[P(h \rightarrow target)], h \in H$. Similarly, we obtain the set $GS_h \quad \forall h \in H$ as the result of linking the clusters described in Section 2.3. Each element $h$ of set $H$ maintains a set of outgoing links $OUT_h$ subset of $GS_h$ to the single destination node either sink or actor. In the next section, we describe the way of building the set $OUT_h \quad \forall h \in H$.

## 3.2   Sensor-Actor Coordination

The main communication paradigm in WSANs is based on the effective sensor-actor coordination. Right actions against the detected events cannot be performed unless event information is transmitted from sensors to actors. Therefore, the ultimate goal of any routing protocol in WSANs is to relay the event readings to the actors within a certain delay limit. In the classical *semi-automated architecture*, there is a central node that is responsible to collect the readings and issue action commands to the actors responsible for the action. Unlike this approach, *automated architecture* has also been realized due to the need of immediate action on the phenomena observed in the sensory field. In the former approach, sink is the destination of events reported by all the sources and is responsible to coordinate with actors. In the latter case, the mobile actors in *automated architecture* are the targets of the event readings observed by the sensor nodes and, hence, the coordination is local.

In order to compute the delay-constrained paths efficiently, we decompose G into a minimized acyclic subgraph $\bar{G} = (\bar{V}, \bar{E})$ constituting a large acyclic region within G. $\bar{V}$ is the set of nodes either in $H$ or belong to the $GS$ sets of cluster-heads i.e. $\bar{V} = H \cup GS_1 \cup GS_2... \cup GS_k$ for $k$ number of clusters. $\bar{E}$ is the set of directed edges such that an edge $\bar{e}(u \to v) \in \bar{E}$ if $u, v \in \bar{V}$. The length of an edge $\bar{e}(u \to v) \in \bar{E}$ may be greater than one because the members in $GS$ may be multi-hop far from heads. For instance, an edge $\bar{e}(u \to v) \in \bar{E}$ might exist due to some member node $w$ such that $u \to w \to v), w \notin \bar{V}, (u, v) \in \bar{V}$. Here, $R(\bar{e})$ is the least available energy of any node visited while traversing the link $\bar{e}(u \to v)$ and $\Delta(\bar{e})$ is the cumulative delay experienced by the data packet on the corresponding link.

The decomposed minimized graph $\bar{G}$ is the backbone to establish the route from the source nodes to either the sink (*semi-automated architecture*) or the actor (*automated architecture*). In the next section, we look into the formation of the graph $\bar{G}$.

**Centralized DEAR (C-DEAR).** In this section, we deal with the centralized *semi-automated architecture*. The sink node is stationary like sensor nodes and the path from cluster-heads to sink is built in proactive way. Sink is the destination for all the source nodes in *semi-automated architecture*. Source to sink path is divided into two phases; source to cluster head and cluster-head to sink. The first phase builds the path from source nodes to cluster-head that is done during the cluster formation in a forward tracking manner. The next phase deals with finding the path from cluster-heads to the sink using backtracking. It is activated initially by the sink during the network configuration phase and is updated periodically. To achieve this, the algorithm visits the graph G and marks all the vertices $h \in H$. A mark is associated with the life of the node, which is deleted as that vertice(node) expires. A vertex can be marked if $h \in H$ has not been already marked or the current path delay $\Delta[P(sink, h)]$ is less than the previously observed path delay. Once all the elements $h$ of set $H$ are marked, we build a path $P(sink, h)$ $\forall h \in H$ in proactive fashion and each element $h \in H$ set its $hops[P(sink, h)] = |P(sink, h)|$.

When $h$ is marked, $h$ adds the incoming link $in(x \to h), x \in V$ to the set $OUT_h$ in reverse-topological order $out(x \to sink)$. The incoming link $in$ may be associated with the last marked element $g \in H$ in the marking process or *null* if $h$ is the first marked item and represents link to the root (sink node). This helps $h$ to extend the set $OUT_h$ by using the pre-determined set $GS_h$. The attributes of the elements of $GS$ set contains the $AdjacentHead$ ID that corresponds to $g$.

For each element $o(m \rightarrow g) \in GS_h$, it searches for the match of $g$ with the attribute *AdjacentHead* of $o$. If there exists such element $o(m \rightarrow g)$ then $h$ adds the link as $o(m \rightarrow g)$ to $OUT_h$ and associate an integral value $H(o)$ apart from the other two real value functions $R(o)$ and $\Delta(o)$. Hence, the edges set $\bar{E}$ of $\bar{G}$ can be obtained as $OUT_1 \cup OUT_2, ..., \cup OUT_k$ for $k$ number of clusters. Fig 2 illustrates the decomposed subgraph $\bar{G}$ with all the possible links to the sink node. We use the term link for set $\bar{E}$ rather than edge because vertices of set $\bar{V}$ may be connected by some intermediate vertices in V. The set $OUT_h$ provides all the possible routes to the sink node and we exploit the multiple entries in $OUT_h$ to provide delay-constrained energy aware routes and implicit congestion control. We describe the criteria of selecting the outgoing link in Section 3.3. The cost of marking process is $O(n)$ and, in fact, it is the actual cost of building route from source nodes to the sink node.

**Implementation.** The marking process is implemented by broadcasting sink *presence* beacon in the network. That is, sink initiates the connection with the sensor nodes by broadcasting its *presence* beacon periodically, where the length of period (life of mark) is larger than the *hello* beacon. This periodic beacon helps to refresh the path because the topology of the sensor nodes is dynamic. A receiving node accepts this beacon if it meets one of the following conditions:

1. It has not already received this beacon or beacon has expired.
2. Delay of this beacon is smaller than the last received beacon.
3. The number of hops traversed by this beacon are small.

When a node receives a packet it calculates the delay and forwards the request in the direction of cluster-head. While the cluster-head forwards it to its neighboring cluster-head. Hence, each cluster-head learns the loop free path to the sink node and gets the delay value and number of hops so far.

**Distributed DEAR (D-DEAR).** The distributed event routing approach is imperative due to the non-existence of central controller. Events detected by the sensor nodes are directly routed to the actor nodes without the intervention of the sink node. To provide the distributed routing in the *automated architecture*, RCR proposes the distributed flavor of DEAR. In D-DEAR, we decompose the graph G into the $m$ number of $\bar{G}$ subgraphs for each of $m$ mobile actors. The idea is similar to C-DEAR described in detail in Section 3.2 except that we have $m$ possible destinations. The marking process is triggered independently by all the $m$ actors to construct $m$ number of $\bar{G}$ representing the paths $P(h, actor_1), (h, actor_2), ..., (h, actor_m) \quad \forall h \in H$. The cost of D-DEAR is $O(mn)$.

In order to optimize the sensor-actor coordination in the distributed environment, the marking process also propagates the current *load* factor of the actor. The *load* represents the number of sources the actor is serving at the moment. The marking criteria in D-DEAR is modified such that $h$ accepts the mark of an actor on the basis of its Eculidian distance. The nearest one is the best candidate for marking the element $h$ of the set H. There might be the possibility that two or more actors reside at the same distance to $h$. In such case, *load* factor breaks the ties among such candidates and less-loaded actor is the winner.

Actors are location aware mobile nodes. Whenever an actor moves, it triggers the construction of graph $\bar{G}$ in addition to the periodic reconstruction of graphs. The periodic update of graphs is required due to the highly dynamic topology of the wireless sensor and actor networks because sensor nodes may be deployed at any time or their energy deplete. Hence, the algorithm updates the path proactively to reduce the chances of path failure like the path establishment.

### 3.3   Alternative Path Selection

Power efficiency has always been an important consideration in sensor networks. Whereas, E2E deadline $\tau$ is another constraint for real-time applications in wireless sensor and actor networks. Real-time event delivery is the main aim of our distributed routing protocol. We have described the process of building the set of outgoing links $OUT$. The selection of a particular link $o(m \rightarrow g) \in OUT_h, g \in H$ by the cluster-head $h \in H$ is based on the criteria to balance the load in terms of delay and energy of its member nodes. The operations of the alternative gateway selection are outlined in the Algorithm 2.

---

**Algorithm 2.** Select Outgoing Link

---

**Ensure:** Delay-constrained energy aware outgoing link $out \in OUT_h$
1: Pseudo-code executed by source cluster-head $h$ to select an outgoing link from the set $OUT_h$.
2: $P = \infty$
3: **for all** $o(m \rightarrow g) \in OUT_h, g \in H$ **do**
4:    **if** $time_{left}/hops[P(sink, s)] < \Delta(o)$ **then**
5:       **if** $R(o) < P$ **then**
6:          $P = R(n)$
7:          $out = o$
8:       **end if**
9:    **end if**
10: **end for**

---

Cluster-head adds the $time_{left}$ field to its data packet that is set to $\tau$ by the source cluster-head. Each intermediate cluster-head looks for this $time_{left}$ field and selects the outgoing link accordingly by executing the above procedure. If the delay constraint can be meet through multiple links then it selects the one according to the criteria as described below:

*"The link, along which the minimum power available (PA) of any node is larger than the minimum PA of a node in any other links, is preferred".*

Every receiving node then updates the $time_{left}$ field as $time_{left} = time_{left} - delay_s$. It can be seen that the link selection criteria implicitly eliminates the congestion by alternating the links towards destination. Whenever a link is congested, the packet delay is increased and this delay is reported to the cluster-head in successive *hello* beacon. The weight of this link is reduced and, eventually, the cluster-head reacts to it by selecting the other available link. Hence, the congestion is avoided in addition to the energy efficiency.

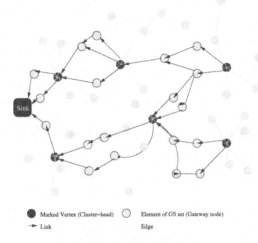

**Fig. 2.** Decomposition of graph G into the minimized acyclic subgraph $\bar{G}$ within the region G

## 4    Performance Evaluation

The performance of $RCR$ is evaluated by using the network simulator ns-2 [20]. The example scenario consists of a sink node, three actors and 100 sensors randomly deployed in $200 \times 200$ meter square area. Three sensor agents developed by NRL are also placed to trigger phenomenon at the rate of 2 events per second each.

The main aim of our proposed framework is to deliver the events triggered in the sensors field to the actors within the given delay bound $\tau$. Fig. 3 illustrates the average delay against the application deadline for the four different configurations; *direct semi-automated*, *indirect semi-automated*, *static-actors automated* and *mobile-actors automated* architecture. The mobility pattern of mobile actors is random walk. The major factor in missed-deadline is mobility of actors as clear by the mobile actors graph in Fig. 3.

Deadlines miss-ratio is an important metric in real-time systems. We measure the miss-ratio for different values of $\tau$. Fig. 4 represents the evaluation of deadlines miss-ratio for all the configurations. The miss-ratio reaches to 0 when $\tau \approx 50ms$ for *direct semi-automated* and *static-actors automated*. This value of $\tau$ can not be set for the other configurations because there is still a significant miss-ratio.

Besides the mobility of actors, network configuration and events congestion are also the factors of missed-deadlines. It happens when the neighbor cluster of actor itself has detected events and at the same time the other clusters

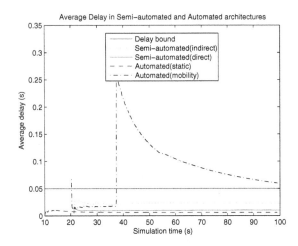

**Fig. 3.** Average delay in *semi-automated architecture* vs *automated architecture*

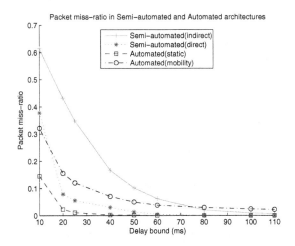

**Fig. 4.** Deadlines Miss-ratio in *semi-automated architecture* and *automated architecture* for $\tau = 10 - 110ms$

are also sending their readings for the same actor through this neighbor cluster. It is possible that the sensors start detecting events while the network is in the configuration state. This scenario not only causes the loss of packets but missed-deadlines as well. Fig. 5(a) represents the delay graph of *semi-automated architecture*. The events start occurring before the network is configured. There are missed-deadlines due to the non-configured network at the beginning of the graph. The same scenario for the *automated architecture* is shown in Fig. 5(b).

Although the missed-deadlines are very small in Fig. 5(b) but the peaks are initially higher than the peaks in stable configuration and causes source of jitter

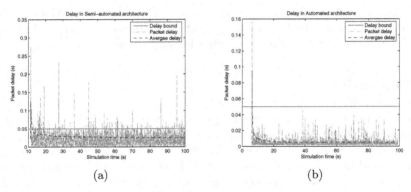

(a)                                    (b)

**Fig. 5.** Event to actor routing starting before configuration of 100 sensor and 3 actor nodes ($\tau = 0.05sec$)

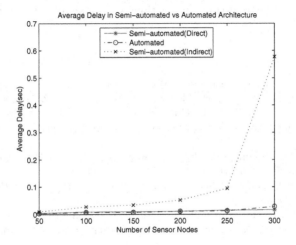

**Fig. 6.** Average delay by increasing number of nodes

in the traffic. The *automated architecture* requires lesser configuration time[3] than the *semi-automated architecture* and therefore, the packet delays are not much affected during the network configuration state. In Fig. 5(b), actors start receiving events at about 6 sec. This value is 12 sec. in the *semi-automated architecture*.

The scalability of the proposed framework is evaluated by increasing the number of sensor nodes in the field. Fig. 6 shows the results up to 300 nodes. The average packet delay in *indirect semi-automated architecture* increases significantly with the number of nodes as compared to the other configurations. The other two configurations (*direct semi-automated architecture* and *static-actors automated architecture*) are not much affected by deploying more sensors. The

---

[3] Configuration time is the time to form clusters, link them and to find out the actors for sending event readings. The network is stable after the configuration.

reason is apparent because the sink node is the only node that provides coordination among clusters and actors in indirect configuration. Whereas, the other two configurations provide direct coordination among sensors and actors.

## 5   Conclusion

There have been a number of routing protocols developed for WSN that claim to provide real-time routing and congestion control. However, none of them has considered the presence of actors. *RCR* addresses the issues in such heterogeneous network and provides a coordination framework for wireless sensor and actor networks. It clusters the sensors and selects a head in each cluster to minimize the energy consumption, estimates the delay budget and makes routing decisions. The communication framework works fine in *semi-automated architecture* as well as *automated architecture*. Simulation results show that *RCR* meets the E2E deadlines for real-time applications with a small value of miss-ratio.

## References

1. I. F. Akyildiz and I. H. Kasimoglu, "Wireless Sensor and Actor Networks: Research Challenges," *Ad Hoc Networks, Vol. 2, Issue 4, pp. 351-367, October 2004.*
2. T. Melodia, D. Pompili, V. C. Gungor and I. F. Akyildiz, "A Distributed Coordination Framework for Wireless Sensor and Actor Networks," *ACM MobiHoc'05, May 2005.*
3. K. Akkaya, M. Younis, "An Energy-Aware QoS Routing Protocol for Wireless Sensor Networks," *23rd ICDCSW, pp. 710-715, 2003.*
4. S. Basagni, "Distributed Clustering for Ad Hoc Networks," *IEEE Proc. Vehicular Technology Conference, 1999.*
5. D. Tian and N.D. Georganas, "Energy Efficient Routing with Guaranteed Delivery in Wireless Sensor Networks," *Mobile Computing and Communications Review, 2001.*
6. Y. Yu, R. Govindan and D. Estrin, "Geographical and Energy Aware Routing (GEAR): a recursive dissemination protocol for wireless sensor networks," *UCLA/CSD-TR-01-0023, Tech. Rep., 2001.*
7. M. J. Handy, M. Haase and D. Timmermann, "Low Energy Adaptive Clustering Hierarchy with Deterministic Cluster-Head Selection," *IEEE International Conference on Mobile and Wireless, 2002.*
8. M. Chatterjee, S.K. Das and D. Turgut, "A Weight Based Distributed Clustering Algorithm for Mobile Ad hoc Networks," *HiPC 2000: pp. 511-521.*
9. A.D. Amis, R. Prakash, T.H.P. Vuong and D.T. Huynh, "Max-Min-D-Cluster Formation in Wireless Ad hoc Networks," *Proc. IEEE Infocom 2000, Tel Aviv, Israel, March 2000.*
10. C. Intanagonwiwat, R. Govindan and D. Estrin, "Directed Diffusion: A Scalable and Robust Communication Paradigm for Sensor Networks," *Proc. ACM MobiCom, March 2000.*
11. T. He, J.A. Stankovic, C. Lu and T. Abdelzaher, "A Real-Time Routing Protocol for Sensor Networks," *Proc. IEEE International Conference on Distributed Computing Systems, May 2003.*

12. V. Rodoplu and T.H. meng, "Minimum Energy Mobile Wireless Networks," *IEEE JSAC, August 1999.*

13. W. R. Heinzelman, A. Chandrakasan and H. Balakrishnan, "Energy Efficient Communication Protocol for Wireless Microsensor Networks," *IEEE Proc. Hawaii International Conf. Sys. Sci. January 2000.*

14. W. R. Heinzelman, J. Kulik and H. Balakrishnan, "Adaptive Protocol for Information Dissemination in Wireless Sensor Networks," *Proc. ACM MobiCom, 1999.*

15. W. R. Heinzelman, A. Chandrakasan and H. Balakrishnan, "An application-specific protocol architecture for wireless microsensor networks," *IEEE Transactions on Wireless Communications, vol. 1, no. 4, pp. 660-670, October 2002.*

16. G. Smaragdakis, I. Matta and A. Bestavros "SEP: A Stable Election Protocol for clustered hetrogenous wireless sensor networks," *2nd International Workshop on Sensor and Actor Network Protocols and Applications (SANPA 2004).*

17. V. Mhatre and C. Rosenberg, "Homogeneous vs. heterogeneous clustered sensor networks: A comparative study," *In Proc. of 2004 IEEE International Conference on Communications (ICC 2004), June 2004.*

18. Q. Sun and H. Langendörfer, "A new distributed routing algorithm for supporting delay-sensitive applications," *Journal of Computer Communications vol. 9 no. 6 May 1998.*

19. R. Gawlick, A Kamath, S. Poltkin and K. G. ramkrishnan "Routing and admission control in general topology networks," *Tech. Report STAN-CS-TR-95-1548, Stanford University, CA, 1995.*

20. UC Berkeley, LBL, USC/ISI and Xerox PARC, "The Network Simulator ns-2," *The VINT Project, http://www.isi.edu/nsnam/ns/.*

# Reliable Data Delivery in Wireless Sensor Networks: An Energy-Efficient, Cluster-Based Approach

Shan-Shan Li, Pei-Dong Zhu, Xiang-Ke Liao, and Qiang Fu

Doctor Team, School of Computer, National University of Defense Technology,
ChangSha, China, 410073
{shanshanli, pdzhu, xkliao, samon.fu}@nudt.edu.cn

**Abstract.** Reliable data delivery in wireless sensor networks is a challenging issue due to the error prone nature of wireless communication and the frangibility of little sensors. However, readings of geographically proximate sensors are strongly correlated, so it's not only unnecessary but also wastes energy to deliver data in all sensors reliably. Sensor nodes could be clustered and only the aggregated data need to be reliably delivered. In this paper, we propose a new reliable, energy-efficient clustering protocol CREED. CREED takes advantage of the multi-rate capabilities of 802.11 a, b, g technologies by reducing energy consumption per bit for shorter transmission range. A dynamic backup scheme EDDS based on nodes' residual energy is presented to enhance the fault tolerance of CREED. EDDS supports quick cluster head selection and failure recovery while only need exchange a few messages. CREED also includes an energy aware multipath inter-cluster routing to guarantee reliable data delivery. Simulation results verify the performance of CREED.

## 1 Introduction

Wireless sensor networks for monitoring environments have recently been the focus of many research efforts and emerged as an important new area in wireless technology [1][2][3]. In a wireless sensor network, a large number of tiny, sensor-equipped and battery-power nodes are often deployed in inhospitable environments and communicate with each other using low data-rate radio. Consequently, sensor networks are subject to high fault rate due to energy constraint, environment noise or malicious destruction. Therefore, some approaches should be taken such as multipath or retransmission in order to guarantee data delivery in the *desired reliability* [4-11]. The essence of these measures is to achieve reliability at the cost of redundancy.

Another critical nature of sensor networks is that reporting of all sensors may generate lots of redundant data because nodes are largely deployed. Furthermore, readings of geographically proximate sensors are strongly correlated. So it's not only unnecessary but also wastes energy to deliver data in all sensors reliably. Clustering is a self-organization technique to optimize energy consuming [12-18]. Nodes can send their data to a central cluster head that can do some data aggregation and then forwards them to the desired recipient. With clustering, only the aggregated data should be reliably delivered without losing fidelity, we can also prolong the network lifetime by reducing channel contention and give nodes the opportunity to sleep while not sensing. The added overhead is the formation and maintenance of clusters.

Y. Koucheryavy, J. Harju, and V.B. Iversen (Eds.): NEW2AN 2006, LNCS 4003, pp. 384–395, 2006.
© Springer-Verlag Berlin Heidelberg 2006

In this paper, we proposed a Cluster based protocol to support Reliable and Energy Efficient data delivery that is completely Distributed (CREED). The formation of cluster is based on a new accurate energy model that breaks down the traditional energy opinion. Cluster heads are selected according to their residual energy.In order to prevent network partitioning due to sudden failures or being maliciously attacked of cluster heads, new backup nodes are always ready instead of having to wait to the next round and resulting in the halting of aggregation and communication. An inter-cluster routing protocol is also proposed to choose multiple energy-efficient and robust paths, aiming for the desired reliability with the minimum number of path.

The rest of this paper is organized as follows. In Section 2, we briefly describe some related work. The communication energy model that CREED is based on is illustrated in Section 3. Section 4 gives an overview of several reliability techniques, especially emphasizing on the multipath routing. In Section 5, we present the detail of CREED protocol, including the intra-cluster and inter-cluster routing and the corresponding reliability scheme. Several simulations described in Section 6 verify the obtained theoretical results. We conclude the paper in section 7.

## 2   Related Work

Much of the existing work related to reliable data delivery in sensor networks deals with multipath and retransmission techniques. In [4], multiple paths from source to sink are used in diffusion routing framework [11] to quickly recover from path failure. The multiple paths provided by such protocols could be used for sending the multiple copies of each packet. [5] proposes a multipath-multipacket forwarding protocol for delivering data packets at required reliability based on data priority using a probabilistic flooding scheme. [6] presents a lightweight end-to-end protocol called ReInForM to deliver packets at desired reliability. Memory less nature of ReInForM is suitable for memory constrained sensor nodes. Similarly, different multipath extensions to well known routing algorithms have been proposed [7, 8].

[9] addresses the problem of efficient information assurance at different assurance levels. It figures out that if nodes have caching capability, then the overhead of end-to-end schemes [6] is significantly higher than hop-by-hop retransmission. However, hop-by-hop retransmission would lead to longer delay. RMST [10] uses retransmission to deliver larger blocks of data in multiple segments. It checks the cache periodically for missing segments and generates NACK messages to ask for retransmission. All techniques discussed above are based on the flat architecture, which has poor scalability and doesn't take advantage of the data aggregation.

A hierarchical network organization will not only facilitate data aggregation, but also help to form the uniform, robust backbones, and give an opportunity for nodes to sleep while not sensing or transmitting. LEACH [12] is an application-specific data dissemination protocol that uses clustering to prolong network lifetime. It does not offer a complete energy optimization solution, as it has no strategy for specifying cluster heads position and distribution. HEED [13] improves LEACH in its generality and energy efficiency. HEED makes no assumption on energy consumption and selects well distributed cluster heads and considers a hybrid of energy and communication cost. In [14], the authors use LEACH-like randomized clustering, but they provide methods to compute the optimal values of the algorithm parameters at prior. [15] develops a clustering protocol DEEP that manages the communication of data while minimizing energy consumption across sensor networks. DEEP starts with an initial cluster head and gradually forms clusters throughout the network by controlling the dimension of clusters and the geographical distribution of cluster heads, which results in a balanced load among cluster heads. [16] presents a new approach for sensor applications that requires coverage for a large area, stationary environment.

To our best knowledge, there is primarily one protocol REED [17] which constructs a cluster head overlay to withstand unexpected failures or attacks on cluster heads. Although REED can be easily incorporated into any clustering protocol, it needs to interleave the construction of k cluster head overlays that seems to be time-consuming and burdensome. In CREED protocol, backup cluster heads and multiple paths are constructed during the formation of clusters and route setup, which need few additional operation and easy to be maintained.

## 3 Energy Model

Traditional opinion often assumed that a reduction of the transmitting (or radiated) energy yields a proportional reduction of total energy consumption. Energy consuming will increase exponentially with the increase of transmission distance. So much work has been done on controlling the hop distance for energy efficiency [19] or analyzing systematic cost based on this energy model [20]. Actually, even without taking into account reception energy, this is not true for any practical power amplifier. In general, all transceiver energy components are summarized as:

$$E(Joule) = \theta + \alpha d^n \qquad (1)$$

Where $\theta$ is a distance-independent term that accounts for the local oscillators overhead, $\alpha$ represents the amplifier factor, d is the transmission distance, n could be a number between 2 and 4. The distance-independent part $\theta$ will dominate the energy consumption [21,22]. Some measurement presented in [22] shows that for the MicaZ and Mica2 hardware, the total power consumption is almost independent of the transmit power. The fraction of power actually transmitted ranges from less than 0.1 percent at the lowest power setting to 1.4 percent for MicaZ and 14 percent for Mica2 at their respective highest output power levels. Therefore, short-hop routing does not yield any substantial energy benefit if more distant relay node can be reached with sufficient reliability.

On the other hand, the rapid advancements in the area of wireless local area networks have led to the development of today's popular IEEE standard such as 802.11a, b, g and many advanced microchips. These standards support multi-rates for different transmission ranges [23]. Table 1 shows the expected data rate for 802.11g wireless technologies. The Atheros 2004 tri-mod chipset [24] is used to monitor the real values for the radio hardware. Table 2 shows its energy model which verify again that despite the fact that the path attenuation energy increases exponentially by the transmission distance, $\theta$ dominates the path loss and therefore,

**Table 1.** Expected data rate of 802.11g technology [23]

| Rate (Mbps) | Max. Range | Rate (Mbps) | Max. Range |
|---|---|---|---|
| 1 | 100m | 18 | 51m |
| 2 | 76.5m | 24 | 41.25m |
| 6 | 64.5m | 36 | 36m |
| 9 | 57m | 48 | 23.1m |
| 12 | 54m | 54 | 18.75m |

**Table 2.** Energy consumption parameters for Atheros 2004 tri-mod chipset [24]

| Mode | Max Output Power, $\alpha d^{2}$ (dBm) | Total Power Consumption (W) | $\theta$ (Watt) | $\eta_{max} \times \alpha d^{n}$ (Watt) |
|---|---|---|---|---|
| 802.11a | +14 | 1.85 / 1.20 | 2.987 | 0.0625 |
| 802.11b | +21 | 1.75 / 1.29 | 2.727 | 0.3125 |
| 802.11c | +14 | 1.82 / 1.40 | 3.157 | 0.0625 |

causes the total power consumption to remain constant as transmission distance increases. Hence we can exploit the multi-rate capabilities of 802.11g and the energy model to decrease the energy consumption for smaller distances by switching to higher data rates and reduces the energy needed per bit as distance shrinks:

$$E(Joule / bit) = \frac{\theta + \alpha d^{n}}{Rate} \tag{2}$$

[15] has presented the energy consumption of Atheros2004 radio for 802.11g technology. It illustrates that single rate communication energy consumption remains constant as transmission distance increases while communication energy consumption for multi-rate transmission decreases for shorter transmission ranges. Therefore, direct transmission consumes less energy than multi-hop transmission when the range is not large since more nodes will take part in transmission. We can compute the most efficient direct transmission range 40 m based on table 1 and table 2. For ranges greater than this distance, minimum energy 2–hop path will be more energy efficient. [15] also made some experiments to get the optimum *d* for direct transmission (37m). In the section 5, we will cluster the network based on this *d*.

# 4  Multipath Routing

Retransmission and multipath are two major techniques to achieve reliability. Retransmission computes the expected number of transmissions. This technique will be used when reliable delivery with low communication overhead is more important than latency in delivery. Multipath Routing allows the establishment of multiple paths between source and destination, which provides an easy mechanism to increase the likelihood of reliable data delivery by sending multiple copies of data along different paths. By introducing such redundancy, the system can compensate for data losses due to local channel error. Multipath routing does not require retransmission so the latency in data delivery would be significantly lower. Hence this approach would be better if the latency in data delivery is a more important factor than communication overhead.

Since applications that need high reliability also demand real time at the same time, we use multipath as a basis to introduce reliability to the inter-cluster routing of CREED. Some routing protocol use primary path to transmit data and fall back to other backup routes when the primary breaks down. In this case, route disruption may lead to long delays at the routing layer, which affects the QoS for delay sensitive applications. Consequently, all data packets are transmitted along the multipath at the same time instead of only using the alternate path when the primary path fails.

## 5   The CREED Protocol

### 5.1   Cluster Formation

Here we want to use single hop routing for intra-cluster data transmission because periodical change of cluster heads will make it burdensome to construct and maintain the multi-hop path. CREED clusters the network based on the distance $d$ discussed in section 3 for the sake of energy efficiency. The intuition is to make the distance between any two nodes in the same cluster not more than $d$, therefore, wherever the cluster head is, the frequent control signal transmission and extra power consumption associated with the head could be avoided. We assume that all nodes are location aware (from GPS). CREED partitioned the network into several grids. Each node uses location information to associate itself with a grid. The diagonal length of a gird is equal to $d$, the relation between the border length $a$ and $d$ is showed in figure 1. Therefore, the distance between any two nodes in one cluster is less than $d$. [15] puts all cluster heads in the centre of the corresponding clusters that result

**Fig. 1.** Grid

$$2a^2 = d \tag{3}$$

in a balanced load among cluster heads. But this requires flooding to choose new cluster heads in every re-clustering round.Even if re-clustering is not so frequent, more and more black holes will be formed since DEEP rotate the cluster head position after current cluster heads are out of energy. Furthermore, [15] doesn't consider any reliability factor.

In general, the closer the sensor nodes are from each other, the more strongly their sensed data are correlated. Therefore, aggregation for adjacent nodes has high cost efficiency. If cluster has too wide range, not only aggregation cannot be well utilized, but also the cluster heads will be with heavy burden. That is to say, cluster size should **NOT** be enlarged with the scale-up of the networks. If the scale of network is too large, hierarchical clustering should be used. This change will make CREED suitable for a wider range of wireless sensor networks.

### 5.2   Cluster Head Selection

Since cluster is fixed due to a designated $a$, cluster member is also fixed. In order to prolong the network lifetime as much as possible, each node in one cluster should be selected as head in turn. Data collected by cluster head are first aggregated and then forward to the sink. This will save much energy while transmission reliability is degraded due to decrease of data redundancy. If some cluster head encounters unexpected failure, data from the whole cluster would be lost. The network has been partitioned for a long time until the next round cluster head selection. CREED considers this issue and uses an Energy Directed Dynamic Sorting backup scheme (EDDS) to support rapid recovery of cluster head failures. Furthermore, unlike LEACH [12] and DEEP[15], EDDS does not assume equivalent energy level for all sensors and uniform energy consumption for cluster heads. EDDS execution can be summarized as follows. By default, nodes referred to below belong to one cluster:

1. At the initialization, all nodes broadcast to exchange their energy information.
2. Each node first sorts the received energy information in descending order, then broadcast the energy list in order to avoid that some node may receive imcomplete information.
3. The node which has the highest energy (on the top of the energy list) becomes cluster head and send cluster head declaration message (DM).
4. If some node hasn't heard any DM in a given time $T_d$, it sends inquiring message(IM). Any node who receives both IM and DM will reply the head information.
5. If it still hasn't received any reply information in another given time $T_i$ after $T_d$, it first deletes the top node in the energy list (which means that node may not work now) and sends reselecting message (RM) to the sub-optimal node that are currently on the top of the energy list. $T_i$ and $T_d$ are set properly so that if the node hasn't received any reply, other nodes are also in the same situation.
6. If the sub-optimal node hasn't received any DM either but received RM, it declares to be the cluster head and goto 3. Other non-cluster head nodes also delete the top node in the energy list when they receive the DM of the sub-optimal node. This process is continued in sequence of the energy list until a cluster head has been successfully selected.
7. In the phase of data transmitting, Cluster head send "alive" message periodically to inform cluster members its state. So if any cluster head failed, its cluster members would soon be aware and it will be deleted from every node's energy list and the sub-optimal node will become the new cluster head. This scheme can remedy the network partition resulted from the failure of cluster head as soon as possible and reduce the loss as much as possible.
8. At the end of this round, the cluster head send its residual energy information to its cluster members so that they can re-sort the energy list. Then re-clustering will begin from 3. Here we assume the energy consumption of non-cluster head nodes is almost uniform because their data collections are quantitatively equivalent since they are adjacent.

Through EDDS, CREED can select the maximal residual energy node to be cluster head while avoiding the energy information exchange of all nodes (only need once at the beginning of clustering). Furthermore, with more energy, the cluster heads are more stable to work. The most important is that EDDS adopts a complete-backup scheme to enhance the fault tolerance of cluster-based communication.

### 5.3  Inter-cluster Routing

In this section, we will present the inter-cluster routing protocol of CREED. This protocol is borrowed the main idea from the energy aware routing (EAR) [25]. EAR keeps a set of good paths and chooses one based on a probabilistic fashion. It means that instead of a single path, a communication would use different paths at different times, thus any single path does not get energy depleted. In order to guarantee the data delivery in desired reliability, we take advantage of the natural relation between EAR and multipath routing. i.e. to allocate number of transmissions to each path proportional to the ratio computed from EAR protocol. What's more, we modify EAR to take some reliability factor into consideration. If the selection ratio of two paths is close, path with less hop number will be selected in preference.

There are three phases in the routing protocol: setup phase, data communication phase and route maintenance. After setup phase, routing nodes will setup a forwarding table. The table conserves an entry for each effective neighbor inside which it includes the selection probability. The setup phase is explained in the following:

1. The destination node initiates the connection by flooding the network in the direction of the source node. Every intermediate node forwards the request only to the neighbors that are closer to the source node than oneself and farther away from the destination node.
2. On receiving the request, the energy metric of the neighbor who sent the request is computed and added to the total cost of the path. Here we use the energy model discussed in section 3 and "J/bit" as measure of energy consumption. Paths that have a very high cost are discarded and not added to the forwarding table.
3. Transmission energy consumption and node residual energy are used to compute the energy cost and added to the forwarding table. Node assigns a probability to each of the neighbors with the probability inversely proportional to the cost.
4. If the probability of some neighbors are close as in equation(4) but with different hops. then the probability of these neighbors will be reallocated inversely

$$\frac{p_i - p_j}{\min\{p_i, p_j\}} <= \alpha \tag{4}$$

$p_i, p_j$ are the probability of neighbor i, j; $\alpha$ is proximate threshold

proportional to their hop number and replace the old value in the forwarding table. In general, the less the hop number, the more reliable the path is (supposed in appropriate power level).

$$p_i^{new} = \frac{h_j}{h_i + h_j}(p_i + p_j), \quad p_j^{new} = \frac{h_i}{h_i + h_j}(p_i + p_j) \tag{5}$$

$h_i, h_i$ are the hop number from neighbor i, j to sink.

5. After the probability is designated. The hop number from a node k to the sink can be computed by equation (6). Supposed the hop number of the sink is 0.

$$h_k = (\sum_{i \in FT_K} p_i * h_i) + 1 \tag{6}$$

After setup phase, CREED enters data communication phase and source cluster head s can transmit data based on the forwarding table. We make some assumptions about the sensor nodes and the underlying network model. Assume that the intra-cluster transmission power and data redundancy are high enough to guarantee information completeness after aggregation. Source nodes always have data to send to sink. Data packets collected from a cluster can be aggregated perfectly into a single data packet For the sake of simplicity, we also assume that each hop has an error rate e, the desired reliability is r. We can use equation (7) to compute the number of paths needed to guarantee the reliability.

$$N_s = \frac{\log(1-r)}{\log(1-(1-e)^{h_s})} \tag{7}$$

s sends the data packet to the neighbors in the forwarding table. The paths number allocated to each neighbor is $N_s * p_i$ ( $i \in FT_S$ ). Here paths number means to transmit $N_s * p_i$ copies of data packet to this neighbor. If $N_s * p_i$ is less than 1, s will forward the data packet to node i with probability $N_s * p_i$ . Each of the inter-mediate nodes forwards the data packet in the similar way. This process continues till the data packet reaches the sink.

Route maintenance is minimal. Localized flooding is performed infrequently from destination to source to keep all the paths alive.

If some cluster head fails and a new cluster head is selected (discussed in section 5.2), the new one will inform its neighborhood of its existence. The upstream neighbors will re-send the routing information to it and let the new cluster head setup the forwarding table. The downstream neighbors either don't differentiate between the old one with the new one or just delete the entry for this cluster from forwarding table if the new one is too far. Neglecting the energy metric with the new cluster head can avoid re-flooding to all downstream nodes. Although it may lead to some inaccuracy, the extra energy consumption may be negligible compared to the re-flooding (since only this cluster re-select head). If some downstream neighbor deletes the entry for this cluster, it should re-allocate the ratio among the entries left. Actually, all routing nodes are cluster head and thus have more energy, so failure may seldom happen.

## 6 Performance Evaluation

In this section, we study the data transport reliability of CREED protocol, as well as its energy-efficiency. In order to get a better understanding on how well CREED works, we present the performance comparison between CREED and LEACH. We will make the evaluation from the following three aspects:

1. The desired reliability and attained reliability (compared with single path routing);
2. Inter-cluster and intra-cluster communication energy consumption and the standard deviation;
3. The timely failure recovery when some cluster heads failed suddenly.

We set border length $a$ of the grid (cluster) to 30m. So for the convenience of computation, we assume that 100 nodes are uniformly dispersed in a square area 90m*90m. The channel error rate is normally distributed across the field. MAC protocol can assign a unique channel for every node to prevent possible collisions.

## 6.1 Attained Reliability

The first two simulations show the attained reliability under different channel error rate and attained reliability. Here we modify LEACH to support EAR routing instead of direct transmission with sink to optimize energy consumption, but LEACH only

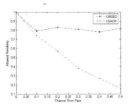

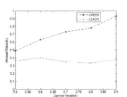

**Fig. 2.** Attained Reliability by LEACH and CREED (Single path vs. Multi-path)

**Fig. 3.** Attained Reliability vs. Desired Reliability by LEACH and CREED

use single path to deliver data. We set the optimum cluster head number to 5 (computed from [12]). The mean channel error is increased from 0 to 50%. Figure 2 shows the average attained reliability of the first 20 round clustering for LEACH and CREED. The desired reliability is 0.8. We can see that CREED provides reliabilities close to the desired reliability. Figure 3 shows the average attained reliability of the first 20 round clustering for LEACH and CREED. The channel error rate is 30%. In these two figures, CREED attain the desired reliability by using multipath routing, but the results for LEACH is dissatisfying because EAR only use single path routing.

## 6.2 Energy Consumption

- Intra-cluster

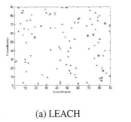

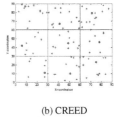

(a) LEACH

(b) CREED

**Fig. 4.** Randomly chosen cluster head distribution for LEACH and CREED

LEACH has no strategy for specifying cluster heads position and distribution so it doesn't offer a complete energy optimization solution. CREED partitions the network into grids so that each head is located inside one grid. Figure 4 presents two randomly chosen cluster head distribution for LEACH and CREED. The corresponding energy consumption for all non-cluster head nodes when they transmit data to their cluster head is shown in figure 5. The energy model is based on the result from section 3. It's obvious that the average intra-cluster data transmission energy consumption for CREED is much less than LEACH. Figure 6 shows the energy consumption and the corresponding standard deviation for non-cluster head nodes in the first 10 rounds for LEACH and CREED. It demonstrates that CREED can perform better in terms of balancing the intra-cluster load among nodes and therefore prolong the network lifetime.

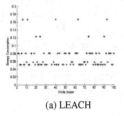
(a) LEACH

(b) CREED

**Fig. 5.** Energy consumption under the two cluster head distributions in figure 4(unit: J/bit)

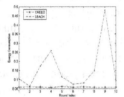

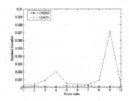

**Fig. 6.** Intra-cluster data transmission energy consumption and the standard deviation

- Inter-cluster

We assume that the source is located in lower left zone (the zone of the lower left cluster) and the sink at the upper right zone (90,90). Figure 7 shows the average inter-cluster data transmission energy consumption of the first 20 round clustering for LEACH and CREED corresponding to figure 2. The channel error rate is 30%. From the figure we can see that with the increase of the desired reliability, the energy consumption of CREED is more than LEACH because of the increase of path number to attain the reliability. It's inevitable because CREED attain the reliability at the cost of data redundancy. But since CREED effectively controls the cluster heads distribution, the increase of inter-cluster energy consumption of CREED is acceptable. The cost of multipath setup and selection is minimal since it's included in the process of route setup. Anyway, this energy consumption is much less than that of a flat architecture in the same desired reliability.

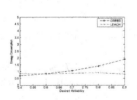

**Fig. 7.** Inter-cluster energy consumption (J/bit)

### 6.3   Robustness to Sudden Cluster Heads Failure

Here we want to evaluate the performance of EDDS. Suppose in a simple situation, some cluster heads that participate in the inter-cluster routing are suddenly failed in the 5<sup>th</sup> round. LEACH doesn't reselect the new cluster heads until the next round so it has to wait a long time to re-flood to reconstruct the route. Figure 8 illustrates the change of attained reliability. The desired reliability is 80% and the channel error rate is 30%. We can see that the desired reliability of CREED only decreases a little (0.71 relative to 0.8) due to the timely cluster head reselection of EDDS. Besides, cost of new cluster head reselection is few because the energy-sorting list facilitates the new cluster head selection.

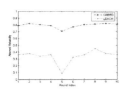

**Fig. 8.** Attained reliability with unexpected failure

## 7   Conclusion

In this paper, we combine the clustering technique with several reliability mechanisms to propose a new reliable, energy-efficient clustering protocol CREED. CREED partitions the network into several grids based on the optimum direct transmission range to control the cluster head distribution. It uses EDDS to support rapid failure recovery and multipath to guarantee reliable data delivery. Experiments show that CREED can attain the reliability with the least energy consumption.

Actually what we are proposing in this paper is a design methodology that can be extended to other cluster techniques. The essence of our method is to provide backup nodes to support quick cluster head selection and failure recovery in intra-cluster communication and redundant data delivery to deal with channel error in inter-cluster transmission. In the future, we will design a framework to easily integrate these reliability mechanisms into common clustering techniques.

**Acknowledgement.** This paper is supported by the National Natural Science Foundation of China (Grant No.90204005), National High-Tech Research and Development Program of China (Grant No.2005AA121570, 2002AA1Z2101 and 2003AA1Z2060) and Grant Project 51415040205KG0151.

## References

[1]   F. Akyildiz et al., "Wireless Sensor Networks: a survey, " *Computer Networks*, Vol. 38, pp. 393-422, March 2002.

[2]   H. Karl and A. Willig, "A short survey of wireless sensor networks," *Technical Report TKN-03-018*, Telecommunication Networks Group,technical University Berlin, October 2003.

[3]   F.L.LEWIS, "Wireless Sensor Networks",Smart Environments:Technologies, Protocol and Applications, 2004.

[4]   D.Ganesan,R.Govindan,S.Shenker,and     D.Estrin,     "Highly-resilient,energy-efficient multipath routing in wireless sensor networks",In Mobile Computing and Communications Review(MC2R) Vol 1.,No.2,2002.

[5]   S.Bhatnagar,B.Deb,and B.Nath, "Service differentiation in sensor networks",In Wireless Personal Multimedia Communication,2001.

[6]   B. Deb, S. Bhatnagar and B. Nath , "ReInForm: Reliable Information Forwarding using Multiple Paths in Sensor Networks", Proc. of IEEE LCN, 2003.

[7]   M. Marina and S. Das, "On-Demand Multipath DistanceVector Routing in Ad Hoc Networks", In Proc. of IEEE ICNP, 2001.

[8]   M. Pearlman, Z. Haas, P. Sholander and S. Tabrizi, "The Impact of Alternate Path Routing for Load Balancing in Mobile Ad-Hoc Networks", In Proc. of the ACM MobiHoc, 2000.

[9]   B.Deb, S.Bhatnagar,and B.Nath, "Information Assurance in Sensor Networks", WSNA,2003.

[10]  Fred Stann,John Heidemann, "RMST:Reliable Data Transport in Sensor Networks",In 1st IEEE International Workshop on Sensor Net Protocols and Applications, 2003.

[11]  C.Intanagonwiwat, R.Govindan, and D.Estrin, "Directed diffusion:a scalable and robust communication paradigm for sensor networks", In MOBICOM,pages 56-57,2000.

[12]  Wendi B. Heinzelman, Anantha P. Chandrakasan, Hari Balakrishnan, "An Application-Specific Protocol Architecture for Wireless Microsensor Networks", IEEE Transactions on Wireless Communications.Vol.1, No.4, October 2002.

[13]  Younis O.,and Fahmy S., "Distributed Clustering in Ad-hoc Sensor Networks: A Hybrid, Energy-Efficient Approach",Proceedings of IEEE INFOCOM. Hong Kong, March,2004.

[14]  S.Bandyopadhyay, E.Coyle, "An Energy Efficient Hierarchical Clustering Algorithm for Wireless Sensor Networks", Proceeding of INFOCOM 2003, Vol.3,pp.1713-1723.

[15]  Marzieh Veyseh, Belle Wei and Nader F.Mir, "An Information Management Protocol to Control Routing and Clustering in Sensor Networks", Journal of Computing and Information Technology - CIT 13,2005, pp.53-68.

[16]  Michael Hempel, Hamid Sharif, Prasad Raviraj, "HEAR-SN: A New Hierarchical Energy-Aware Routing Protocol for Sensor Networks", Proceedings of the 38th Hawaii International Conference on System Sciences – 2005.

[17]  Ossama Younis, Sonia Fahmy, Paolo Santi, "An Architecture for Robust Sensor Network Communications", International Journal of Distributed Sensor Networks, 1:1-23,2005.

[18]  Y.Xu,J.heidemann, D.Estrin, "Geography-informed Energy Conservatino for Ad-hoc Routing",In Proceedings of the Seventh Annual ACM/IEEE International Conference on Mobile Computing and Networking 2001, pp.70-84.

[19]  L.C. Zhong, J.M. Rabaey and A. Wolisz, "An integrated data-link energy model for wireless sensor networks", ICC 2004, Paris, France, June 20-24, 2004.

[20]  Vivek Mhatre, Catherine Rosenberg, "Design guidelines for wireless sensor networks: communication, clustering and aggregation", Ad Hoc Network 2 (2004), p45-63.

[21]  MIN, R., et. al., "Top Five Myths about the Energy Consumption of Wireless Communication", Mobile Comput. and Commun. Review, vol. 1, Num. 2, 2003.

[22]  Martin Haenggi, Daniele Puccinelli, "Routing in Ad Hoc Networks: A Case for Long Hops", IEEE Communication Magazine, October,2005.

[23]  Broadcom white paper, The New Mainstream Wireless LAN Standard, softcopy at http://whitepapers.zdnet.co.uk/0,39025945,60072368p-39000522q,00.htm.

[24]  MEHTA, S., et al., A CMOS Dual-Band Tri-Mode Chipset for IEEE 802.11a/b/g Wireless LAN, 2003 IEEE Radio Frequency Integrated Circuits (RFIC)Symposium, pp. 427–430, June 2003.

[25]  SHAH, R.C., et. al., "Energy Aware Routing for Low Energy Ad Hoc Sensor Networks", IEEE Wireless Commun. Networking Conf., vol. 1, pp. 350–355, March 2002.

# Experimental Analysis and Characterization of Packet Delay in UMTS Networks

Jose Manuel Cano-Garcia, Eva Gonzalez-Parada, and Eduardo Casilari

Electronic Technology Department, University of Malaga
cano@dte.uma.es

**Abstract.** This paper presents the results of a set of experiments aiming to characterize the behaviour of the IP data service over a real UMTS network. According to our empirical measurements, packet losses are not frequent, meaning that UMTS link-level retransmission mechanisms are adequate to cope with data corruption through the wireless link. However, ARQ loss recovery mechanisms produce a high variability of the packet delay, which can also affect the performance of the IP data service. By analysing actual UMTS traces, we explore the behaviour of the packet delay through the UMTS network, explaining the effects of the underlying mechanisms. Finally, we derive a simple model to imitate the behaviour of UMTS Internet access. This model could be very useful in simulation or emulation experiments.

## 1 Introduction

Third generation Universal Mobile Telecommunication Systems (UMTS) have been specifically designed to face up the challenges of providing high performance mobile access to the Internet. Wireless links have singular characteristics that affect the performance of transport protocols, such as a variable bandwidth and delay, corruption losses, asymmetry, etc. It is well known that the congestion control of the Transport Control Protocol (TCP) reacts to packet losses on the assumption that they are the result of network congestion, which reduces its performance when this is not the case.

The last decade has witnessed a massive research effort devoted to design solutions both to improve TCP behaviour over wireless links [1, 2], and to optimize the protocols of cellular systems for mobile Internet access ([3, 4]). To assure a better controllability of the scenario, these studies are often carried out by simulations. Since UMTS topology is complex, simulation studies usually focus on accurately modeling the wireless interface, because it is assumed to be the bottleneck link. However, despite of this simplification, simulation models involving all the protocols of the UMTS radio interface are still complex and comprise a high number of parameters. Therefore, the research community on transport protocols have become more and more interested in developing simple models of the data transmission over mobile cellular networks based on the behaviour observed in real links [5]. Experimental testing of data transport over operative cellular networks is now possible, and several practical studies, mainly focused on the evaluation of the performance of TCP over GPRS, can be found

Y. Koucheryavy, J. Harju, and V.B. Iversen (Eds.): NEW2AN 2006, LNCS 4003, pp. 396–407, 2006.

in the literature [6, 7]. In addition, there is a solid research background on the empirical characterization and modeling of Internet (see for example [8, 9, 10] for delay characterization). Models of wireless links and practical measurements are not only useful for transport protocol performance evaluation, but also for the consideration and design of new enhancements and protocols.

This paper presents an empirical evaluation and characterization of the Internet access over UMTS. The study is not only aimed at the evaluation of end-to-end performance of the Internet access, but also at developing a simple and general UMTS link model that can be used in the simulation of transport protocols. The analysis included in this paper is mainly focused on delay variation, because of the low loss rate and the negligible packet reordering that have been observed.

This paper is structured as follows: Section 2 provides an overview of UMTS architecture. The utilized testbed is presented in section 3. Section 4 discusses the measurement results and their implications for the link characterization. A model for UMTS link simulation is subsequently proposed in section 5. Finally, section 6 summarizes the main conclusions.

## 2   UMTS Overview

UMTS is a standard for the deployment of third generation cellular networks based on Wideband Code Division Multiple Access (WCDMA). The UMTS network is divided into the Core Network (CN), the Radio Access Network (UTRAN) and the User Equipment (UE), in order to facilitate the migration from 2G to 3G networks by allowing the coexistence of different access techniques and core network technologies [11]. The User equipment (UE) is connected to the UTRAN through the Uu radio interface, which supports both FDD and TDD mode operation, although only the FDD mode is usually used on currently deployed 3G networks. For both modes the same network architecture and the same protocols are used. The radio interface protocol architecture is divided into the physical, the data link (L2) and the network layers [12]. L2 is further divided into the Medium Access Control (MAC), the Radio Link Control (RLC) and the Packet Data Convergence Protocol.

In the physical layer of the Uu Interface a WCDMA multiple access scheme is used to divide the available bandwidth into physical channels on which transport channels information is mapped. Physical layer operation requires a temporization, which is set by establishing 10 ms frames. Due to block-coding and interleaving, transport channels are clocked at multiples of this period, known as Transmission Time Interval (TTI). Every TTI, information blocks (transport blocks, TB) are delivered by the upper protocols to the physical layer for transmission over the wireless link. Usually, the TTI can be 1, 2 or 4 frames, depending on the service configuration. The size of the data that can be transferred every TTI varies depending on the configured data rate. As we explore in the next section, the TTI temporization clearly affects the end-to-end delay experienced by IP packets transferred over UMTS.

The MAC sublayer maps logical channels onto the transport channels offered by the physical layer, while the RLC sublayer [13] comprises all the mechanisms needed to the fragmentation and reassembly of upper layer Data packets (Service Data Units, SDUs) into smaller data blocks (Packet Data Units, PDUs) so that they can be transmitted through the radio interface by the lower layers. This sublayer provides transparent, unacknowledged (UM) or acknowledged (AM) mode data transfer to the upper layers. For Internet data transport, RLC AM mode with in-order delivery is used. The acknowledged mode uses a sliding window protocol with a selective repeat mechanism to guarantee error free data transmissions. According to this mechanism, the receiving RLC entity, under certain conditions, sends back status reports to the transmitting RLC entity, pointing out which PDUs have been correctly received and which ones are missing. Basing on this information, the transmitting RLC entity decides either the retransmission of previously sent PDUs or the transmission of new ones. In order to avoid stalling, the RLC transmitting entity can also poll the peer receiving entity and request a status report. The RLC AM mode hides radio packet corruption from upper layers, but at the cost of introducing packet delay variability.

# 3   Measurement Infrastructure

The main goal of our testbed setup is to analyse the end-to-end packet delay and losses. To carry out the characterization, we have developed a software tool able to periodically transmit IP probe packets through the UMTS access to measure the delay and losses that they experience. In order to do this in a simple way, our measurement software is not divided between a transmitter side and a sender side, but it is located in a probe laptop with 2 networks interfaces, as shown in fig. 1. This probe laptop has access to UMTS through a 3G Sony Ericsson V800 handheld device connected to the laptop by mean of a USB interface. The laptop is also equipped with an Ethernet network card connected to our University corporate network. The UMTS Internet access provided by the network operator sets up a dedicated channel with up to 64 kbps and 384 kbps for uplink and downlink transmissions respectively, which is the QoS most commonly offered by current UMTS providers. By using this configuration, our measurement software is capable of injecting packets in the UMTS interface and receiving them through the Ethernet link, and viceversa. This allows us to accurately measure one-way packet delay and analyze the loss process for uplink and downlink directions independently.

It should be highlighted that it is not possible to send packets through one of the interfaces (for example the UMTS) directly addressed to the other (for example the Ethernet link), because the network protocols of the operating system running in the laptop PC (Linux) would detect that the packet is addressed to the local machine, and, therefore, it would deliver them immediately to the upper layers instead of sending them to the network. To solve this problem, we have used an intermediate Network Address Translation (NAT) router. The Laptop addresses the packets to the NAT router, so they are sent through the

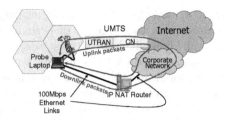

**Fig. 1.** Measurement testbet

appropriate interface by the network protocols and received by the NAT, which is configured to alter the source and destination addresses of the received packets, so they can be redirected to the probe laptop (see fig. 1). The NAT router is simply a Linux desktop computer with two Ethernet interfaces, one attached to our corporate network and the other directly connected to the probe laptop. Source and destination NAT is provided by the Linux kernel and configured through the `iptables` command.

The advantage of the proposed schema over a two-side measurement strategy (i. e. using two different desktops, one connected through UMTS and other attached to our corporate network) is that only one clock (the probe laptop clock) is employed to estimate the delay, which avoids any problem of synchronization between the origin and the target nodes of the packets. It should be noted that the additional delay produced by the NAT processing and redirection at the intermediate router is negligible when compared to the total end-to-end delay. The major error source of our measurement testbed, which would also be present if we had used a two-side measurement schema, is provoked by the path between the GGSN and our corporate network. Packet journey through this path is unavoidable, since we have no direct access to the GGSN, and may not only increase the measured packet delay, but can also generate delay fluctuations, packet losses and reordering. However, observations outlined in the next section indicate that the these effects can also be neglected, allowing an adequate observation of different features of the packet delay process. In order to evaluate how much the packet delay is increased in the segment between the GGSN and our network, we have used the `traceroute` tool to determine the number of hops (routers) in the path and the median round trip time (RTT) to each router in the path. The number of hops detected is 14. The RTT between the probe laptop and the first node seen, which is presumably the GGSN or other node belonging to the access provider (nodes of the core network are transparent to the user) was estimated to be 119 ms. On the other hand, the RTT between the probe laptop and the last node (our NAT router) is 140 ms. According to the trace presented by `traceroute`, the 20 ms difference in round trip delay between these nodes are mainly spent in the Spanish Universities backbone (RedIris) between Madrid and Malaga.

The probe laptop is also a Linux computer, in order to take advantage of the bunch of free network testing tools available, including packet sending tools like `ping` or `hping2`, packet capture tools such as `tcpdump`, and bandwidth estimation tools (`pathchar`, `pathrate`, `pathchirp`, etc.). Using available open source code, we have developed a customized tool capable of periodically sending UDP or fake-TCP packets at small intervals (up to 10ms) to characterize end to end packet behaviour. Each transmitted packet is marked with a sequence number, and its departure time is stored in order to measure one-way delay and losses. Packet size can be configured to test its effect on end-to-end performance.

Although in this paper we have particularly focused on studying end-to-end packet behaviour, our testbed setup is also valid to study other traffic aspects such as the available bandwidth or TCP performance over UMTS. Our intention is to tackle these issues in further work, with the goal of expanding the study presented in this paper.

# 4 Results and Discussion

Using the testbet setup and the software tool described in the previous section, we have accomplished several set of measurements in a commercial UMTS network to study end-to-end packet behaviour. The data traces analysed in this paper were captured in three different days of November and December 2005. For all three data set a fixed interval of 20 ms between probe packets was configured, while the packet size was set up to 40 bytes, thus yielding a data rate of about 16 kbps, which is below both the uplink and the downlink maximum data rate (64 and 384 kbps respectively). The two first data sets were acquired using fake TCP packets, whereas the third data set corresponds to UDP data traffic. Both trace 2 and 3 are 10 hour long (1,800,000 packets) and were captured overnight. Trace 1 is 5 hour long and was captured in the morning. Data sets features and measurements in both uplink and downlink directions are summarized in table 1. These data sets and the measurement software will be publicly available in our web site [14].

**Table 1.** Dataset features and basic measured parameters

| Data Set | Downlink | | | Uplink | | |
|---|---|---|---|---|---|---|
| | 1 | 2 | 3 | 1 | 2 | 3 |
| Mean Packet Delay (ms) | 101.3 | 94.2 | 108.3 | 105.9 | 98.1 | 101.10 |
| Packet Delay Standard Deviation (ms) | 61.17 | 63.2 | 66.5 | 77.4 | 72.9 | 70.3 |
| Packet Delay Median (ms) | 81.4 | 72.4 | 83.1 | 79.9 | 74.2 | 78.2 |
| Number of Detected Spikes | 11508 | 26244 | 29076 | 14043 | 25859 | 26118 |
| Mean Spike duration(ms) | 217.5 | 216.3 | 217.4 | 232.7 | 232.7 | 235.9 |
| Standard Deviation of spike duration(ms) | 82.8 | 78.8 | 77.6 | 101.5 | 95.6 | 93.8 |
| Median of spike duration (ms) | 200 | 199.7 | 199.4 | 210 | 207.9 | 212.9 |
| Mean Time between spikes (s) | 1.56 | 1.37 | 1.19 | 1.28 | 1.4 | 1.3 |
| Standard Deviation of Time between spikes(s) | 1.5 | 1.35 | 0.94 | 1.46 | 1.45 | 1.35 |
| Median of Time between spikes (s) | 1.16 | 1.0 | 1 | 0.76 | 0.82 | 0.88 |

A first glance of the results can be obtained in figs. 2.a-d, where selected parts of the captured data are represented for a better view of different details. In all the four plots, each dot represents the estimated delay of a single packet.

Packet delay is represented in the Y-axis versus the packet departure time represented in the X-axis. Figure 2.a shows a 10 minute length segment of dataset 2 (30,000 packets), whereas plots b-d display shorter periods to magnify certain details. From these plots we can observe that packet delay is highly variable. Although an important fraction of the packets experience a delay around 80 ms, there is a significant amount of packets that exhibit a higher latency. Similar results (not included in fig. 2) are obtained for uplink. If we take a closer look (fig. 2.b-c), we can appreciate that the delay variation is mainly due to the presence of delay spikes, which are present in both uplink and downlink directions. As it can be clearly appreciated in the figures, each packet in a spike experiences a delay 20 ms lower than the previous one, which is precisely the periodicity of the packet emission. This suggests that the spikes are produced by the ARQ mechanisms used by the RLC sublayer to assure reliable and in-order delivery. When a block is lost over the wireless link, no data is delivered by the RLC to the upper layers until the ARQ mechanism has recovered the lost frames. When the loss is recovered, all data stored in the receiver-side reconstruction buffer of the RLC are delivered in one go to the upper layers. Since a packet is sent every 20 ms, each packet affected by the recovery process has spent in the reconstruction buffer 20 ms less than the previous one, which explains the shape of the delay spike when represented in the plot.

Besides, a sawtooth pattern can be observed for packets not affected by the delay spikes (fig. 2.d). A slow drift and a periodic hop of 20 ms can be observed for uplink packets, whereas downlink packets follow a similar pattern, but with an amplitude of 10 ms. This pattern is a measurement artifact caused by the drift of the measurement computer timer with respect to the UMTS physical layer temporization, and should not be modelled. As explained in previous section, the physical layer takes data from the upper layer every TTI for transmission over the wireless link. TTI is a configurable parameter, but it can be set only to 1, 2 or 4 radio frames of 10 ms. Probe Packets are periodically sent through the network every 20 ms. Since the network clock and the measuring computer timer are not synchronous, their frequency can be slightly different and therefore the phase difference between them may change linearly, yielding a linear variation of the delay. When the phase difference reaches a certain threshold, the data transmission skips one idle TTI, thus producing the one-TTI "step" or abrupt change in the sawtooth pattern. By the observation of the amplitude of these "steps" in both directions we can deduce that the TTI is set to 2 frames for the uplink and 1 frame for the downlink, which is consistent with the default configuration suggested by [15] for background traffic over a dedicated channel.

From the measurements, we can also conclude that packet reordering is very rare. For example, for the second trace only 20 of the almost 2 million uplink packets are out of order, while no reordered downlink packets were found. In fact, evidence suggests that the observed packet reordering occurs in the journey through the Internet, and not in the UMTS segment. As previously commented, RLC ARQ mechanisms can deliver to upper layers several consecutive packets in one go when a lost PDU is recovered. When this burst of packets has to traverse

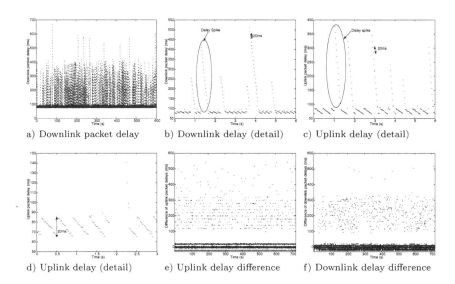

a) Downlink packet delay    b) Downlink delay (detail)    c) Uplink delay (detail)

d) Uplink delay (detail)    e) Uplink delay difference    f) Downlink delay difference

**Fig. 2.** Delay plots for uplink and downlink directions

the Internet, which is the case of the uplink transmission, they are prone to be reordered under certain circumstances. With regard to packet loss, we can conclude that UMTS ARQ mechanisms are quite efficient, since the percentage of lost packets is lower than 0.5%. Given these facts, and considering that about a 15 % of the packets are affected by delay spikes in all the presented data sets, we have decided to focus the study on the characterization of this phenomenon rather than on packet loss.

A plausible solution to model packet delay behaviour is to characterize the spike arrival process (i. e., the length of loss-free periods) and the spike duration (i.e., the time required to recover from a transmission loss). In order to characterize both parameters, we can compute the difference of the delay of consecutive packets (i.e. $diff(d[n]) = d[n + 1] - d[n]$, where $d[n]$ is the delay of the $n - th$ packet) and set up a threshold to discriminate between the delay increments corresponding to a spike and those produced by other causes. Figs. 2.e-f show this difference for a subset of consecutive uplink and downlink packets from one of the datasets. Figure 2.e (uplink) reveals that packets not affected by delay spikes are usually clustered around 0 ms, except for the packets suffering a 20 ms delay increment provoked by the sawtooth pattern, which are clustered around 20 ms. Packets affected by delay spikes are clustered at -20 ms, with the exception of the first packet of each spike, which are the dots with bigger positive values. It should be noted that these values are a good estimate of the spike duration (i.e. the time needed by the RLC mechanisms to recover from a loss). These points can be easily discriminated from the others by setting a reasonable threshold , since in both figures a dot-free area exists between 20 and 120 ms. Results in figure 2.f (downlink) are similar but, as the TTI is 10 ms, packets affected by the sawtooth pattern are clustered around this value.

If we analyse the spike duration by computing its histogram, we can easily observe that its nature is clearly discrete (see fig. 3.a-b). This discretization is due to the temporization of the UMTS link, because the RLC layer always requires an integer number of TTIs to recover from a loss. The detected discretization is 10 ms for downlink and 20 for uplink, which is consistent with previous inferences. It should be highlighted that the fact of being able to neatly observe the discretization originated by the physical layer temporization in uplink packets suggests that the path between the GGSN and our corporate network has a minor impact on the delay fluctuation.

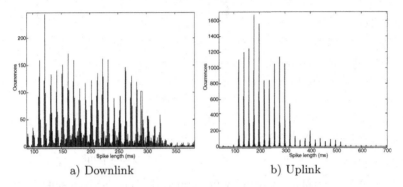

a) Downlink                              b) Uplink

**Fig. 3.** Histogram of the spike duration (RLC loss-recovery time)

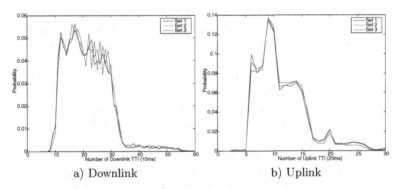

a) Downlink                              b) Uplink

**Fig. 4.** Probability function of the spike duration (RLC loss-recovery time)

Given the natural discretization of the RLC loss recovery time, measured values for these durations were rounded to a multiple of TTIs to compute the probability function of the spike duration for all the three data series, which are shown in figs 4.a-b. The results are similar for all the series, which suggests that this process is quite stable. The amount of time needed by the RLC to recover from a loss depends on its configuration and is variable because different mechanisms to trigger and inhibit status reports are simultaneously active in the RLC. In any case, the minimum time required to recover from a loss seems to be 6 TTI (120 ms) for the uplink, and 10-11 TTIs (100-110 ms) for the downlink. These limits are approximately consistent with theoretical values derived by [16] using the delay budgets suggested in [17]. In addition to the probability function of the RLC recovery time, we have also explored the autocorrelation of this process (i.e. if there is a dependence between the duration of two consecutive loss episodes), and we have found that each recovery can be assumed to be independent from the previous one, since no correlation is observed.

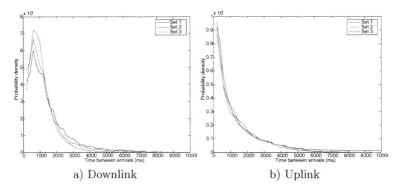

a) Downlink                          b) Uplink

**Fig. 5.** Probability density function of the time between spikes

If we now focus on the spike arrival process, time between spikes can be estimated with a resolution given by the packet sending period used to acquire the data series. Figures 5.a-b show the estimated probability density function for this parameter. As we can see, this process is less stable in comparison with the spike duration, and some minor differences appear between the different datasets. This is logical if we consider that this statistic is more dependent on the test conditions, which can vary from one set to other. Similarly, by using the autocorrelation estimation, we concluded that no correlation can be observed for this parameter.

To complete the end-to-end delay characterization, we have also performed further tests to verify possible interdependence on both the uplink and downlink processes. To do so, we have computed the autocorrelation of uplink and downlink aggregated delay signal, and their cross correlation, for different aggregation levels. Only short-range dependence, provoked by the presence of the delay spikes, is found in the delay signal, while it disappears in the aggregated signal. On the other hand, no significant cross-correlation can be found between the uplink and the downlink delay processes. Finally, we have computed aggregated signals representing the number of spikes observed in consecutive periods of a given duration. This signals have been used to analyse the autocorrelation and cross correlation of uplink and downlink spike interarrival process. No evidence of significant correlation have been found using this analysis.

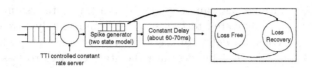

**Fig. 6.** Proposed model for UMTS simulation/emulation

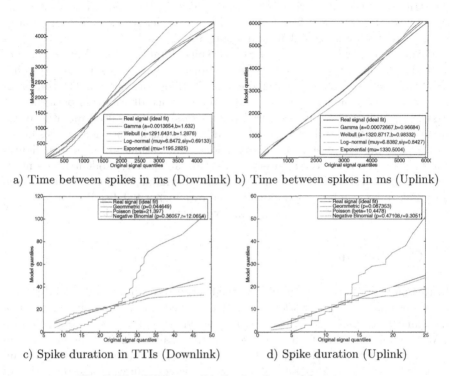

a) Time between spikes in ms (Downlink) b) Time between spikes in ms (Uplink)

c) Spike duration in TTIs (Downlink)         d) Spike duration (Uplink)

**Fig. 7.** Fit of different distribution for model parameters

## 5    Proposal of an UMTS Delay Model

In this section we propose a simple simulation/emulation model for UMTS based
on the observations performed in the previous section. This model is intended to
be used in preliminary simulation or emulation experiments to evaluate the performance of transport and application protocols and can be easily implemented
for both environments. The model structure is summarized in figure 6, and is
valid for both uplink and downlink directions (only one direction is represented
in the figure). Uplink and downlink models are independent from each other.
The first stage of the model is a constant bit rate server with a temporization
based on the TTI. Packets arriving at the system are queued, and every TTI a
given amount of data are served depending on the link bitrate. When all data
in the packet have been served, the packet is delivered to the next stage, which
is introduced to reproduce the delay spikes observed in the previous section.

The spike generation stage can be modelled as a two state system. In the loss-free state, all packets delivered to this stage are immediately sent to the next, which is simply a constant delay. In the loss-recovery state, all packets arriving to this stage are stored in a buffer, until the state changes to loss-free. When this occurs, all packets stored in the buffer are sent in one go to the next stage. The amount of time spent in each state is random and statistically modelled by a distribution function derived from the analysis of the characterization of the time between spikes (loss free state) and the spike length (loss recovery state). Both processes are independent and identically distributed.

Given the evident discrete nature of the spike duration (it is a limited number or TTIs), we have tested discrete distributions to characterize it. On the other hand, we have considered continuous distributions for the characterization of the time between spikes. In order to determine which distribution performs the best fit of the empirically measured parameters, we have used quantile plots. In these plots, the ideal fit is represented by a straight line. The curves closer to this line are the best fitted distributions. The results are shown in figures 7.a-d. Spike duration (loss-recovery period) are adequately modelled by a negative binomial distribution both in uplink and downlink. Similarly, time between downlink spikes can be reasonably adjusted by a lognormal distribution, whereas the best fit for the uplink process is an exponential distribution.

## 6   Conclusions

In this paper we have presented a testbed setup to evaluate end-to-end performance of real UMTS networks. Using this testbed, we have conducted a set of measuments to analyse the end to end behaviour of Internet access based on UMTS. Our study has mainly focused on the analysis and characterization of packet delay. By periodically sending packets and monitoring their end-to-end delay, we are able to infer several aspects of the configuration and operation of the UMTS network. Using these relatively simple measurements we have developed a parsimonious model of UMTS end-to-end behaviour that can be used in network simulation or emulation experiments.

The work presented in this paper can be expanded in different ways. Firstly conclusions derived from the analysis of the measurements and the proposed model, should be validated by further and more extensive tests. Additional experiments using different packets sizes and sending periods should be executed to determine the influence of these parameters in the end-to-end behaviour. Also, the study of the effect of the link quality and handovers on the observed results should be considered. On the other hand, the measurement testbed can be used to extend the analysis to other parameters. A deeper study of packet losses, and an analysis of the available bandwidth behaviour should be also accomplished to enhance the characterization of the UMTS Internet access. Tests involving TCP connection performance analysis can also be interesting in order to validate the conclusions derived by using the methodology proposed in this paper.

# Acknowledgements

This paper has been partially supported by Spanish public research fund program under project TIC2003-07953-C02-01.

# References

1. C. Casetti et al., "TCP westwood: end to end bandwidth estimation for enhanced transport over wireless links," *Wireless Networks*, vol. 8, pp. 467–479, 2002.
2. M. Chan and R. Ramjee, "Improving TCP/IP performance over third generation wireless networks," in *IEEE INFOCOM*, May 2004.
3. F. Vacirca, A. D. Vendictis, A. Todini, and A. Baiocchi, "On the effect of ARQ mechanisms on TCP performance in wireless enviroments," in *IEEE Globecom*, December 2003.
4. C. Chiasserini and M. Meo, "Impact of ARQ protocols on QoS in 3GPP systems," *IEEE Transactions on Vehicular Technology*, vol. 52, no. 1, pp. 205–215, January 2003.
5. A. Gurtov and S. Floyd, "Modeling wireless links for transport protocols," *ACM Computer communication Review*, vol. 34, no. 2, pp. 85–96, April 2004.
6. P. Benko, G. Malicsko, and A. Veres, "A large-scale, pasive analysis of end to end TCP performance over GPRS," in *IEEE INFOCOM*, May 2004.
7. R. Chakravorty, J. Cartwright, and I. Pratt, "Practical experience with TCP over GPRS," in *IEEE GLOBECOM*, November 2002.
8. Y. Zhang, N. Duffield, V. Paxson, and S. Shenker, "On the constancy of internet path properties," in *Proceedings of ACM SIGCOMM Internet Measurement Workshop (IMW'2001)*, November 2001.
9. L. Carbone, F. Coccetti, P. Dini, R. Percacci, and A. Vespignani, "The spectrum of internet performance," in *Pasive and Active Measurements (PAM2003)*, 2003.
10. H. Jiang and C. Dovrolis, "Passive estimation of tcp round-trip times," *SIGCOMM Comput. Commun. Rev.*, vol. 32, no. 3, pp. 75–88, 2002.
11. "Network architecture," 3GPP, Technical Specification 25401, 2000.
12. "Radio interface protocol architecture," 3GPP, Technical Specification 25.301-v4.3.0 Release 4, June 2002.
13. "RLC protocol specification," 3GPP, Technical Specification 25.322-v4.5.0 Release 4, June 2002.
14. "Diana research group web page." [Online]. Available: http://pc21te.dte.uma.es
15. "Common test environments for user equipment (UE)," 3GPP, Technical Specification 34.108-Release 4, May 2002.
16. A. Mutter, M. C. Necker, and S. Lck, "IP packet service time distributions in UMTS radio access networks," in *EUNICE 2004*, 2004.
17. "Delay budget within the access stratum," 3GPP, Technical Specification 25.853-Release 4, March 2001.

# End-to-End QoS Signaling for Future Multimedia Services in the NGN

Lea Skorin-Kapov[1] and Maja Matijasevic[2]

[1] R&D Center, Ericsson Nikola Tesla, Krapinska 45, HR-10000 Zagreb, Croatia
[2] FER, University of Zagreb, Unska 3, HR-10000 Zagreb, Croatia
lea.skorin-kapov@ericsson.com, maja.matijasevic@fer.hr

**Abstract.** The paper presents a model for Quality of Service (QoS) signaling for complex multimedia services in the next generation network (NGN), from establishing necessary QoS support at session setup, to further QoS modifications in response to dynamic changes during the session. The model aims to maximize user perceived quality, while taking into account constraints imposed by the user, network, and service itself. While the model is independent of the particular service and network scenario, we take a Networked Virtual Reality (NVR) service as a fairly good representative of intricate media-rich services in the NGN. We further propose a mapping onto the IP Multimedia Subsystem (IMS), as the most prominent NGN-driven part of the current Universal Mobile Telecommunications System (UMTS) architecture; and identify its possible functional enhancements for supporting services such as NVR.

## 1 Introduction

One of the key requirements for the next-generation network (NGN) is to provide seamless Quality of Service (QoS) for converged mobile and fixed multimedia services "anywhere-anytime" [6]. The provisioning of end-to-end (E2E) QoS, in general, involves signaling and dynamic negotiation/renegotiation between involved parties in order to agree on a common and feasible set of QoS parameters at session set-up, as well as reacting to possible changes during the session. In this paper, we present a model supporting the signaling, negotiation, and adaptation of QoS for media-rich interactive services, addressing issues arising due to the heterogeneous environment of NGNs. We focus on control signaling at the application/session level, rather than on the underlying mechanisms and architecture that provide actual service delivery. For reference purposes, we use the model proposed in our previous work [12], which takes into account the following issues: (1) user terminal and access network constraints; (2) way(s) of expressing user preferences in terms of application components (media elements); (3) dynamic resource availability and cost; and, (4) mapping of user/application requirements to transport QoS parameters. While the proposed model is service independent, we discuss it in the context of Networked Virtual Reality (NVR), looking to address one of the most "complex multimedia service" scenarios. Characterized by

Y. Koucheryavy, J. Harju, and V.B. Iversen (Eds.): NEW2AN 2006, LNCS 4003, pp. 408–419, 2006.

3D graphics, integrated multimedia components, and user interaction (perceived as) in real-time, examples of NVR services include networked 3D games, virtual worlds for training and collaborative work, and many more.

The paper is organized as follows. Related work on E2E QoS signaling and negotiation for multimedia services is discussed in Section 2. A generic QoS model is described in Section 3, with sequence diagrams provided to illustrate session QoS setup and (re)negotiation. Section 4 presents a case study involving the mapping of model entities to the IMS architecture. In Section 5, we give concluding remarks and identify open issues for ongoing and future work.

## 2   Related Work

Issues of E2E QoS signaling and negotiation regarding the NGN architecture have been addressed in ITU-T, 3GPP, IETF, and ETSI/TISPAN, as well as in recent research literature. The ITU-T defines an NGN as a packet-based network able to provide telecommunication services and make use of multiple broadband, QoS-enabled transport technologies [3]. A generic architecture for E2E QoS control and signaling for multimedia services is defined by the ITU-T in [4]. In parallel to ITU-T, ETSI/TISPAN has embraced the 3GPP IP Multimedia Subsystem (IMS) architecture, which has become an internationally recognized standard for offering multimedia services in the packet switched domain. With regards to application level QoS signaling, the IETF Session Initiation Protocol (SIP) [10] has been adopted by 3GPP as the key session control protocol for IMS. The common idea of all mentioned approaches is a horizontally layered architecture, with QoS-signaling interfaces between service-level call/session functionality and the underlying transport/connectivity functions. The relationship and efficient mapping of QoS parameters across layers, however, remains a challenge, especially for complex, higly interactive, and media-rich services. As such, it has been investigated from various points of view, with those mentioned next the most relevant to our approach. An evaluation of scenarios involving relationships between application-level and network-level QoS signaling during session negotiation, renegotiation, and handover can be found in [9]. The work [8] proposes an End-to-End Negotiation Protocol (E2ENP) for active negotiation of QoS for multimedia, based on SIP and SDPng (new generation), the successor of the Session Description Protocol (SDP) [7]. The authors assume an end-user application deriving valid QoS Contracts (service configurations) negotiated between users and enforced based on dynamic resource availability and user expectations. Further dealing with the issue of convergence between user preferences and expectations and network resource constraints, the authors in [5] propose an adaptive QoS control architecture for multimedia wireless applications. The referenced approaches have been analyzed with the goal of inspiring and contributing to the proposed model supporting signaling and dynamic QoS adaptation for complex multimedia such as NVR. Our model aims to provide a general framework within which various methods for QoS negotiation and/or parameter matching may be applied.

# 3   QoS Control Model

The presented model represents an adaptive QoS system supporting the process of E2E QoS control based on negotiation/renegotiation and service adaptation from the moment a user accesses a service until service termination. The model, shown in Fig. 1, is composed of a number of functional entities grouped logically into three components: *Client*, *Access and Control*, and *Application Server*. In the figure, solid arrows between model entities represent data flows, while dashed lines represent control (signaling) flows.

Upon a user's initiated service request, the *Client* sends a request to the *Access and Control* entities. The request contains user preferences and terminal capabilities in a client profile, or, a reference to a profile in a *client profile repository*. The *Access and Control* entities are responsible for identifying the client and service requirements, and authorizing necessary network resources. Negotiated and authorized parameters serve as input to an optimization process designed to dynamically calculate the resource allocation and *application operating point*. The application operating point refers to the final application configuration (included media components, corresponding codecs/formats, etc.) to be delivered to the user from the *Application Server*. After calculation, reservation mechanisms are invoked to reserve network resources, hence we are assuming an underlying network with implemented QoS mechanisms. A high-level view of the QoS control and negotiation process is given in Fig. 2. Once the service starts, further optimization and renegotiation/adaptation procedures may be invoked during service lifetime in response to changes in resource availability and/or resource cost; the client profile (user preferences, terminal capabilities, access network); and service requirements.

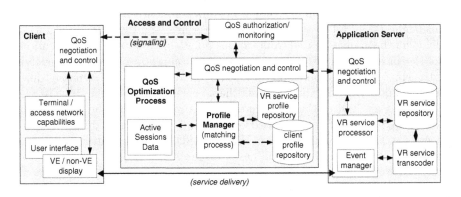

**Fig. 1.** Model for dynamic negotiation and adaptation of QoS for NVR services

## 3.1   Access and Control

The *Access and Control* component, shown in Fig. 1, represents a logical grouping of service control functionalities. In an actual network scenario, proposed functionalities may be distributed among different entities in the network.

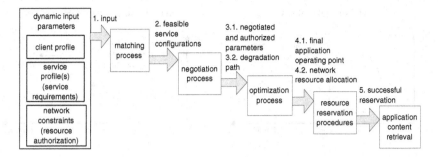

**Fig. 2.** QoS negotiation and control process

**Client profile determination.** A user's initial service request is constructed by way of a user interface enabling a user to request a service and set (possibly predefined) preferences. Options include enabling a user to specify desired quality (e.g. high, medium, low) of media components, preferences such as "audio has priority over video", budget constraints, minimum acceptable framerate, and maximum acceptable download time. Preferences, together with terminal hardware, terminal software, and access network characteristics are incorporated into a client profile. The proposed generic client profile is shown in Table 1. For the purposes of this paper, we assume that communication end points are capable of QoS-related signaling. It should be noted, though, that in a general case, where the user terminal does not support QoS signaling mechanisms, QoS requests may be issued through network session control entities.

**Table 1.** Generic client profile

| Client Profile parameters | | | |
|---|---|---|---|
| Terminal hardware | Network Characteristics | Terminal Software | User Preferences |
| - Model<br>- Display size<br>- Processor type<br>- Processor MIPS<br>- Memory<br>- Color depth<br>- Sound option<br>- Input character set<br>- Output character set | - Current bearer service<br>- Supported bearers<br>- Downlink (bandwidth, delay, loss, jitter, BER)<br>- Uplink (bandwidth, delay, loss, jitter, BER) | - OS (name, vendor, version)<br>- Browser info<br>- Supported software<br>- Supported media types | - Acceptable service format(s)<br>- Textures (accept, desired quality)<br>- Audio (accept, desired quality)<br>- Video (accept, desired quality)<br>- Text (accept, desired quality)<br>- Data (accept, desired quality)<br>- Max. download time<br>- Budget<br>- Min. acceptable framerate<br>- Degradation code |

**Profile Manager.** The client request, including or referencing the client profile, passes via the *QoS authorization/monitoring* module and is received by the *QoS Negotiation and Control* (QNC) module of the *Access and Control* component. The profile, or a reference to it, is passed to the *Profile Manager* (PM) module. The PM matches the restrictive parameters of the client profile with parameters of the retrieved VR service profile(s) corresponding to the requested service in order to determine feasible service configurations.

The *VR service profile repository* contains profiles specifying supported configurations (versions) of services. Multiple application configurations may be

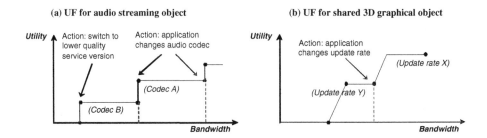

**Fig. 3.** Example Utility Functions for VE objects

feasible in order to address issues of heterogeneity stemming from diverse end user capabilities and preferences. For example, an application may be implemented in two ways: with media streaming (configuration 1), and without media streaming (configuration 2). The repository may be located on the application server hosting the actual service content, or on a different server in the network. We specify a generic service profile as shown in Table 2.

By considering the NVR service to be comprised of a collection of virtual world objects (e.g. 3D graphic objects, sound, video, etc.) integrated into a virtual environment (VE), communication requirements may be expressed as a combination of the requirements of particular objects mapped down to transport level QoS parameters. A detailed discussion of the parameters and a proposed mapping to QoS parameters and classes for UMTS, is given in [11]. In addition, factors such as a user's level of interest in an object inside the VE may be considered. Hence, we specify particular object requirements by combining object utility functions (UFs) and the notion of their relative importance. An example of UFs defined as functions of bandwidth for an audio streaming object (a) and a shared 3D graphical object (b) within a VE is shown in Fig. 3. Minimum requirements are indicated by those values at which utility drops to zero, while maximum requirements correspond to the marginal utility of utility value approaching one. It should also be noted that UFs may vary over time and accross different configurations. Assuming that the application can adapt its configuration to available bandwidth, Fig. 3 also illustrates actions to be taken when certain bandwidth

**Table 2.** Generic service profile

| Service Profile parameters | | | |
|---|---|---|---|
| Metadata | Network Requirements | Processing Requirements | Adaptation Policy |
| - Service designation | - General | - Display size | - User interest adaptation policy |
| | - downlink  (bandwidth, delay, loss, jitte  r, BER) | - Polygons | |
| - Service description | - uplink (bandwidth, delay, loss, jitter, BER) | - Textures | |
| | | - Sound | - Network resource adaptation policy |
| - Service identification | - Object $i$ profile | - Video | |
| | - name, object url, media type, formats | - Lighting | |
| - Service format | - downlink: UF(s)/loss/delay, jitter, weight factor | - Text | - Frame rate adaptation |
| | - uplink: UF(s)/loss/delay, jitter, weight factor | - Data | |
| | - media quality preference mapping to traffic class | - File size | |
| | | - Software support | |
| | [ ... profiles for other objects ...  ] | | |

thresholds are reached. Application adaptation based on multi-dimensional UFs has been addressed previously in literature [13].

In Table 2, the *Network Requirements* parameters refer to E2E QoS requirements as perceived by a user. The first subset (labelled *General*) corresponds to the overall minimum network requirements. The second subset contains particular object requirements (labelled *Object* n *profile*), specified in the form of UFs. The relative importance of objects (media components) to the user is taken into account by multiplying utility values with weight factors (WF) ranging from 0 to 1. One way of determining object WFs is by specifying user perceived Level of Interest (LoI) as *high, medium,* and *low*. A user indicating that audio is more important than video can lead to an LoI value of *high* for audio (e.g. a WF of 1 for the audio object UF), and an LoI value of *low* for video (e.g. a WF of 0.25 for the video object UF). The *Adaptation Policy* options specify the actions to be taken as a result of change in a) User interest (LoI); b) Network resources; and, c) Frame rate. For example, in response to decrease in bandwidth, a lower bitrate codec may be selected; or, to maintain a stable frame rate, the order in which to degrade the quality of objects may be specified. Parameters of the service profile relating to network requirements are updated dynamically over the course of the service lifetime (e.g. media streaming object is started/stopped; a new user joins a shared 3D environment; certain WFs have changed). Based on the given service profile and the particular client profile, the matching process is conducted by the PM to determine *feasible* service configurations, such that: (1) A client's terminal capabilities can support the service processing requirements; and (2) The client's access network can support the minimum requirements for all (required) VE objects; and (3) The client's preferences in terms of acceptable media components and maximum acceptable download time can be met.

After the matching process, the PM extracts a set of potential session parameters (may include multiple potential media formats, codec types, etc.) from those service configurations that are feasible. These parameters are passed to the QNC module, which sends a session offer to the *Client* with a set of session description parameters. The *Client* may then accept/deny/modify offered parameters. A message is returned indicating the subset of offered parameters agreed to by the *Client*. Network entities authorize resources based on the agreed parameter subset. The returned session parameter subset and authorization is passed back to the PM, which then orders the feasible service configurations based on achievable user perceived quality into a so-called *degradation path* from the highest to the lowest quality configuration. Establishment of a degradation path is determined by user preferences (e.g. a user considers audio to be more valuable than video). This is used when service degradation or upgrading is necessary. Finally, the service profile corresponding to the highest quality feasible configuration is passed on to the *QoS Optimization Process* (QOP) module.

**QoS Optimization Process.** The *QOP* is responsible for calculating the optimal resource allocation and application operating point. The goal is to maximize user perceived quality, by combining the user's notion of relative VE object "importance" and UF based adaptation, and taking into account the following constraints: (1) Requested network resources must be less than or equal to authorized resources; (2) Allocated resources must fall within threshold values for service requirements across all VE objects; (3) Total price of allocated resources must be less than or equal to specified user budget; (4) The configured operating point must satisfy negotiated parameters (terminal capabilities). The function to be maximized is a linear combination of UFs multiplied by WFs defined across all VE objects. A detailed description of the objective function and constraint formulation is given in [12]. After calculation, the $QOP$ passes the final profile (via the $QNC$ module) specifying the calculated application operating point and required resources to the $QNC$ module of the *Application Server*.

## 3.2 Application Server

The *Application Server* is responsible for retrieving and adapting stored VR service content based on the negotiation and calculation conducted by the *Access and Control* entities. The $QNC$ module receives the final profile and initiates resource reservation mechanisms according to it. The final service profile is then passed to the *VR service processor*, which retrieves the service content from the VR service repository (possibly modified by the *VR service transcoder*), and forwards the content to the end user.

A user's interest in VE objects is subject to dynamic changes. A change in user's LoI , a user interaction, or a change detected by the application cause an event to be passed from the application to the *Event Manager* (EM). The *EM* then determines wheter the change is "significant" enough to require recalculation of the optimal application operating point and reallocation of network resources to meet new QoS requirements. It maps the event to an LoI for a given VE object, and sends to the $QOP$, which consults the *user interest adaptation policy* (of the service profile) and updates object profiles.

## 3.3 Session Establishment and Modification

The process of initial session establishment is shown in Fig. 4. At the top of the diagram we see the modules of the proposed model involved in the process. For multimedia services, the model is not tied to any particular session setup protocol, but they could be mapped, for example, to SIP messages such as INVITE, SESSION PROGRESS, PRACK, OK, ACK, etc. We outline three general scenarios supported by the proposed model involving optimization and renegotiation procedures invoked during the course of the service lifetime, due to changes in: (1) resource availability/cost; (2) client profile (preferences, terminal capabilities, access network); and (3) service requirements (object profiles).

**Scenario 1: Change in resource availability/cost**
Procedures related to a change occurring in resource availability are shown in Fig. 5. When available bandwidth decreases (or cost increases), the QOP runs the optimization process and searches for an optimal operating point and resource allocation (arrow *3*). As long as no solution is found within given constraints, service degradation is requested, based on the previously defined degradation path,

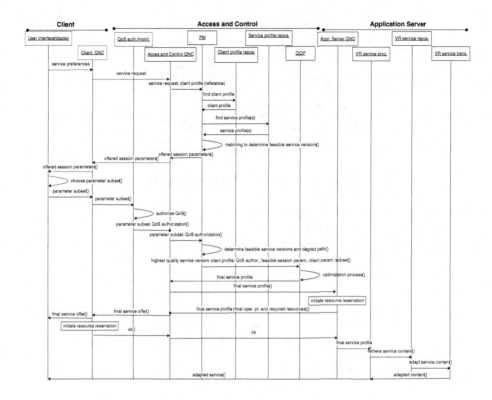

**Fig. 4.** Initial VR session establishment

and the optimization process is run (arrows *4–6*). Once a solution is found, the new final service profile is sent to the QNC module. A session description update is sent to the client indicating newly determined session parameters, and the final service profile is passed to the Application Server responsible for updating session parameters accordingly. If no solution is found, the client is informed of session establishment failure. In case of increased resource availability (or, cost decrease), the QOP sets the current version to the highest quality version in the degradation path and then runs the optimization process again.

**Scenario 2: Change in client profile**
The client profile may change due to dynamic changes in the access network, in software configurations (e.g. new codec), in terminal resource usage, in user preferences, as well as in the terminal itself. In addition, a service or network provider may update client profile parameters such as those relating to service subscription data. The signaling related to changes in client profile is shown in Fig. 6. The PM conducts a new matching process to determine the feasible session parameters (arrow *3*). If the new client profile parameters and the old client profile parameters yield the same matching results, session data is updated with the new client profile and the optimization process is invoked (arrows *14–15*). Procedures following this process correspond to those already shown in

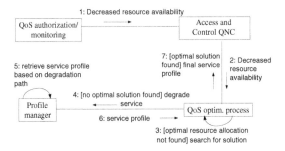

**Fig. 5.** Change in resource availability

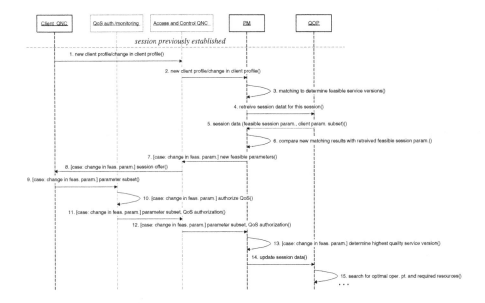

**Fig. 6.** Change in client profile

Fig. 4. On the other hand, if the result of the new matching process is not the same as the one obtained from the previously conducted matching, this means that a change has occurred in the media parameters that are considered feasible for this session (For example, due to decreased bandwidth in the client access network, video streaming may no longer be supported.) In this case, negotiation procedures are invoked (arrows *7–13*).

### Scenario 3: Change in service requirements

Changes in service requirements, such as change of an object's UF (LoI), or changes in an object's assigned WF, are signaled to the QOP by the EM module of the Application Server, as shown in Fig. 7 (arrows *2–6*). The QOP updates the necessary object profiles in the corresponding session data (arrow *7*), and re-formulates the objective function based on the current UF states and WFs

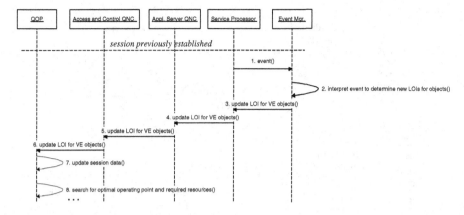

**Fig. 7.** Change in service requirements

(arrow *8*). Finally, it should be noted that re-calculation of the optimal operating point at each and every change in constraining factors is not desired (nor viable). Rather, thresholds may be established indicating modifications "important enough" to do so. Determining such thresholds, and effects thereof onto user perceived value, requires extended study and is considered for future work.

## 4   Case Study: Mapping to IP Multimedia Subsystem

While the discussed model is independent of the particular network, we conduct a case study to determine model applicability in UMTS networks by mapping it onto the IMS, as shown in Fig. 8. The IMS session establishment begins with an IMS terminal, shown as User Equipment (UE) (e.g. mobile phone, PC), obtaining access to the IMS through an access network (e.g. GPRS). The next steps involve the allocation of a Proxy-Call Session Control Function (P-CSCF) to serve as an outbound/inbound SIP proxy, and SIP application level registration to the IMS network. The P-CSCF also interfaces with a Policy Decision Function (PDF) which authorizes the use of bearer and QoS resources within the access network [2]. The central node of the signaling plane that is responsible for session establishment, modification, and release is the Serving-CSCF (S-CSCF), located in the user's home network. The S-CSCF interrogates the Home Subscriber Server (HSS) to access user profile information, fetch subscription data, and for authentication/authorization/accounting purposes. Additionally, the S-CSCF plays an important role in service provision by invoking one or more Application Servers (AS). The (IMS) Application and Services domain hosts both application/content, and reusable common functions that provide value-added services (e.g. messaging, presence). Various other ASs (e.g. gaming server) may be invoked along the signaling path. Having briefly described the IMS, we now present the mapping of the proposed model functionality onto it.

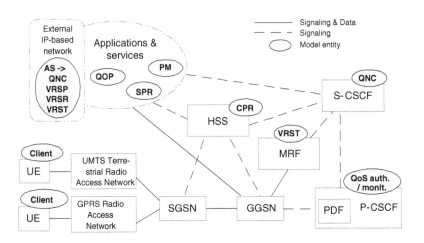

**Fig. 8.** Mapping of model entities to the IMS architecture

The *Client* is located in the *User Equipment* (UE). We map the functionality of the *Access and Control QoS authorization/monitoring* entity to the P-CSCF and the PDF. The *Access and Control QNC* entity is mapped to the S-CSCF. The *client profile repository* (CPR) storing user profile information is considered to be a part of the HSS. We place the *VR Service Profile Repository* (SPR) in the Application and Services domain, where it is interrogated by the S-CSCF. The *PM*, responsible for matching client profile parameters with service requirements, and the *QOP*, which calculates the optimal service operating point and resource allocation, are both mapped to a SIP AS in the Application and Services domain. With the specification of generic client and service profiles, the implementation of these functions is independent of the actual application content being delivered to the end user. From an IMS perspective, introducing such enhanced QoS support in the network, as a generic "reusable service", could lead to quicker time-to-market for new services, with service providers only being required to provide the network with a service profile stating service requirements. The network would then take care of determining the QoS parameters that would maximize user perceived quality. The *Application Server* (AS) is composed of four modules: *QNC*, *VR service processor* (VRSP), *VR service repository* (VRSR), and *VR service transcoder* (VRST). These modules may reside in the IMS network or on a server in an External IP Based Network (ExIPBN). Transcoding functionality of the *VRST* module may be mapped to the IMS *Media Resource Function* (MRF), responsible for the manipulation of multimedia streams. The proposed mapping indicates possible upgrades to some mentioned IMS entities, but also in most cases identifies existing functionality that should be used. The main enhancement to IMS would be the AS providing advanced QoS support in a reusable way. In addition, our approach stresses the need for standardization of QoS parameter specification in the form of client and service profiles.

# 5  Conclusions and Future Work

In this work, we presented a proposed QoS control model supporting the signaling, negotiation, and adaptation of QoS for multimedia services, intended for use in NGNs, and mapped onto the 3GPP IMS. Two issues may be identified as future work. The foremost is the question of scalability, considering the overhead resulting from QoS related signaling, both in terms of cost and time. The second important issue for further study is security. With regards to scalability issues relating to model deployment, it is clear that for a large number of users, running the optimization procedure separately for each user and when dynamic changes occur is definitely time consuming and costly. Besides the establishment of re-calculation thresholds, a solution may be to offer a set of discrete solutions calculated in advance for particular combinations of constraints. Investigation of such issues is considered for future work. A key requirement is for dynamic service adaptation and QoS renegotiation to occur with minimal user perceived disruptions.

# References

1. –, 3GPP TS 23.228: IP Multimedia Subsystem (IMS); Stage 2. June (2005)
2. –, 3GPP TS 23.803: Evolution of Policy Control and Charging. September (2005)
3. –, ITU-T Recommendation Y.2001: General Overview of NGN. December (2004)
4. –, ITU-T Recommendation H.360: An architecture for end-to-end QoS control and signalling. March (2004)
5. Araniti, G., De Meo, P., Iera, A., Ursino, D.: Adaptively Controlling the QoS of Multimedia Wireless Applications through 'User Profiling' Techniques. *IEEE J. on Selected Areas in Communications*, Vol. 21, No. 10, Dec. (2003) 1546-1556
6. Gao, X., Wu, G., Miki, T.: End-to-end QoS Provisioning in Mobile Heterogeneous Networks. *IEEE Wireless Communications Magazine*, June (2004) 24-34
7. Handley, M., Jacobson, V.: SDP: Session Description Protocol. IETF RFC 2327 (1998)
8. Guenkova-Luy, T., Kassler, A. J., Mandato, D.: End-to-End Quality-of-Service Coordination for Mobile Multimedia Applications. *IEEE J. on Selected Areas in Communications*, Vol. 22, No. 5, June (2004) 889-903
9. Prior, R., Sargento, S., Gomes, D., Aguiar, R. L.: Heterogeneous Signaling Framework for End-to-end QoS support in Next Generation Networks. Proc. of *38th Hawaii International Conf. on Systems Sciences* (HICSS-38 2005), CD-ROM/ Abstracts Proceedings, Big Island, HI, USA January (2005)
10. Rosenberg, J., et al.: SIP: Session Initiation Protocol. IETF RFC 3261 (2002)
11. Skorin-Kapov, L., Mikic, D., Vilendecic, D., Huljenic, D.: Analysis of end-to-end QoS for networked virtual reality services in UMTS. *IEEE Communications Magazine*, Vol. 42 No. 4, April (2004) 49-55
12. Skorin-Kapov, L., Matijasevic, M.: Dynamic QoS Negotiation and Adaptation for Networked Virtual Reality Services Proc. of *IEEE WoWMoM 2005*, Taormina, Italy, June (2005) 344-351
13. Wang, X., Schulzrinne, H.: An Integrated Resource Negotiation, Pricing, and QoS adaptation framework for multimedia applications *IEEE J. on Selected Areas in Communications*, Vol. 18, No. 12, Dec. (2000) 2514-2529

# Optimization of Charging Control in 3G Mobile Networks

Farah Khan and Nigel Baker

Centre for Complex Cooperative Systems, UWE Bristol UK, BS16 1QY,
Tel.: (+44) 07769782368
Farah.Khan@uwe.ac.uk, Nigel.Baker@uwe.ac.uk

**Abstract.** The charging predicament in communication networks is complex and multifaceted because it involves issues such as customer requirements whilst ensuring compatibility with existing and planned protocols and a diverse set of desired services. Therefore, it is essential to understand and evaluate the usefulness of available and emerging new techniques that provide improvement and extend the capability of charging schemes in order to reduce congestion, ensure and deliver quality of service to the users who need it most and hence achieve the most efficient consumption of resources available. This paper describes a charging simulation model based on Opnet UMTS model for the dimensioning of Charging Data Record (CDR) generation process by analysing the real network load, to estimate how many CDRs are generated at various points in the network and to realise optimisation techniques to assess the reasonable time duration for trigger settings and optimal rate of CDR generation.

**Keywords:** Charging data, Tariff Models, UMTS, Opnet.

## 1 Introduction

"It is said that the electronic-economy, based upon communication networks that provide businesses with new ways to access their customers, is destined to be much more than a simple sector of the economy. It will someday be the economy." [1] The convergence of cellular networks with other technologies like Bluetooth, WIMAX and Internet; has a strong impact on the evolution of services offered by the mobile network operators. There will be a wide variety of application and services offered limited only by the imagination and marketing capability of the service providers and third party vendors. This means that the converged network technologies and communication infrastructures should cooperate seamlessly and congruously to ensure consistent, transparent and ubiquitous service provision to end-users.

Information transport services typically segregated different media types but digital convergence and associated packetization of information now enables transport of integrated multimedia flows over 3G mobile systems. Due to the differentiation of services according to user's requirements; network operators and service providers face a greater challenge, beyond the technological issues, related to implementation and deployment of more effective and logical charging mechanisms. In the past with national monopoly telecommunication companies there was no pressure for charges to mirror costs, but this pressure has been growing with the impact of competition. The involvement of new players in the service provision process and the introduction

Y. Koucheryavy, J. Harju, and V.B. Iversen (Eds.): NEW2AN 2006, LNCS 4003, pp. 420–432, 2006.

of a flexible enterprise model which enables the service deployment without the need of additional subscriptions for the user, mean new charging requirement and solution. Until recently, the network elements and support systems used closed proprietary systems to capture the usage, which does not provide a solution for the differentiated charging schemes as required by Third Generation (3G) applications and service provisioning.

Paradigms for the charging of information transport services are intimately related to the technological foundations on which the respective services are based. For example, in the Public Switched Telephone Network (PSTN), the networking paradigm is one of hierarchical logical topologies with standardised fixed bandwidth access channels interconnected via circuit-switched facilities, which can be assigned on a dedicated basis to particular user sessions. The associated generic charging model involves flat (monthly), measured rates for local calls, and metered toll charges levied on the basis of time and/or distance for non-local calls. There is now a significant need for suitable systems, which can quickly establish any type of charging scheme for the use of any service. It should be possible to set up flexible, customised usage based charging schemes for any type of service elements. Therefore it is important to analyse various service scenarios according to customer behaviour. In previous work described under Section 3, the authors has performed Usage based study based on charging data generated for different service scenarios and comparison of Charging Data Records (CDRs) generated from a test GPRS network. This paper describes the next phase of this research as the OPNET based Charging simulation to abstract the behaviour of a real system that is amenable to experimentation in a safe and efficient manner. Section 2 describes the high-level charging principles and mechanisms for mobile networks. The simulation model is explained in section 4 and section 5 concludes the paper with direction of future work.

## 2 Charging and Billing Mechanisms

The principles and requirements related to charging and billing of mobile communication has been evolving according to the level of services offered and depending on the demands of the consumers. With all mobile networks, subscriber charging is associated with unique user identification. The basic structure of the charging model is usually consistent with the type of services offered by the serving networks; however the subscriber is given various options to select the charging model which bests suits their needs. These options vary according to the type of the call/session. For example, whether it is an incoming or outgoing call, whether it is a request for the service for which the user is already subscribed for or it is a request for new services. However, the main requirements and high-level charging and billing principles relate to two major forms of charging mechanisms; Online and Offline charging.

GSM/UMTS networks provide functions that implement offline and/or online charging mechanisms on the bearer (e.g. GPRS), subsystem (e.g. IMS) and service (e.g. MMS) levels. In order to support these charging mechanisms, the network performs real-time monitoring of resource usage on the above three levels in order to detect the relevant chargeable events. Typical examples of network resource usage are a voice call of certain duration, the transport of a certain volume of data, or the submission of a MM of a certain size. The network resource usage requests may be initiated by the UE (MO case) or by the network (MT case). Offline and online charging may be performed simultaneously and independently for the same chargeable event. [4]

## 2.1  Offline Charging

Offline Charging is a process where charging information for network resource usage is collected concurrently with that resource usage. The charging information is then passed through a chain of logical charging functions and at the end of this process, the network creates CDR files. These files are then transferred and processed offline by the Billing Domain for the purpose of subscriber billing, inter-operator account settlement and determination of charges, generally seen as itemised charging based on Call Event Recording (CER). Thereby offline charging is a mechanism where charging information does not affect, the service rendered, in real time.

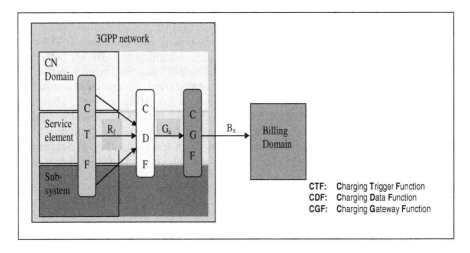

**Fig. 1.** Logical ubiquitous offline charging architecture

## 2.2  Online Charging

Online Charging is a process where charging information for network resource usage is collected concurrently with that resource usage in the same fashion as in offline charging. However, authorization for the network resource usage must be obtained by the network prior to the actual resource usage to occur. This authorization is granted by the Online Charging System (OCS) upon request from the network. When receiving a network resource usage request, the network assembles the relevant charging information and generates a charging event towards the OCS in real-time. The OCS then returns an appropriate resource usage authorization. The resource usage authorization may be limited in its scope (e.g. volume of data or duration) and may have to be renewed from time to time for as long as the user's network resource usage persists. Therefore in online charging, a direct interaction of the charging mechanism with the control of network resource usage is required. [4]

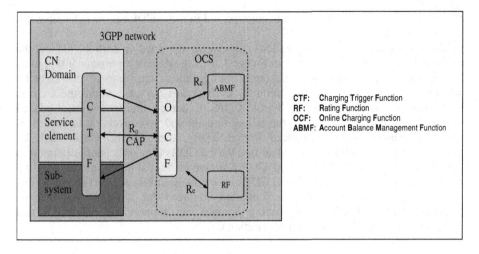

**Fig. 2.** Logical ubiquitous online charging architecture

## 2.3   Charging Data Analysis

The GSM and UMTS PLMN support a wide range of packet-based services by means of the General Packet Radio Service (GPRS). The charging functions for GPRS include the generation of CDRs by the Serving GPRS Support Node and the Gateway GPRS Support Node as well as the transport of these CDRs to a Billing System (BS) through a CGF. CDRs in GPRS includes fields identifying the user, the session and the network elements as well as information on the network resources and services used to support a subscriber session in the PS domain. The CDR fields are divided into two types; Mandatory (a field that shall be always present in the CDR) and Conditional (a field that shall be present in a CDR if certain conditions are met, these conditions are specified as part of the field definition).

Since CDR processing and transport consume network resources, operators may opt to eliminate some of the fields. This operator provisionable reduction is specified by the field category. A logical diagram showing the possible field categories is shown in Figure 4. The content of the CDRs shall be specified on all the open network interfaces that are used for CDR transport. They include the GSN to CGF interface and the outward interface from the core network to the billing system. The CDRs are categorised into following types as functional in the ISS cluster over these interfaces:

**Partial CDR:** A CDR that provides information on part of a subscriber session. A long session may be covered by several partial CDRs.

**Full-Qualified Partial CDR** (CPC): A partial CDR that contains a complete set of the fields. This includes all the mandatory and conditional fields as well as those fields that the PLMN operator has provisioned to be included in the CDR. The first Partial CDR shall be a Full Qualified Partial CDR.

**Reduced Partial CDR** (RPC): Partial CDRs that only provide mandatory fields and information regarding changes in the session parameters relative to the previous CDR.

For example, location information is not repeated in these CDRs if the subscriber did not change its location.

The possible charging configurations that can be supported on both the Ga and the outbound interfaces are illustrated in Figure 3. A is the default arrangement that must be supported by all systems. The other configurations are optional that may be supported in addition to configuration A. B illustrates the case where the CGF is converting Reduced to Complete Partial CDRs. C depicts the case where Reduced Partial CDRs can be received in the billing domain and no conversion is needed. If the CGF cannot support Reduced Partial CDRs, then all the GSNs shall be provisioned to generate only Full Qualified Partial CDRs (i.e., only configuration A is possible). On the other hand, if the CGF can convert Partial CDRs format then the GSNs may generate Reduced Partial CDRs based on the rules specified above. In this case configurations B can also be supported. Reduced Partial CDRs may also be generated by the GSNs if the billing domain can support the reduced format regardless of the CGF features (configuration C).

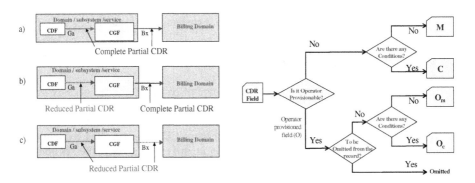

**Fig. 3.** Possible Configurations of Ga and Bx CDR Formats

**Fig. 4.** Logical Diagram illustrating the different CDR Field Categories

## 3 Charging Data Dimensioning

The role of charging is not only to cover the costs of service provision and generate income for the service provider, but also to influence the way customers use network services. This happens as each individual customer reacts to tariffs and seeks to minimise charges. The tariff structure should provide the right incentives for users to use network resources efficiently. This is the key idea of incentive compatibility. Tariffs should guide customers to select service and use the network in ways that are good for overall network performance. Tariffs, which are not incentive compatible, give wrong signals and lead users to use the network in very inefficient ways. A desirable property of a charging scheme would be to encourage the cooperative sharing of information and characterisation effort between the user and the network. [2]

### 3.1 Scenario Analysis

3G mobile systems will be distinctive in providing a variety of services with varying characteristics independent of the underlying network infrastructure. Due to the differentiation of services according to user's need, network operators and service providers face a great challenge to provide better Quality of service (QoS) and employ more effective and logical charging and billing mechanisms. Therefore it is important to analyse various service scenarios according to customer behaviour, which we have categorized as follows:

- High Fliers: Professional Workers (Applications assisting in their work).
- Trend Setters: Younger people (Teenagers to mid twenties mostly using data-light services).
- Assured: Mature generation (use technology to enhance their lifestyle).

These consumer types are analysed further according to the service package they choose for example the Trend Setters may either choose flat rate or prepaid packages to keep control on their expenditure whereas the High Fliers who will be using the service for critical business reasons may not worry about the cost so we assume they will be post-paid (usage based monthly payments) These Scenarios are then compared with the number of CDRs generated for a particular service or service mix with the relevant tariff models, to define the validation criteria of the results expected from the simulation. Overall the results are forecasted to give an idea of how efficient and incentive compatible a particular tariff model is as compared to the level of access and type of services offered within the subscription package.

### 3.2   CDR Flows Versus Trigger Settings

The authors have performed a usage related study based on the number of CDRs generated for different scenarios. It is a spreadsheet-based model, which makes assumptions regarding the proportion of time, which a particular type of consumer will spend on various types of services. According to the proportion of time which the users spend on individual services, the authors have calculated the total data transferred over the network for each individual type of service as shown in Table 1. Table 1 show that a user transfers 862 Kbytes of data by browsing the web for 72 minutes (40% of the PDP context duration which is assumed as 3hrs).

**Table 1.** Parameters and total data transferred by the user during a web browsing session

| Web browsing Service Parameters | | |
|---|---|---|
| Session traffic (kbytes) | 59.875 | |
| Mean session duration (seconds) | 300 | |
| | | |
| Transfer rate per minute (kbytes) | 11.975 | |
| | | |
| number of packet calls per session | 5 | |
| number of packets in each packet call | 25 | |
| packet size (bytes) | 479 | |
| | | |
| Total Web data (kbytes) | 11.975 x 72 | 862.2 |

The volume of data for web browsing session as shown in Table1 is calculated by considering the typical parameters which include the number of packet calls and within each packet call the number of packets transferred and the typical packet size. These parameters provide the total session traffic volume, which in the present case is 59.9 Kbytes (See table 1). Assuming that the mean web session duration is 300 seconds, the authors have derived the data transfer rate per minute as 11.9 Kbytes and therefore for the user who spent 72 minutes, it is 862 Kbytes. Adding the data for all service derives the total data volume which is 8.2 Mb, transferred by various users in 3 hr duration. This total data is then used to calculate the number of CDRs generated.

### 3.3 Results and Evaluation

Considering the data volume, the authors calculate the peak transfer rate per hour per context. With 8.2 Mb data transferred by 24k parallel PDP contexts in 3 hr duration, the peak transfer rate comes to be 90 Kb. Assuming that the mean to peak loading ratio is 50% and the GSN throughput is set fixed at 6k (100 byte) packet per second.

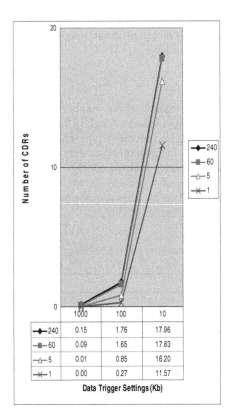

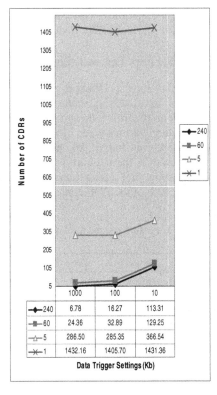

Chart1. Number of CDRs generated according to Data Volume Trigger Settings for 20% of PDP Contexts generating 80% of traffic

Chart2. Number of CDRs generated according to Data Volume and Open Time Trigger Settings for total traffic

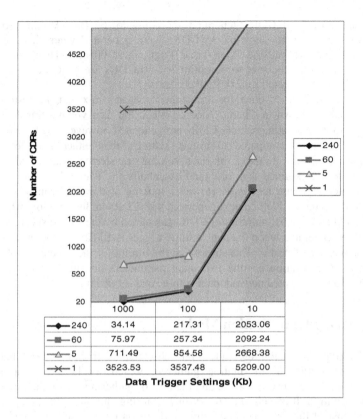

Chart3. Number of CDRs generated according to GPRS Test Data dimensioning statistics

The authors have modelled the CDR generation for the following two traffic profiles; one bursty and one non-bursty.

- 20% of contexts generate 80% of the traffic, active for 10 minutes each hour.
- 80% of contexts generate 20% of the traffic, active for 5 minutes every two hours.

Outside the active periods, the users will transfer no data at all, but the PDP Context will remain open. For simplification, the authors assume that only one type of CDR is used; either S-CDR from SGSN or G-CDR from GGSN. The authors then take the partial CDR triggers according to either Data Volume or Open Time to analyse the number of CDRs generated for various trigger combinations.

Chart 1 show the number of CDRs generated according to the Data Volume triggers, which are set at 1000 Kbytes, 100 Kbytes and 10 Kbytes intervals. When the Data Volume Trigger is 100 Kbytes, as compared to 180 Kbytes of traffic, generated by 20% of the contexts, it is likely that only two or less CDRs will be generated when moving from time settings of 240 minutes down to 1 minute. However when the Data Volume Trigger is quite small as 10 Kbytes, it is likely to get 180k/10k = 18 CDRs per hour. As the Time Limit trigger is reduced, it is more likely that some of the CDRs will be triggered by Time Limit, and not by Data Volume. Chart 2 however shows the number of CDRs generated according to the Data Volume triggers, for 100% of PDP contexts showing 33 CDRs were generated when the Data Volume trigger was 100 Kb and the Open Time Trigger was 60 min. Therefore the rate of CDR generation can be said to be optimal if 30 CDRs are generated every hour for the average traffic created by 24k PDP contexts.

The same analysis is done for a customer scenario mix and the results follow a similar trend as shown in Charts 1 and 2. When the data volume transferred is high compared to trigger settings, more CDRs are generated however when the data volume transferred is small compared to trigger settings, then either 1 or less CDR are generated. Therefore in terms of data volume transferred, triggers should be set according to the assumed data rate (and essentially the data volume generated) for a particular service, customer type. However looking at the data from the Test GPRS network, the same 24 PDP context generated 408 CDRs in 3hr duration, which is about 30% more than the ideal numbers of CDRs calculated by the spread sheet model as 257 CDRs to be generated when the data volume trigger is 100kb and the open time trigger is 60 minutes (see Chart3). This shows that the timing for the generation of CDRs and rate of CDR generation are the two main parameters that need to be set optimally according to the expected network data volume and service mix.

## 4   Charging Optimisation Model

The charging dimensioning study highlights that a trade-off exist between the amounts of statistical information gathered, which would allow more accurate characterisation of a user's traffic, and the cost of gathering, storing and manipulating such information. It is well known fact that charging and billing is a major portion of the total cost of the networks and if this trade-off is not realised, avalanche of CDRs will possibly be generated from the network.

Since bandwidth is a limited and scarce resource within a mobile network, avalanche of CDRs would cause the network congestion and can cause signalling overload for some applications requiring online charging. Therefore it would benefice the network to realise the optimal setting for CDR triggering and transfer functions. The authors have performed a charging simulation using Optimized Network Engineering Tools (OPNET) [9] UMTS model for the dimensioning of CDR generation process to estimate and forecast the possibility of CDR avalanche. The main goal/focus to design this simulation is to predict CDR generation based on various traffic load conditions and system parameters.

### 4.1   Simulation Modelling

Proper modelling is very critical for any system-level simulator as the objective is to build a model which has similar functionality and behaves similar to the real system. OPNET provides a comprehensive development environment supporting the modelling of UMTS networks and incorporates tools for all phases of a simulation study, including model design, simulation, data collection and analysis. Both behaviour and performance of modelled systems can be analyzed by performing discrete event simulations [9].

For practical reasons the charging simulation only implements the mandatory network features required for UMTS charging data trigger and generation. The Opnet UMTS core network model is based on 3rd Generation Partnership Project (3GPP) Release 1999 standards. Charging information in the packet domain core network is collected by two network nodes: the Serving GPRS Support Nodes (SGSNs) and Gateway GPRS Support Nodes (GGSNs), which are serving that UE. The SGSN collects local radio network usage charging information whilst the GGSN collects external data network usage charging information for each UE. The charging simulation has enhanced the functionality of both SGSN and GGSN to generate the relevant CDRs (S-CDR at SGSN and G-CDR at GGSN). In addition the GTP process model is programmed to monitor the data packets for the calculation of traffic volume necessary to implement the charging triggers.

### 4.2 Simulation Scenarios

Charging simulation modelling adopts a user centric approach for driving the simulation environment. Based on this approach a number of "User Profiles" are created, each using a different mix of applications like e-mail, multimedia messaging, voice, internet access and file download as categorised below:

- Leisure 1: Email(light) and Web-browsing (light)
- Leisure 2: Multimedia Messaging (light) and File Transfer (light)
- Corporate: Email (heavy) and Web-browsing (heavy)
- IMS: Voice over IP, Push to talk and Database access (light), always on profile.

These user profiles are represented by User Equipment (UEs) densities in various cells with their own sets of service preferences and usage patterns The current stage of simulation generates charging data for 100 UEs supporting one of the above profiles with the data trigger setting of 10KB, which means that a partial SCDR is generated after every 10KB of data traffic volume (uplink plus downlink) being transferred over the network.

The model generates initial CDRs for each UE at the PDP activation and verifies that no traffic is transmitted before successful PDP activation has taken place. The transmission of data volume occurs at the start time of each profile which is kept random to represent an unpredictable network activity as depicted in the real system.

### 4.3 Results and Discussion

The total charging data mounts as the simulation progresses, however it is interesting to note how charging data can vary according to the different user profiles characterised by application demands on traffic data volume resulting into few or more CDRs. Therefore charging trigger function has to be intelligent according to the resource usage of the network. Graph 1 shows the traffic sent for various applications used as part of the User profiles. As the simulation progress, UEs use a combination of these applications during a single PDP context and therefore different applications are showing traffic in parallel; however the bytes per seconds are varying per applications reflecting the mix of User profiles using either data heavy or data light version of these applications.

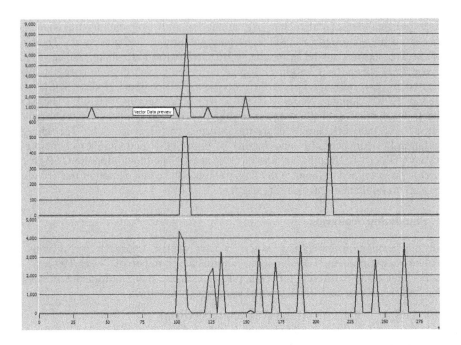

Graph 1. Traffic Volume (bytes/second) for Email, FTTP and HTTP applications

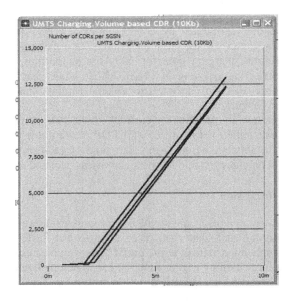

Graph 2. Number of CDRs generated per SGSN along the simulation time

Graph 2 shows the number of CDRs generated per SGSN for the simulation run with typical busy hour load. Each SGSN is assigned with a random mix of UEs with one of the four profiles using the above shown applications, however about 20% of

UEs were using the corporate profile transmitting data volume according to data heavy services, very similar to the bursty scenario used in the usage based study in section 3 where 20% of context created 80% of traffic over the network. It is evident from the graph that a CDR avalanche condition can be created with 10Kb or even 100Kb of data volume trigger under such scenario.

Graph 3 however shows a scenario where 30% of UEs are connected with always on session with the IMS profiles using applications like Presence and VOIP. Under such scenario even the 10kb trigger setting is quite high as the data volume created over the network is quite low. A time trigger setting for CDR generation would be better suited for this scenario.

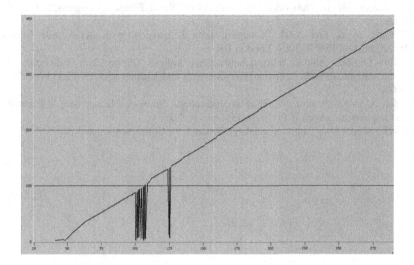

Graph 3. Number of CDRs generated by User Profiles using Data Light applications

## 5   Conclusions and Future Work

The simulation realises the importance of optimal generation and transport of CDRs according to controlled input in terms of user profiles, data volume and time trigger settings. In specific terms CDRs measurements should be performed according to the thresholds or limits of bandwidth on network nodes and interfaces. A desirable ratio of CDRs versus the traffic data volume transferred over the network in parallel is 2% under every scenario. Since GPRS was initially designed for always on PDP session, this ratio can be achieved by formalising the rules of charging data generation under various scenarios. However during the last few years, since GPRS is used under a dialup mode and a new PDP context is created for every new service API, a higher percentage of CDRs is also generated due to repeated PDP activation and deactivation of the session.

The authors would extend the simulation to test further scenarios for the possibility of CDR avalanche leading to the formalisation of CDR generation rules.

# References

[1] C Courcoubetis et al. 2003, "Pricing Communications Networks, Economics, Technology and Modelling", John Wiley & Sons Ltd, England.

[2] D. J. Songhurst, 1999, "Charging Communication Networks", Elsevier Science B.V, Amsterdam.

[3] Maria et al. 2001, "Subscription management and Charging for Value-added services in UMTS networks", Proceedings of IEEE VTC01 Pg2160

[4] 3GPP Technical Specification, June 2006, "Charging Management – Charging architecture and principles" Rel6, (3GPP TS 32.240 V6.2.0).

[5] F Khan et al. May 2002, "Impact of Service Mix and Tariff Models on Charging in Third Generation Mobile Networks)" Proceedings of World Wireless Congress, San Francisco USA.

[6] F Khan et al. Oct 2004, "Charging Data Dimensioning in 3G Mobile Networks" Proceedings of IEE 3G2004, London UK.

[7] 3GPP Digital cellular telecommunications system (Phase 2+). Universal Mobile Telecommunication System (UMTS) Charging and Billing. (3G TS 22.115 V3.2.0), Jan 2000.

[8] Salah Aidarous et al, 1998, Telecommunications Network Management – Technologies and Implementations", IEEE Press, Piscataway, NJ.

[9] UMTS Model User Guide, Specialized Model User Guide, www.opnet.com.

# Asymptotic Performance Analysis of a TCM System over Fading Multiple-Access Channel

Felix Taubin

St.-Petersburg State University on Aerospace Instrumentation
Bolshaya Morskaya 67, 190000 St.-Petersburg, Russia
ftaubin@yahoo.com

**Abstract.** Trellis coded modulation (TCM) with maximum likelihood (ML) decoding for multiple-access transmission over Rayleigh fading channel is considered. An upper bound for the bit-error rate is derived by considering a super-trellis, i.e. the combination of the user's trellis codes. This upper bound leads to the definition of design criteria for the selection of preferable user's codes. It is shown also that at high signal-to-noise ratio, the logarithm of the bit error probability goes to zero with the same slope as for the single-user transmission. The asymptotic energy loss due to other-users interference is completely determined by the factor that accounts for an increase the number of nearest (in terms of a minimum distance) neighbors in the supertrellis with respect to the code trellis of specific user.

**Keywords:** Trellis coded modulation, multiple-access transmission, fading channels, maximum likelihood decoding, bit-error rate, other-users interference.

## 1 Introduction

Conventional multiple-access techniques rely on bandwidth expansion in some form, and conventional single-user detection techniques are then usually applied. In code-division multiple-access (CDMA) systems, each user is assigned $N$ times the resources normally required for a single-user. The same resources are then shared by multiple users, where a total load of $N$ users are considered good applying conventional techniques. The performance of CDMA systems depends on the correlation properties of the spreading codes. If all codes are orthogonal, in a perfectly synchronous communication, multiple access interference does not arise and the number of served users $N$ is limited only by the dimension of the global signal space from which the signatures are chosen. When $N$ exceeds this dimension, the system is oversaturated and the signatures can no longer remain orthogonal. In this case, multiple access interference becomes unavoidable and multiuser detection becomes necessary to combat [1].

Oversaturation strategies are estimated to play important role in future wireless communication systems due to the ever increasing requirement for system flexibility. Indeed, in systems where a conventional DS-CDMA approach is adopted, e.g. W-CDMA, it may happen that irregular and sudden increase in the traffic demand, e.g. in downlink due to Web browsing services, can not be supported, the limiting factor being the length of the employed orthogonal spreading codes [2,3].

Y. Koucheryavy, J. Harju, and V.B. Iversen (Eds.): NEW2AN 2006, LNCS 4003, pp. 433–440, 2006.
© Springer-Verlag Berlin Heidelberg 2006

In contrast with conventional multiple-access techniques, one can consider a nar-rowband multiple-access scheme, where multiple users share the same bandwidth as required by a single-user. The narrowband multiple-access scheme is based on band-width-efficient trellis coded modulation (TCM) and is termed trellis code multiple-access [4,5]. Instead of using a unique spreading sequence for each user as in CDMA, the multiple-access is provided entirely by some user-specific feature.

In the following, the TCM multiple-access system over a fading channel is consid-ered, and maximum likelihood (ML) decoding is examined. For this scenario, an up-per bound for the bit-error rate is derived. Asymptotic analysis of the upper bound provides one to suggest design criteria for the selection of preferable user's codes, and to get the asymptotical energy loss due to other-users interference.

## 2  System Model

Consider a multiple-access communication system in which $N$ statistically inde-pendent sources transmit information to $N$ separate data sinks. Each source is connected to a separate encoder. The multiple-access channel has $N$ input ports and single output port connected to a single (common) decoder.

Binary information stream $\mathbf{a}_i$ of $i$ th source is encoded by encoder of a trellis code $C_i$ with the rate $R_i = k_i / n$ bit per coded symbol and constraint length $v_i$. The $i$ th coded stream $\mathbf{c}_i$ is the sequence of complex-valued $n$-tuples: $\mathbf{c}_i = (\underline{c}_{i1}, \underline{c}_{i2}, \dots)$, where $\underline{c}_{ij} = (c_{ij1}, \dots c_{ijn})$. It is assumed that i) an identical $q-$ary signal set $S$ is used by all encoders, i.e. $c_{ijm} \in S$, and  ii) the set $S$ is nor-malized in such a way that variance of code sequence is equal to 1, i.e. $\overline{\left| c_{ijm} \right|^2} = 1$.

The code symbols of $i$ th user are ideally interleaved, multiplied by the coefficient $\sqrt{R_i E_{bi}}$, where $E_{bi}$ is the energy per bit, and transmitted over a $i$ th discrete-time Rayleigh fading channel (a $i$ th subchannel of the multiple-access channel ). Hereaf-ter it is assumed that i) $n$-tuples of all users are synchronously transmitted, and  ii) the $N$ Rayleigh fading channels (employing by $N$ users ) are mutually independent. The output symbols of the $N$ Rayleigh fading channels are added, generating the re-ceived sequence $\mathbf{r}$. After deinterleaving the symbols $r_l$, $l \geq 1$, of the received se-quence are given by

$$r_l = \sum_{i=1}^{N} \sqrt{R_i E_{bi}} \mu_{il} c_{il} + n_l. \tag{1}$$

In (1) $c_{il}$ denotes $l$ th symbol of the $i$ th coded stream $\mathbf{c}_i$ (obviously, $c_{il} = c_{ijm}$ for $j = \lfloor l/n \rfloor$, $m = l - n \cdot j$ ), $\left\{ \mu_{il} \middle| 1 \leq i \leq N, l \geq 1 \right\}$ is the set of mutually independent complex-valued Gaussian random variables with zero mean and variance $\overline{|\mu_{il}|^2} = \alpha_i$, and $\left\{ n_l \middle| l \geq 1 \right\}$ is the complex-valued AWGN with variance $N_0$.

## 3  ML Decoding and Performance Bound

Let us denote a superstream $\mathbf{c}$ as $\mathbf{c} = (\mathbf{c}_1, \dots \mathbf{c}_N)$. Assuming the exact values of $\left\{ \mu_{il} \middle| 1 \leq i \leq N, l \geq 1 \right\}$ are available for the receiver, the ML superstream's decoder has to maximize the metric

$$\Gamma(\mathbf{c}) = \sum_{l \geq 1} (2\mathrm{Re}[r_l^* \cdot \sum_{j=1}^{N} \sqrt{R_j E_{bj}} \, \mu_{jl} c_{jl}] - \left| \sum_{j=1}^{N} \sqrt{R_j E_{bj}} \, \mu_{jl} c_{jl} \right|^2) \qquad (2)$$

by searching through a supertrellis (combined trellis ) with $\sum_{j=1}^{N} v_j$ states. In (2) $*$ stands for complex conjugation, and $\mathrm{Re}[x]$ denotes the real part of $x$.

An upper bound on the average bit error probability of $i$ th user is obtained as

$$P_{bi} \leq \frac{1}{k_i} \sum_{\mathbf{c}} p(\mathbf{c}) \sum_{\tilde{\mathbf{c}} \in E_i(\mathbf{c})} wt_i(\mathbf{c}, \tilde{\mathbf{c}}) P(\mathbf{c} \rightarrow \tilde{\mathbf{c}}), \qquad (3)$$

where $wt_i(\mathbf{c}, \tilde{\mathbf{c}})$ is the number of $i$ th user's bit errors that occur when the superstream $\mathbf{c}$ is transmitted and the superstream $\tilde{\mathbf{c}} \neq \mathbf{c}$ is chosen by decoder, $p(\mathbf{c})$ is the probability of transmitting $\mathbf{c}$, $E_i(\mathbf{c})$ denotes the set of all possible detours (error events) $\tilde{\mathbf{c}} = (\tilde{\mathbf{c}}_1, \dots \tilde{\mathbf{c}}_N)$, $\tilde{\mathbf{c}}_i \neq \mathbf{c}_i$, in the supertrellis with respect to the superstream $\mathbf{c}$ starting at time 0, and $P(\mathbf{c} \rightarrow \tilde{\mathbf{c}})$ represents the pairwise error probability. The upper bound of (3) is efficiently evaluated using the transfer function bound approach (see for example Biglieri [6] ).

The pairwise error probability

$$P(\mathbf{c} \rightarrow \tilde{\mathbf{c}}) = \mathrm{Pr}\{\Gamma(\tilde{\mathbf{c}}) - \Gamma(\mathbf{c}) \geq 0\} =$$

$$\mathrm{Pr}\{\sum_{l \geq 1} (2\mathrm{Re}[n_l^* \cdot \sum_{j=1}^{N} \sqrt{R_j E_{bj}} \, \mu_{jl}(\tilde{c}_{jl} - c_{il})] - \left| \sum_{j=1}^{N} \sqrt{R_j E_{bj}} \, \mu_{jl}(\tilde{c}_{jl} - c_{jl}) \right|^2) \geq 0\}. \qquad (4)$$

Applying Chernoff bound for upper bounding the pairwise error probability (4) conditioned on the matrix $\mathbf{M} = \left\{ \mu_{jl} \middle| 1 \leq j \leq N, l \geq 1 \right\}$, one obtains:

$$P(\mathbf{c} \rightarrow \tilde{\mathbf{c}} | \mathbf{M}) \leq \min_{\lambda > 0} \exp((-\lambda + \lambda^2 N_0) d^2(\mathbf{c}, \tilde{\mathbf{c}})) = \exp(-d^2(\mathbf{c}, \tilde{\mathbf{c}})/4N_0), \qquad (5)$$

where

$$d^2(\mathbf{c},\tilde{\mathbf{c}}) = \sum_{l \geq 1} \left| \sum_{j=1}^{N} \sqrt{R_j E_{bj}} \, \mu_{jl}(\tilde{c}_{jl} - c_{jl}) \right|^2 .$$

Averaging (5) over the probability density function of $M$ gives

$$P(\mathbf{c} \rightarrow \tilde{\mathbf{c}}) \leq \prod_{l \geq 1} \left\{ 1 + \frac{1}{4N_0} \sum_{j=1}^{N} \alpha_j R_j E_{bj} |c_{jl} - \tilde{c}_{jl}|^2 \right\}^{-1} \tag{6}$$

Let us denote

$$D = \left\{ d \| c - \tilde{c} \| = d; \ c, \tilde{c} \in S \right\},$$

$$\Delta = \left\{ \underline{d} = (d_1, \ldots, d_N) \mid d_i \in D, \ 1 \leq j \leq N \right\}$$

$$J(\mathbf{c}, \tilde{\mathbf{c}}, \underline{d}) = \left\{ l \geq 1 \| c_{jl} - \tilde{c}_{jl} \| = d_j, 1 \leq j \leq N, \underline{d} \in \Delta \right\},$$

$$w(\mathbf{c}, \tilde{\mathbf{c}}, \underline{d}) = \text{card} \, (J(\mathbf{c}, \tilde{\mathbf{c}}, \underline{d})).$$

Then the upper bound (6) can be represented in the form

$$P(\mathbf{c} \rightarrow \tilde{\mathbf{c}}) \leq \prod_{\underline{d} \in \Delta} \left\{ 1 + \frac{1}{4N_0} \sum_{j=1}^{N} \alpha_j R_j E_{bj} d_j^2 \right\}^{-w(\mathbf{c}, \tilde{\mathbf{c}}, \underline{d})} . \tag{7}$$

Substituting (7) in (3), we obtain the desired performance upper bound as follows

$$P_{bi} \leq \frac{1}{k_i} \sum_{\mathbf{c}} p(\mathbf{c}) \sum_{\tilde{\mathbf{c}} \in E_i(\mathbf{c})} wt_i(\mathbf{c}, \tilde{\mathbf{c}}) \prod_{\underline{d} \in \Delta} \left\{ 1 + \frac{1}{4N_0} \sum_{j=1}^{N} \alpha_j R_j E_{bj} d_j^2 \right\}^{-w(\mathbf{c}, \tilde{\mathbf{c}}, \underline{d})} . \tag{8}$$

## 4 Asymptotic Analysis of Bit Error Rate

Clearly, the upper bound of (8) will be dominated by the term in summation which has the slowest rate in descent with $E_i / N_0$. This term represents a weighted sum of products each consisting of

$$\min_{\mathbf{c}; \tilde{\mathbf{c}} \in E_i(\mathbf{c})} \sum_{\underline{d} \in \Delta} w(\mathbf{c}, \tilde{\mathbf{c}}, \underline{d}) \tag{9}$$

factors. It is easy to verify that (9) coincides with the minimum Hamming distance $\delta_i$ of the code $C_i$. Let us denote $E_i(\mathbf{c}, \delta_i)$ the subset of $E_i(\mathbf{c})$, such that

$$E_i(\mathbf{c},\delta_i) = \Big\{ \tilde{\mathbf{c}} \mid \sum_{\underline{d}\in\Delta} w(\mathbf{c},\tilde{\mathbf{c}},\underline{d}) = \delta_i \Big\}.$$

Then the dominant term in the sum (8) has the form

$$\frac{1}{k_i}\sum_{\mathbf{c}} p(\mathbf{c}) \sum_{\tilde{\mathbf{c}}\in E_i(\mathbf{c},\delta_i)} wt_i(\mathbf{c},\tilde{\mathbf{c}}) \prod_{\underline{d}\in\Delta} \Big\{1+\frac{1}{4N_0}\sum_{j=1}^{N}\alpha_j R_j E_{bj} d_j^2\Big\}^{-w(\mathbf{c},\tilde{\mathbf{c}},\underline{d})}, \tag{10}$$

where each product consist of $\delta_i$ factors. It is significant that i) the subset $E_i(\mathbf{c},\delta_i)$ contains, in particular, the superstreams $\tilde{\mathbf{c}}$ of the form $\tilde{\mathbf{c}} = (\mathbf{0},...,\mathbf{0},\tilde{\mathbf{c}}_i,\mathbf{0},...,\mathbf{0})$ with $\tilde{\mathbf{c}}_i$ such that the Hamming distance between $\mathbf{c}_i$ and $\tilde{\mathbf{c}}_i$ is equal to $\delta_i$, and ii) the stream $\tilde{\mathbf{c}}_i$ of each superstream $\tilde{\mathbf{c}} \in E_i(\mathbf{c},\delta_i)$ has the same feature, namely the Hamming distance between $\mathbf{c}_i$ and $\tilde{\mathbf{c}}_i$ is equal to $\delta_i$.

Let $q_i(\delta_i)$ be the cardinality of the subset $E_i(\mathbf{c},\delta_i)$, and let $m_i(\delta_i)$ be the number of error events paths in the code $C_i$'s trellis that have the Hamming weight $\delta_i$. The ratio $q_i(\delta_i)/m_i(\delta_i)$ accounts for an increase of the number of nearest neighbors (in terms of $\delta_i$) in the supertrellis with respect to the code $C_i$'s trellis.

The product in (10) can be upper bounded in the form:

$$\prod_{\underline{d}\in\Delta}\Big\{1+\frac{1}{4N_0}\sum_{j=1}^{N}\alpha_j R_j E_{bj} d_j^2\Big\}^{-w(\mathbf{c},\tilde{\mathbf{c}},\underline{d})} \le (\alpha_i R_i E_{bi}/4N_0)^{-\delta_i} d_p^{-2}(\mathbf{c}_i,\tilde{\mathbf{c}}_i)\lambda_i(\mathbf{c},\tilde{\mathbf{c}}). \tag{11}$$

In (11) $d_p^2(\mathbf{c}_i,\tilde{\mathbf{c}}_i)$ denotes the product of the squared Euclidean distances along the error event path of the code $C_i$ [7], and the distance coefficient $\lambda_i(\mathbf{c},\tilde{\mathbf{c}})$, allowing for the influence of other users, is given by

$$\lambda_i(\mathbf{c},\tilde{\mathbf{c}}) = \prod_{\underline{d}\in\Delta}\Big\{\sum_{j=1}^{N}\frac{\alpha_j R_j E_{bj} d_j^2}{\alpha_i R_i E_{bi} d_i^2}\Big\}^{-w(\mathbf{c},\tilde{\mathbf{c}},\delta_i)}.$$

It is easy to see that

$$\lambda_i(\mathbf{c},\tilde{\mathbf{c}}) \le 1. \tag{12}$$

Applying the bound (11) to (10), one obtains that the dominant term is upper bounded by

$$\frac{1}{k_i}\Big(\frac{\alpha_i R_i E_{bi}}{4N_0}\Big)^{-\delta_i} d_{pi}^{-2}\sum_{\mathbf{c}} p(\mathbf{c}) \sum_{\tilde{\mathbf{c}}\in E_i(\mathbf{c},\delta_i)} wt_i(\mathbf{c},\tilde{\mathbf{c}})\lambda_i(\mathbf{c},\tilde{\mathbf{c}}),$$

where $d_{pi}^2$ denotes the minimum squared product distance of the code $C_i$.

Therefore, asymptotically with high signal-to-noise ratio, the average bit error probability of $i$ th user is approximately given by

$$P_{bi} \cong \frac{1}{k_i} \left( \frac{\alpha_i R_i E_{bi}}{4N_0} \right)^{-\delta_i} d_{pi}^{-2} \sum_{\mathbf{c}} p(\mathbf{c}) \sum_{\tilde{\mathbf{c}} \in E_i(\mathbf{c}, \delta_i)} wt_i(\mathbf{c}, \tilde{\mathbf{c}}) \lambda_i(\mathbf{c}, \tilde{\mathbf{c}}). \tag{13}$$

Taking into account the inequality (12), we get more simple (than (13)) approximation for the average bit error probability of $i$ th user:

$$P_{bi} \cong \frac{\chi_i(\delta_i) d_{pi}^{-2}}{k_i} \left( \frac{\alpha_i R_i E_{bi}}{4N_0} \right)^{-\delta_i}, \tag{14}$$

where $\chi_i(\delta_i)$ is the total number of bit errors for all error paths $\tilde{\mathbf{c}}_i$ in $\tilde{\mathbf{c}} \in E_i(\mathbf{c}, \delta_i)$.

Let us compare now the average bit error probability of $i$ th user (14) with the performance of the single-user transmission. Assuming the only $i$ th user transmits information stream $\mathbf{a}_i$ over the fading channel, i.e. $\alpha_j = 0$, $1 \le j \le N$, $j \ne i$, asymptotically the average bit error probability of $i$ th user is approximately given by

$$P_{bi}^{(su)} \cong \frac{\beta_i(\delta_i) d_{pi}^{-2}}{k_i} \left( \frac{\alpha_i R_i E_{bi}}{4N_0} \right)^{-\delta_i}, \tag{15}$$

where $\beta_i(\delta_i)$ is the total number of bit errors for all error paths $\tilde{\mathbf{c}}_i$ in the code $C_i$'s trellis that have the Hamming weight $\delta_i$.

Comparing (14) with (15), one can see that $P_{bi}/P_{bi}^{(su)} \cong \chi_i(\delta_i)/\beta_i(\delta_i)$, i.e. other-users interference asymptotically leads to increase of the error coefficient by a factor $\chi_i(\delta_i)/\beta_i(\delta_i)$. By this is meant that the logarithm of the bit error probability $P_{bi}$ goes to zero with the same slope as the single $i$ th user bit error rate $P_{bi}^{(su)}$. In other words, for multiple-access transmission over Rayleigh fading channel we have the asymptotic efficiency $\eta = 1$. It is rather surprising result, when taking into account that for multiple-access transmission over AWGN channel, the asymptotic efficiency $\eta < 1$ [1].

The asymptotic energy loss due to other-users interference

$$\theta_i = (10/\delta_i) \lg(\chi_i(\delta_i)/\beta_i(\delta_i)). \tag{16}$$

Notice that the ratio $\chi_i(\delta_i)/\beta_i(\delta_i)$ in (16) can be approximated by the ratio $q_i(\delta_i)/m_i(\delta_i)$, which, as mentioned above, accounts for an increase of the number of nearest neighbors in the supertrellis with respect to the code $C_i$'s trellis; clearly, the ratio $q_i(\delta_i)/m_i(\delta_i)$ increases with a rise of the number of served users $N$.

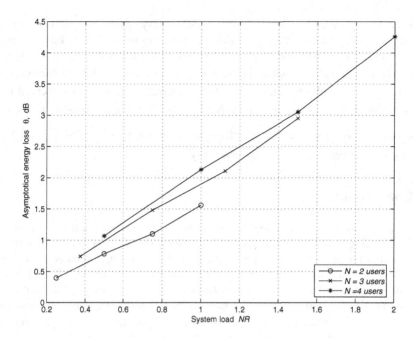

**Fig. 1.** Asymptotic energy loss due to other-users interference versus the system load

Fig.1 shows the asymptotic energy loss versus the system load. All users employ BPSK and binary convolutional 4-state codes with rate $R$ ; the system load is defined as $NR$ information bits per channel symbol. One can see that the asymptotic energy loss due to other-users interference is roughly linear with the system load. For an oversaturated scenario, over the range of the system load from 1 to 2 bits per channel symbol, the energy loss is approximately doubled: from ~2 dB to 4.2 dB. We find also that given the system load, performance loss increases with a rise of the number of users $N$ .

The asymptotic performance estimate (14) indicates that in addition to the well known code design criteria for single-user transmission - the minimum Hamming distance $\delta_i$ and product distance $d_{pi}^2$, the factor $\chi_i(\delta_i)$ has to be taken into account in the multiuser scenario. It should be pointed out that the factors $\chi_i(\delta_i)$, $1 \le i \le N$, are interrelated, and hence the joint minimization of $\chi_i(\delta_i)$, $1 \le i \le N$, is troublesome. At the same time, for transmission scheme with identical user's trellis codes, $\forall i \ \delta_i = \delta$, $\chi_i(\delta_i) = \chi(\delta)$, and the specified problem disappears.

*Remark.* For typical downlink scenario, ML decoding is now too complex to implement. A lot of iterative decoding schemes were therefore suggested as an alternative to the ML decoding. Among these schemes, recently proposed interleave-division multiple-access (IDMA) [8] with an iterative multiuser decoding procedure, based on combining the successive interference cancellation and turbo concept employing soft values, is highly attractive. To a first approximation, asymptotic analysis of the IDMA scheme is reduced to estimation of residual other-users interference. For considered transmission scheme with identical user's trellis codes, it can be shown that asymptotical variance $\zeta$ of residual other-users interference goes to zero as the signal-to-noise ratio goes to infinity. This results holds, provided the number of served users $N$ is bounded by a certain threshold $N_{max}$, depending on the specific trellis code employed. It follows that under specific conditions ( $N \leq N_{max}$ and $\zeta \sim o(N_0)$ ), near-single-user performance is asymptotically attained.

## 5 Conclusions

In this paper, we have considered the narrowband multiple-access transmission over Rayleigh fading channel. The transmission scheme is based on bandwidth-efficient trellis coded modulation, and it has no bandwidth expansion compared to a single-user scenario. The upper bound for the bit-error rate is obtained. The asymptotic analysis of this bound shows that at high signal-to-noise ratio, the logarithm of the bit error probability goes to zero with the same slope as for the single-user transmission, i.e. the asymptotic efficiency $\eta = 1$. The asymptotic performance analysis indicates that in addition to the well known code design criteria for single-user transmission - the minimum Hamming distance and product distance, the factor that accounts for an increase of the number of nearest neighbors in the supertrellis with respect to the code trellis of specific user has to be considered.

## References

1. S. Verdu, "Multiuser Detection", Cambridge University Press, 1998.
2. J.A.F. Ross, D.P. Taylor, "Multiuser signaling in the symbol-synchronous AWGN channel", *IEEE Trans. Inform. Theory*, 41(4), July 1995, pp. 1174-1178.
3. J. Paavola, V. Ipatov, "Oversaturating synchronous CDMA systems on the signature per user basis", *Proceedings of 5th European Personal Mobile Communications Conference*, April 2003.
4. T. Aulin and R. Espineira, "Trellis coded multiple access (TDMA)", *Proceedings of IEEE Int. Conf. Commun. (ICC'99)*, Vancouver, Canada, June 1999, pp. 1177-1181.
5. F. Brännström, T. Aulin and L. Rasmussen, "Iterative detectors for trellis-code multiple access", *IEEE Trans. Commun.*, 50 (9), Sept. 2002, pp. 1478-1485.
6. E. Bigliery, "High level modulation and coding for nonlinear satellite channels", *IEEE Trans. Commun.*, 37(7), July 1989, pp. 669-676.
7. D. Divsalar and M. K. Simon, "The design of trellis coded MPSK for fading channels: Performance criteria and set partitioning for optimum code design", *IEEE Trans. Commun.*, 36(9), Sept. 1988, pp. 1004-1022.
8. L. Ping, L. Liu, K. Wu, and W.K. Leung, "Interleave division multiple access (IDMA) communication systems", *Proceedings of $3^{rd}$ International Symposium on Turbo Code &Related Topics*, 2003, pp. 173- 180.

# A MIMO Architecture for IEEE 802.16d (WiMAX) Heterogeneous Wireless Access Using Optical Wireless Technology

Neda Cvijetic[1] and Ting Wang[2]

[1] University of Virginia, Charlottesville, VA, 22904, USA
[2] NEC Laboratories America, Princeton, NJ, 08540, USA

**Abstract.** Recently, there has been much interest in convergence in the communications industry: convergence of fixed and mobile terminals, convergence of optical and wireless networks, and convergence of wireless cellular and broadband platforms. At the core of the push for mobile broadband convergence is a user-centric model for next-generation networks that are envisioned to provide an array services "anytime, anywhere" reliably and cost-effectively. Solutions for this problem will undoubtedly require a cross-layer approach that considers everything from physical layer parameters, to scheduling and routing protocols.

In this paper, we focus on heterogeneous fixed broadband topologies and study physical-layer convergence of optical and wireless networks. We adopt a point-to-point architecture and present a performance analysis of a multiple-input, multiple output (MIMO) hybrid RF/optical wireless system that implements the physical layer of the IEEE 802.16d WiMAX standard. The key technique of interest is multi-subcarrier modulation of OFDM signals on the optical carrier. Average power efficiency, atmospheric turbulence, adaptive modulation, receiver noise, as well as cross-layer protocol design are considered. Expressions for average symbol error rate are derived and evaluated to determine system performance. We thus assess optical wireless technology as a potential distributor of WiMAX traffic and highlight the MIMO advantage in such a hybrid system.

**Keywords:** network convergence, optical wireless communication, orthogonal frequency division multiplexing (OFDM), WiMAX, subcarrier modulation, multiple-input multiple output (MIMO) systems, cross-layer protocol design.

## 1 Introduction

The "anytime, anywhere" mobile broadband network of the future may be regarded as the ultimate goal for the telecommunications and networking industries. Since both the mobility and availability requirements for this next-generation network are extremely high, it will require seamless and efficient convergence of emerging wireless and traditional wired technologies. On the wireless end, what is envisioned is a heterogeneous, all-IP architecture that incorporates

Y. Koucheryavy, J. Harju, and V.B. Iversen (Eds.): NEW2AN 2006, LNCS 4003, pp. 441–451, 2006.

everything from wireless LANs to cellular and Bluetooth [1]. Incorporating wired networks into this vision also comes with its own set of challenges due to some all-encompassing differences between the wired and wireless media. The growing field of cross-layer protocol design seeks to address these incompatibilities, namely the strict layered architecture of wired networks that becomes inefficient for time-varying wireless channels transmitting heterogeneous traffic. In the cross-layer scenario, rather than having protocol layers function independently, protocols are designed to exploit dependence between layers and improve system performance [2].

Several important challenges and questions emerge from this cross-layer approach to network convergence. The re-definition of the traditional protocol layers has placed great importance on the architecture design for converged networks – without proper care for the architecture, cross-layer design may result in performance losses rather than gains [2]. Secondly, this re-definition has raised questions about the role of the physical layer (PHY) in the wireless portion of converged networks. While the PHY has played a relatively small role in wired networks, signal-processing advances and the time-varying nature of the wireless channel allow it to have a greater contribution to overall system performance.

In this paper, we seek to address these fundamental questions by proposing a novel architecture for a converged, heterogeneous network that effectively uses cross-layer protocol design as well as an elaborate physical layer. Specifically, we consider the distribution of WiMAX traffic over a multiple-input, multiple-output (MIMO) wireless channel that exploits optical wireless technology. The motivation behind this approach is reliably maximizing traffic throughput in a wireless channel with a high degree of multipath fading. The following section provides a brief overview of the key technologies involved, while Sect. III presents a detailed model of the proposed architecture. Section IV overviews performance analysis. Summary and conclusions are provided in Sect. V.

## 2     Technology Overview

### 2.1     IEEE 802.16d (WiMAX) Standard

The IEEE 802.16d standard (WiMAX), also commonly referred to as 802.16REVd or 802.16-2004, was published in October 2004 for fixed wireless access to customers within a 10-15 mile radius of a WiMAX base station. It is designed to allow up to 100 Mbps transmission on a 20 MHz bandwidth at potential carrier frequencies between 2 – 60 GHz (covering both licensed and unlicensed spectrum.) [3] In October 2005, the standard was updated and extended to include mobile access via 802.16e or Mobile *WiMAX* [4]. The WiMAX PHY interface of interest here is orthogonal frequency division multiplexing (OFDM) that uses multiple orthogonal data carriers in parallel, such that the total bandwidth occupied is the same as for the SC system, but the bandwidth occupied by each carrier is only a small part of the total bandwidth [5]. By thus subdividing the frequency spectrum, a frequency-selective channel is essentially converted into a collection of adjacent frequency-flat channels, which combats multipath fading and greatly simplifies the receiver. Since OFDM also allows low-complexity multiuser access schemes (such as OFDMA) and can be employed in tandem with MIMO to boost capacity over multipath wireless channels, it has emerged as the prominent broadband wireless access scheme and has generated much recent research interest [6].

A simplified MIMO-OFDM system can be described as follows [6]: incoming data from a digital source first undergoes forward error correction and interleaving before being mapped to a digital constellation (WiMAX allows for adaptive modulation and coding.) After digital modulation, the data is encoded by a MIMO encoder via either spatial multiplexing or space-time coding. Finally, each of the parallel outputs of the MIMO encoder undergoes OFDM modulation: pilot insertion, inverse FFT (IFFT) modulation, cyclic prefix attachment, and preamble insertion for accurate timing. At the receiver, the received symbol streams undergo synchronization, OFDM demodulation, MIMO decoding, and finally digital demodulation and decoding to recover the original data.

## 2.2   Optical Wireless Technology

Optical wireless technology (free-space optics) is generating renewed interest as a viable means for fixed broadband connectivity [7]. Among its primary advantages are the unlicensed spectrum of operation, quick and inexpensive setup, as well as seamless integration with legacy fiber-optic networks. Principal challenges for optical wireless communication is various weather-related fading effects, including clear-air scintillation: random changes in the air's index of refraction distort the optical wave's phase front, causing received power to fluctuate [8]. Although the physical model for scintillation fading is different than the multipath fading model in RF systems, the principle of spatial diversity can be applied to optical wireless channels nonetheless. Multiple uncorrelated transmitters can be introduced into the random fading channel to improve the received signal quality [9].

In optical fiber systems, subcarrier modulation (SCM) is currently a popular transmission technique, whereby a number of electrical signals are multiplexed in the RF domain and transmitted by a single wavelength [10]. With the advent of dense wavelength division multiplexing (DWDM) this technique yields promise of great bandwidth efficiency in next-generation fiber networks. The SCM technique, referred to as MSM in optical wireless communication, has recently attracted much attention in optical wireless systems as well. Its greatest benefits are the high data speeds obtained without the need for equalization, increased bandwidth efficiency over single-carrier transmission with on-off-keying (OOK) or pulse-position modulation (PPM), and finally, the ability to modulate MSM electrical signal onto the optical carrier using either intensity, frequency, or phase modulation [10].

The main drawback of MSM is the poor average optical power efficiency. Since most of such systems use intensity modulation and direct detection of the optical signal (IM/DD) and the MSM electrical signal can take on large instantaneous negative values, a dc bias must be added to the electrical signal to prevent clipping and distortion in the optical domain (since optical intensity cannot take on negative values.) This is particularly detrimental for OFDM MSM, where the large number of carriers creates unfavorable peak-to-average-power ratios (PAPR). Several techniques have been proposed to mitigate these effects, including time-varying dc bias, block codes that improve power efficiency, as well as subcarrier signal point sequences that distribute a single symbol across multiple carriers [11]. A recent study showing that subcarrier phase-shift keying (PSK) system outperforms OOK in an optical wireless system in the presence of atmospheric turbulence may be found in [12].

# 3   Proposed Architecture

In the proposed architecture, we make use of the MIMO-OFDM model described in Sect. II but we make the additional proposition that the OFDM-modulated WiMAX traffic is used as subcarrier modulation of optical wireless carriers rather than traditional RF antenna elements [13]. A practical application of this system might be metropolitan areas where large data throughput is required but the physical landscape causes a high degree of multipath. Thus, what is needed is a wireless solution to act as a reliable bridge between the optical core network and wireless WiMAX fringes. The optical wireless approach is effective because its channel is both non time-dispersive (flat fading) and linear, making it very well suited for broadband transmission of analog signals. Furthermore, the time and cost efficiency of installing optical wireless terminals allows a flexible and scalable architecture that provides carrier-class availability and reliability through prudent link lengths and link margins.

The most common form of providing transmitter and receiver diversity in optical wireless literature are appropriately-spaced laser and photodetector arrays, where array elements are separated by the channel correlation distance, often cited as $\sqrt{\lambda L}$, where $\lambda$ is the optical wavelength and $L$ is the link length [14]. This ensures that the fading coefficients along the optical paths are reasonably independent. However, this approach is employed in all-digital optical systems where pulses of light are non-coherently combined in the optical domain to reap the MIMO advantage. However, with the MSM approach, the transmitted optical signal is analog, requiring preservation of its frequency and phase information and thus a highly-sophisticated form of coherent detection that is currently not feasible in practical systems.

Consequently, we propose an alternative method to ensure spatial diversity. On the transmitter end, we propose a laser array operating at different optical wavelengths, such that the channel fading experienced by the wavelengths is uncorrelated (i.e. we obtain transmitter diversity via wavelength diversity.) This then provides an optical wireless complement to fiber-optic WDM systems. Through optical filtering on the receiver end, we ensure that each photodetector receives only the signal from its 'intended' wavelength so that the coherence properties of the analog signal are preserved. After photodetection (optoelectronic conversion) the OFDM streams are demodulated and sent to the MIMO decoder, so that the MIMO processing is restricted to the electrical domain.

We may restate the above description as follows: each optical transmitter in the transmitter array is intensity modulated through multisubcarrier modulation by an electrical MIMO-encoded OFDM analog signal. Since the optical wireless channel is frequency flat-fading and linear, the analog properties of the signal are preserved upon transmission. The receiver array contains one photodetector for each transmitter wavelength and includes optical filtering to tune the device to the wavelength of interest (discarding undesired wavelengths and background radiation.) Through direct detection, the received optical signals are converted to the electrical domain, where both OFDM demodulation and MIMO decoding take place.

Since we assume a metropolitan area application as the motivation for this architecture, we restrict ourselves to short to moderate link lengths over which the fading channel is modelled as lognormal. In lognormal fading, the p.d.f. for the optical *amplitude* path gain, $A$, in the turbulent channel is

$$f_A(a) = \frac{1}{(2\pi\sigma_X^2)^{1/2}a} \exp(-(\log_e a - \mu_X)^2/2\sigma_X^2), \quad a > 0 \qquad (1)$$

where $A = e^X$ and $X$ is normal with mean $\mu_X$ and variance $\sigma_X^2$. The optical power gain $A^2$ is also lognormal, and we define $\mathbf{G} = \mathbf{A} \odot \mathbf{A} = \{a_{jk}^2, j = 1, ..., J, k = 1, ..., K\} = \{g_{jk}, j = 1, ..., K, j = 1, ..., K\}$ as the channel fading matrix with $J$ transmitters and $K$ receivers, where $a_{jk}$ denotes the path gain from transmitter $j$ to detector $k$. To make a fair comparison with the non-fading case, we set the mean path gain $E[G] = 1$, which requires that $\mu_X = -\sigma_X^2$. To characterize the degree of fading, we use the 'scintillation index' $(S.I.)$ defined as

$$S.I. = \frac{E[A^4]}{E^2[A^2]} - 1 = \frac{E[G^2]}{E^2[G]} - 1 \qquad (2)$$

and related to the variance $\sigma_X^2$ by $S.I. = e^{4\sigma_X^2} - 1$. The typical values for $S.I.$ range from 0.4-1.0 in the literature. We treat the lognormal fading as constant over an OFDM symbol block but slowly-varying over successive blocks. The channel's effect is thus equivalent to a real (rather than complex) scale factor, since it will be multiplicative in intensity.

The $S.I.$ thus emerges as a natural channel metric in this setup that may be used for opportunistic communication over this channel via interaction between the PHY and medium-access control (MAC) layers, resulting in a novel cross-layer protocol design. Since we have assumed MIMO encoding in the electrical domain prior to OFDM modulation and optical-wireless transmission, we assume that the PHY of this wireless LAN is capable of multipacket reception (i.e. receiving multiple packets simultaneously from multiple data streams.) We also assume that the upper layers of the system give a hierarchical order to the data streams, ordering them in terms of importance and QoS requirements. Consequently, by using OFDM pilot signals to monitor the channel $S.I.$, we may set up a $S.I.$ threshold to adaptively load the channel and use it in an opportunistic fashion depending on channel conditions. For low $S.I.$, the incoming data streams may be transmitted simultaneously through the MIMO channel, while priority may be given to certain streams in cases of high *S.I.* Through interaction between PHY and MAC layers we may thus form a simple, flexible multi-user/MIMO technique that actively adapts to traffic demands and channel conditions.

## 4    Performance Analysis

In previous work, we simulated average symbol error rates for single-input, single-output non-fading and fading FSO OFDM systems using multi-subcarrier modulation with direct detection and QPSK, 16 QAM and 64 QAM signalling [13]. Here, we derive analytical average symbol error probability expressions that supports our previous work, and we extend it to the MIMO case.

The laser input current may be expressed as

$$i_L(t) = I_{dc}[1 + \mu s(t)] \tag{3}$$

where $I_{dc}$ is the required dc bias, $\mu$ reflects the modulation index, and $s(t)$ is the baseband OFDM signal, normalized to a maximum value of unity. If we choose the bias current $I_{dc}$ as the midpoint of the input current range $\Delta I$ that produces a linearly-dependent optical power output, we impose $|\mu s(t)| < \Delta I/2$ by clipping the peaks of $s(t)$ to a limit $\gamma s_{rms}$, a multiple of its root-mean-square value. Given our choice for $\mu$ ($0 \leq \mu \leq 1$), the output optical intensity is

$$P_s(t) = P_T[1 + \mu s(t)] \tag{4}$$

where $P_T$ is the average optical power at the transmitter. We may express the total photodetected electrical signal as

$$r(t) = \rho A^2(t) P_s(t) + n(t) \tag{5}$$

where $A(t)$ is a log-normally distributed random process, $P_s(t)$ is the optical intensity in absence of atmospheric turbulence, and $n(t)$ is the noise signal that captures both photodetection shot noise and post-detection thermal noise. Filtering $r(t)$ to a bandwidth $B$ (Hz) yields a new zero-mean Gaussian noise process whose one-sided power spectral density is given by

$$N_0 = 2q\rho P_0 R^2 + 4k_b T_{sys} R \quad (v^2/Hz); \tag{6}$$

where $k_b$ is Boltzmann's constant, $T_{sys}$ is equivalent system temperature (Kelvin), $R$ is the resistance ($\Omega$), and $q$ is electronic charge (Coulombs). The resulting electrical SNR of the recovered OFDM signal conditioned on the *power* fading gain, $A^2 = a^2 = g$, is

$$SNR_{|G=g} = \frac{g\rho^2\mu^2 P_0^2 R^2}{\gamma^2 B(2k_b T_{sys} R + \rho P_0 q R^2)} \tag{7}$$

A tight upper bound on symbol error probability (conditioned on $g$) is a function of $E_s$, the symbol energy in the given constellation, and $N_0$, such that

$$P_{s|G=g} \leq \beta Q\left(\sqrt{\frac{E_s}{N_0}\frac{3}{M-1}}\right) = \beta Q\left(\sqrt{SNR\frac{T}{N}\frac{3}{M-1}}\right) \tag{8}$$

where $T$ is OFDM symbol time, $N$ is number of active OFDM subcarriers, $M$ is the constellation size and $\beta = 2, 4, 4$ for QPSK, 16 QAM and 64 QAM respectively. We may then obtain the average symbol error probability by averaging $P_{s|g}$ with respect to $f_G(g)$. We also note that if $g = 1$, we obtain symbol error probability for the non-fading channel.

We now extend the above analysis to two MIMO scenarios with $J$ transmitters and $K$ transmitters (we assume $J = K$.) For an uncoded, non-fading channel with different information transfer across the $J$ transmitters, we obtain

$$P_s \leq \beta Q \left( \sqrt{\frac{E_s}{JN_0} \frac{3}{M-1}} \right) \tag{9}$$

Finally, in an uncoded, fading channel with repeated information transfer across the $J$ transmitters, we obtain

$$P_s \leq \beta Q \left( \sqrt{\sum_j g_j \frac{E_s}{JN_0} \frac{3}{M-1}} \right) \tag{10}$$

Equations (9)–(10) were used to evaluate average error symbol probability in an optical wireless channel under these two MIMO scenarios. Unless otherwise stated, the physical parameter values assumed throughout are given in Table 1.

Table 1. Physical parameter set

| Parameter | Value |
|-----------|-------|
| $\rho$ | 0.8 |
| $\mu$ | 1 |
| $P_0$ | -40 → -10 dBm |
| $B$ | 26 MHz |
| $T_{sys}$ | 300 K |
| $R$ | 100 $\Omega$ |
| $T$ | 10 $\mu$sec |
| $\gamma$ | 4 |
| $N$ | 200 |

Figure 1 shows the average symbol error probability in a non-fading FSO OFDM channel for QPSK and 64 QAM signalling and several different transmitter/receiver array sizes (i.e. $J = K = 1, 2, 4, 8$.) We note that QPSK outperforms 64 QAM by almost 7 dB at $P_s = 10^{-9}$ for all $J$. Since the laser driver current is normalized so that its peak value is fixed, the laser is biased so as to operate in its linear region, and the channel is non-fading, the difference in $P_s$ performance between QPSK and 64 QAM for a fixed $J = K$ may be attributed to the underlying baseband modulation formats. The larger constellation sizes

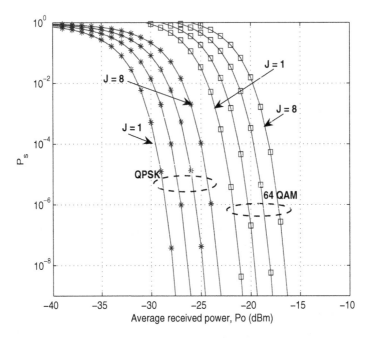

**Fig. 1.** Symbol error probability vs. average received optical power in non-fading FSO channel for QPSK and 64 QAM for $J = 1, 2, 4, 8$

require larger average received power to accurately discriminate among transmitted symbols since symbol energy, $E_s$, is fixed with increasing constellation size. We also note the steepness of the $P_s$ curves, indicating that a large gain in $P_s$ results from a small increase in $P_0$. For 64 QAM and $J = 8$, for example, Fig. 1 indicates that a 2 dB increase in $P_0$ (from -18 dBm to -16 dBm) reduces $P_s$ by almost seven orders of magnitude. The steepness of the $P_s$ curves in Fig. 1 is also an immediate indication that the system is thermal-noise-limited even at $P_0 = -10$ dBm. Thus, its thermally-noise-limited SNR $\propto P_0^2$; as a result, the system requires a $P_0$ increase of only 1.5 dB for a 3 dB electrical SNR improvement. This also explains why the penalty for increasing $J$ by a factor of 2 is 1.5 dB for both QPSK and 64 QAM. Since we fix the total transmitted optical power in order to have a fair comparison with the $J = 1$ case, the optical power per transmitter is halved each time the transmitter array size doubles. Consequently, we incur a 6 dB penalty in $P_0$ at $P_s = 10^{-9}$ by changing $J = 1$ to $J = 8$. However, we note that this uncoded, non-fading MIMO system transmits different information across transmitters so that trade-off for this $P_0$ penalty is an increase in the channel rate (and thus transmitted symbol rate) by a factor of $J$. In other words, going from $J = 1$ to $J = 8$ incurs a 6 dB penalty in $P_0$ but for a given $P_s$, but allows transmitting at a symbol rate $JR_s$ for that same symbol error probability. In short to moderate links where increasing $P_0$ by this amount is feasible, this is an effective way to increase data throughput.

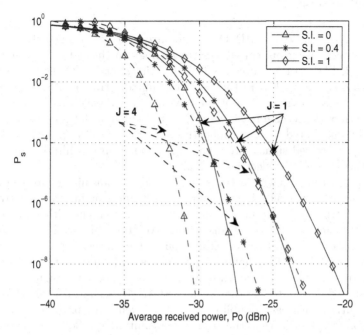

**Fig. 2.** Symbol error probability vs. average received optical power for QPSK in fading FSO channel for several *S.I.* values

Figure 2 plots the average symbol error probability in a lognormally-fading FSO OFDM channel with QPSK signalling, for two transmitter/receiver array sizes and a range of scintillation indexes. (In this case, since we are repeating information across the transmitter array, the symbol rate $R_s$ is the same as in the non-diverse case.) For a single transmitter-receiver pair ($J = 1$), the $P_0$ penalty at $P_s = 10^{-9}$ due to channel fading over the non-fading ($S.I. = 0$) case is 4 dB and 7 dB for $S.I. = 0.4$ and $S.I. = 1$ respectively. In the MIMO channel with $J = 4$, increased atmospheric turbulence incurs also incurs $P_0$ penalty of 4 dB and 7 dB for $S.I. = 0.4$ and $S.I. = 1$ respectively over the non-fading case at the same $P_s$. However, for all the cases considered, the $J = 4$ system performs almost 3 dB better than the analogous $J = 1$ setup, indicating the MIMO diversity advantage. Note that we could assume a Rayleigh fading channel and average expression (10) with respect to the square of this distribution in order to get a more 'quantitative' measure of the diversity magnitude as this would produce $P_s$ curves with asymptotic slopes. However, since the Rayleigh model is the limiting case for optical wireless channels and is not as common in practice as the lognormal model, we employed the latter one in this work. By focusing on the $J = 4$ system, we may also select a $S.I.$ threshold based on $P_s$ and $P_0$ criteria. If at $P_0 = $ -25 dBm, we require $P_s \leq 10^{-9}$, for example, $S.I. = 0.4$ is an appropriate threshold that determines how the upper-layer protocols will use the channel.

# 5    Conclusions

In this paper, we present a MIMO architecture for distributing IEEE 802.16d (WiMAX) traffic using optical wireless technology to through multi-subcarrier modulation (MSM). In the presented model, the electrical OFDM waveforms are multiplexed and applied to the laser transmitter array as MSM. Transmitter diversity was provided via optical wavelength diversity of the lasers. The turbulent atmospheric channel is modelled as frequency-flat, contributing a a lognormally-distributed real, multiplicative scale factor to the transmitted optical intensity. Received signals are separated through optical filtering, reserving MIMO processing for the electrical domain.

Two MIMO scenarios with $J$ transmitters and $K$ transmitters are studied in this work: an uncoded, non-fading channel with different information transfer across the $J$ transmitters, and an uncoded, fading channel with repeated information transfer across the $J$ transmitters. We derive and evaluate analytical average symbol error probability expressions for each case. Finally, the scintillation index, $S.I.$, of the channel is chosen as a channel metric that allows opportunistic use of the channel through cross-layer protocol design.

The obtained data indicates that in the first MIMO scenario, a 1.5 dB penalty in $P_0$ is incurred with each doubling of $J$ at a given error probability $P_s$. The trade-off for this loss is that the overall transmitted symbol rate potentially increases from $R_s$ to $JR_s$, increasing data throughput $J$-fold. In the second MIMO scenario, the symbol rate $R_s$ remains unchanged, and an almost 3 dB gain in $P_0$ is obtained through diversity techniques with $J = 4$. The values of $S.I. = 0.4$ is chosen as the opportunistic channel threshold for the studied case.

# References

1. H. Jiang and X. Shen, "Cross-layer design for resource allocation in 3G wireless networks and beyond," *IEEE Comm. Mag.*, pp. 120–126, Dec 2005.
2. V. Srivastava and M. Motani, "Cross-layer design: a survey and the road ahead," *IEEE Comm. Mag.*, pp. 110–119, Dec 2005.
3. A. Henley, "WiMAX– a standard air interface for broadband wireless channels," *Microwave Jnl*, April 2005.
4. T. Kwon, H. Lee, J. Kim, D.-H. Cho, S. Cho, S. Yun, W.-H. Park, and K. Kim, "Design and implementation of a simulator based on a cross-layer protocol between MAC and PHY layers in a WiBRO compatible IEEE 802.16e OFDMA system," *IEEE Comm. Mag.*, pp. 136–146, Dec 2005.
5. W. Zou and Y. Wu, "COFDM: an overview," *IEEE Trans. Broadcasting*, vol. 41, no. 1, pp. 1–6, March 1995.
6. H. Yang, "A road to future broadband wireless access: MIMO–OFDM-based air interface," *IEEE Comm. Mag.*, pp. 53–60, Jan 2005.
7. A. Acampora, "Last mile by laser," *Scientific American*, June 2002.
8. S. Wilson, M. Brandt-Pearce, J. Leveque, and Q. Cao, "Free-space optical MIMO transmission with Q-ary PPM," *IEEE Trans. Commun.*, vol. 53, no. 8, pp. 1402–1406, Aug 2005.

9. N. Cvijetic, S. Wilson, and M. Brandt-Pearce, "Optimizing system performance of free-space optical MIMO links with APD receivers," Proc. 2006 Optical Fiber Conference (OFC).  Anaheim, CA, Mar. 5-10 2006.

10. T. Ohtsuki, "Multiple-subcarrier modulation in optical wireless communications," *IEEE Commun. Mag.*, pp. 74–79, Mar 2003.

11. S. Teramoto and T. Ohtsuki, "Multiple-subcarrier optical communication systems with subcarrier signal-point sequence," *IEEE Trans. Commun.*, vol. 53, no. 10, pp. 1738–1743, Oct 2005.

12. Q. Lu and G. Mitchell, "Performance analysis for optical wireless communication systems using subcarrier PSK intensity modulation through turbulent atmospheric channel," *In Proc. IEEE Globecom*, 2004.

13. N. Cvijetic and T. Wang, "WiMAX over free-space optics – evaluating OFDM multi-subcarrier modulation in optical wireless channels," Proc. 2006 IEEE Sarnoff Symposium.  Princeton University, Mar. 26-28 2006.

14. J. Strohbehn, *Laser Beam Propagations in the Atmosphere.*  Springer, 1978.

# Cooperative Transmission Through Incremental Redundancy

Vladimir Bashun and Konstantin Dubkov

St.Petersburg State University of Aerospace Instrumentation, 190000, Bolshaya
Morskaya st., 67, St.Petersburg, Russia
{bashun, kostia}@vu.spb.ru

**Abstract.** Rapid development of wireless networks leads to unceasing
appearance of new techniques and methods. Recently cooperative relaying emerged as viable and tempting option for future wireless networks.
Cooperative transmission attempts to provide spatial diversity for hardware devices that have only one antenna through the joint use of antennas
that belong to several nodes. An overview of approaches in this area is
presented. For proposed coded cooperative strategy, regions for which
cooperation results in a gain in Frame Error Rate (FER) are analyzed.

**Keywords:** cooperative transmission, coded cooperation, incremental
redundancy, error-correcting codes.

## 1 Cooperative Transmission Overview

Wireless data transmission is broadcast by nature. It means, that a signal transmitted by source to the destination can be "overheard" at neighboring nodes.
Even usage of directional antenna allows data to be received by all stations that
are located within the transmission range. This effect does not cost anything
since transmitter side needs no additional power to make other radio nodes receive the signal. The neighboring nodes can attempt to help destination to receive
the message. Cooperative communication refers to processing of this overheard
information at the surrounding nodes and retransmission to the destination to
create spatial diversity, thereby to obtain higher throughput and reliability.

The main idea is to obtain spatial diversity. Generally, transmit diversity requires more than one antenna at the transmitter. Using multiple antennas (at
both transmitter and receiver) can provide robustness against channel variations and improve performance. However, main constraints for wireless networks
are limitations in size and complexity of mobile node and power limitations.
Many wireless devices are limited to one antenna. Cooperative communication
attempts to provide spatial diversity through the joint use of antennas that belong to several nodes. At the same time, the relay node (cooperator) is usually
closer to the destination than the source. Using of relay allows exploiting decreasing in path loss between relay and destination (widely used in multi-hop
networks). So, cooperative transmission attempts to benefit from:

- spatial diversity (for nodes with one antenna)
- path loss savings

Y. Koucheryavy, J. Harju, and V.B. Iversen (Eds.): NEW2AN 2006, LNCS 4003, pp. 452–460, 2006.

It is shown [2], [3], [5] that cooperation with neighboring nodes results in robustness to channel variations, higher throughput (by increasing achievable rate regions [1]), extended coverage, extended battery life.

Cooperative relaying comes from meshed multi-hop networks communications, where the information from the source to the destination is relayed via other mobiles. It is natural to use relay as cooperator in ad hoc networks that work without base station. But cooperative communication can be also used for networks that work with base stations (e.g. 802.11, 802.16). Cooperative relaying can be used for improving of uplink in such systems (since mobile nodes usually have one antenna, and base station may have several antennas).

## 1.1  Cooperative Transmission Strategies

When cooperator detects transmission of its partner, it should somehow make decision, what exactly to transmit and when it needs to transmit.

All cooperative schemes may be grouped by the level of data processing (demodulating, decoding, etc.) at cooperator. The main question is how cooperator should process information, received from its partner, and what additional data it should transmit in order to improve the received signal or decrease error probability at the destination.

Currently two basic approaches exist. The operator can either forward its received signal without processing (or with minimal processing), or it can process received signal (e.g. decode and re-encode) before retransmitting it. First strategy is usually called amplify-and-forward [3]. Second is further referred to as coded cooperation [1], [4]. Both approaches are explained in detail below.

**Amplify-and-forward.** The concept of amplify-and forward strategy is perhaps the simplest among the cooperative signaling methods. The relay node in this case simply retranslates incoming signal [3]. Each cooperator in this method receives a noisy version of the signal transmitted by the source node. The cooperator simply amplifies and retransmits this noisy signal (Figure 1).

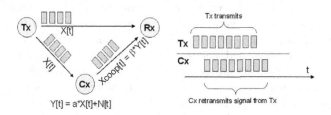

**Fig. 1.** Amplify-and-forward cooperation scheme

So cooperator acts as simple repetition. This scheme can be viewed as repetition coding from two separate transmitters, except that relay transmitter (cooperator) amplifies its own receiver noise [3]. The receiver can hear both signals — from source node and from relay (cooperator). It will combine the information sent by the user and partner and will make a final decision on the transmitted bit. In such systems, synchronization of source and cooperator is required.

The performance is constraint by the quality of source-cooperator channel.

**Coded cooperation.** In this case cooperator does not simply forward the received noisy signal, but tries to correct errors. Various cooperative protocols were proposed in [3]. In simplest Decode-and-Forward scheme the transmitter encodes packets using some FEC code and sends it to destination. Cooperator detects the signals from transmitter, demodulates and decodes it (correcting errors occurred), then encodes message again using the same code (restores original signals) and transmits it to the receiver [3]. Having two copies of original signals at receiver side it is possible to extract data from it more reliable (Figure 2).

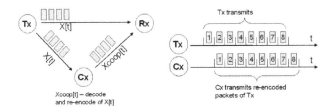

**Fig. 2.** Simple decode-and-forward scheme

More advanced Decode-and-Re-encode schemes are based on calculating **incremental redundancy** on cooperator. This approach is usually called coded cooperation. Incremental redundancy schemes were studied in [1], [4], [5].

It is possible for several FEC techniques to form different number of redundant blocks for original data. So source node uses one set of redundant bits and cooperator after decoding the message forms another set of redundant blocks that are sent to the destination. In ideal situation, the more redundancy destination receives, the better decoder works. Having more parity check symbols is more efficient than just having several copies of message, so generally this method is more efficient then previous two. But efficiency costs additional complexity.

In order to use this method, cooperator generally needs to obtain source message without errors first, decoding the message that it received from the source. This means that cooperator needs to fully decode received signal. Decoding procedure may be hard computational task that requires some time. It can be impossible to do this in real time. One possible way to overcome it is using systematic codes, when first $K$ bits of codeword for $(N, K)$ code shall contain message bits, and the remaining $N - K$ symbols will contain parity check bits. If cooperator gets first $K$ bits of the codeword without errors, it can immediately encode received message bits using some code and send only appropriate check bits. First part of the message is usually additionally protected by CRC for error detection [1], [5].

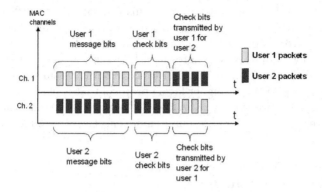

**Fig. 3.** Coded cooperation through time division

In order to realize coded cooperative scheme, one should use special code constructions that will allow separating creation of parity check bits between source node and cooperator and allow computing incremental redundancy on cooperator. Some additional constraints for code constructions may be given in order to optimize code performance (see e.g. [1]).

A coded cooperative scheme that used time division between relay and cooperator and assumed existence of feedback channel between source and relay was proposed in [4]. Proposed code construction was based on rate-compatible punctured convolutional codes (RCPC-codes) and also involved cyclic redundancy check (CRC) for error detection.

In [5] code construction based on turbo codes and iterative decoding was proposed. This construction employed two consistent recursive systematic convolutional codes with interleaving. Also, modification of basic time division framework called space-time cooperation was proposed. In this scheme, user transmits first part of the message in the first frame, and in the second frame user transmits his own parity bits as well as parity bits for his partner.

Coded cooperation can be implemented using error-correcting codes, e.g. block or convolutional codes. In order to partition code bits between source and cooperator, one can use puncturing, products codes, or some other methods. One can use block codes (RS, etc.), rate-compatible punctured convolutional codes [4], turbo codes [5] and other code constructions. LDPC codes work well for wireless channel in general, but it is still an uncompleted task to construct LDPC codes that will allow getting benefit from incremental redundancy.

Some mixed schemes of the above mentioned strategies may also exist. For example, cooperator can switch from coded cooperation to amplify-and-forward and back depending on channel state. Additional research may reveal some other hybrid (combined) schemes.

Besides level of processing, all cooperative transmission schemes can be divided by two groups basing on their organization method. First method supposes absence of any agreement between source and cooperator side, i.e. transmission is made in usual way and cooperator may either assist in it or not depending on its own decision. The main advantage of this method is above-PHY protocol simplicity: transmitter side doesn't care about cooperator presence and receiver processes cooperator's signal on PHY only if it's assisted.

Another method supposes some agreement between cooperator and transmitter and/or receiver side about further assistance in data transmission. This leads to more complicated MAC (or higher layer) protocol that is used for negotiation between radio nodes. However, a priori knowledge about cooperator presence allows transmitter to use its energy resources more effectively or to increase system data throughout.

## 2    The Proposed Scheme

Coded cooperation schemes seem to be more promising since having some parity check information is more efficient than just having several copies of message. Generally, coded cooperation method is more efficient. At this stage of research, principal scheme of cooperation needs to be selected and its performance and usage limits need to be analyzed. In this chapter, a simple modification of existing coded cooperation schemes is presented. As a starting solution, Reed-Solomon codes were selected as the basis for cooperation scheme. In future some advanced code constructions using codes with soft decoding (e.g. LDPC) need to be taken.

### 2.1    Scheme Description

This coded cooperation scheme is a modification of scheme presented in [4]. The Reed-Solomon $(N, K)$ codes are used. The transmitter encodes $K$ symbols into a codeword with $N$ symbols using RS-code. After that it additionally protects first $K$ symbols of message by CRC. The code is systematic (where first $K$ symbols of codeword presents uncoded source message bits).

Transmission of message is divided into two periods, so the time slot for transmission is divided into two frames. For the first frame, node transmits a block consisting of the message part (first $K$ symbols) of codeword with corresponding checksum and half of parity check symbols.

If there is free cooperator nearby source, it can receive the message word and if CRC is correct, it sends to the source and destination the acknowledgement that it will take part in cooperation, and then sends the second half of code word checks. Figure 4 illustrates this process.

If the source hasn't received acknowledgement, it sends remaining parity check symbols, otherwise it relies on cooperator. Destination receives code word parts and decodes them correctly if number of errors less than or equal to $t$.

We consider a Rayleigh slow fading channel and also include the path loss effect. It is assumed that the average received SNR per coded bit is proportional to $D^{-\alpha}$, where $\alpha$ is the path loss component which is determined by the environment and $D$ represents the normalized distance between the source and the destination. Typically the path loss exponent $\alpha$ is between 2 and 4 [7]. A simulation pipeline for 802.16 PHY level was developed by Intel CTL Russia group. Results of Intel CTL Russia group for 802.11 PHY modulations are presented in [8]. BER/SNR results were used for calculating bit error rate for given SNR on receiver, and this obtained level of BER was used to generate errors in transmitted packets.

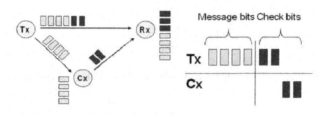

**Fig. 4.** Incremental redundancy principle

A simple cooperative diversity scenario with one source, one relay and one destination is considered. Let's denote a channel from source to destination as *direct channel*, a channel from source to cooperator as *internal channel* and a channel from cooperator to destination as *cooperation channel*. We consider two scenarios: in first scenario, cooperator is used when possible. In second scenario, cooperative transmission is never used. Note that overall transmitter power usage is assumed equal in both cases (with and without cooperation), since amount of transmitted data is also equal.

In simulations, RS-codes with rate 0.5 were used. This scenario assumes code length $N$ is equal to 12, information word length, $K$ is equal to 6, number of corrected errors, $t = K/2$ is equal to 3.

## 2.2   Simulation Results

We want to investigate the situations, in which cooperation gives benefit (in terms of frame error rate, FER). It is clear that the share of frames where cooperator is used in the total number of frames depends on internal channel quality (cooperator will not be used if it can't receive first $K$ bits correctly). At the same time, efficiency of using cooperator is determined by correlation of SNR in direct channel and cooperation channel.

In first experiment we consider the situation when cooperator is closer to destination than the source (e.g. it can be located between the source and destination). Assume that SNR in internal and cooperation channel is fixed and the channel is good (SNR is equal 13 dB); SNR in direct channel is changing (from bad to good).

Source and cooperator observe different independent channels to destination. Errors are added into transmitted packets for different channels independently, depending on SNR on receiver. Destination node combines both parts of message and attempts to decode it. If total number of errors is less than $t$, decoding will be successful. If cooperator is used (scenario 1) and cooperation channel is better

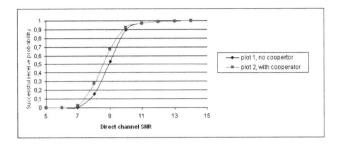

**Fig. 5.** Successful receive probability against SNR in direct channel

than direct channel, cooperator can help to decrease number of errors in second part of the message, and so it can increase successful decode probability (see Figure 5).

The regions (in Euclidian space) where it is reasonable to use cooperator will be found out further. It is interesting to know how these regions depend on distance between the nodes (source, cooperator and destination). For this purpose, the map of effective cooperator usage region has been constructed.

The summary of results is presented in Figure 6. The source is shown $S$ point, and destination is shown as $D$ point. From the previous experiment one can see that for good direct channel, using cooperator does not give significant gain. We consider that direct channel is bad (the normalized distance between them is 5 units that correspond to 7 dB in SNR). The distance can be interpreted in terms of normalized distance units, or in terms of received SNR (since received SNR is proportional to $D^{-\alpha}$).

The main areas where cooperator is used and where using of cooperator is efficient are presented as shaded areas. The area 2 denotes the region, where cooperator has the best efficiency. Area 1 shows region where using cooperator is also efficient, but gives smaller gain. Moving cooperator closer to a source increases the distance between cooperator and destination. Hence it leads to degradation of cooperation channel quality. For both areas, cooperators have almost similar internal channel. But cooperation channel for cooperator in area 2 is better than in area 1. Moving cooperator closer to destination decreases the quality of internal channel, and the share of frames where cooperator is used in the total number of frames decreases (and benefit from using cooperator also decreases). Area 3 represents the region where the rate of cooperator usage is very small and cooperator almost never used. At some distance from the source (affected by path loss exponent and coding scheme) cooperator stops giving any benefit.

Several bounds we found. "Gain boundary" divides the region where cooperator may decrease FER from the region in which cooperator gives no gain. This bound depends only on distance between the source and the destination and cannot be moved by changing of coding scheme. "Cooperator usage boundary" limits the region where cooperator can still receive first $K$ message symbols without errors, from the region, where it always receives first part of the message with errors (cooperator is never used in the second case). Cooperator is used only inside "cooperator usage boundary" region. This bound can be moved by using more efficient coding scheme, because better BER for the same SNR can be obtained. This bound also depends on cooperative strategy. Path loss exponent (parameter $\alpha$) also influences on "cooperator usage boundary". Increasing $\alpha$ moves this bound closer to the source.

One can see from the figure that even for Reed-Solomon codes (those are not so efficient for wireless media) using simple coded cooperation can be beneficial. At the same time, the regions of efficient cooperator usage are limited by certain areas. Efficient code constructions for calculating incremental redundancy need to be invest⁻⁻⁺⁻ᵈ

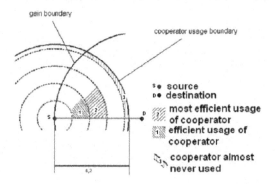

**Fig. 6.** Effective cooperator usage regions

## 3   Conclusion

Cooperative transmission is an interesting and tempting technique that may be exploited in future wireless systems. An overview of currently discussed cooperative transmission methods and strategies were presented. Coding cooperation seems to be more promising compared to simple amplify-and-forward or detect-and-forward cooperating schemes. A simple coded cooperative scheme based on RS-codes was proposed. For this scheme, the regions (in Euclidian space) of effective cooperator usage are analyzed and the spatial maps for effective usage of cooperators are presented.

## References

1. Elza Erkip, Andrew Sendonaris, Andrej Stefanov, and Behnaam Aazhang. Cooperative Communication in Wireless Systems, DIMACS Workshop on Network Information Theory, Piscataway, New Jersey, March 2003
2. Ernesto Zimmermann, Patrick Herhold, Gerhard Fettweis. On the Performance of Cooperative Diversity Protocols in Practical Wireless Networks, in European Transactions on Telecommunications (ETT), January 2005

3. J.N. Laneman, D.N.C. Tse, and G.W. Wornell. Cooperative diversity in wireless networks: Efficient Protocols and Outage Behavior. IEEE Trans. Inform. Theory, vol.50 pp. 3062-3080, Dec 2004
4. T. Hunter and A. Nosratinia. Cooperation diversity through coding. In IEEE ISIT, June 2002
5. M. Janani, A. Hedayat, T. Hunter and A. Nosratinia. Coded Cooperation With Space-Time Transmission and Iterative Decoding, in Proceedings WNCG Wireless Networking Symposium, October 2003
6. F.J. MacWilliams, and N.J.A. Sloan. The Theory of Error-Correcting Codes, North-Holland, Amsterdam, 1977
7. H.L. Bertoni. Radio Propagation for Modern Wireless System, Upper Saddle River, NJ:Prentice Hall, 2000.
8. E.A.Krouk et al. Research of LDPC codes and 10G Base-T. Ethernet physical layer development. Intel Technical report, CTL Russia, December 2004

# Comparison of Effective SINR Mapping with Traditional AVI Approach for Modeling Packet Error Rate in Multi–state Channel

Martti Moisio[1] and Alexandra Oborina[2]

[1] Nokia Research Center, P.O. Box 407, FIN-00045 NOKIA GROUP, Finland
martti.moisio@nokia.com
[2] University of Joensuu, P.O. Box 111, FI-80101 Joensuu, Finland
ext-alexandra.oborina@nokia.com

**Abstract.** This paper provides an overview and comparison of traditional and advanced Link-to-System (L2S) interface mappings used in system level simulations to evaluate link performance in terms of packet error rate (PER). An Actual Value Interface (AVI) as traditional L2S interface is compared with an Effective SINR Mapping (ESM) as advanced L2S interface. This comparison is able to highlight the main differences in prediction of multi-state channel performance effecting on final data throughput, delay of data packets and required CPU time. Furthermore, this paper proposes an improved version of AVI method in order to improve the accuracy of system performance estimation. The comparison results show that in case of multi-state OFDM system, highest accuracy is achieved with Mutual Information ESM (MIESM) method and relatively good accuracy with the exponential ESM (EESM) method.

**Keywords:** System simulator, L2S interface, EESM, MIESM, AVI, OFDM.

## 1 Introduction

Within the scope of International Telecommunication Union–Radio [1] vision for the future development of wireless systems, the new radio access interfaces will have to support up to 1 Gbit/s for mobile wireless access and up to 100 Mbit/s for nomadic/local wireless access. Due to these very high data rate requirements, technologies that work well over broadband frequency–selective radio channel are needed. The most suitable techniques in this category are multiple antenna systems, OFDM modulation and OFDMA multiple access. To analyze the changes in system performance with varying system parameters – such as traffic, channel model and coding scheme – link level and dynamic system level simulations need to be performed in several typical environments.

Traditionally, a link level simulator is used to evaluate the physical layer performance. On link level a continuous radio link is modeled, involving specific characteristics like spatial pre- and post–processing, synchronization, channel estimation and small–scale fading. The simulation results typically provide user

Y. Koucheryavy, J. Harju, and V.B. Iversen (Eds.): NEW2AN 2006, LNCS 4003, pp. 461–473, 2006.
© Springer-Verlag Berlin Heidelberg 2006

quality in terms of bit error rate (BER) as a function of averaged signal–to–interference plus noise ratio (SINR).

Instead, the objective of system level simulations is to study higher layer events and interactions. Its main target is to explore the behavior of the whole network with predefined number of base stations (BS) and (typically) uniformly distributed mobile stations (MS). On the system level all relevant network-level radio resource management (RRM) operations are simulated (for example power control, handover, link adaptation).

Link level and the system level simulators interact with each other by means of the so–called Link–to–System (L2S) interface. In practice, this interface is realized through a set of mapping (look-up) tables. These mapping tables are constructed on the link level and they represent tabulated BER functions of instantaneous system level SINR. At each step the system level simulator computes channel quality measures for all of the resource elements and maps them onto a certain BER using the mapping tables. After receiving all bits of a packet, mapping from BER to packet error rate is performed.

In order to evaluate the performance of a new OFDM radio access system with various coding and modulation schemes and channel characteristics, a large number of resource elements should be measured. Since multidimensional mapping onto packet error rate is often too complex, it is desirable to have a simple and at the same time generic link performance model to determine PER using a limited number of quality value parameters [2]. Thus, the objective of the L2S interface is in accurate compression from instantaneous channel state – where instantaneous quality measures of all resource elements are utilized – to one or two specific measures that are later used for mapping. In this paper different methods realizing such compression are evaluated and compared. Also, an improvement of traditional L2S mapping is proposed.

This paper is organized as follows. An overview of advanced and traditional L2S interface approaches for system performance estimation is given in section 2. In section 3 a comparison of presented methods from information measure point of view is presented, together with an improvement proposal for the traditional AVI approach. Sections 4 and 5 describe the system simulator parameters and simulation results, respectively. Finally, the conclusions can be found in section 6.

## 2  Link to System Interface: Traditional and Advanced Mapping Approaches

To estimate the system performance, link performance model is required to produce the results for evaluating the link quality, which is based on resource and power allocation in a certain radio propagation environment. Figure 1 introduces a potential realization of such link performance model binding link and system level simulators for performance prediction [2].

At each step of the system level simulator, resources are allocated between users and power levels are determined for each transmitting link. The individual instantaneous channels involving large- and small–scale fading, path loss and

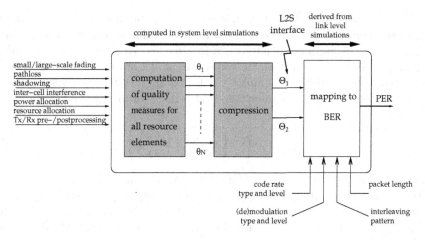

**Fig. 1.** Link performance model [2]

interference are modeled. Furthermore, a set of quality measures are derived for all resource elements. This study concentrates on a multi–state OFDM system, where information bits of a single user can be transmitted over several OFDM symbols (time domain) and over several OFDM subcarriers (frequency domain) which all can have different quality. Hence, the number of time and frequency resource elements is typically too large to be used directly, and a compression from instantaneous channel state to a smaller number of significant quality measures is needed. Typically, one or two scalar parameters are enough to predict packet error probability of a specific channel state. The key problem is to find these suitable parameters.

Different proposals for measuring quality of resource elements can be found from literature, but herein we focus on signal–to–interference plus noise ratio as a quality measure. For multi–state OFDM channel with $N$ subcarriers, the subcarrier specific SINR samples $\{\gamma_j\}$ are simply calculated as

$$\gamma_j = \frac{C_j}{I_j + N_0} \qquad 1 \leq j \leq N \tag{1}$$

where $C_j$ is received signal power in $j^{\text{th}}$ subcarrier, $I_j$ is interference power in $j^{\text{th}}$ subcarrier, and $N_0$ is background noise power.

When applying the conventional L2S interface called Actual Value Interface (AVI) [3] to OFDM(A), SINR sample $\gamma_{avi}$ is the averaged SNR over all subcarriers assuming one specific channel model.

$$\gamma_{avi} = \frac{1}{N} \sum_{j=1}^{N} \gamma_j \tag{2}$$

This SINR sample is then mapped onto BER or PER. For this mapping methodology the link level simulator has to produce a specific mapping table for all channel realizations. The main drawback of this method is the use of direct arithmetic average and the need of a new mapping table for every channel model.

A general realization of advanced L2S interface is given by an Effective SINR Mapping (ESM) that maps an instantaneous multi–state channel - described by a set of subcarrier SINR samples – onto instantaneous scalar value. This is called an effective signal–to–interference plus noise ratio $\gamma_{eff}$, and it is defined as follows

$$\gamma_{eff} = \alpha_1 I^{-1} \left( \frac{1}{N} \sum_{j=1}^{N} I \left( \frac{\gamma_j}{\alpha_2} \right) \right) \tag{3}$$

where function $I(\cdot)$ is referred to "information measure" function, and $I^{-1}(\cdot)$ is its inverse. The scaling parameters $\alpha_1$ and $\alpha_2$ are used to match the model to related modulation and coding scheme (MCS). The effective SINR is later mapped onto a single bit error rate via mapping table constructed for additive white Gaussian noise (AWGN) channel on the link level. The accuracy of the ESM is validated through adjustment of the predicted BER($\gamma_{eff}$) for AWGN channel to the measured instantaneous BER($\gamma_j$) derived from link level simulations. This equalization of BER samples for all instantaneous channel states allows utilizing AWGN mapping table for various channel models in system level simulations.

## 3    Comparison of L2S Mappings from Information Measure Point of View

As proposed in [4] the "information measure" function $I(\cdot)$ characterizing the capacity of a multi–state channel has the following property

$$I \left( \frac{\gamma_{eff}}{\alpha_1} \right) = \sum_j p_j I \left( \frac{\gamma_j}{\alpha_2} \right) \overset{(3)}{=} \frac{1}{N} \sum_{j=1}^{N} I \left( \frac{\gamma_j}{\alpha_2} \right). \tag{4}$$

Here, $p_j = \frac{1}{N}$ is the probability mass function for $\gamma_j$ sample on $j^{th}$ subcarrier since the total number of subcarriers equals to $N$. Notice that the function $I(\cdot)$ only characterizes the channel capacity as it may not determine well–defined channel capacity in informational theory. Application of various information measures into (4) yields specific L2S interfaces with corresponding accuracy and efficiency in performance modeling. Hence, it has significant influence on the system level performance.

In order to obtain AVI performance (2), the quality measure itself should be utilized as the information measure, with $\alpha_1 = \alpha_2 = 1$.

$$I(\gamma) = \gamma \tag{5}$$

For advanced L2S interface we focus on two known information measures – exponential and mutual information – referred to an Exponential Effective SINR Mapping (EESM) and Mutual Information SINR Mapping (MIESM), . The exponential information measure function has a scaling parameter $\beta = \alpha_1 = \alpha_2$ and it is defined as

$$I(\gamma) = A - Be^{\frac{\gamma}{\beta}} \qquad A, B \in \mathbb{R} \tag{6}$$

introducing EESM given in [4]:

$$
\begin{aligned}
I(\gamma_{eff}) &\overset{(4)}{=} \tfrac{1}{N} \sum_{j=1}^{N} A - Be^{\frac{-\gamma_j}{\beta}} \\
e^{\frac{-\gamma_{eff}}{\beta}} &= \tfrac{1}{N} \sum_{j=1}^{N} e^{\frac{-\gamma_j}{\beta}} \\
\gamma_{eff} &= -\beta \ln\left( \tfrac{1}{N} \sum_{j=1}^{N} e^{-\frac{\gamma_j}{\beta}} \right)
\end{aligned}
$$

The mutual information is specified by Bit Interleaved Coded Modulation (BICM) capacity defined as [6]

$$I_{m_j}(\gamma) = m_j - E_Y \left( \frac{1}{2^{m_j}} \sum_{i=1}^{m_j} \sum_{b=0}^{1} \sum_{z \in \Gamma_b^i} \log \frac{\sum_{\hat{\gamma} \in \Gamma} exp(-|Y - \sqrt{\gamma}(\hat{\gamma} - z)|^2)}{\sum_{\tilde{\gamma} \in \Gamma_b^i} exp(-|Y - \sqrt{\gamma}(\tilde{\gamma} - z)|^2)} \right) \tag{7}$$

where $m_j$ is number of bits per symbol $j$, $\Gamma$ is the set of $2^{m_j}$ symbols, $\Gamma_b^i$ is the set of symbols for which bit $i$ equals to $b$, $Y$ is a zero mean, unit variance, complex Gaussian variable, and $E(\cdot)$ is the expectation value. This mutual information formula is the basis for the MIESM model with scaling parameter $\beta = \alpha_1 = \alpha_2$.

In Figure 2, the three presented information measure functions are plotted for QPSK$\frac{1}{2}$ and 16QAM$\frac{1}{2}$ modulation and coding schemes. In order to compare the shapes of the information measure curves $A$ and $B$, the EESM information measure (6) parameters are chosen as follows

$$
\begin{aligned}
A = B = 1 &\quad \text{if} \quad \text{MCS} = \text{QPSK}\tfrac{1}{2}, \\
A = B = 2 &\quad \text{if} \quad \text{MCS} = \text{16QAM}\tfrac{1}{2}.
\end{aligned}
$$

Since the scaling factor $\beta$ is adjusted to match the effective SINRs to a specific modulation scheme, we represent two curves of information measures for EESM and MIESM. In plots "without beta shifting", the parameter $\beta$ is equal to 1.0 for both information measure functions, while in plots "with beta shifting" the parameter $\beta$ is tuned for the system level simulator. It can be observed that the $\beta$–shifted EESM and MIESM curves approach each other in sense of Mean Square Error (MSE). $\beta$ shifting improves QPSK MSE from 0.221 to 0.02 and 16QAM MSE from 3.034 to 0.2955.

These functions have two different shape characteristics: convexity and sigmoid (so called S–shape). The information measure of AVI is a convex function, but EESM and MIESM information measure curves are sigmoidal. This difference partially occurs due to the modulation format being involved in EESM and MIESM, and it has a great impact on how various SINR samples are relatively weighted in certain information measure function. Since the amount of information that a channel can pass with certain modulation should follow sigmoidal curves [4], AVI method will overestimate the carried information by overweighting the high SINR samples. EESM and MIESM have sigmoidal shape and they will better describe the information transferred by individual SINR samples.

AVI information measure weights all SINR samples equally, and the end result $\gamma_{avi}$ describes the performance for a given channel "on average". Taken into

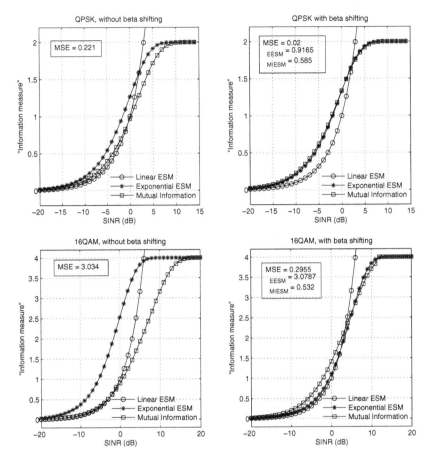

**Fig. 2.** Information measures for QPSK$\frac{1}{2}$ and 16QAM$\frac{1}{2}$

account the used modulation, this quality measure is then used to predict the BER/PER performance.

Instead, EESM information measure puts higher weight on lower SINR samples and less weight on higher SINR samples. Thus, the critical (in terms of final error probability) lower SINR samples dominate in the mean value of weighted SINRs: $\frac{1}{N} \sum_{j=1}^{N} e^{-\frac{\gamma_j}{\beta}}$. Applying the inverse information measure to the derived mean value yields higher result for higher mean. Hence, only the meaningful SINR samples are represented in the effective SINR value.

MIESM in general behaves similarly as EESM, but the weight of lowest and highest SINR samples are limited by the information measure functions lower and upper limits. Such evaluation of SINR samples allows crucial SINR samples to dominate on performance quality estimation. Comparison of EESM and MIESM in terms of channel capacity given in [4] instead of SINR weighting shows that MIESM information measure better characterizes the system performance – however the EESM function is also able to reach good accuracy.

After analyzing the behavior of relative weighting for convex and sigmoidal functions we can propose an improvement for the convex AVI information measure. Due to bounding of sigmoid we may also bound the convex AVI information measure so that crucial SINR samples will have more contribution in final quality measure. Since EESM and MIESM information measures are bounded by the modulation format and the bandwidth, these characteristics should define the thresholds for AVI information measure to indicate crucial SINRs. In AVI approach the modulation structure and the bandwidth are involved in performance estimating through coded BER calculations. So, the bounds for AVI information measure may be derived from link level look–up tables. Based on lower- and upper–bounded AVI information measure this paper proposes an improved AVI method.

## 4   Simulation Assumptions and Parameters

A sophisticated, dynamic, time–driven system level simulator was used to study the effect of different L2S mapping methods on final system performance. The simulator was able to simulate performance of cellular OFDM(A) system with high accuracy. Every OFDMA symbol was simulated, one simulation time step was set to 22 microseconds. The simulated macro cellular network was covered by 10 base stations and 200 mobile stations uniformly distributed over the simulation area. Mobile station speed was 3 km/h.

The simulator had TDMA/TDD multiple access on downlink. In frequency domain, OFDM modulation/waveform was used in 20.0 MHz bandwidth at 5.0 GHz carrier frequency. In the simulations, downlink noise power was -94.0 dBm, and base station transmit power was 43.0 dBm. Selective ARQ was used to correct the packet transmission errors. Traffic model was a Full Buffer FTP model and only downlink direction was studied. On reverse channel only control data such as acknowledgements were sent.

To simulate correlated fast fading on system level, the simulator generated complex fading samples from the used tapped delay line model, where each tap faded according to well known Jakes model. Lognormally distributed samples with 6dB standard deviation were added to the signal to simulate slow fading. The distance pathloss model was similar to the classical OkumuraHata model.

In simulations two different modulation and coding schemes were used without link adaptation: QPSK$\frac{1}{2}$ and 16QAM$\frac{1}{2}$. The utilized scaling parameter $\beta$ for both MCSs is given in Table 1. Based on link level look–up tables the upper

**Table 1.** Settings for $\beta$ and AVI bounds in system simulator

| Modulation | Code Rate | $\beta$ value EESM | $\beta$ value MI–ESM | low bound improved AVI | upper bound improved AVI |
|------------|-----------|-----------|-----------|------------|------------|
| QPSK | 1/2 | 0.9165 | 0.585 | -15 (dB) | 0 (dB) |
| 16QAM | 1/2 | 3.0787 | 0.532 | -15 (dB) | 5 (dB) |

and low limits for improved AVI were specified as shown in Table 1. Two types of simulations were run: "error–free" simulations, where all transmitted packets were received correctly; and normal simulations with errors and retransmissions. The first type of simulations was run with time period of 50000 steps (1.1 seconds real time), the second one with 10 millions steps (220 seconds real time).

## 5   Simulation Results

In order to highlight the differences between the selected L2S approaches applied to exactly same set of SINR samples (i.e. excluding the effect of retransmissions), a set of error–free simulations were carried out. Simulated coded BER values and their linear regressions are presented in Figure 3 and Figure 4. Numerical results for the BER dispersion are summarized in Table 2.

It can be noticed that both effective SINRs and final coded BERs collected for EESM and MIESM have no significant difference. However, higher deviation of coded BER for $16QAM\frac{1}{2}$ can be observed. These results are predictable since adjusted information measure curves do not follow exactly the same shape (see Figure 2), and MSE error in the case of 16QAM modulation is approximately 15 times higher. The exponential ESM relatively underweights meaningful SINRs especially in the case of $16QAM\frac{1}{2}$ leading to overestimation of system level performance. The AVI simulation results in Table 2 as well as the corresponding subplots in Figure 3 indicate a considerable distinction between ESM simulation results. Such a high variance in effective SINR values may be justified by the fact that given modulation format is not taken into $\gamma_{avi}$ calculations (2). But large deviations that occurred in simulated coded BER values show that AVI approach can not accurately estimate the performance. Our proposal for AVI may slightly improve the accuracy of performance estimation as less deviations of coded BER values obtained through the improved AVI can be observed in Table 2. The presented results demonstrate that improved AVI approach produces more reliable quality measures and provides better estimation of the performance, but still can not be equivalent to EESM or MIESM approaches.

Linear regressions for simulated coded BER values depicted in Figure 4 were examined through MSE, and the results are summarized in Table 3. The better performance of improved AVI (lower MSE) can be seen.

The presented "error–free" simulation results indicate that the differences in information measure curves lead to essential differences in link quality metrics. Finally, this leads to differences in such key user quality metrics as PER and throuhgput. The EESM and MIESM methods produce quite similar results in

**Table 2.** Deviation of coded BER values

| | AVI | | improved AVI | | EESM | |
|---|---|---|---|---|---|---|
| | $QPSK\frac{1}{2}$ | $16QAM\frac{1}{2}$ | $QPSK\frac{1}{2}$ | $16QAM\frac{1}{2}$ | $QPSK\frac{1}{2}$ | $16QAM\frac{1}{2}$ |
| EESM | $67.48{\cdot}10^{-4}$ | $96.80{\cdot}10^{-4}$ | $64.50{\cdot}10^{-4}$ | $91.38{\cdot}10^{-4}$ | – | – |
| MIESM | $68.17{\cdot}10^{-4}$ | $98.97{\cdot}10^{-4}$ | $65.07{\cdot}10^{-4}$ | $92.62{\cdot}10^{-4}$ | $1.57{\cdot}10^{-4}$ | $8.94{\cdot}10^{-4}$ |

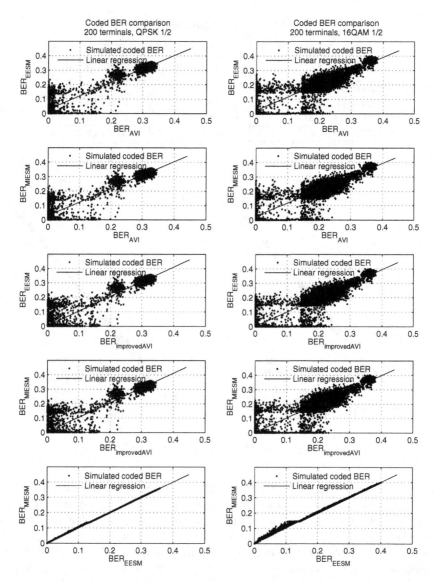

**Fig. 3.** BER results for QPSK$\frac{1}{2}$ and 16QAM$\frac{1}{2}$ obtained in error–free simulations

the case of QPSK modulation, but slight difference is observed in the case of 16QAM modulation. Evaluation of the performance through the AVI approach can cause the overestimation, and its improved version can only slightly increase the accuracy.

Next, normal system simulations with packet errors and retransmissions were conducted to study the real cellular system performance. The cumulative density functions of effective SINR (as output from system simulations) are depicted in Figure 5. In Table 4 average user throughput and PER are shown as final key

**Table 3.** Linear regression MSE

|  | QPSK$\frac{1}{2}$ | 16QAM$\frac{1}{2}$ |
|---|---|---|
|  | EESM vs MIESM | EESM vs MIESM |
| improved AVI vs MIESM | $4.07\times10^{-3}$ | $1.22\times10^{-3}$ |
| improved AVI vs EESM | $3.84\times10^{-3}$ | $2.39\times10^{-3}$ |
| AVI vs MIESM | $4.58\times10^{-3}$ | $3.43\times10^{-3}$ |
| AVI vs EESM | $4.35\times10^{-3}$ | $3.52\times10^{-3}$ |

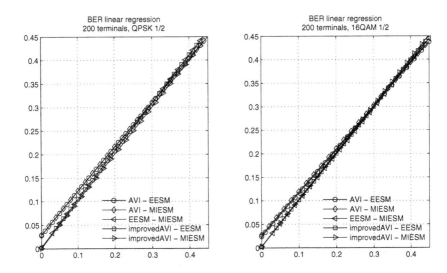

**Fig. 4.** BER results for QPSK$\frac{1}{2}$ and 16QAM$\frac{1}{2}$

performance indicators. The plots in Figure 5 reflect the similarities in MIESM and EESM mappings. The distributions of effective SINRs obtained through MIESM and EESM for QPSK modulation have the same behavior until the MIESM effective SINR reaches the upper limit. As stated earlier, for 16QAM modulation the effective SINR samples computed by EESM and MIESM are noticeably different, which is explained by differences in the information measure functions. The distribution of SINR samples produced by AVI approach shows significant density shifting to higher quality measures which comes from overestimating the performance due to simple subcarrier averaging and because modulation format is not taken into account. The improved AVI dampens this overestimation compared to conventional AVI, resulting into enhancements in estimation accuracy. Note, that AVI and improved AVI SINR distribution should not follow the shape of MIESM and EESM SINR distributions since the modulation format is not included in $\gamma_{avi}$ computation (2).

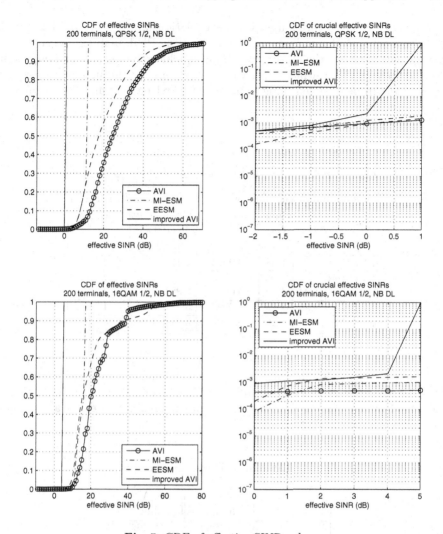

**Fig. 5.** CDF of effective SINR values

When analyzing the amount of erroneously received packets during the simulation time, it is seen PER increased when more accurate L2S interface is used. The results presented in Table 4 exhibit maximum percentile of failed packets for MIESM and the minimum percentile of failed packets for AVI. It is anticipated since the AVI produces relatively higher link quality measures that later causes lower packet error probability. The improved AVI affects similarly, but here insignificant SINR samples are upper bounded. Hence, relatively lower quality measures in comparison are produced. Thus, higher packet error probability will be achieved in the case of improved AVI. Since EESM and MIESM weight crucial SINR values more accurately, the probability of packet error is not underestimated in these cases and larger amount of failed packets is detected.

The user perceived throughput has clear relationship with utilized L2S interface approach and the findings are in-line with the PER analysis. According to the simulation results the higher throughput is achieved using AVI and improved AVI for link quality estimation. Two main reasons for throughput reductions with ESM can be outlined: larger amount of failed packets and larger amount of retransmissions.

**Table 4.** Statistics of quality of service measures

| | QPSK$\frac{1}{2}$ | | | 16QAM$\frac{1}{2}$ | | |
|---|---|---|---|---|---|---|
| | User Throughput | | % of Failed Packets | User Throughput | | % of Failed Packets |
| | Mean Mb/s | Dev Mb/s | % | Mean Mb/s | Dev Mb/s | % |
| AVI | 1.22 | 1.58 | 0.0073 | 4.11 | 4.76 | 2.86 |
| improved AVI | 1.04 | 1.32 | 0.0081 | 3.32 | 4.55 | 3.78 |
| EESM | 0.98 | 1.19 | 0.0097 | 3.21 | 4.22 | 4.24 |
| MIESM | 0.94 | 1.20 | 0.0102 | 3.21 | 4.24 | 4.25 |

In order to consider the usability of the L2S interface in practice the simulation CPU time is an essential indicator. Table 5 shows CPU time statistics collected from normal simulations running for explored L2S interface approaches for QPSK$\frac{1}{2}$ and 16QAM$\frac{1}{2}$.[1]

**Table 5.** CPU time results

| | QPSK$\frac{1}{2}$ | 16QAM$\frac{1}{2}$ |
|---|---|---|
| | dd:hh:mm:ss | dd:hh:mm:ss |
| AVI | 7:00:54:47 | 4:19:49:37 |
| improved AVI | 7:19:49:05 | 4:23:10:56 |
| EESM | 10:13:54:47 | 6:13:31:19 |
| MIESM | 10:05:51:16 | 7:21:53:54 |

Clearly, the main drawback of ESM mapping in comparison with AVI approach is its high simulation time. Moreover, MIESM approach also needs additional offline time to construct BICM function for desirable modulation format. However, in these simulations no effort was done to optimize the execution speed of ESM. Considerable CPU time savings can be achieved by e.g. reducing the amount of calculated OFDMA subcarriers. With typical channel models

---

[1] The simulations were run under Linux 2.4 compatible operating system with 3 GHz processors. The time period of simulations was 10 millions steps.

and speeds, the correlation of adjacent sub-channel SINR is high in both time and frequency. Consequently, the nearby subcarriers can be grouped into larger chunks, leading into reduced amount of calculation.

## 6  Conclusions

In this paper, both traditional AVI and advanced ESM-based link-to-system interface methods were investigated, including a proposal for AVI improvement. The investigation was focused on final system level performance, in terms of BER, PER and simulated user throughput.

The analyzed L2S interface methods were compared from information measure point–of–view and via sophisticated system level simulations. A dynamic OFDM/TDMA/TDD system level simulator was used to carry out the two required types of simulations: "artificial" error–free simulations where all transmitted packets were received correctly; and "normal" simulations with real packet errors and retransmissions.

The artificial simulations showed that EESM and MIESM provided good accuracy in case of QPSK modulation. For 16QAM modulation the EESM was less accurate and slightly overestimated the performance. Nevertheless, the link quality measures in terms of BER values were sufficiently similar in order to yield reasonable accuracy. The AVI approach clearly provided low accuracy of multi–state channel performance estimation for both modulations. In the proposed AVI improvement lower and upper bounds are introduced for the SNIR samples which clearly improved the accuracy. The normal simulations showed the same trend and revealed that AVI can result into 30 percent overestimation in final simulated average user throughput. With improved AVI, the error was only 6 percent. Unfortunately, the CPU time required for the advanced L2S mapping is high compared to the traditional mapping. Therefore, some speed improvements are needed for the ESM methods in order to reduce the running time.

## References

1. ITU-R: Framework and overall objectives of the future development of IMT–2000 and systems beyond IMT–2000. Rec.ITU-R M.1645 (2003)
2. Brueninghaus, K., Astély, D., Sälzer, T., Visuri, S., Alexiou, A., Karger, S., Seraji, G.: Link performance models for system level simulations of broadband radio access systems. PIMRC 2005 (September, 2005)
3. Hämäläinen S., Slanina, P., Hartman, M., Lappeteläinen, A., Holma, H., Salonaho, O.: A novel interface between link and system level simulations. Proceedings of the ACTS Mobile Telecommunications Summit'97, Aalborg, Denmark (October, 1997) 599–604
4. Ericsson: Effective SNR mapping for modeling frame error rates in multiple–state channels. 3GPP2-C30-20030429-010 (April, 2003)
5. Ericsson: System level evaluation of OFDM — further considerations. 3GPP TSG-RAN WG1 35, R1-031303, Lisbon, Portugal (November, 2003)
6. Caire, G., Taricco, G., Biglieri, E.: Capacity of bit–interleaved channels. Electronic Letters **32 N12** (June, 1996) 1060–1061

# Privacy Homomorphism for Delegation of the Computations

Mikhail Stepanov[1], Sergey Bezzateev[1], and Tae-Chul Jung[2]

[1] Saint-Petersburg State University of Airspace Instrumentation
mike@catalina.spb.ru,
bsv@aanet.ru
[2] Samsung Advanced Institute of Technology
tc.jung@samsung.com

**Abstract.** Privacy homomorphisms (PHs) are encryption functions mapping a set of operations on plaintext to another set of operations on ciphertext. In this paper we present new PH scheme that is based on the theory of the finite fields.

## 1 Introduction

The idea of privacy homomorphism (PH) was firstly introduced by Rivest, Adleman and Dertouzos [1]. They proposed several PHs to process encrypted data without decrypting. A PH is an encryption function which allows the processing the encrypted data without decryption. Formally privacy homomorphism can be defined as an encryption function $e$ which has efficient algorithms to compute $e(xy)$ and $e(x+y)$ from $e(x)$, $e(y)$ without revealing x and y. Simple scheme that was offered in [1] is PH where the plaintext was represented in the system of the residue classes.

The cryptanalysis of [1] was done by Brickell and Yacobi in [3] using known plaintext attack.

Later Domingo-Ferrer proposed two algebraic PHs in 1996 and 2002 [4],[5]. Cryptanalysis of these schemes was done in works [7],[6],[8].

Delegation of computing and data is a major field of application for PHs. If the computations performed on encrypted data have arithmetical nature, the PH presented in this paper will be especially useful. The delegation computing problem may appear in the following cases:

- software application operates with data on the untrusted framework (mobile agents [10], java applets [9]),
- processor capabilities are too low to compute result of some hard processing algorithm (calculations of the signature in sensor networks [12], ubiquitous computing [11])

Proposed PH is based on theory of the finite fields. Then it is possible to apply proposed PH to the algorithms that operate with data in finite field. Particular cases of these algorithms are the public key based signature, encryption algorithms. More precisely:

Y. Koucheryavy, J. Harju, and V.B. Iversen (Eds.): NEW2AN 2006, LNCS 4003, pp. 474–480, 2006.

- cryptographic algorithms on elliptic curves,
- cryptographic algorithms on error correcting codes.

Here we propose the encryption homomorphism based on theory of finite fields. Main idea is hiding the set of elements that belong to finite field in the more powerful set of elements that belongs to the finite ring.

The paper is organized as follows: in Section 2 we give the required definitions and set up homomorphism between finite field and some ring, and show construction of the efficient backward mapping, in section 3 we describe the PH scheme, in section 4 we give simple example, in section 5 we describe some notes about security of the proposed scheme, the conclusion is done in section 6.

## 2   Required Algebraic Notes

Let $g(x)$ be the primitive polynomial over $F_p$ with $\deg(g) = l$. We can easily prove that the set of the residues modulo $g(x)$ over $F_p$ forms the field. We will denote the set of the residues as $F_p[x]/g(x)$ and denote obtained field as $F_{p^l}$. All elements in the set $F_{p^l}$ will be of the form $\sum_{i=0}^{l-1} a_i x^i$, where $a_i \in F_p$. It is well known from theory of finite fields [2], that

$$x^{p^l - 1} \equiv 1 \bmod g(x). \tag{1}$$

We will use this property in the following.

**Definition 1.** *By substitution we define* $x := y^k$, *where* $k$ *is arbitrary chosen so that* $\gcd\left(p^l - 1, k\right) = 1$.

Let's define $k^{-1} : k \cdot k^{-1} \equiv 1 \bmod p^l - 1$. Let's set function $j(i)$ as

$$j(i) := i \cdot k \bmod p^l - 1. \tag{2}$$

Using Definition (1) and (2) we define the polynomial $\tilde{g}(y) = \sum_{i=0}^{\deg g'(x)} g_i' \cdot y^{j(i)}$, where $\{g_i'\}_{i=1..\deg g'}$ are coefficients of the $g'(x) = g(x) \cdot f(x)$, $f(x)$ is some arbitrary polynomial.

Now we can define the mapping.

**Definition 2.** *The encryption mapping is defined as:*

$$e : \begin{cases} F_{p^l} \to \Omega \\ \sum_{i=0}^{l-1} a_i \cdot x^i \longmapsto \sum_{i=0}^{l-1} a_i \cdot y^{j(i)} \quad \bmod \tilde{g}(y) \end{cases}$$

It is clear that $\Omega \subseteq F_p[y]/\tilde{g}(y)$. In other words, we define the embedding of the field $F_{p^l}$ in the $F_p[y]/\tilde{g}(y)$. Since we don't know either $\tilde{g}(y)$ is primitive or not then we can conclude that $F_p[y]/\tilde{g}(y)$ is at least the ring. Now let's show that mapping $e$ is the homomorphism by operations $\{+, -, \cdot\}$.

**Theorem 1.** *Mapping e is homomorphic by "±":*

$$e\left(\sum_{i=0}^{l-1} a_i x^i \pm \sum_{i=0}^{l-1} b_i x^i\right) = e\left(\sum_{i=0}^{l-1} a_i x^i\right) \pm e\left(\sum_{i=0}^{l-1} b_i x^i\right)$$

*and homomorphic by " · ":*

$$e\left(\sum_{i=0}^{l-1} a_i x^i \cdot \sum_{i=0}^{l-1} b_i x^i\right) = \left(e\left(\sum_{i=0}^{l-1} a_i x^i\right) \cdot e\left(\sum_{i=0}^{l-1} b_i x^i\right)\right) \quad \bmod \tilde{g}(y).$$

*Proof.* Let's express the right side of first equality as

$$e\left(\sum_{i=0}^{l-1} a_i x^i\right) \pm e\left(\sum_{i=0}^{l-1} b_i x^i\right) = \sum_{i=0}^{l-1} a_i y^{j(i)} \bmod \tilde{g}(y) \pm \sum_{i=0}^{l-1} b_i y^{j(i)} \bmod \tilde{g}(y).$$

Since sum or differences of the polynomial residues can't exceed the $\deg \tilde{g}(y)$, then

$$\left(\sum_{i=0}^{l-1} a_i y^{j(i)} \pm \sum_{i=0}^{l-1} b_i y^{j(i)}\right) \bmod \tilde{g}(y) = e\left(\sum_{i=0}^{l-1} a_i x^i \cdot \sum_{i=0}^{l-1} b_i x^i\right).$$

So the homomorphism by $\pm$ is follows.

Let's now express the right side of second equality.

$$e\left(\sum_{i=0}^{l-1} a_i x^i\right) \cdot e\left(\sum_{i=0}^{l-1} b_i x^i\right) =$$
$$\left(\left(\sum_{i=0}^{l-1} a_i y^{j(i)}\right) \bmod \tilde{g}(y) \cdot \left(\sum_{i=0}^{l-1} b_i y^{j(i)}\right) \bmod \tilde{g}(y)\right) \bmod \tilde{g}(y),$$
$$\left(\left(\sum_{i=0}^{l-1} a_i y^{j(i)}\right) \cdot \left(\sum_{i=0}^{l-1} b_i y^{j(i)}\right)\right) \bmod \tilde{g}(y) = e\left(\sum_{i=0}^{l-1} a_i x^i \cdot \sum_{i=0}^{l-1} b_i x^i\right).$$

So the homomorphism by multiplication is follows.

Division can not be carried out in general because the set $F_p[y]/\tilde{g}(y)$ may not be a field. A good solution is to leave division unchanged and handle divisions in rational format by considering the field of rational functions.

With the help of mapping from Definition 2 encryption function can be obtained. Now we need to define the decryption function. Lets define the following mapping.

**Definition 3.** *The decryption mapping d is defined as:*

$$d: \begin{cases} F_p[y]/\tilde{g}(y) \to F_{p^l} \\ \sum_{v=0}^{\deg(\tilde{g})-1} a_v \cdot y^v \longmapsto \sum_{v=0}^{\deg(\tilde{g})-1} a_v \cdot (x^{k^{-1}})^v \quad \bmod g(x) \end{cases}$$

Correctness of the decryption follows from the theorem.

**Theorem 2.** *The mappings e and d, is satisfied the property:*

$$d\left(e\left(\sum_{i=0}^{l-1}a_ix^i\right)\right)=\sum_{i=0}^{l-1}a_ix^i$$

*Proof.* Let's express the right side of the expression:

$$d\left(e\left(\sum_{i=0}^{l-1}a_ix^i\right)\right)=d\left(\left(\sum_{i=0}^{l-1}a_iy^j(i)\right)\bmod \tilde{g}(y)\right)=$$

$$\left(\left(\sum_{i=0}^{l-1}a_ix^{k^{-1}j(i)}\right)\bmod \tilde{g}(x^{k^{-1}})\right)\bmod g(x)=$$

$$\left(\left(\sum_{i=0}^{l-1}a_ix^{i\cdot k\cdot k^{-1}}\right)\bmod g(x)f(x)\right)\bmod g(x)=\left(\sum_{i=0}^{l-1}a_ix^{i\cdot k\cdot k^{-1}}\right)\bmod g(x).$$

using (1), we obtain

$$\left(x^{k\cdot k^{-1}}\sum_{i=0}^{l-1}a_ix^i\right)\bmod g(x)=\sum_{i=0}^{l-1}a_ix^i$$

Theorem 2 proves that with the help of mapping from Definition 3 the decryption function may be obtained.

To make our scheme applicable we should point our attention to right choice of the polynomial $\tilde{g}(y)$. We should take in mind that our scheme should work in large fields. Then arbitrary choice of the $\tilde{g}(y)$ leads to the very high degree of $\tilde{g}(y)$ that makes our scheme impractical.

Let's define the polynomial $\hat{g}(y)=\sum_{i=0}^{l}g_i\cdot y^{j(i)}$, where $\{g_i\}_{i=0..l}$ are coefficients of the $g(x)$. Now we show how to reduce the highest degree term of the polynomial $\hat{g}(y)$.

**Definition 4.** *Set $\Gamma=\{x^{i1},x^{i2},...,x^{il}\}$ forms reduction basis of the $F_{p^l}$ if:*

- *residuals $x^{i1}\bmod g(x),x^{i2}\bmod g(x),...,x^{il}\bmod g(x)$ are linearly independent*
- *each $i_v<p^l-1$ for $v=1..l$,*
- *$j(i_v)<\deg\hat{g}$ for $v=1..l$.*

Let $g_ny^{j(i_n)}$ be the term with highest degree of $\hat{g}(y)$. We can express $g_nx^{i_n}$ in the basis $\Gamma$ so that $g_nx^{i_n}\equiv\gamma_1x^{i1}+\gamma_2x^{i2}+...+\gamma_lx^{il}\bmod g(x)$, where $\gamma_v\in F_p$, $v=1..l$. Then we can construct polynomial $g'(x)=g(x)-g_nx^{i_n}+\gamma_1x^{i1}+\gamma_2x^{i2}+...+\gamma_lx^{il}=g(x)f(x)$, where $f(x)=\frac{g'(x)}{g(x)}$, and obtain $\tilde{g}(y)$. It is easy to see that $\deg g'(x)\geq\deg g(x)$ and $\deg\tilde{g}(y)<\hat{g}(y)$ as desired. Described procedure can be applied several times until the desired degree of $\tilde{g}(y)$ is obtained.

## 3   Scheme Description

### 3.1   Setup

- Find arbitrary primitive polynomial $g(x)$ over $F_p$, set $l=\deg g(x)$.
- Choose $k$ arbitrary so that $\gcd\left(k,p^l-1\right)=1$, and compute $k^{-1}\bmod p^l-1$.

– Generate basis $\Gamma$ following to Definition 4.
– Compute appropriate $\tilde{g}(y)$ from $g(x)$. Public Keys - $\tilde{g}(y)$.
  Secret Keys - $g(x), k$.

## 3.2   Encryption

Let's represent message $m = (m_1, m_2, ..., m_l)_p$ as $m(x) = \sum_{i=0}^{l-1} m_i \cdot x^i$.

Choose arbitrary $\omega \in F_{p^l}$ with $\deg e(\omega) < \deg \tilde{g}(y)$. Represent $\omega$ in basis $\Gamma$ such that $\omega = \gamma_1 x^{i_1} + \gamma_2 x^{i_2} + ... + \gamma_l x^{i_l}$ and let $\delta(y) = e\left(\omega - \gamma_1 x^{i_1} - \gamma_2 x^{i_2} - ... - \gamma_l x^{i_l}\right)$. The encryption is:

$$c = \text{Enc}\ (m(x)) = e\left(\sum_{i=0}^{l-1} m_i \cdot x^i\right) + \delta(y) =$$
$$\sum_{i=0}^{l-1} m_i \cdot y^{j(i)} \bmod \tilde{g}(y) + \delta(y).$$

## 3.3   Computations

Computations are processed on the images of the field elements in the set $F_p[y]/\tilde{g}(y)$. The correctness of the computations follows from Theorem 1.

## 3.4   Decryption

Decryption is done by applying mapping from Definition 3

$$m = \text{Dec}\ (c) = d(c) + d(\delta(y)) = d\left(e\left(\sum_{i=0}^{l-1} m_i \cdot x^i\right)\right) + d(\delta(y)) =$$
$$\sum_{i=0}^{l-1} m_i \cdot x^i + \delta(x^{k^{-1}}) \bmod g(x) = \sum_{i=0}^{l-1} m_i \cdot x^i$$

# 4   Example

**Setup**

1. $g(x) = x^3 + x + 1, p = 2, p^l = 2^3 = 8$
2. $k = 2, k^{-1} = 4$
3. $\Gamma = \{x^6 + x^5 + 1\}$
4.

$$g(x) = x^3 + x + 1$$
$$x^3 \equiv x^6 + x^5 + 1 \bmod g(x)$$
Let's substitute $y^2$ in $(x^3 + x^6 + x^5 + 1)$ then $y^6 + y^5 + y^3 + 1$
$$\hat{g}(y) = y^6 + y^2 + 1$$
$$\tilde{g}(y) = y^5 + y^3 + y^2$$

**Encryption:**

$$m' = (111)_2 = x^2 + x + 1$$
$$\delta_1(y) = y^4 + y^3 + y^2 + 1$$
$$\text{Enc } (m') = y^4 + y^2 + 1 \text{ mod } \tilde{g}(y) + \delta_1(y) = y^3$$
$$c' = (01000)_2$$

$$m'' = (101)_2 = x^2 + 1$$
$$\delta_2(y) = y^4 + y^2 + y$$
$$\text{Enc } (m'') = y^4 + 1 \text{ mod } \tilde{g}(y) = y^4 + 1 + \delta_2(y) = y^2 + y + 1$$
$$c'' = (00111)_2$$

**Computations:**
$$c' \cdot c'' = c'(y) \cdot c''(y) = (y^3) \cdot (y^2 + y + 1) \text{ mod } \tilde{g}(y) = y^4 + y^2 = (10100)_2.$$

**Decryption:**
Dec $(y^4 + y^2) = (x^4)^4 + (x^4)^2 \text{ mod } g(x) = x^2 + x = (110)_2.$
Check the correctness: $m' \cdot m'' = (x^2 + x + 1)(x^2 + 1) = x^2 + x.$

## 5   Security Analysis

**Definition 5.** *([5],Definition 3) A privacy homomorphism is said to be secure against a known plaintext attack if, for any fixed number n of known plaintext-ciphertext pairs, the probability of successful secret key extracting can be made arbitrary small by properly choosing the security parameters of the homomorphism.*

Suppose we have plaintext-ciphertext pair $m(x), c(y)$. Then brute force attack consists of finding of $x^v, g(x) : c(x^v) \text{ mod } \tilde{g}(x) = m(x)$, where $v := 1..p^l - 1$, $g(x) \in F_p[x]/x^{l+1}$, and $|F_p[x]/x^{l+1}| = p^{l+1}$ . So, if $x^v, g(x)$ are randomly chosen, then $\Pr\{c(x^v) \text{ mod } \tilde{g}(x) = m(x)\} \approx \frac{1}{p^{2l+1}}.$

## 6   Conclusion

In this paper we propose new PH scheme based on the theory of finite fields. We may summaries the features of the proposed scheme.

- In the proposed scheme addition, subtraction, multiplication and division can be carried out on encrypted data at an unclassified level.
- Encryption and decryption can be implemented efficiently since all operations are executed modulo some $p$ and require execution of only polynomial time algorithms. The degree of elements bounded by polynomial $\tilde{g}(y)$.
- Plaintext can be encrypted to the different ciphertexts.

# References

1. R.L. Rivest, L. Adleman and M.L. Dertouzos, "On Data Banks and Privacy Homor-phisms," In Foundataions of Secure Computataion, pp. 169-179, Academic Press, 1978.
2. F. J. MacWilliams and N. J. A. Sloane, The Theory of Error-Correcting Codes, Elsevier Science Publishers B.V. New York,1997.
3. E. Brickell and Y. Yacobi, "On Privacy Homoorphisms," In Advances in Cryptology-Eurocrypt'87, pp. 117-125, Springer-Verlag, 1988.
4. J. Domingo-Ferrer,"New Privacy Homomorphism and Applications," Information Processing Letters, Vol. 60, no. 5, pp. 277-282, Dec.1996.
5. J. Domingo-Ferrer,"Provably Secure Additive and Multiplicative Privacy Homo-morphism," ISC2002, LNCS. Vol. 2443, pp.471-483, 2002.
6. F. Bao, "Cryptanalysis of a Provable Secure Additive and Multiplicative Privacy Homomorphism," ICSD2003, 2003
7. D. Wagner."Cryptanalysis of an Algebraic Privacy Homomophism," ISC2003, 2003.
8. Jung Hee Cheon, Hyun Soo Nam "A Cryptanalysis of the Original Domingo-Ferrer's Algebraic Privacy Homomorphism", http://eprint.iacr.org/
9. K. C. Hopson, Stephen E. Ingram ,"Developing Professional Java Applets", Sams.net, June 1, 1996
10. Danny B. Lange, Mitsuru Oshima, "Programming and Deploying Java Mobile Agents with Aglets",Addison-Wesley Professional; 1st edition (September 15, 1998)
11. M. Weiser, "Hot topics-ubiquitous computing", Computer, vol.26, issue 10, pp.71-72, oct 1993
12. Ian F. Akyildiz, WellJan Su, Yogesh Sankarasubramaniam, Erdal Cayirci, "A Sur-vey on Sensor Networks", IEEE Communications Magazine,vol.40,issue 8, pp.102-114, aug 2002

# Enhanced MIMO Transmission Scheme for MB-OFDM System with Spatial Mode Selection

Jae-Seon Yoon, Myung-Sun Baek, So-Young Yeo,
Young-Hwan You, and Hyoung-Kyu Song

uT Communication Research Institute,
Sejong University, Seoul, Korea

**Abstract.** The transmission of huge multimedia data and software components has been increased in recent years in home and this explosive growth of wireless communication is creating demand for high speed, reliable, and spectrally efficient communication in home networking environments. Therefore, dynamic data transmission with proposed scheme for MB-OFDM system is considered in this paper. The proposed scheme is investigated for the most suitable adaptation in home networking environments. And this scheme includes STBC, VBLAST, and STBC-VBLAST in all. By properly selecting their MIMO schemes agreeable to estimated SNR, the proposed scheme has the advantage of each technique; diversity gain or high transmission rate relative to general scheme.

## 1 Introduction

Recently, the orthogonal frequency division multiplexing (OFDM) is commonly used for high data rate in wireless communication due to its inherent error susceptibility in a multipath environment and has been chosen for several broadband WLAN standards like IEEE802.11a, European HIPERLAN/2 and Japanese MMAC, and WPAN proposal like IEEE802.15.3a multi-band orthogonal frequency division multiplexing (MB-OFDM) ultra wideband (UWB) [1]-[2].

And multiple input multiple output (MIMO) structures have been used for realizing high speed and reliability data transmission. The space time coding (STC) system increases the error performance of the communication systems and obtains the space-time diversity gain by coding over the different transmitting antennas [3], while space division multiplexing (SDM) system provides very high data-rate communication over wireless channels without increasing the total transmitting power and bandwidth [4]. And the STC-SDM system can achieve the high-speed and reliable communication [5]. Then, these are considered for high-speed and reliable communication.

In home networking environment, users will utilize various communication systems anytime and anywhere at multipath fading channel condition. Therefore the MB-OFDM system with MIMO structure for mobile communication environment should change their system reliability and data rate depending on dynamically changing user's demand and channel condition.

Y. Koucheryavy, J. Harju, and V.B. Iversen (Eds.): NEW2AN 2006, LNCS 4003, pp. 481–488, 2006.

In this paper, the space-time block coding (STBC) for STC, the Vertical Bell Labs layered space time (VBLAST) for SDM, and STBC-VBLAST for STC-SDM are used for MB-OFDM with MIMO structures. We propose an enhanced MIMO transmission scheme for MB-OFDM system with spatial mode selection agreeable to estimated SNR [6]. By the proposed scheme, we get both the diversity gain and the improvement of transmission rate. The performance of the proposed scheme is evaluated in terms of bit error rate (BER) and bits per subcarrier (BPS) throughput.

The remainder of this paper is organized as follows. Section 2 describes the system model. The SNR estimation is described in Section 3. In Section 4, enhanced MIMO transmission scheme for MB-OFDM System is proposed. In Section 5, simulation results are presented. Finally, we conclude the paper in Section 6.

## 2   System Model

### 2.1   MB-OFDM System with MIMO Structure

We consider MB-OFDM system that the whole available UWB spectrum between 3.1-10.6GHz is divided into several sub-bands with smaller bandwidth, whose bandwidth is approximately 500MHz [7]. In each sub-band, a normal OFDM modulated signal with $K = 128$ subcarriers and QPSK is used. The main difference between the MB-OFDM system and other narrowband OFDM systems is in the way that different sub-bands are used in the system. In this system, the transmission is not done continually on all sub-bands. Different patterns of sub-band switching are chosen for different users (different piconets) such that the multiuser interference is minimized [7].

Consider the MB-OFDM system link comprising $N_t$ transmitting antennas and $N_r$ receiving antennas. The received signals are corrupted by additive noise that is statistically independent among the $N_r$ receivers. Let $\{X_i^s(k)|k = 0,...,K-1\}$ denote the $K$ subcarrier symbols where $k$ and $i$ represent the corresponding subcarrier and transmitting antenna in $s$-th sub-band, respectively. At the receiver, the output in the frequency domain is

$$\mathbf{R}^s = \mathbf{H}^s\mathbf{X}^s + \mathbf{W}^s, \tag{1}$$

where $\mathbf{R}^s$ and $\mathbf{W}^s$ is an $N_r \times 1$ matrix of an received MIMO-OFDM symbol and an $N_r \times 1$ matrix of an additive white Gaussian noise (AWGN), respectively and $\mathbf{H}^s$ is an $N_r \times N_t$ matrix of propagation coefficient which is statistically independent in $s$-th sub-band.

### 2.2   MIMO Transmission Techniques

In this subsection, the MIMO transmission techniques could provide higher spectral efficiency and increase potentially the system capacity. Since the MIMO transmission techniques with 4 transmitting antennas($N_t = 4$) are used in this paper as described on Section 4, the maximum transmission rate is 4 for the BLAST scheme and the minimum transmission rate is 1 for the STBC. A detailed explanation is as follows. Generally, MIMO systems are classified into two large groups. Firstly, MIMO diversity transmission technique exists. Particularly, STBC is a representative diversity transmission technique and obtains gain of transmitting diversity by transmitting the same data for multiple transmitting antennas. Therefore, STBC technique brings the improvement of error performance through the diversity gain [3]. However, MIMO diversity technique has a waste of BPS performance in a high SNR environment by giving up the transmission of data despite of link capability. The transmission matrix of the STBC for $N_t = 4$ can be represented as

$$\mathbf{X}_{\text{STBC}} = \begin{pmatrix} \mathbf{X}_1 & -\mathbf{X}_2^* & \mathbf{X}_3 & -\mathbf{X}_4^* \\ \mathbf{X}_2 & \mathbf{X}_1^* & \mathbf{X}_4 & \mathbf{X}_3^* \\ \mathbf{X}_3 & -\mathbf{X}_4^* & \mathbf{X}_1 & -\mathbf{X}_2^* \\ \mathbf{X}_4 & \mathbf{X}_3 & \mathbf{X}_2 & \mathbf{X}_1 \end{pmatrix}. \tag{2}$$

This STBC technique transmits 4 different data sequences four times using 4 transmitting antennas. Therefore, the transmission rate is 1.

On the other side, MIMO multiplexing technique (known also as BLAST) transmitting the different data at multiple transmitting antennas supports the high speed transmission rate without increasing of system bandwidth. According to the transmission mode at the transmitter, BLAST is divided into the diagonal BLAST (DBLAST) and vertical BLAST (VBLAST). To diagonally transmitting data, D-BLAST uses a specific block coding among the transmitting. Therefore, it has the advantage of high efficiency in frequency and the disadvantage of high embodiment complexity. On the contrary, by independently transmitting data for each transmitting antennas, VBLAST has a low complexity in the process of embodiment. The VBLAST can theoretically offer a nearly linear increase in capacity. But the VBLAST has the loss of BER performance at a low SNR environment [4]. The transmission matrix of the VBLAST for $N_t = 4$ can be represented as

$$\mathbf{X}_{\text{VBLAST}} = \begin{pmatrix} \mathbf{X}_1 \\ \mathbf{X}_2 \\ \mathbf{X}_3 \\ \mathbf{X}_4 \end{pmatrix}. \tag{3}$$

This BLAST scheme transmits 4 different data simultaneously using 4 transmitting antennas. Therefore, the transmission rate is 4.

Finally, the STBC-VBLAST can achieve the high-speed and reliable communication, because it has the advantages of both STBC and VBLAST. However, this scheme has the disadvantage of both STBC and VBLAST; the transmission efficiency of STBC-VBLAST is only one half of VBLAST [5]. Therefore, the STBC-VBLAST is not an optimum solution. The STBC-VBLAST transmission matrix is presented as follows

$$\mathbf{X}_{\text{STBC-VBLAST}} = \begin{pmatrix} \mathbf{X}_1 & -\mathbf{X}_2^* \\ \mathbf{X}_2 & \mathbf{X}_1^* \\ \mathbf{X}_3 & -\mathbf{X}_4^* \\ \mathbf{X}_4 & \mathbf{X}_3^* \end{pmatrix}. \tag{4}$$

## 3    SNR Estimation for Spatial Mode Selection

### 3.1    Basic Theory of Adaptive Transmission

Assuming perfect channel quality estimation, the instantaneous channel SNR is measured by the receiver and the transmission mode selector determines the mode used in the next transmission. The information for the selection of transmission mode is delivered into the transmitter, using the system's control channel. This side-information is named mode-selecting feedback information. The different transmission modes are used according to the available mode-selecting feedback information. Specifically, a transmission mode is selected as shown in eqn.(5), if the instantaneous channel SNR perceived by the receiver exceeds the corresponding switching levels [6].

$$T_l = \begin{cases} \text{Mode}_1 & \text{if } \gamma < \mu_1 \\ \text{Mode}_2 & \text{if } \mu_1 \leq \gamma < \mu_2 \\ \vdots & \vdots \\ \text{Mode}_L & \text{if } \gamma \geq \mu_{L-1} \end{cases}, \tag{5}$$

where $\gamma$ and $\mu_l$ denote the channel quality value(value of the estimated SNR) and $l$-th mode-switching level, respectively and $L$ is transmission mode level. Generally, the mode-switching level $\mu_l$ is determined such that the average BPS throughput is maximized, while satisfying the average target BER requirement.

And mode selection probability $Pr(T_l)$ is defined as the probability of selecting the $l$-th mode from the set of available transmission modes,

$$Pr(T_l) = Pr[\mu_{l-1} \leq \gamma < \mu_l] \tag{6}$$

$$= \int_{\mu_{l-1}}^{\mu_l} f(\gamma)d\gamma,$$

where $f(\gamma)$ represents the probability density function (PDF) of $\gamma$.

### 3.2    SNR Estimation for MB-OFDM with MIMO Structure

In the proposed MIMO transmission scheme with spatial mode selection for MB-OFDM system, the SNR estimation is executed in all sub-band, respectively. And every sub-band(e.g. bandgroup 1, 2, 3, 4 ,5) is independent each other. Therefore, the estimated SNR at $k$-th subcarrier in each sub-band from $i$-th transmitting antenna to $j$-th receiving antenna, which specify the power ratio of received pure signal and that of noise, is defined in the form of

$$\text{SNR}_j^k[\text{dB}] =$$

$$10 \log_{10} \frac{1}{\varepsilon_j^2(k)} \sum_{i=1}^{N_t} |H_{ji}(k)|^2 \cdot (X_i(k)/\sqrt{N_t})^2, \quad (7)$$

where

$$\varepsilon_j^2(k) = (R_j(k) - \sum_{i=1}^{N_t} H_{ji}(k) \cdot X_i(k)/\sqrt{N_t})^2, \quad (8)$$

where $\varepsilon_j^2(k)$ is a noise power of $k$-th subcarrier at $j$-th receiving antenna, $R_j(k)$ is received symbol at $j$-th antenna, $H_{ji}(k)$ is a propagation coefficient of channel from $i$-th transmitting antenna to $j$-th receiving antenna, and $X_i(k)$ is transmitted symbol at $i$-th transmitting antenna.

In this paper, since the proposed transmission scheme use multiple antennas, the average SNR value which is estimated and averaged at all receiving antennas is required. That is

$$\text{SNR}^k[\text{dB}] = \frac{1}{N_r} \sum_{j=1}^{N_r} \text{SNR}_j^k . \quad (9)$$

For proposed scheme, the overall average SNR which is averaged for all subcarriers can be represented as

$$\text{SNR}[\text{dB}] = \frac{1}{K} \sum_{k=0}^{K} \text{SNR}^k . \quad (10)$$

## 4   Enhanced MIMO Transmission Scheme with Spatial Mode Selection

In this proposed scheme, the transmission modes correspond to the STBC, VBLAST, and STBC-VBLAST. In the VBLAST, the number of the receiving antennas is equal to or larger than that of the transmitting antennas [6]. Therefore, the number of transmitting and receiving antennas is the same in this scheme ($N_t = N_r = 4$). In this case, the transmission mode selector among STBC, VBLAST and STBC-VBLAST decides its MIMO transmission mode according to the estimated SNR of channel in eqn.(10). The transmission mode is defined by the following rules,

$$T_l = \begin{cases} \text{STBC} & \text{if } \gamma < \mu_1 \\ \text{STBC-VBLAST} & \text{if } \mu_1 \leq \gamma < \mu_2 \\ \text{VBLAST} & \text{if } \gamma \geq \mu_2 \end{cases} . \quad (11)$$

By properly selecting each spatial mode, this scheme satisfies the diversity gain and very high-rate transmission according to the channel condition in home networking environment.

# 5    Simulation Results and Discussions

In this section, we examine the performance of the MB-OFDM system apply-ing the proposed scheme. To evaluate the performance of this system, the MB-OFDM system with FFT size of 128 is considered in a bandgroup 1 of CM2 UWB channel environment. And this system is modulated by QPSK and has 4 trans-mitting, 4 receiving antennas. In this simulation, the values of SNR thresholds for switching the transmission mode are $\mu_1 = 10$ dB and $\mu_2 = 20$ dB, respec-tively. These SNR thresholds are determined to employ the proposed scheme by several experiments.

Fig. 1 shows the selection probability of each transmission mode according to the SNR threshold. In this figure, it shows that the STBC scheme is changed into the STBC-VBLAST scheme when the probability of mode selection in transmis-sion modes is reached at about 50 percent on the neighborhood of $\mu_1$. Also it shows that the STBC-VBLAST scheme is changed into the VBLAST scheme when the probability of mode selection in transmission modes is reached at about 50 percent on the neighborhood of $\mu_2$. Therefore, it shows that the pro-posed scheme could achieve the higher data rate or higher reliability by changing spatial mode in compliance with SNR of varying channel environments.

The performances of BER and BPS are shown on Fig. 2 and Fig. 3, respec-tively which are also shown the performances of STBC, STBC-VBLAST and VBLAST to compare the proposed scheme. As depicted in Fig. 2, the BER curve of proposed scheme is started from the STBC curve at low SNR and is reached to VBLAST curve at high SNR through the STBC-VBLAST curve. Therefore, it shows that the performance of BER is overall improved by using the proposed

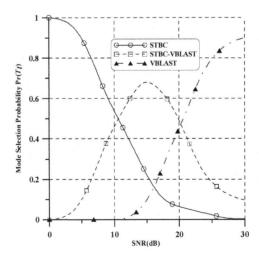

**Fig. 1.** Spatial mode selection probability of enhanced MIMO transmission scheme for MIMO-OFDM with spatial mode selection

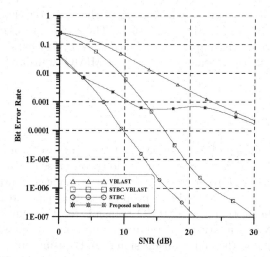

**Fig. 2.** BER performance of enhanced MIMO transmission schemes for MB-OFDM system with spatial mode selection

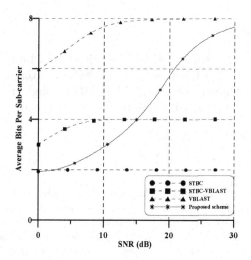

**Fig. 3.** BPS performance of enhanced MIMO transmission schemes for MB-OFDM system with spatial mode selection

scheme, compared with each conventional MIMO transmission technique, from low SNR to high SNR.

Fiq. 3 shows that the BPS curve of the proposed scheme is located between STBC and VBLAST. This curve is started from the STBC curve and is gradually approached to VBLAST curve through the STBC-VBLAST curve according to increasing SNR. At SNR > 20 dB, the proposed scheme attains to 8 BPS like a throughput performance of VBLAST.

Consequently, the proposed scheme has all the advantages of STBC, STBC-VBLAST , and VBLAST in Fig. 2 and Fig. 3. The performances of BER and BPS of proposed scheme are adaptively changed in varying channel environments, while the performance of each conventional MIMO transmission technique is fixed.

## 6   Conclusions

The main disadvantage of MB-OFDM system using each conventional MIMO transmission technique is a waste of BER and BPS to result from varying channel environments. In order to solve this problem, enhanced MIMO transmission scheme with spatial mode selection is proposed in this paper. It is introduced for the proposed scheme that three MIMO transmission techniques are adaptively used with the SNR estimation. Therefore, the proposed scheme can satisfy both the diversity gain and the high transmission rate. By using this proposed scheme, the reliable and high data-rate communications which are well adapted to the various user's demands and channel conditions are achieved in home network environments.

## Acknowledgments

This work is financially supported by the Ministry of Education and Human Resources Development (MOE), the Ministry of Commerce, Industry and Energy (MOCIE) and the Ministry of Labor (MOLAB) through the fostering project of the Lab of Excellency and is supported by MIC Frontier R&D Program in KOREA.

## References

1. M. Minani, H. Morikawa, and T. Aoyama, "The design of naming-based service composition system for ubiquitous computing applications," *IEEE Applications and the Internet Workshops*, pp. 304-312. January 2004.
2. P.-L. Tsai, C.-L. Lei, and W.-Y. Wang, "A remote control scheme for ubiquitous personal computing," *IEEE Networking, Sensing and Control Conference*, vol. 2, pp. 1020-1025. March 2004.
3. S. M. Alamouti, "A simple transmit diversity technique for wireless communications," *IEEE*, vol. 16, no. 8, pp. 1451-1458. October 1998.
4. G. J. Foschini, "Layered space-time architecture for wireless communications in a fading environment when using multi-element antennas," *Bell Labs Technical Journal*, vol. 1, no. 2, pp. 41-59, Autumn 1996.
5. A. F. Naguib, N. Seshadri, and A. R. Calderbank, "Increasing data rate over wireless channels," *IEEE Signal Processing Megazine*, vol. 17, no. 3, pp. 76-92, May 2000.
6. L. Hanzo, C. H. Wong, M. S. Yee, "Adaptive wireless Transceivers," Wiley, 2002.
7. "IEEE 802.15 WPAN high rate alternative PHY Task Group 3a (TG3a) [Online]". Available: http://www.ieee802.org/15/pub/TG3a.html

# Improving Energy Efficiency and Responsiveness in Bluetooth Networks: A Performance Assessment

Laura Galluccio, Alessandro Leonardi, and Antonio Matera

Dipartimento di Ingegneria Informatica e delle Telecomunicazioni (DIIT)
University of Catania, Italy
name.surname@diit.unict.it

**Abstract.** Bluetooth is a widespread solution for personal area networks (PAN). However, it currently lacks in spontaneous and energy efficient device communication which could make it more successful. To allow spontaneous networking, communication nodes should have the ability to timely discover each other. Moreover, in order to improve energy efficiency and increase battery lifetime the energy spent during discovery procedures should be minimized while guaranteeing high probability of successfully completing the discovery and the information transfer procedure. In this paper we derive performance considerations on the behavior of the system through experiments in a physical testbed where devices move while discovering nodes in their neighborhood and then exchange information. These considerations can be used for accurately model neighbor discovery process and information transfer in Bluetooth networks.

**Keywords:** Bluetooth, Markov model, Analysis, Experimental Results.

## 1 Introduction

The support of spontaneous networking [2, 3, 1] will allow mobile devices to autonomously look for other devices in their proximity every-time they enter a new area and then, if necessary, exchange information. This will make the Bluetooth technology more successful. To allow spontaneous networking, the discovery of neighbor nodes must be timely. Moreover, in order to improve energy efficiency and increase battery lifetime the energy spent during the discovery procedures should be minimized while guaranteeing high probability of successfully completing the procedures. Energy efficiency and responsiveness are antagonist to each other and, consequently, an appropriate trade-off has to be sought. In a previous work [4] we developed an analytical framework for neighbor discovery procedures modelling in self-organizing networks which can be applied also to Bluetooth networks. More specifically, we found out that a trade-off between energy efficiency and responsiveness can be identified; in particular, we understood how the parameters of the neighbor discovery process can be tuned to satisfy these two requirements in a static scenario where nodes do not move and only

Y. Koucheryavy, J. Harju, and V.B. Iversen (Eds.): NEW2AN 2006, LNCS 4003, pp. 489–500, 2006.
© Springer-Verlag Berlin Heidelberg 2006

discover other nodes and services available in their neighborhood. In this paper, instead, we want to pursue two targets: the first one is validating the analytical model through a comparison to a real Bluetooth 1.1 testbed; then, we want to study a mobile Bluetooth testbed where devices have to successfully discover each other and perform information exchange during the limited neighborhood time. This will help us to elaborate a comprehensive analytical model to take into account a mobile scenario where nodes velocity impacts on the capability of the network to efficiently work. The remainder of this paper is organized as follows. In Section 2 the discovery model proposed in [4] is summarized and the QoS performance parameters defined; in Section 3 some significant analytical results are reported. In Section 4 two testbed scenarios, the static and the mobile one, are investigated and used to satisfy the proposed targets. Finally, in Section 5, concluding remarks are drawn.

## 2    Discovery Model

Bluetooth [5] is a short range communication standard allowing wireless data communications at a maximum net rate of about 700 Kbps for Bluetooth 1.0 and from 1 Mbps up to 3 Mbps for Bluetooth 2.0. When a Bluetooth device enters a network, it should spontaneously search for other devices in its neighborhood. Once discovered, before activating a connection, the Bluetooth device has to retrieve the 48-bit Bluetooth device address, clock, class-of-device information (which characterizes the transmission power level being used), as well as other information required for the establishment of a connection. Devices in the neighborhood are usually referred to as discoverable. A discoverable device in range periodically listens on a certain physical channel and responds to possible inquiries received on that channel. Discoverable devices are usually also connectable; this means that such a device periodically listens and responds in an appropriate channel used to establish a connection. Once discovered, two or more Bluetooth devices are able to exchange data. When Bluetooth devices are mobile, it is necessary to periodically reinvoke the device discovery procedure so as to update the list of devices in the neighborhood and, thus, services and resources available in the closest proximity. Neighbor discovery process is thus preliminary to connection establishment. In order to characterize the neighbor discovery process, in the following sections we will recall the basic aspects of the analytical Markov model we proposed in [4].

### 2.1    Hunting Process

A single node which wants to discover possible neighbor nodes runs a set of procedures which we call *hunting* process.

Suppose that at time $t = 0$, a mobile Bluetooth node enters the radio coverage area of another mobile Bluetooth node and, therefore, they become neighbors. Let $T$ be the random variable representing the time needed for the two nodes to discover each other. Responsiveness of the discovery process can be represented by the probability distribution function of the random variable $T$, $F_T(t)$, which is defined as follows

$$F_T(t) = P\{T \le t\} \quad \forall t \ge 0 \tag{1}$$

Calculation of $F_T(t)$ is based on a Markov model of the neighbor discovery process which results from the interactions between the hunting processes run by the two nodes.

The *hunting process* run by a node can be in three possible states: 1) **Inquiry (I)** where the node transmits a beacon message to advertise its presence to neighbor nodes; 2) **Inquiry Scan (S)** where the node listens in order to receive beacon messages from possible neighbor nodes; 3) **Doze (D)** where the node is not executing any functionality to discover or being discovered by its neighbors (in this state the node's energy consumption is negligible).

We will model this behavior through a Markov chain in which each of the above states is a macro-state. More specifically, let us call $M$ the number of frequency channels which can be used for the discovery and $I_m$ and $S_m$, with $m \le M$, the state of the Markov chain in which the node transmits and listens to the $m-th$ channel, respectively. Let $Q^{(H)}$ represent the transition rate matrix of the hunting process. This matrix can be used to easily evaluate the array of the steady state probabilities of the hunting process in a single node, $\Pi^{(H)}$ as follows [6]:

$$\begin{cases} \pi^{(H)} \cdot Q^{(H)} = 0 \\ \sum_{\sigma \in F^{(H)}} \left[\pi^{(H)}\right]_{[\sigma]} = 1 \\ \left[\pi^{(H)}\right]_{[\sigma]} = P\left\{S^{(H)}(t) = \sigma\right\} \end{cases} \tag{2}$$

where $\left[\pi^{(H)}\right]_{[\sigma]}$ represents the generic element of the array $\pi^{(H)}$, $F^{(H)}$ is the state space of the hunting process and $S^{(H)}(t) \in F^{(H)}$ is the state of the hunting process of a node in a time instant $t$.

## 2.2   Neighbor Discovery Model

In this section we will derive the *neighbor discovery process* resulting from the interactions between the hunting processes of two nodes. Suppose that at time $t$, where $t \ge 0$, two mobile nodes, $N_1$ and $N_2$, become neighbors, i.e., they are in radio coverage of each other. The state of the *discovery process*, $S^{(P)}(t)$, can be represented by the pair

$$S^{(P)}(t) = \left(S^{(H_1)}(t), S^{(H_2)}(t)\right) \quad \text{with } S^{(H_1)}(t) \text{ and } S^{(H_2)}(t) \in F^{(H)} \tag{3}$$

Consequently, the state space of the neighbor discovery process, $F^{(P)}$, is given by the cartesian product of the state spaces identified in Section 2.1 for the two nodes. Moreover, an array $\pi^{(P)}(t)$ can be defined, in which the generic element represents the probability that the state of the discovery process is $(\sigma_1, \sigma_2)$ at time $t$:

$$\left[\pi^{(P)}(t)\right]_{[\sigma_1, \sigma_2]} = P\left\{S^{(P)}(t) = (\sigma_1, \sigma_2)\right\} \tag{4}$$

Observe that $\pi^{(P)}(t)$ can be calculated as [6] $\pi^{(P)}(t) = \pi^{(P)}(0) \cdot e^{-Q^{(P)} \cdot t}$ where:

- $\pi^{(P)}(0)$ represents the array of the state probabilities when the two mobile nodes become *neighbors*, i.e., at time $t = 0$. Since the hunting states of the two nodes involved in the discovery process were independent of each other before time $t = 0$, $\pi^{(P)}(0)$ can be calculated as the cartesian product of the arrays of the steady state probabilities of the two mobile nodes' hunting processes, $\pi^{(P)}(0) = \pi^{(H_1)} \times \pi^{(H_2)}$

- $Q^{(P)}$ is the transition rate matrix of the discovery process.

The discovery between $N_1$ and $N_2$ occurs if one of the two mobile nodes transmits a beacon message in the channel where the other mobile node is listening. Therefore, if at time $t$ the state of the discovery process is $S^{(P)}(t) = (\sigma_1, \sigma_2)$, then the discovery occurs in the time interval $(t, t + \Delta t)$ with probability $P^{(DISC)}_{(\sigma_1, \sigma_2)}(t, t + \Delta t)^1$ which depends on $Q^{(H_1)}$ and $Q^{(H_2)}$ that are the state transition matrixes of the hunting process of $N_1$ and $N_2$. Accordingly, it follows that the array of the discovery rate, $\Lambda^{(DISC)}$, can be defined, in which the generic element $\left[\Lambda^{(DISC)}\right]_{[(\sigma_1,\sigma_2)]}$ represents the discovery rate when the state of the discovery process is $(\sigma_1, \sigma_2)$ and can be straightforwardly calculated as

$$\left[\Lambda^{(DISC)}\right]_{[(\sigma_1,\sigma_2)]} = \frac{P^{(DISC)}_{(\sigma_1,\sigma_2)}(t, t + \Delta t)}{\Delta t} \tag{5}$$

Obviously, the arrays $\pi^{(P)}(t)$ and $\Lambda^{(DISC)}$ can be used to evaluate the overall discovery rate at time $t$, $\rho(t)$, i.e.,

$$\rho(t) = \pi^{(P)}(t) \cdot \Lambda^{(DISC)*} \tag{6}$$

where $\Lambda^{(DISC)*}$ denotes the transpose of $\Lambda^{(DISC)}$.

Now, let us focus the attention on the evolution of the discovery process when the two nodes do not succeed in discovering each other.

Let $\pi^{(DISC)'}_{T \geq t}(t)$ be the array in which the generic element is the probability that the state of the pair of neighbors is $(\sigma_1, \sigma_2)$ given that the two nodes have not discovered each other, i.e.,

$$\left[\pi^{(DISC)'}_{T \geq t}(t)\right]_{(\sigma_1,\sigma_2)} = P\left\{S^{(P)}(t) = (\sigma_1, \sigma_2)|T \geq t\right\} \tag{7}$$

In order to evaluate $\pi^{(DISC)'}_{T \geq t}(t)$, it must be considered the matrix $Q^{(P)}_{NODISC}$ of the state transition rates, given that the mobile nodes do not discover each other. The matrix $Q^{(P)}_{NODISC}$ can be calculated from matrix $Q^{(P)}$ as follows.

First, it is set $Q^{(P)}_{NODISC} = Q^{(P)}$; then, the elements of the matrix which result in the discovery of the two mobile nodes are set to zero. Accordingly, the array $\pi^{(P)}_{T \geq t}(t)$ can be calculated as $\pi^{(P)}_{T \geq t}(t) = \pi^{(P)}(0) \cdot e^{-Q^{(P)}_{NODISC} \cdot t}$

The array $\pi^{(P)}_{T \geq t}(t)$ can be used for the computation of the discovery rate at time $t$, given that the mobile nodes have not yet discovered each other, $\rho_{T \geq t}(t)$. Thus,

$$\rho_{T \geq t}(t) = \pi^{(P)}_{T \geq t}(t) \cdot \Lambda^{(DISC)*} \tag{8}$$

---

[1] Details on the derivation of $P^{(DISC)}$ can be found in [4].

## 2.3 Responsiveness Estimation

The results in Section 2.2 are used to evaluate the probability distribution function, $F_T(t)$, of the random variable $T$ which represents the time when the nodes $N_1$ and $N_2$ discover each other.

To this purpose the following lemmas can be demonstrated[2]:

**Lemma 1:** *The probability distribution function of the random variable $T$ is given by:* $F_T(t) = 1 - k \cdot e^{-A(t)}$ *being* $k = 1 - \sum_{m=1}^{M} \left( [\pi^{(P)}]_{(S_m,I_m)} + [\pi^{(P)}]_{(I_m,S_m)} \right)$ *and* $A(t) = \int_0^t \rho_{T \geq t}(\tau) d\tau$.

Before illustrating Lemma 2, some notation needs to be introduced. Let $L$ be the number of columns of $Q_{\text{NODISC}}^{(P)}$ and apply the spectral decomposition to the matrix $Q_{\text{NODISC}}^{(P)}$:

$$Q_{\text{NODISC}}^{(P)} = \Gamma^{-1} \cdot D \cdot \Gamma \qquad (9)$$

where $D$ is the diagonal matrix of the eigenvalues of $Q_{\text{NODISC}}^{(P)}$ and $\Gamma$ is the matrix consisting of the eigenvectors of matrix $Q_{\text{NODISC}}^{(P)}$. It is now possible to introduce the following lemma:

**Lemma 2:** *The value of $A(t)$ is given by:*

$$A(t) = \pi^{(P)}(0) \cdot \Gamma^{-1} \cdot B \cdot \Gamma \cdot \Lambda^{(DISC)*} \qquad (10)$$

where the diagonal matrix $B$ is written as

$$[B]_{[(i,j)]} = \begin{cases} \left[ \frac{1-e^{-d_i \cdot t}}{d_i} \right] & \text{if } i = j \text{ and } i \leq L \\ t & \text{if } i = j, \ L < i \leq H \\ 0 & \text{otherwise} \end{cases} \qquad (11)$$

## 2.4 Energy Efficiency Estimation

In order to take into account the energy efficiency of the discovery process, an energy cost, $c$ is considered. This energy cost depends on the percentage of time spent by the mobile node performing operations which require energy consumption. Specifically, the cost is given by two terms: the power consumption due to the time required for the transmission of beacon messages and the power consumption due to the time required for scanning the signaling channels to listen to beacon messages.

To calculate this parameter, the array $\Pi^{(H)}$ is used, as follows:

$$c = \sum_{m=1}^{M} \left[ PW_{\text{INQ}} \cdot [\Pi^{(H)}]_{[I_m]} + PW_{\text{SCAN}} \cdot [\Pi^{(H)}]_{[S_m]} \right] \qquad (12)$$

where $PW_{\text{INQ}}$ and $PW_{\text{SCAN}}$ represent the power consumption while the mobile node is in Inquiry and Inquiry Scan mode, respectively. Observe that the energy

---

[2] Proofs of these lemmas can be found in our previous work [4].

cost ranges in the interval $(0, max\{PW_{INQ}, PW_{SCAN}\})$. Therefore, given that the power utilized to transmit is generally larger than the power used to receive, $0 < c < PW_{INQ}$. Once we have defined the analytical Markov model used for evaluating the responsiveness and energy efficiency parameters, we will estimate the maximum $F_T(t)$, provided that a constraint on the energy consumption $c$ is given.

## 3  Analytical Results

In this section we will perform an analytical study on the performance of the system modelled as described in the previous sections. More specifically, we will maximize the probability that the discovery occurs within a certain time $t^*$, provided that the energy consumption is lower than a given threshold. Let us now focus on the Bluetooth standard.

According to it, a node in the Inquiry state transmits $J$ trains of *Inquiry Messages*, where:

- The parameter $J$ is set equal to $J = (256 \cdot K \cdot (M/16))$, with $K \in \{2, 4, 6\}$ and $M$ is the number of frequency channels which can be used for the discovery.
- Each train consists of 16 Inquiry Messages, each transmitted in a different channel according to a pseudo-random sequence, called the *Inquiry Hopping Sequence* [5].
- The average time between the transmission of two Inquiry Messages is $T_{\text{SLOT}} = 625\mu s$, although only $68\mu s$ are utilized for the actual transmission of the Inquiry Message.

Accordingly, the average time spent in the Inquiry macro-state, $T_{\text{INQ}}$, is equal to $T_{\text{INQ}} = 16 \cdot 256 \cdot K \cdot \left(\frac{M}{16}\right) \cdot T_{\text{SLOT}}$.

A BT node entering the Inquiry Scan mode executes the following two steps:

- *Step 1:* The BT node listens to a channel $f_k$ belonging to the Inquiry Hopping Sequence, for a time interval $T_{\text{RX}}$, with $T_{\text{RX}} \geq 16 \cdot T_{\text{SLOT}}$.
- *Step 2:* The hunting process of the BT node spends a time period $T_{\text{WR}}$, with $T_{\text{WR}} \leq 2.56$ s, without listening or sensing beacon messages.

The aforementioned two steps are repeated according to a pseudorandom sequence over different channels for the entire duration of the Inquiry Scan procedure. A BT node in Inquiry Scan mode, $N_1$, receiving an Inquiry Message from node $N_2$ leaves the Inquiry Scan mode to notify its presence to $N_2$. However, in order to avoid collisions, this is not done immediately. Rather, $N_1$ waits for a random time interval $T_{\text{WAIT}}$, where $0 < T_{\text{WAIT}} \leq 0.64$ s, before entering the Inquiry Scan mode again. In this new Inquiry Scan mode, $N_1$ notifies its presence as soon as it receives an Inquiry Message again. This concludes successfully the discovery process.

When the BT node is not involved in any of the above procedures, we already said that it is considered to be in Doze mode. The average duration of the time period spent in the Doze mode is denoted as $T_{\text{DOZE}}$. Observe that the values of

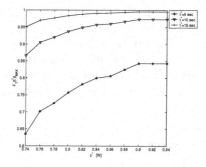

**Fig. 1.** Values of $F_T(t^*)_{\text{MAX}}$ provided that $c \leq c^*$ for different values of time $t^*$ and assuming $T_{\text{WR}} = 1.28$ s (Analytical Results)

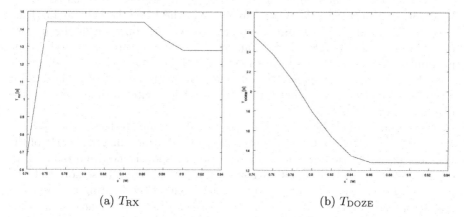

(a) $T_{\text{RX}}$                    (b) $T_{\text{DOZE}}$

**Fig. 2.** Optimal values of $T_{RX}$ and $T_{DOZE}$ vs. $c^*$ (Analytical results)

the parameters $K$, $T_{\text{RX}}$, $T_{\text{WR}}$, $T_{\text{WAIT}}$, and $T_{\text{DOZE}}$ are not specified by the BT standard [5]. However, note that:

- the value of $K$ depends on the number of ongoing synchronous connections, $N_{\text{SCO}}$. In the following it is assumed that $N_{\text{SCO}} = 0$ and, accordingly, $K = 2$.
- a typical value of $T_{\text{WR}}$ is $T_{\text{WR}} = 1.28$ s.
- the value of $T_{\text{WAIT}}$ is randomly chosen in the interval $(0, 0.64)$ s according to a uniform distribution. Therefore, the average value of $T_{\text{WAIT}}$ is equal to 0.32 s.

For what concerns the values of $T_{\text{RX}}$ and $T_{\text{DOZE}}$ they are chosen so as to satisfy the target of maximizing $F_T(t^*)$, given a constraint on the energy cost, while complying with the standard specifications, i.e., $T_{\text{RX}} \geq 16 \cdot T_{\text{SLOT}}$.

In Figure 1 we show the value of $F_T(t^*)_{\text{MAX}}$, which can be obtained analytically provided that the energy cost is not higher than $c^*$. In the same figure, $F_T(t^*)_{\text{MAX}}$ is shown for different values of $t^*$, assuming that $M = 16$

channels are used. Observe that the three curves have similar behavior, i.e., $F_T(t^*)_{\text{MAX}}$ increases as $c^*$ increases, whereas the slope decreases. The above values of $F_T(t^*)_{\text{MAX}}$ can be obtained using the values of the parameters $T_{\text{RX}}$ and $T_{\text{DOZE}}$ given in Figures 2 and derived through the use of the proposed analytical framework. Observe that in Figures 2 the optimal values of $T_{\text{RX}}$ and $T_{\text{DOZE}}$ do not depend on $t^*$, i.e., the three curves (obtained for $t^* = 5$, 10 and 15 s) overlap.

# 4   Experimental Results

In this section we will illustrate the experimental results obtained using a Bluetooth 1.1 testbed consisting of two laptop devices employing Bluetooth DIGI-COM Palladio adapters and using the RedHat 9- 2.4.20 Linux distribution.

The Bluetooth implementation being used in the testbed has been the Bluez Bluetooth protocol stack for Linux, included in the Linux kernel starting from version 2.4. Using the `hcitool` command we set the duration of the inquiry phase in multiples of 1.28 s as recommended by the standard [5], as well as the maximum number of devices to be discovered before the device exits the inquiry procedure. In the following we have chosen to have an inquiry phase lasting for a typical time of 10.24 s so that the device is able to switch 3 times between the two frequency trains.

Two scenarios have been considered for the Bluetooth testbed. The QoS parameters being estimated are the probability distribution function $F_T(t^*)$ of the neighbor discovery time which estimates system responsiveness and the average energy cost $c_M$ which takes into account energy efficiency. Please note that $c_M$ is obtained averaging the energy cost at each cycle of the hunting process. The two scenarios are: 1)*Scenario 1: Static Neighbor discovery scenario* where the two devices do not move and only want to discover each other in the shortest possible time, provided that the required energy consumption is not that high; 2)*Scenario 2: Mobile Neighbor discovery scenario with file transfer* where the two devices move with respect to each other and, once discovered, should also perform information exchange.

The experimental results derived in these two scenarios aim at two targets. The first one is validating the analytical model through a comparison to a real Bluetooth 1.1 testbed. The second target is studying a mobile Bluetooth 1.1 testbed where devices have to successfully discover each other and then make a connection so as to exchange information. This second study will be used to extend the analytical framework to generalize it to the case of a mobile environment where devices are in proximity only for a limited time and need to exchange information over a wireless unreliable interface.

## 4.1   Scenario 1

In this scenario the devices are static and alternate between inquiry, inquiry scan and doze. Using the `hci_commands` the behavior of the two devices is controlled

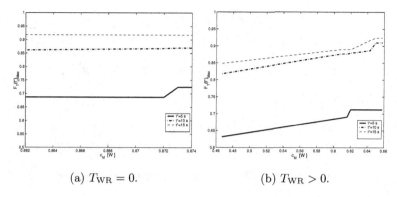

(a) $T_{WR} = 0$.                    (b) $T_{WR} > 0$.

**Fig. 3.** Scenario1: $F_T(t^*)_{MAX}$ vs. $c_M$ for different values of $T_{WR}$ (Experimental results)

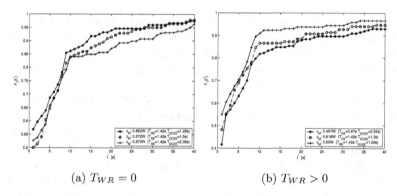

(a) $T_{WR} = 0$                    (b) $T_{WR} > 0$

**Fig. 4.** Scenario1: $F_T(t^*)_{MAX}$ vs. $t^*$ for different values of $(T_{RX}, T_{DOZE})$ pairs

and the $F_T(t^*)$ function can be calculated. The initial state of the device hunting process is randomly chosen.

By using the Bluez hciconfig command inqparms (win:int), we set the win parameter for the inquiry scan window duration equal to $T_{RX}$ whose values are given in Figure 2(a); similarly, we set the doze time value, $T_{DOZE}$, by using the command noscan whose values are given in Figure 2(b). As a consequence we derived the corresponding values of the average energy cost $c_M$. In the calculation of $c_M$ we distinguished two cases: $T_{WR} > 0$ and $c_M \in [0.46, 0.66]$ W and $T_{WR} = 0$ and $c_M \in [0.862, 0.875]$ W. This is because a different choice of $T_{WR}$ has a different impact on network performance. In Figures 3(a) and 3(b) lower values of the $F_T(t^*)_{MAX}$ are obtained for both $T_{WR} > 0$ and $T_{WR} = 0$ with respect to analytical results shown in Figure 1. This is due to having $M = 32$ frequency channels with respect to the analytical case when only 16 channels were considered due to the complexity of the mathematical model. Accordingly, the probability that two nodes discover each other in the experimental testbed is reduced since frequency hopping is performed among a larger set of frequencies.

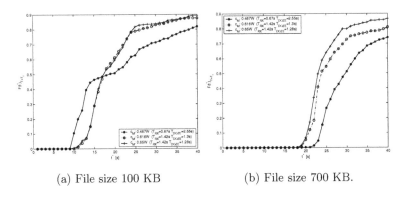

(a) File size 100 KB    (b) File size 700 KB.

**Fig. 5.** Scenario 2: $F_{T+T_T}(t^*)$ vs. $t^*$, for different values of $(T_{\mathrm{RX}}, T_{\mathrm{DOZE}})$ pairs

Please observe that comparing analytical and experimental results, Figure 1 gives a maximum bound on the cost since the $F_T(t^*)_{\mathrm{MAX}}$ values are obtained provided that $c \leq c^*$ while Figure 3 is obtained considering the average cost, $c_M$, which satisfies the condition of being lower than the corresponding values of the cost shown in Figure 1.

In Figures 4(a) and 4(b) we show the values of $F_T(t^*)$ vs. time, $t^*$, for different pairs $(T_{\mathrm{RX}}, T_{\mathrm{DOZE}})$ of the inquiry scan window and the doze time obtained from our analytical framework and shown in Figures 2. Each pair is associated to a corresponding average energy cost $c_M$.

By keeping unchanged the pairs $(T_{\mathrm{RX}}, T_{\mathrm{DOZE}})$, we have also evaluated the case with $T_{\mathrm{WR}} = 0$, shown in Figure 4(a). By setting $T_{\mathrm{WR}}=0$, the total time $T_{\mathrm{SCAN}}$ spent in Inquiry Scan mode decreases, while the scanning period, $T_{\mathrm{RX}}$, remains unchanged. Consequently, lowering $T_{\mathrm{SCAN}}$ and thus $T_{\mathrm{Cycle}}$ defined as $T_{\mathrm{Cycle}} = T_{\mathrm{DOZE}} + T_{\mathrm{SCAN}} + T_{\mathrm{INQ}}$, increases the probability of Inquiry Scan on the $m-th$ channel , $[\Pi^{(H)}]_{[S_m]}$, as well as the cost, because $[\Pi^{(H)}]_{[S_m]}$ is proportional to the reciprocal of $T_{\mathrm{Cycle}}$. In Figure 4(a), as expected, there is not a significant increase in the $F_T(t^*)$ values, despite of a growth in costs with respect to Figure 4(b). To explain this behavior, we observe that, on the one hand a continuous scanning is expected to perform better in terms of the discovery probability, on the other hand inquiry windows tend to overlap more frequently than in the case with a non zero $T_{\mathrm{WR}}$. Accordingly, setting $T_{\mathrm{WR}} = 0$ does not represent a good choice in terms of both responsiveness and energy consumption.

In the next section we will observe that, also in case of neighbor discovery and information transfer with mobile devices, choosing $T_{\mathrm{WR}} > 0$ is more appropriate in order to improve system performance.

### 4.2  Scenario 2: Neighbor Discovery with File Transfer

In this scenario, two phases are considered:1) the devices move while discovering other nodes in their neighborhood 2)after the discovery has happened, a connection should be established and then information transfer can be performed.

Accordingly, the time when the two devices are in proximity of each other is limited due to nodes movement and should be efficiently used for both performing discovery, connection establishment and information transfer. Target of our investigation is understanding how to tune neighbor discovery parameters in the analytical model and evolve the model so as to guarantee good performance in terms of both neighbor discovery and information transfer. For simplicity we assumed that one of the two devices is static and the other one moves over a straight trajectory with a constant velocity, $v$, equal to about 2 Km/h (i.e. a slowly walking person). The process is considered to be successfully completed when the two devices discover each other, establish a connection and complete the information transfer. In order to perform an information transfer, it is necessary to explicitly set up a connection. This can be done through a paging procedure. Paging is executed together with the discovery procedure, using the PAND module to create a connection. The Secure Copy Protocol (SCP) is used to perform a secure information transfer. The probability of discovering devices in the neighborhood and successfully performing an information transfer in a time lower than $t^*$ is defined as $F_{T+T_T}(t^*) = P\{T + T_T \leq t^*\}$      $\forall t^* \geq 0$

where $T$ is the random variable representing the time needed for the two nodes to discover each other and $T_T$ is the variable representing the time needed for the information transfer.

For the hunting process, we used the same input parameters employed in Scenario 1 and we tested the performance for various file size transmissions. We assumed a maximum allowed bit-rate of 720 Kbps, compliant with Bluetooth 1.1 specifications. By performing the experiments, we observed that using the Bluez drivers there is no way for a discovered device to notify the discovery event at the higher levels; only the inquiring device can do that. Only if the inquiry scan command is invoked together with the *page scan* command, this will allow, once a scanning device is discovered, to establish a connection.

The Page Scan procedure works during the time $T_{WR}$, i.e. the time interval between two Inquiry Scan windows; page scan allows to detect page requests, thus starting the procedure for accepting a connection by a paging device. Accordingly, it is now evident that choosing a $T_{WR}$ value greater than zero, allows to improve the performance of the system since we can devote enough time slots to the page procedure to let it work successfully.

Let us observe that the energy consumption due to the Paging procedure, where the page scan window has been chosen to last 18 slots (i.e. about 10 ms) as suggested by standard specifications [5], does not represent a relevant overhead in terms of the average energy cost associated to the neighbor discovery and information transfer process.

Figure 5(a) shows the values of $F_{T+T_T}(t^*)$ vs. the time, $t^*$, calculated for a file size of 100 KB and for three different pairs $(T_{RX}, T_{DOZE})$ chosen according to Figure 2.

Looking at this figure it can be observed that, initially, at least for a time equal to 10 s, $F_{T+T_T}(t^*)$ is basically zero. This is due to the fact that, the non deterministic time needed for connection set-up and SCP authentication, as well as the deterministic transfer time which depends on the file size, give a minimum bound on the time required for successfully completing both neighbor discovery and information transfer. Observe that, initially, the curve with the lowest average cost exhibits better performance with respect to the other curves. Looking at the corresponding pair $(T_{RX}, T_{DOZE})$ we observe that this curve exhibits the lowest value of $T_{RX}$. This is an advantage for the page process since it allows to speed up the connection process and also reduce the energy cost.

Moreover, this behavior can be justified considering that for low values of time $t^*$ the impact of the page process is predominant over the neighbor discovery process as evident comparing Figures 4 and 5 where the difference between the three curves in Figure 5 for low values of $t^*$ is larger than for Figure 4. This means that the impact of the paging process cannot be neglected. For higher file sizes, the neighbor discovery process has a higher impact with respect to the page process leading to a better performance in case of higher $c_M$, as shown in Figure 5(b). As expected, transmission of small size files exhibits a better performance in terms of $F_{T+T_T}(t^*)$ with respect to larger file sizes because the time required for information exchange (i.e. file transfer) is lower.

## 5   Conclusions

In this paper we studied the problem of finding a trade-off between energy efficiency and responsiveness of the neighbor discovery process in a Bluetooth system. Starting from the Markov model of the neighbor discovery process in a static ad hoc network scenario derived in our previous paper, we compared analytical and experimental results derived in a Bluetooth 1.1 testbed to both assess the effectiveness of the analytical model to represent the physical behavior of a Bluetooth discovery process and evolve the analytical model to cope with limited neighborhood time due to nodes movement. Moreover, we aim at extending our analytical framework so as to model both the neighbor discovery process and the information transfer process.

## References

1. C. Prehofer and C. Bettstetter. Self-organization in communication networks: principles and design paradigms. *IEEE Communications Magazine.* Vol. 43, No. 7.
2. J. P. Hubaux et al. Toward Self-Organized Mobile Ad Hoc Networks: The Terminodes Project. *IEEE Communications Magazine.* Vol. 39 No. 1.
3. A. Westerlund L. M. Feeney and B. Ahlgren. Spontaneous Networking: An application-oriented approach to ad hoc networking. *IEEE Communications Magazine.* Vol. 39 No. 6.
4. L. Galluccio, G. Morabito, and S. Palazzo. Analytical Evaluation of a Trade-off between Energy Efficiency and Responsiveness of Neighbor Discovery in Self-Organizing Ad Hoc Networks. *IEEE JSAC.* Vol. 22, No. 7.
5. J. Haartsen. Bluetooth Baseband Specification, http://www.bluetooth.com/.
6. M.F. Neuts. Matrix-Geometric Solutions in Stochastic Models. *The John Hopkins Press.* Baltimore, MD, 1981.

# Adaptive Backoff Exponent Algorithm
# for Zigbee (IEEE 802.15.4)

Vaddina Prakash Rao and Dimitri Marandin

Chair of Telecommunications,
Department of Electrical Engineering and Information Technology,
Technische Universität Dresden, D-01062, Dresden, Germany
Prakash-Rao.Vaddina@rsd.rohde-schwarz.com,
marandin@ifn.et.tu-dresden.de

**Abstract.** The IEEE 802.15.4 is a new wireless personal area network standard designed for wireless monitoring and control applications. In this paper a study of the backoff exponent (BE) management in CSMA–CA for 802.15.4 is conducted. The BEs determine the number of backoff periods the device shall wait before accessing the channel. The power consumption requirements make CSMA–CA use fewer Bes which increase the probability of devices choosing identical Bes and as a result wait for the same number of backoff periods in some cases. This inefficiency degrades system performance at congestion scenarios, by bringing in more collisions. This paper addresses the problem by proposing an efficient management of Bes based on a decision criterion. As a result of the implementation potential packet collisions with other devices are restricted. The results of NS-2 simulations indicate an overall improvement in effective data bandwidth, validating our claim.

## 1 Introduction

The wireless market has been traditionally dominated by high end technologies, and so far Wireless Personal Area Networking (WPAN) products have not been able to make a significant impact on the market. While some technologies like the Bluetooth have been quite a success story, in the areas like computer peripherals, mobile devices, etc, they could not be expanded to the automation arena.

This led to the invention of the wireless low datarate personal area networking technology, Zigbee (IEEE 802.15.4), for the home/Industrial automation. It has received a tremendous boosting among the industry leaders and critics have been quick enough to indicate that no less than 80 million Zigbee products will be shipped by the end of 2006[14].

ZIGBEE is a new wireless technology guided by the IEEE 802.15.4 Personal Area Networks standard. It is primarily designed for the wide ranging automation applications and to replace the existing non–standard technologies. It currently operates in the 868MHz band with a datarate of 20Kbps in Europe, 914MHz band at 40Kbps in the USA, and the 2.4GHz ISM bands Worldwide at a maximum data–rate of 250Kbps. Some of its primary features being:

- Standards–based wireless technology
- Interoperability and worldwide usability

Y. Koucheryavy, J. Harju, and V.B. Iversen (Eds.): NEW2AN 2006, LNCS 4003, pp. 501–516, 2006.
© Springer-Verlag Berlin Heidelberg 2006

- Low data–rates
- Ultra low power consumption
- Very small protocol stack
- Support for small to excessively large networks
- Simple design
- Security, and
- Reliability

The IEEE 802.15.4 standard can be defined with a set of primitives. Primitives are services of each layer built over the services offered by the next lower layer. These services are offered to the next higher layer or sublayer. There are 14 physical and 35 MAC layer primitives supported by Zigbee. Based on the number of primitives supported by these nodes two types of devices are defined: the Full Function Devices (FFDs) and the Reduced Function Devices (RFDs). An FFD is a full functional device that supports all the primitives, whereas the RFD is designed to support a subset of them[2]. Also an FFD is capable of acting as a coordinator, or as a network node routing data to peer nodes or as a simple network node communicating only with the coordinator, whereas, the RFD is only capable of being an end node with communication to the coordinator.

The low rate WPAN supports two types of topologies. They can form a star topology where the nodes can only talk to the coordinator and also as in a peer–to–peer topology where capable network nodes can route data. Several peer–to–peer networks can work together to form a mesh/cluster tree topologies. In this paper a star network topology is considered.

The IEEE 802.15.4 compliant devices use CSMA–CA to access the media for their data transmissions. The algorithm is implemented in units of time called 'backoff periods' where one backoff period is equal to aUnitBackoffPeriod (= 20 Symbols, for IEEE 802.15.4). Before trying to sense a channel a device shall wait for a specified number of backoff periods determined by the variable Backoff Exponent (BE) (0 – $(2^{BE}-1)$). Each device uses a minimum BE value before the start of a new transmission and increment it after every failure to access the channel. BE determines the number of backoff periods the device shall wait before trying to access the channel.

A study of the system performance at collision scenarios reveals an exponential increase in the number of packet drops, for higher datarate operation. The drop in system performance can be attributed to the numerous poor link quality packets. Link Quality is an attribute to characterize the quality of each incoming packet. This measurement is implemented as a Signal–to–Noise Ratio (SNR) estimation in most cases. Often, poor link quality is a direct consequence of the hidden node problem. However, in the current scenario it has been assumed there is no hidden node problem. Because of the power consumption constraints the BE of devices is never allowed to vary outside the range of 3–5. As a result, devices often choose identical number of backoff periods and detect an idle channel. Thus, the devices on choosing an identical number of backoff periods transmit data/MAC–command packets without being aware that another node has also detected an idle channel. This result in frequent confrontations among nodes which result in collisions, packet drops and as a consequence retransmissions, which affect the throughput of the network. It is this cause that is addressed in this paper.

Here we present a brief study of the CSMA–CA mechanism used in IEEE 802.15.4, with emphasis on the improper BE distribution which result in frequent packet collisions and a loss in systems performance. We later move on to provide an algorithm called the Adaptive Backoff Exponent (ABE) which reduces the probability of devices choosing identical number of backoff periods at collision rates, thus improving the systems performance considerably at these rates. NS–2[13] Simulation results are provided to validate our claim of better performance.

## 2  CSMA-CA

The CSMA–CA algorithm is used to access the channel to transmit data or MAC command frames. The devices first need to ensure the channel is free before attempting to transmit their data. This is achieved by checking the idleness of the channel at random intervals of time. The units of time for which the devices shall wait for data transmission are called the backoff periods. One backoff period is equal to aUnitBackoffPeriod (=20 Symbols, for IEEE 802.15.4). Based on weather beacons are used/not used, the CSMA–CA will choose either a slotted/unslotted procedure. The unslotted version is also used in cases when the beacons could not be detected in the PAN. In the slotted mechanism, the backoff period boundaries of every device in the PAN will be aligned with the start of the beacon transmission. Hence all the devices will have their backoff periods aligned with each other. Where as in the unslotted mechanism there is no such restriction. In the following scenario only the slotted mechanism is considered.

Each device shall maintain three variables for each transmission attempt: NB, CW and BE.

*NB*: It is the number of times the CSMA–CA algorithm was required to backoff while attempting the current transmission; this value shall be initialized to 0 before each new transmission attempt.

*CW*: It is the contention window length, defining the number of backoff periods that need to be clear of channel activity before the transmission can commence; this value shall be initialized to 2 before each transmission attempt and reset to 2 each time the channel is assessed to be busy.

*BE*: It is the variable which determines the number of backoff periods a device shall wait before attempting to assess a channel's status. The number of backoff periods that a device shall wait is chosen as a random number in the range of 0 to ($2^{BE}-1$).

The following steps are executed to access the channel:

*STEP–1*: The MAC sublayer shall initialize NB (to 0), CW (to 2) and BE (to macMinBE, 3), and then locate the boundary of the next backoff period. If the Battery Life Extension variable is set, the maximum value of BE can be only 2.

*STEP–2*: The MAC sublayer shall wait for a random number of complete backoff periods in the range of 0 and ($2^{BE}-1$).

*STEP–3*: After the completion of the backoff periods, the MAC shall request the PHY to perform a Clear Channel Assessment (CCA).

*STEP–4*: If the channel is assessed to be busy the MAC sublayer shall increment both NB and BE by one. Because of restrictions due to limited power supply it is ensured

BE is never greater than the maximum allowed value (aMax BE=5). CW is also reset to 2. If the value of NB is less than or equal to the maximum allowed backoff attempts (macMaxCSMABackoffs =4), the mechanism shall proceed with STEP–2 again. If not, the CSMA mechanism shall terminate with a Channel Access Failure status.

*STEP–5*: If the channel is determined to be idle, the mecha-nism shall ensure that CW is equal to zero. If not it shall go back to STEP–3 to the Clear Channel Assessment phase. Else if the contention window is 0, the MAC shall start transmission immediately.

The Fig. 1 presents the above steps graphically.

## 3   The Adaptive Backoff Mechanism

### 3.1   The Inefficient Backoff Management

As had been already discussed in the Introduction, there is an exponential increase in the number of packet drops at collision rates. Zigbee devices are severely constrained in terms of power. Hence every effort is made to save as much energy as possible. During channel accesses the CSMA–CA algorithm is allowed to use only a very small range of backoff exponents (macMinBE – aMaxBE), where the minimum BE a device can support is indicated by macMinBE (=3) and the maximum by aMaxBE (=5). Since the variable BE determines the number of backoff periods the devices shall wait to access the channel, the

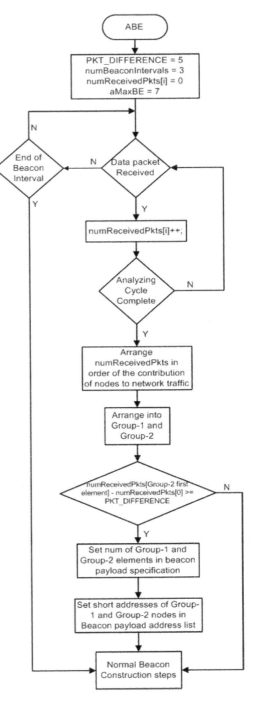

**Fig. 1.** CSMA-CA Mechanism

higher the value of BE, the longer the device will spend trying to access the channel in some cases. The longer wait adds up to the power consumption of the device. Therefore a lower BE range will ensure the devices will never spend too much time waiting for channel access. This makes two or more devices using the same number of backoff periods. As a result they detect an idle channel simultaneously and proceed with their transmissions which results in frequent collisions. An example scenario is described in the following passage:

Assume a large set of devices are present in a star formation with a PAN coordinator and the others acting as end nodes. And assume a subset of these devices has data to be transmitted. And assume of all these transmitting devices three nodes labeled 4, 11 and 23 are of our interest. Let us assume at time t1, node–4 and node–11 has data to be transmitted and have set their BE to the minimum value (macMinBE) of 3. They shall determine the number of backoff periods it needs to use to be in the range of $0-(2^3-1)$. The choice of the correct value is out of scope of the IEEE 802.15.4 standard. It could possibly be determined by the application developer. Let us assume the devices choose to use 2 backoff periods. After the completion of the 2 backoff periods they shall try accessing the channel's status. Let us assume other devices are transmitting in this time and the channel is busy. So during this try they find the channel is busy. Therefore, they increment their BE by 1. And again at time t2, they determine the number of backoff periods they need to wait in the range of $0-(2^4-1)$. Let us assume the nodes 4 and 11 this time choose an unequal number of backoff periods of 4 and 6 respectively.

They wait for their number of backoff periods and node–4 exhausts its backoff periods first. Again let us assume a busy channel and node–4 shall increment it's BE

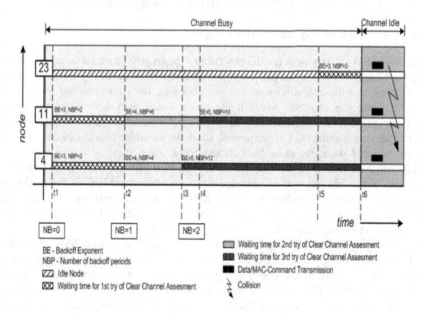

**Fig. 2.** An Example Collision Scenario with Inefficient Backoff Exponent

to 5, at time t3. Similarly node–11 after completion of its backoff periods of 6 finds the channel is busy. So it increments it's BE to 5 at t4. Now they shall use the backoff periods in the range of $0-(2^5-1)$. Let us assume this time the nodes 4 and 11 use backoff periods of 12 and 10 respectively. This implies at any point of time (say t5) when node–4 has completed 9 backoff periods, node–11 shall have completed 7. Let us again assume at this point of time (t5) node–23 is ready for its transmission and shall start off with a minimum backoff exponent of 3, which might lead it to use 3 backoff periods. Hence all the three devices, 4, 11 and 23 have 3 backoff periods to be completed, before they try to access the channel. And let us finally assume an idle medium after t6, when the nodes complete their backoff periods. Hence all the three devices shall detect an idle medium at the end of their backoff periods, and shall attempt to transmit their data which would result in collision.

The seriousness of the problem is clearly evident when many nodes transmitting at higher datarates is considered. The probability of atleast two devices using the same number of backoff periods is increased. And the problem is magnified several fold when the inefficient backoff exponent management is considered. Thus the odds of packet collisions are high at higher datarates which impacts the throughput of the network at these rates adversely. The insufficient backoff exponents of 3(macMinBE) – 5(aMaxBE), is the primary cause to blame for the poor performance at these rates. However, allowing the devices to use higher backoff exponents would impact the energy performance of the whole PAN as devices are often subjected to use exorbitantly higher number of backoff periods to sense the channel.

Therefore we have proposed an algorithm which not only supports higher backoff exponents but also provides nearly same or better energy consumption while providing considerable improvement in the system's throughput performance.

### 3.2 Adaptive Backoff Exponent (ABE) Algorithm

The adaptive backoff exponent (ABE) algorithm is primarily based on three important principles. Firstly is the idea of providing a higher range of backoff exponents to the devices, to reduce the probability of devices choosing the same number of backoff periods to sense the channel. Secondly is to do away with a constant minimum backoff exponent (macMinBE) value as used in the standard CSMA–CA. In this algorithm, the minimum backoff exponent shall be variable; hence devices are not likely to start off with the same backoff exponent when they wish to start a data transmission. And thirdly is the way the minimum backoff exponent (macMinBE) is maintained. Since the algorithm implements a variable macMinBE, the variation factor is each node's contribution to the network traffic. Only devices that are involved in a transmission are taken into consideration. And devices that are not transmitting do not come under the purview of the algorithm.

As can be observed, the algorithm does not interfere with CSMA–CA, but compliments it with an improved Backoff exponent management. According to this algorithm, all devices that are contributing more to the network traffic are slapped with higher macMinBE's, and devices which contribute less to the network congestion will use lower minimum backoff exponents. Therefore devices with longer macMinBE are likely to wait longer than devices with lower macMinBE. At regular intervals (called, Analyzing Cycles) the coordinator decides if it needs to implement a change in the macMinBE of the transmitting nodes. During the period of the analyzing cycle the coordinator shall keep track of each node's contribution to the network traffic, which shall be its decision criteria. If it finds an uneven distribution in network traffic contribution, it shall implement a change in the macMinBE of the involved nodes. Also the devices are now allowed to use a higher backoff exponent (aMaxBE = 7). At the beginning of the transmission the devices shall start off with a macMinBE of 3. However as time progresses and the node in question contributes unevenly to the network traffic, its macMinBE can be either decremented or incremented. Hence on the next transmission it shall use the new macMinBE. A detailed explanation of the algorithm follows. A star network formation with each transmission directed to the coordinator is assumed.

The algorithm is implemented in three different phases.

- The analyzing phase
- The decision phase, and
- The implementation phase

The analyzing and the decision phases are implemented at the coordinator while the implementation phase is carried out at the end nodes involved in data transmissions.

### 3.2.1  The Analyzing Phase

The analyzing phase is repeated after every analyzing cycle. The analyzing cycle is the time duration during which the coordinator observes the contribution of each node to the network traffic. It is a variable that can be used to fine tune the algorithm to produce the best results. In the current simulations the analyzing cycle is taken as 3 Beacon Intervals.

At the initiation of data transmissions the devices shall start off by setting their macMinBE to 3 as in the standard CSMA–CA for 802.15.4. After the initialization, the devices shall continue with their data transmissions. During this period the coordinator counts the number of packets contributed by each node until it reaches the end of the analyzing cycle.

### 3.2.2  The Decision Phase

The decision phase starts at the end of the analyzing phase. Thus, at the end of the analyzing phase, the coordinator has enough information on each node's contribution to the network traffic. It shall now proceed to see if it is time to apply a change in the macMinBE. If yes, which nodes shall increment their macMinBE and which nodes shall decrement it.

Before the steps involved in the Analyzing and Decision phases are explained a brief introduction of the involved parameters is given.

1. *Group–1*: It is a group of devices/nodes whose macMinBE (minimum backoff exponent) is to be decreased by 1. These are the nodes which contribute less to the network traffic compared to the other nodes.
2. *Group–2*: It comprises of devices/nodes whose macMinBE (minimum backoff exponent) is to be increased by 1. These are the nodes which contribute more in comparison to the other nodes. When, the number transmitted packets (a measure of traffic) by Group–2 nodes differ from Group–1 nodes by at least

PKT_DIFFERENCE number of packets, the decision algorithm is completely applied. Else the algorithm goes back to the next analyzing cycle.

3. *PKT_DIFFERENCE*: It is the variable which is the deciding factor for the application of the decision algorithm. If the difference between first element of Group–2 and the first element of Group-1 is more than PKT_DIFFERENCE number of packets, the decision algorithm is applied. If not the Analyzing Phase is invoked again. To apply the decision algorithm further the coordinator will indicate its decision in the beacon payload and transmits it.

4. *numBeaconIntervals*: The number of beacon intervals which comprises one analyzing cycle. It is the time period after which the need for an adjustment of macMinBE is analyzed. This is determined as a number of beacon intervals after which this analysis is to be conducted. Example: If the beacon interval is 0.384secs (for BO=3), and the numBeaconIntervals=3, then one analyzing cycle would be 1.152 seconds. Thus the decision phase will be applied after every 1.152 seconds.

5. *numReceivedPkts[i]*: The array used to maintain the count of the number of packets contributed by each transmitting source node–i to the network traffic.

6. *nodeIndexes[i]*: The array holding the node indexes of the sources (i) which are transmitting to the coordinator. The order of arrangement of the nodes in this array and their contributions as indicated by the array numReceivedPkts[i] is always matched. Arranging the contributions array in any order should make a corresponding adjustment to this array.

7. *'BeaconPayloadFields' ABE Specification Fields*: The specification field of the Beacon Payload with 1 octet in size, indicating the number of nodes whose macMinBE needs to be decremented and the number of nodes whose macMinBE is incremented. See Fig. 3.

8. *addrList[8]*: The list of short addresses (2 octets) of devices whose macMinBE is incremented or decremented. The first four node addresses indicate nodes whose minimum BE is decreased by 1, while the end 4 node addresses indicate nodes whose minimum BE is increased by 1..

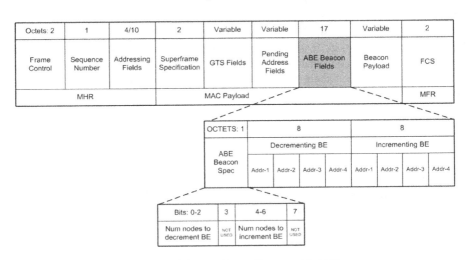

**Fig. 3.** Beacon Frame Format with ABE

The analyzing phase and the decision phase can be broadly explained with the following steps (Fig. 4). Before the steps are carried out the following variables are initialized to their respective values. Note that the values are not absolute and the correct tuning of these values are required to produce efficient results. The indicated

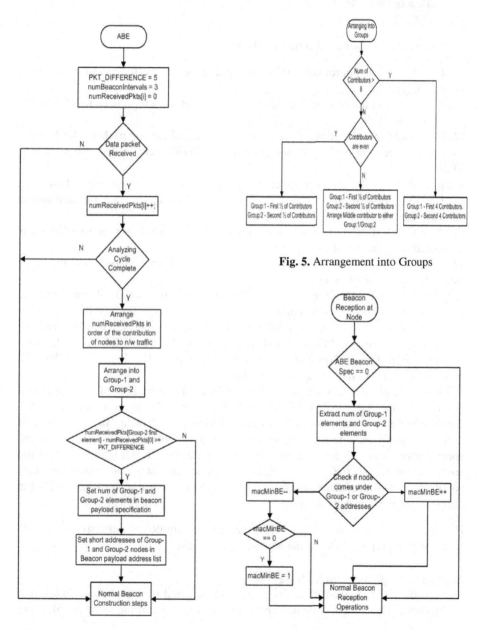

**Fig. 5.** Arrangement into Groups

**Fig. 4.** ABE Analysis and Decision Phases          **Fig. 6.** Implementation Phase

values have been used in the simulations that follow, and have been observed to be producing good results as indicated.

– PKT_DIFFERENCE: 5
– numBeaconIntervals: 3
– aMaxBE: 7

These steps are conducted at the coordinator:

*STEP–1*: Firstly the coordinator waits for a packet reception. If a packet is received proceed to STEP–2.

*STEP–2*: Check if the received packet is a data packet. If not, go back to STEP–1. If yes, follow STEP–3.

*STEP–3*: Increment the packet count for the source node of the packet. Check if the analyzing cycle is complete. If not, go back to STEP–1 else follow the next step STEP–4.

*STEP–4*: Arrange the nodes in order of the number of packets contributed by each node, so that the least contributor stays in the first place and the major contributor is placed at the last.

*STEP–5*: Check if the number of contributors is more than 8. If yes, go to STEP–6. If the contributors are less than 8, go to STEP–7.

*STEP–6*: The first four elements of the contributors list form Group–1. And the last 4 elements come under Group–2.

*STEP–7*: The first half of the nodes form Group–1 and the second half the Group–2. Check if the number of contributors is even. If odd arrange the middle element to either Group–1 or Group–2. If the difference in the packet contribution of this node to the packet contribution of the first node is greater than PKT DIFFERENCE, the node is listed under Group–2 else it comes under, Group–1. The Fig. 5 describes the division of the flows into groups in more detail.

*STEP–8*: Check if the difference between the first elements of Group–2 and Group–1 is more than PKT_DIFFERENCE. If yes, include Group–1 elements in the decrement fields of the beacon payload and Group–2 elements in the increment fields. Similarly include the number of elements in Group–1 and Group–2 in the specification fields. If the difference in the first elements is not significant, set the specification fields to 0. This indicates small traffic rates and hence the little variation in traffic contribution by nodes and hence algorithm proceeds with the normal beacon reception operation.

Having done the above steps, reset the arrays, numReceivedPkts[i] and node Indexes[i], to prepare them for the next analyzing cycle.

### 3.2.3 The Implementation Phase

The coordinator has taken its decision and has indicated it in the beacon fields, and the beacon is transmitted. Now it is up to the node to act according to the coordinators indication. The following steps, specific to ABE implementation are followed at every node that receives the beacon. These steps are indicated in Fig. 6.

*STEP–1*: Upon beacon reception, the node extracts the ABE Beacon Payload Specification fields.

*STEP–2*: See if the Beacon Payload Specification field is set to 0. If yes, continue with normal operation. If no, extract the number of decrements and number of increments.

*STEP–3*: Check if its address matches to the address list in the first four fields of the addrList. If yes, decrement its macMinBE by 1, and check if its equal to 0. If yes, set it to 1. If its address is not listed in the first four address list, go to STEP–4.

*STEP–4*: Check if its address matches to the address list in the last four fields of the addrList. If yes, increment its macMinBE by 1. If its address is not listed, continue with the normal beacon operation.

# 4  Beacon Modification

The algorithm utilizes additional space within the beacon payload to convey the coordinator's decision on the macMinBE adjustment. Hence additional space is required to indicate the number of incrementing/decrementing nodes and their respective node addresses. The algorithm uses 1 octet to indicate the number of increments and decrements and a maximum of 16 octets to indicate the short addresses of the devices. Depending on the number of decrementing and incrementing nodes a maximum of 17 octets are used for this mechanism.

The Fig. 3 highlights the beacon fields concentrating on the ABE fields in detail. Thus, applying a dynamically adjustable backoff mechanism would increase the size of the beacon by 17 octets. However, since every beacon is transmitted in the first slot of the superframe, independent of the contention access and the free periods, there is no effect on the bandwidth utilization and thus would bring in no overhead. However, when speaking of beacon overhead it cannot be entirely ruled out, since this has an impact on networks that operate at low beacon orders with the coordinator performing GTS management, as indicated in the following calculation.

The maximum size of a beacon can be calculated with the superframe duration as in equation (1).

The algorithm comes with an inherent disadvantage. Using the Fig. 3 we can see that including the implementation of the Adaptive Backoff Exponent, we need a total of $(2 + 1 + 10 + 2 + 17 + 2) = 34$ octets. However when the network is operating at Low Superframe Order (1 or 2), the slot duration is (for SO=2) $60x2^2 = 60x4 = 240$ symbols = 240bits = 30octets. Thus a very low superframe order indicates that the above algorithm cannot be applied. And when operating at a SO=3, $60x2^3 = 60x8 = 480$symbols = 480bits = 60octets. Thus, at superframe orders greater than 2, the algorithm can be applied successfully. However, there can be still restrictions to support a complete set of guaranteed time slots.

$$Max\_Beacon\_Size = \frac{SuperFrameDuration}{16}$$
$$= \frac{aBaseSuperFrameDuration \times 2^{SO}}{16}$$
$$= \frac{aBaseSlotDuration \times aNumSuperframeSlots \times 2^{SO}}{16} \quad (1)$$
$$= \frac{60 \times 16 \times 2^{SO}}{16}$$
$$= 60 \times 2^{SO}$$

## 5  Performance Analysis

Several metrics can be defined to grade the performance of a technology against the elements of wireless networking. Some of these metrics have been carefully chosen to validate the performance improvement brought about by the implementation of the Adaptive Backoff Exponent algorithm. They are described here.

- Network Throughput
- End–to–End Delay
- Delivery Ratio
- Energy Consumption

The simulations are conducted on a star network topology with 15 nodes and 8 traffic flows with CBR traffic and one way communication from the node to the coordinator. Each packet is assumed to be of 70Bytes. A Drop Tail queue is used. Two-Ray ground propagation model with an OmniAntenna is used. The nodes are all placed at a distance of 10m from the coordinator. All simulations are conducted with the Network Simulator–2 using the software modules provided by [4].

The graphs in Fig. 7 highlight a comparative analysis of the two methods of backoff management. They indicate the improvement brought about by the application of the Adaptive Backoff Exponent algorithm. The performance metrics are measured as a function of the traffic load. The results are produced with 95% confidence levels, to express the accuracy of the simulations.

With reference to the graphs, it is stated that using the Adaptive Backoff Exponent algorithm effectively moves the congestion area farther. The areas, where extreme numbers of collisions are detected, result in drop in performance of the system. It is these areas which mark the beginning of network congestion. At traffic rates below the congestion areas, the performance of the system without ABE is similar to that of the system applying ABE. At lower rates the number of nodes contending for channel access at any given time is smaller than the number of nodes accessing it at higher rates. At higher traffic rates, more number of nodes try to access the channel and the insufficient distribution of backoff exponents increases the probability of nodes using the same number of backoff periods, making them prone to transmit simultaneously with other nodes. Thus applying the ABE will implement an efficient backoff exponent maintenance by allocating a higher backoff exponent for devices contributing more to the traffic load and lower backoff exponents for devices contributing less. The graphs indicate a peak throughput of 6.734kbps can be achieved for a system applying ABE to that of 5.3Kbps for a system working without ABE. It indicates an improvement in the peak throughput by, ((6.73–5.3)/5.3) X100 = 27%. Similarly the improvement obtained at 1% PER or 99% of delivery ratio is ((6.554–5.25)/5.25)X100 = 25%. Similarly under the indicated simulation conditions, we can see that the system is able to maintain a very low delay until 1.5pkts/sec for a system applying ABE compared to that of only until 1.2pkts/sec for a system without ABE. And as can be observed from the energy analysis, this improvement is achieved with better or near similar power consumption.

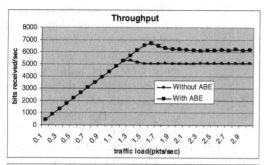

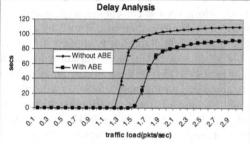

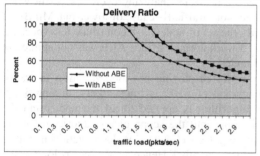

The improvement in the performance can be attributed to lower number of drops due to bad link quality. As can be viewed from Fig. 8, the number of packets dropped due to bad link quality decrease as we apply higher datarates. However, due to extreme poor system performance in terms of the packet error ratio and the delay performance at rates higher than 1.5pkts/sec, the improvement brought about at these rates remain far beyond being realistic.

The algorithm although brings about much needed performance improvement at higher traffic rates, there does appear

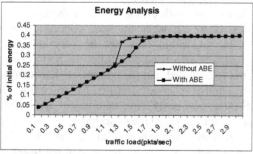

**Fig. 7.** Performance Analysis with and Without ABE

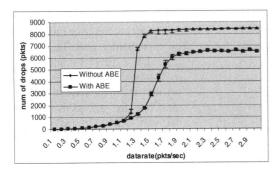

**Fig. 8.** LQI Drops with and without ABE

some inherent difficulties in applying the algorithm to all type of scenarios. Some of them are discussed here.

The current implementation of the algorithm is limited to analyze eight different traffic sources. However there can be more than eight traffic sources, but only eight of them can be effected with ABE in one analyzing cycle. The other transmitting nodes can be influenced in the next analyzing cycle by virtue of the imbalance caused in the traffic contribution of each node, by applying ABE. Consider the following scenario. Let us assume 15 different nodes are transmitting in a PAN. However the current implementation can only be effective on eight nodes: Four nodes to decrease their macMinBE by one and four other nodes to increment their macMinBE. The algorithm is applied for all the nodes, but let us assume the nodes 1–8 display a rather more variation in traffic contribution compared to the other set of nodes (9–15). Of these nodes, let us assume nodes 1–4 have been contributing less compared to the nodes 5–8. As a result the macMinBE of nodes 1–4 is decremented because of lower contribution to network traffic compared to the other set of nodes (5–8). Hence the nodes 9–15 remain out of scope of the algorithm for the current analyzing cycle. Let us move on to the next analyzing cycle. The effect of ABE in the previous analyzing cycle is evident when the nodes 1–4, contribute more to the network traffic (because of lower macMinBE's) in the current analyzing cycle. Similarly the set of nodes, 5–8 contribute less (because of higher macMinBE's) and nodes 9–15 work normally since ABE was not applied in the previous analyzing cycle for these nodes.

As a result the new set of eight nodes displaying a wider variation in traffic contribution, for the current implementation of ABE, is 'most likely' among 5–15. However, it is stated again that the algorithm is applied to all the nodes and the probability that the set of nodes 1–4, come into the scope of the algorithm for the current analyzing cycle is bleak. Again eight nodes with a wider variation in traffic contribution are selected and the algorithm is applied. Hence, as we see, as long as the nodes are reasonably low, the algorithm might work efficiently even though it has been implemented for only eight nodes. However, when the node concentration is more, the current implementation of the algorithm may not be very effective. Probably, the algorithm can produce a spiraling effect and destabilize the performance and produce even worse results. But it is to be noted that the current implementation can be easily extended to support more number of nodes. But this has to be done at the expense of other features like GTS (discussed next). Thus an efficient application of ABE shall depend on the type of application, and the type of scenario.

An important assumption for the current implementation of the algorithm is that the coordinator can account for every transmitted packet, irrespective of the result of the transmission. Since the contribution of each node to the network traffic is measured at the coordinator by summing the number of packets received from each node, we assume that it is 'always' possible to determine the source of the packet, be the transmission a success or a failure.

The implementation of ABE requires the use of the beacon payload. Since the size of the beacon is limited, the implementation of ABE often limits the use of Guaranteed Time Slots (GTS). Also in order to extend ABE to support more number of nodes, the GTS's should be restricted or better eliminated. Similar constraints apply when there is a two way communication. If transmission from the coordinator to a node is restricted, the Pending Address Fields can also be added to the available ABE fields. Thus it is completely depended on the type of application and scenario. Assuming a large PAN, where providing GTS's is not a priority, and if there is no transmission from the coordinator to the node, the ABE can be utilized with an extended version supporting more number of nodes, using the additionally available addressing fields of GTS and Pending addresses. Similarly a relatively smaller PAN with a few nodes authorized to use guaranteed time slots, the current implementation can be successfully used to achieve a higher throughput.

Even though ABE can be extended to support more number of nodes, it is realized that with the current addressing technique (2 octets to address a node) it is not possible to support extremely large PAN's (say more than 50 nodes). Thus there is a pressing need for an efficient addressing technique for such type of application scenarios. Considering the possible throughput gains, the task can be a very good addition for further investigation.

## 6   Conclusions and Future Work

Zigbee is a promising new wireless technology in the home/industrial automation field. With its promises of reliable short range communications at low datarates, and ultra low power consumption, it has created a market for itself. However, minor inefficiencies do not allow it to display its ultimate capability. One such problem (the inefficient backoff management) is addressed here, and a workaround (ABE) has been introduced. Also results indicate the percentage of improvement that can be obtained by applying ABE.

It is observed that the Analyze and Decision Phase of the algorithm can be applied for an efficient active/inactive period management based on the traffic rate to achieve better throughput and energy performance. It can be applied to check the network load and change the superframe duration appropriately so as to vary the active and inactive parts of the superframe to suit the requirements of the network under the current traffic conditions. An efficient addressing technique to address the nodes in the beacon, after the decision has been made by the coordinator, can help ABE support much larger PAN's. These issues can be ideal topics for further research.

## References

1. IEEE computer society, "Wireless MAC and PHY Specifications For LR–WPANS" IEEE, 2003.
2. Jianliang Zheng, and Myung J. Lee, "Will IEEE 802.15.4 Make Ubiquitous Networking a Reality?: A Discussion on a Potential Low Power Low Bit Rate Standard" IEEE Communications magazine, pp. 140–146, June 2004.

3. Gang Lu, Bhaskar Krishnamachari, Cauligi S. Raghavendra, "Performance Evaluation of the IEEE 802.15.4 MAC for Low–Rate Low–Power Wireless Networks" IEEE, pp. 701–706, 2004.
4. http://ees2cy.engr.ccny.cuny.edu/zheng/pub/, "Zigbee Software Modules", City College of NewYork.
5. http://www.energizer.com/, "Energizer Engineering DataSheet", Energizer Battery Co.
6. Chris Evans–Pughe, "Bzzzz .. Is the ZigBee wireless standard, promoted by an alliance of 25 firms, a big threat to Bluetooth?", IEEE Review, March 2003.
7. "What exactly is ... ZigBee?", IEE communirations Engineer, August/September 2004.
8. Jeff Grammer, CEO, Ember Inc, "Zigbee Starts To Buzz", IEE Review, November 2004.
9. Sinem Coleri Ergen, "ZigBee/IEEE 802.15.4 Summary", September 10, 2004.
10. "ZigBee. Wireless Technology Made Simple", RF & WIRELESS, May, 2005.
11. Nick Baker, "ZigBee and Bluetooth, Strengths and Weaknesses for Industrial Applications", IEE Computing and Control Engineering, April/May, 2004.
12. Pete Cross, "Zeroing in on ZigBee", Circuit Cellar, Issue 176, March 2005.
13. The network simulator (Version 2), "http://www.isi.edu/nsnam/ns/"
14. ABI Research, "The Zigbee Wave: Some Hesitate, Some Plunge In", September 2004.

# Analysis of TCP-AQM Interaction Via Periodic Optimization and Linear Programming: The Case of Sigmoidal Utility Function⋆

K. Avrachenkov[1], L. Finlay[2], and V. Gaitsgory[2]

[1] INRIA Sophia Antipolis, France
K.Avrachenkov@sophia.inria.fr
[2] University of South Australia, Australia
{luke.finlay, vladimir.gaitsgory}@unisa.edu.au

**Abstract.** We investigate the interaction between Transmission Control Protocol (TCP) and an Active Queue Management (AQM) router, that are designed to control congestion in the Internet. TCP controls the sending rate with which the data is injected into the network and AQM generates control signals based on the congestion level. For a given TCP version, we define the optimal strategy for the AQM router as a solution of a nonlinear periodic optimization problem, and we find this solution using a linear programming approach. We show that depending on the choice of the utility function for the sending rate, the optimal control is either periodic or steady state. Main attention is paid to a problem with a sigmoidal utility function, in which the evolution of the optimal sending rate resembles a "saw-tooth" behavior of the "instantaneous" TCP sending rate.

## 1 Introduction

Most traffic in the Internet is governed by the TCP/IP protocol [3], [10]. Data packets of an Internet connection travel from a source node to a destination node via a series of routers. Some routers, particularly edge routers, experience periods of congestion when packets spend a non-negligible time waiting in the router buffers to be transmitted over the next hop. The TCP protocol tries to adjust the sending rate of a source to match the available bandwidth along the path. During the principle Congestion Avoidance phase the current TCP New Reno version uses Additive Increase Multiplicative Decrease (AIMD) binary feedback congestion control scheme. In the absence of congestion signals from the network TCP increases sending rate linearly in time, and upon the reception of a congestion signal TCP reduces the sending rate by a multiplicative factor. Thus, the instantaneous AIMD TCP sending rate exhibits a "saw-tooth"

---

⋆ The work was supported by the Australian Research Council Discovery-Project Grants DP0346099, DP0664330, Linkage International Grant LX0560049 and France Telecom R&D Grant "Modélisation et Gestion du Trafic réseaux Internet" no. 42937433.

Y. Koucheryavy, J. Harju, and V.B. Iversen (Eds.): NEW2AN 2006, LNCS 4003, pp. 517–529, 2006.

behavior. Congestion signals can be either packet losses or Explicit Congestion Notifications (ECN) [17]. At the present state of the Internet, nearly all congestion signals are generated by packet losses. Packets can be dropped either when the router buffer is full or when an Active Queue Management (AQM) scheme is employed [7]. In particular, AQM RED [7] drops or marks packets with a probability which is a piece-wise linear function of the average queue length. Given an ambiguity in the choice of the AQM parameters (see [4] and [15]), so far AQM is rarely used in practice. In the present work, we study the interaction between TCP and AQM. In particular, we pose and try to answer the question: What should be the optimal dropping or marking strategy in the AQM router? For the performance criterion, we choose the average utility function of the throughput minus either the average cost of queueing or the average cost of losses. This performance criterion with a linear utility function was introduced in [1]. We have analyzed not only the currently used AIMD congestion control, but also Multiplicative Increase Multiplicative Decrease (MIMD) congestion control. In particular, MIMD (or Scalable TCP [11]) is proposed for congestion control in high speed networks. However, since it turns out that the results for MIMD and AIMD congestion control schemes are similar, we provide the detailed analysis only for AIMD TCP.

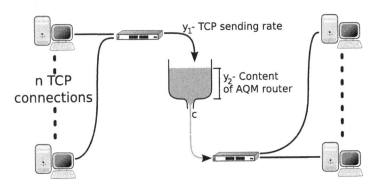

**Fig. 1.** Fluid model for data network

We restrict the analysis to the single bottleneck network topology (see Figure 1). In particular, we suppose that $n$ TCP connections cross a single bottleneck router with the AQM mechanism. We take the fluid approach for modeling the interaction between TCP and AQM [13], [14], [19]. In such an approach, the variables stand for approximations of average values and their evolution is described by deterministic differential equations. Since we consider long-run time average criteria, our TCP-AQM interaction model falls into the framework of the periodic optimization (as outlined in the next section).

This paper is an updated version of the earlier authors' work [2]. The main difference from [2] is that in the current paper we consider (and present numerical results for) a practically important case of a sigmoidal utility function.

## 2  Statement of the Problem

Consider the control system

$$\dot{y}(t) \; = \; f(u(t), y(t)), \quad t \in [0, T], \quad T > 0 \,, \tag{1}$$

where the function $f(u, y) : U \times \mathbb{R}^m \to \mathbb{R}^m$ is continuous in $(u, y)$ and satisfies Lipschitz conditions in $y$; the controls are Lebesque measurable functions $u(t) : [0, T] \to U$ and $U$ is a compact subset of $\mathbb{R}^n$.

Let $Y$ be a compact subset of $\mathbb{R}^m$. A pair $(u(t), y(t))$ will be called *admissible* on the interval $[0, T]$ if the equation (1) is satisfied for almost all $t \in [0, T]$ and $y(t) \in Y \;\; \forall t \in [0, T]$. A pair $(u(t), y(t))$ will be called *periodic admissible* on the interval $[0, T]$ if it is admissible on $[0, T]$ and $y(0) = y(T)$.

Let $g(u, y) : U \times \mathbb{R}^m \to \mathbb{R}^1$ be a continuous function. The following problem is commonly referred to as the *periodic optimization problem*:

$$\sup_{(u(\cdot), y(\cdot))} \frac{1}{T} \int_0^T g(u(t), y(t)) dt \overset{def}{=} G_{per} \,, \tag{2}$$

where *sup* is over the length of the time interval $T > 0$ and over the periodic admissible pairs defined on $[0, T]$.

A very special family of periodic admissible pairs is that consisting of constant valued controls and corresponding steady state solutions of (1):

$$(u(t), y(t)) \; = \; (u, y) \in \mathcal{M} \overset{def}{=} \{(u, y) \mid (u, y) \in U \times Y \,, \; f(u, y) = 0 \}. \tag{3}$$

If *sup* is sought over the admissible pairs from this family, the problem (2) is reduced to

$$\sup_{(u, y) \in \mathcal{M}} g(u, y) \overset{def}{=} G_{ss} \tag{4}$$

which is called a *steady state optimization problem*. Note that

$$G_{per} \; \geq \; G_{ss} \tag{5}$$

and that, as can be easily verified, $G_{per} = G_{ss}$ if the system (1) is linear, the sets $U$, $Y$ are convex and the function $g(u, y)$ is concave. Note also that in a general case (e.g., the dynamics is non-linear and/or the integrand is not concave), (5) can take the form of a strict inequality (examples can be found in [5],[8],[9] and in references therein).

We formulate the problem of optimal control of TCP-AQM interaction as a periodic optimization problem, in which the state space is two dimensional

$$y = (y_1, y_2) \,, \quad f(u, y) = (f_1(u, y_1), f_2(u, y_1)) \,; \tag{6}$$

and the control $u$ is a scalar: $u(t) \in U$, with

$$U \overset{def}{=} \{u : 0 \leq u \leq 1\} \,. \tag{7}$$

We consider two congestion control schemes: Additive Increase Multiplicative Decrease (AIMD) scheme and Multiplicative Increase Multiplicative Decrease (MIMD) scheme. In both cases the first state component $y_1(t)$ is interpreted as a sending rate at the moment $t$, while the second state component $y_2(t)$ represents the size of the queue in the router buffer. In the AIMD scheme, the evolution of $y_1(t)$ is defined by the equation

$$\dot{y}_1(t) = f_1(u(t), y_1(t)) \stackrel{def}{=} \alpha(1 - u(t)) - \beta y_1^2(t)u(t), \tag{8}$$

where $\alpha = \alpha_0 n/\tau^2$ and $\beta = 1 - \beta_0/n$. Here $n$ is the number of competing TCP connections, $\tau$ is the round trip time. Typical values for $\alpha_0$ and $\beta_0$ are 1 and 0.5, respectively. In the MIMD scheme, the evolution of $y_1(t)$ is defined by the equation

$$\dot{y}_1(t) = f_1(u(t), y_1(t)) \stackrel{def}{=} \gamma y_1(t)(1 - u(t)) - \beta y_1^2(t)u(t), \tag{9}$$

where $\gamma = \gamma_0/\tau$ and $\beta$ as in the AIMD case. A typical value for $\gamma_0$ is 0.01. The control $u(t)$ is interpreted as the dropping/marking probability. A detail derivation of equation (8) can be found for instance in [13] and [19]. Note also that if the control is not applied ($u(t) = 0$), the sending rate grows linearly in time if AIMD is used, and the sending rate grows exponentially in time if MIMD is used.

We study active queue management with and without explicit congestion notifications. When AQM with ECN is used, the packets are not dropped from the buffer when control is applied and the buffer is not full. In the case of AQM with ECN (AQM-ECN scheme), the evolution of the router buffer content $y_2(t)$ is described by

$$\dot{y}_2(t) = f_2(u(t), y_1(t)) = f_2(y_1(t)) \stackrel{def}{=} \begin{cases} y_1(t) - c, & 0 < y_2(t) < B, \\ [y_1(t) - c]_+, & y_2(t) = 0, \\ [y_1(t) - c]_-, & y_2(t) = B, \end{cases} \tag{10}$$

where $c$ is the router capacity, $B$ is the buffer size, $[a]_+ = \max(a, 0)$ and $[a]_- = \min(a, 0)$. In the case of AQM without ECN (AQM-non-ECN scheme), AQM signals congestion by dropping packets with rate $u(t)y_1(t)$. Consequently, the dynamics of the router buffer content $y_2(t)$ is described by

$$\dot{y}_2(t) = f_2(u(t), y_1(t)) \stackrel{def}{=} \begin{cases} (1 - u(t))y_1(t) - c, & 0 < y_2(t) < B, \\ [(1 - u(t))y_1(t) - c]_+, & y_2(t) = 0, \\ [(1 - u(t))y_1(t) - c]_-, & y_2(t) = B. \end{cases} \tag{11}$$

The function $g(u, y)$ in the objective (2) will be defined as follows

$$g(u, y) = \psi(y_1) - \kappa y_2 - Me^{K(y_2 - B)}y_1, \tag{12}$$

where $\psi(\cdot)$ is the utility function for the sending rate value, $\kappa y_2$ is the cost of delaying the data in the buffer, and $Me^{K(y_2 - B)}y_1$ is the penalty function for losing data when the buffer is full. Examples of the utility functions we will be dealing with are:

$$\psi(y_1) = y_1 \ , \quad \psi(y_1) = \log(1 + y_1) \ , \quad \psi(y_1) = \frac{y_1^2}{y_1^2 + a} \ . \qquad (13)$$

The linear utility function corresponds to the throughput maximization. The concave utility function such as logarithm is the conventional utility function for elastic applications in the Internet. Finally, the sigmoidal utility function corresponds well to the audio and video streaming applications [6, 12, 18]. The sigmoidal utility function has a single inflexion point which separates a convex part for low sending rates and a concave part for high sending rates.

The rest of the paper is organized as follows: In Section 3 we give an overview of the linear programming approach to periodic optimization. Then, in Sections 4 and 5 we apply the general technique of Section 3 to the problem of interaction between TCP and AQM. We show that depending on the utility function for the sending rate, we obtain either periodic or steady state optimal solution. We conclude the paper with Section 6.

## 3   Linear Programming Approach

In [9] it has been shown that the periodic optimization problem (2) can be approximated by a family of finite dimensional Linear Programming Problems (LPPs) (called in the sequel as approximating LPP). This approximating LPP is constructed as follows.

Let $y_j$ ($j = 1, ..., m$) stand for the $j$th component of $y$ and let $\phi_i(y)$ be the monomomial:

$$\phi_i(y) \stackrel{def}{=} y_1^{i_1} ... y_m^{i_m} \ ,$$

where $i$ is the multi-index: $i \stackrel{def}{=} (i_1, ..., i_m)$. Let us denote by $I_N$ the set of multi-indices

$$I_N \stackrel{def}{=} \{i \ : \ i = (i_1, ..., i_m) \ , \quad i_1, ..., i_m = 0, 1, ..., N \ , \quad i_1 + ... + i_m \geq 1 \ \}.$$

Note that the number of elements in $I_N$ is $(N + 1)^m - 1$. Assume that, for any $\Delta > 0$, the points $(u_l^\Delta, y_k^\Delta) \in U \times Y$ , $l = 1, ..., L^\Delta$ , $k = 1, ..., K^\Delta$ , are being chosen in such a way that, for any $(u, y) \in U \times Y$, there exists $(u_l^\Delta, y_k^\Delta)$ such that $||(u, y) - (u_l^\Delta, y_k^\Delta)|| \leq c\Delta$ , where $c$ is a constant.

Define the polyhedral set $W_N^\Delta \subset \mathbb{R}^{L^\Delta + K^\Delta}$

$$W_N^\Delta \stackrel{def}{=} \left\{ \gamma = \{\gamma_{l,k}\} \geq 0 \ : \ \sum_{l,k} \gamma_{l,k} = 1 \ , \right.$$

$$\left. \sum_{l,k} (\phi_i'(y_k^\Delta))^T f(u_l^\Delta, y_k^\Delta) \gamma_{l,k} = 0 \ , i \in I_N \right\}, \qquad (14)$$

where $\phi_i'(\cdot)$ is the gradient of $\phi_i(\cdot)$. Define the approximating LPP as follows

$$\max_{\gamma \in W_N^\Delta} \sum_{l,k} \gamma_{l,k} g(u_l^\Delta, y_k^\Delta) \overset{def}{=} G_N^\Delta ,\tag{15}$$

where $\sum_{l,k} \overset{def}{=} \sum_{l=1}^{L^\Delta} \sum_{k=1}^{K^\Delta}$ .

As shown in [9] (under certain natural and easily verifiable conditions), *there exists the limit of the optimal value $G^{N,\Delta}$ of the LPP (15) and this limit is equal to the optimal value $G_{per}$ of the periodic optimization problem (2)*:

$$\lim_{N \to \infty} \lim_{\Delta \to 0} G_N^\Delta = G_{per} .\tag{16}$$

*Also, for any fixed $N$,*

$$\lim_{\Delta \to 0} G_N^\Delta \overset{def}{=} G_N \geq G_{per} .\tag{17}$$

Thus, $G_N^\Delta$ can be used as an approximation of $G_{per}$ if $N$ is large and $\Delta$ is small enough.

Let $(u^*(\cdot), y^*(\cdot))$ be the solution of the periodic optimization problem (2) defined on the optimal period $T = T^*$ (assuming that this solution exists and is unique) and let $\gamma^{N,\Delta} \overset{def}{=} \{\gamma_{l,k}^{N,\Delta}\}$ be an optimal basic solution of the approximating LPP (15). From the consideration in [9] it follows that an element $\gamma_{l,k}^{N,\Delta}$ of $\gamma^{N,\Delta}$ can be interpreted as an estimate of the "proportion" of time spent by the optimal pair $(u^*(\cdot), y^*(\cdot))$ in a $\Delta$-neighborhood of the point $(u_l, y_k)$, and in particular, the fact that $\gamma_{l,k}^\Delta$ is positive or zero can be interpreted as an indication of that whether or not the optimal pair attends the $\Delta$-neighborhood of $(u_l, y_k)$.

Define the set $\Theta$ by the equation

$$\Theta \overset{def}{=} \{(u,y) : (u,y) = (u^*(\tau), y^*(\tau)) \ \ for \ some \ \ \tau \in [0, T^*] \} .\tag{18}$$

This $\Theta$ is the graph of the optimal feedback control function, which is defined on the optimal state trajectory $\mathcal{Y} \overset{def}{=} \{y : (u,y) \in \Theta\}$ by the equation $\psi(y) \overset{def}{=} u \ \forall \ (u,y) \in \Theta$ . For the definition of $\psi(\cdot)$ to make sense, it is assumed that the set $\Theta$ is such that from the fact that $(u', y) \in \Theta$ and $(u'', y) \in \Theta$ it follows that $u' = u''$ (this assumption being satisfied if the closed curve defined by $y^*(\tau)$ , $\tau \in [0, T^*]$ does not intersect itself).

Define also the sets:

$$\Theta_N^\Delta \overset{def}{=} \{(u_l^\Delta, y_k^\Delta) : \gamma_{l,k}^{N,\Delta} > 0\},\tag{19}$$

$$\mathcal{Y}_N^\Delta \overset{def}{=} \{y : (u,y) \in \Theta_N^\Delta\} ,\tag{20}$$

$$\psi_N^\Delta(y) \overset{def}{=} u \ \forall \ (u,y) \in \Theta_N^\Delta,\tag{21}$$

where again it is assumed that from the fact that $(u', y) \in \Theta_N^\Delta$ and $(u'', y) \in \Theta_N^\Delta$ it follows that $u' = u''$. Note that the set $\Theta_N^\Delta$ (and the set $\mathcal{Y}_N^\Delta$) can contain no more than $(N+1)^m$ elements since $\gamma^{N,\Delta}$, being a basic solution of the LPP (15), has no more than $(N+1)^m$ positive elements (the number of the equality type constraints in (15)).

As mentioned above, the fact that $\gamma_{l,k}^{N,\Delta}$ is positive or zero can be interpreted as an indication of that whether or not the optimal pair attends the $\Delta$-neighborhood of $(u_l^\Delta, y_k^\Delta)$, and thus, one may expect that $\Theta_N^\Delta$ can provide some approximation for $\Theta$ if $N$ is large and $\Delta$ is small enough. Such an approximation has been formalized in [9], where it has been established that:

*(i) Corresponding to an arbitrary small $r > 0$, there exists $N_0$ such that, for $N \geq N_0$ and $\Delta \leq \Delta_N$ ($\Delta_N$ is positive and small enough),*

$$\Theta \subset \Theta_N^\Delta + r\mathcal{B} . \tag{22}$$

*(ii) Corresponding to an arbitrary small $r > 0$ and arbitrary small $\delta > 0$, there exists $N_0$ such that, for $N \geq N_0$ and $\Delta \leq \Delta_N$ ($\Delta_N$ being positive and small enough),*

$$\Theta_N^{\Delta,\delta} \subset \Theta + r\mathcal{B} , \tag{23}$$

*where* $\Theta_N^{\Delta,\delta} \overset{def}{=} \{ (u_l^\Delta, y_k^\Delta) \; : \; \gamma_{l,k}^\Delta \geq \delta \, \}$.

Note that in both (22) and (23), $\mathcal{B}$ is the closed unit ball in $\mathbb{R}^{n+m}$.

The fact that $\Theta_N^\Delta$ "approximates" $\Theta$ for $N$ large and $\Delta$ small enough leads to the fact that $\mathcal{Y}_N^\Delta$ approximates $\mathcal{Y}$ and to the fact that $\psi_N^\Delta(y)$ approximates (in a certain sense) $\psi(y)$. This gives rise to the following algorithm for construction of near-optimal periodic admissible pair [9]:

1) Find an optimal basic solution $\gamma_N^\Delta$ and the optimal value $G_N^\Delta$ of the approximating LPP (15) for $N$ large and $\Delta$ small enough; the expression "$N$ *large and $\Delta$ small enough*" is understood in the sense that a further increment of $N$ and/or a decrement of $\Delta$ lead only to insignificant changes of the optimal value $G_N^\Delta$ and, thus, the latter can be considered to be approximately equal to $G_{per}$ (see (16)).

2) Define $\Theta_N^\Delta$, $\mathcal{Y}_N^\Delta$, $\psi_N^\Delta(y)$ as in (19). By (22) and (23), the points of $\mathcal{Y}_N^\Delta$ will be concentrated around a closed curve being the optimal periodic state trajectory while $\psi_N^\Delta(y)$ will give a point wise approximation to the optimal feedback control.

3) Extrapolate the definition of the function $\psi_N^\Delta(y)$ to some neighborhood of $\mathcal{Y}_N^\Delta$ and integrate the system (1) starting from an initial point $y(0) \in \mathcal{Y}_N^\Delta$ and using $\psi_N^\Delta(y)$ as a feedback control. The end point of the integration period, $T^\Delta$, is identified by the fact that the solution "returns" to a small vicinity of the starting point $y(0)$.

4) Adjust the initial condition and/or control to obtain a periodic admissible pair $(u^\Delta(\tau), y^\Delta(\tau))$ defined on the interval $[0, T^\Delta]$. Calculate the integral $\frac{1}{T^\Delta} \int_0^{T^\Delta} g(u^\Delta(\tau), y^\Delta(\tau)) d\tau$ and compare it with $G_N^\Delta$. If the value of the integral proves to be close to $G_N^\Delta$, then, by (16), the constructed admissible pair is a "good" approximation to the solution of the periodic optimization problem (2).

In conclusion of this section, let us consider the following important special case. Assume that, for all $N$ large and $\Delta$ small enough, the optimal basic solution $\gamma^{N,\Delta}$ of the approximating LPP (15) has the property that

$$\gamma_{l^*,k^*}^{N,\Delta} = 1 , \qquad \gamma_{l,k}^{N,\Delta} = 0 \quad \forall \, (l,k) \neq (l^*, k^*) , \tag{24}$$

which is equivalent to that the set $\Theta_N^\Delta$ consists of only one point

$$\Theta_N^\Delta = \{(u_{l^*}^\Delta, y_{k^*}^\Delta)\} . \tag{25}$$

Note that that the indexes $l^*, k^*$ in (24) and (24) may depend on $N$ and $\Delta$.
    Assume that there exists a limit

$$\lim_{\Delta \to 0} (u_{l^*}^\Delta, y_{k^*}^\Delta) = (\bar{u}, \bar{y}) , \tag{26}$$

(the same for all sufficiently large $N$). Then, as follows from results of [9], the pair $(\bar{u}, \bar{y})$ is the steady state solution of the periodic optimization problem (2) and, in particular,

$$G_{per} = G_{ss} = g(\bar{u}, \bar{y}) . \tag{27}$$

## 4    Optimal Periodic Solution for Sigmoidal Utility Function

In this and the next sections it is always assumed that

$$Y = \{(y_1, y_2) \mid y_i \in [0, 4], \ i = 1, 2\} \tag{28}$$

and that $U$ is defined by (7); it is also assumed everywhere that $c = 1$ and $B = 4$ (see the equations describing the dynamics of the buffer's content (10) and (11)).
    Let us consider the interaction between AIMD TCP (8) and the AQM-ECN router (10), the former being taken with $\alpha = 1/98$ and $\beta = 1/2$ (such a choice of these parameters corresponds to the case of a single TCP connection and a typical value of the round trip time).
    Let us use the objective function (12), with the following values of the parameters: $\kappa = 0$, $M = 20$ and $K = 5$; and with the sigmoidal utility function being defined by the equation

$$\psi(y_1) = \frac{y_1^2}{y_1^2 + a}, \tag{29}$$

with $a = 12$. As mentioned in the Introduction section the choice of the sigmoidal utility function correspond to streaming video and audio applications in the Internet (see more details on sigmoidal utility functions in [6, 12, 18]). The conventional concave utility functions are analyzed in the next section.
    Define the grid of $U \times Y$ by the equations (with $U$ and $Y$ mentioned as above)

$$u_i^\Delta \overset{def}{=} i\Delta, \quad y_{1,j}^\Delta \overset{def}{=} j\Delta, \quad y_{2,k}^\Delta \overset{def}{=} k\Delta. \tag{30}$$

Here $i = 0, 1, \ldots, \frac{1}{\Delta}$ and $j, k = 0, 1, \ldots, \frac{4}{\Delta}$ ($\Delta$ is chosen in such a way that $\frac{1}{\Delta}$ is an integer). The approximating LPP (15) can be written in this specific case as

$$G_N^\Delta \overset{def}{=} \max_{\gamma \in W_N^\Delta} \sum_{i,j,k} \left( \frac{(y_{1,j}^\Delta)^2}{(y_{1,j}^\Delta)^2 + 12} - 20 e^{5(y_{2,k}-4)} y_{1,j}^\Delta \right) \gamma_{i,j,k}, \tag{31}$$

where $W_N^\Delta$ is a polyhedral set defined by the equation

$$
W_N^\Delta \stackrel{def}{=} \left\{ \gamma = \{\gamma_{i,j,k}\} \geq 0 \; : \; \sum_{i,j,k} \gamma_{i,j,k} = 1 \; , \right.
$$

$$
\left. \sum_{i,j,k} (\phi'_{i_1,i_2}(y_{1,j}^\Delta, y_{2,k}^\Delta))^T f(u_i^\Delta, y_{1,j}^\Delta, y_{2,k}^\Delta)\gamma_{i,j,k} = 0, \; (i_1, i_2) \in I_N \right\}, \quad (32)
$$

in which $\phi_{i_1,i_2}(y_1, y_2) \stackrel{def}{=} y_1^{i_1} y_2^{i_2}$.

The problem (31) was solved using the CPLEX LP solver [20] for $N = 5$ and $N = 7$ with $\Delta$ varying from 0.00625 to 0.05. We have obtained the following optimal values of the LPP (31):

$G_5^{0.05} \approx 0.07755, G_5^{0.025} \approx 0.07757, G_5^{0.0125} \approx 0.07757, G_5^{0.00625} \approx 0.07758,$
$G_7^{0.05} \approx 0.07755, G_7^{0.025} \approx 0.07755, G_7^{0.0125} \approx 0.07756, G_7^{0.00625} \approx 0.07756.$

From this data one may conclude that $G_7 = \lim_{\Delta \to 0} G_7^\Delta \approx 0.07756$. Since $G_7 \geq G_{per}$, it follows that, if for some admissible periodic pair $(u(\tau), y(\tau))$,

$$
\frac{1}{T} \int_0^T \left( \frac{y_1^2(\tau)}{y_1^2(\tau) + 12} - 20 e^{5(y_2(\tau)-4)} y_1(\tau) \right) d\tau \approx 0.07756 \; , \quad (33)
$$

then this pair is an approximate solution of the periodic optimization problem (2).

Let $\left\{ \gamma_{i,j,k}^{N,\Delta} \right\}$ stand for the solution of (31) and define the sets

$$
\Theta_N^\Delta \stackrel{def}{=} \left\{ (u_i, y_{1,j}, y_{2,k}) \; : \; \gamma_{i,j,k}^{N,\Delta} \neq 0 \right\} ,
$$

$$
\mathcal{Y}_N^\Delta \stackrel{def}{=} \left\{ (y_{1,j}, y_{2,k}) \; : \; \sum_i \gamma_{i,j,k}^{N,\Delta} \neq 0 \right\} . \quad (34)
$$

Let us mark with dots the points on the plane $(y_1, y_2)$ which belong to $\mathcal{Y}_N^\Delta$ for $N = 7$ and $\Delta = 0.00625$. The result is depicted in Figure 2. The points are represented with $\Box$ or $\bullet$ and have an associated $u = 1$ or $u = 0$, respectively. It is possible to construct a feedback control by using two thresholds for the buffer content. As the queue length $y_2$ is decreasing (in the region where $y_1 < 1$), we have a certain threshold for when the control should be dropped and allow the data rate $y_1$ to grow. The same can be said for the opposite case when the queue is increasing and $y_1 \geq 1$. The threshold values in our numerical example can be chosen as 2.071 and 2.1, respectively. Thus, the feedback control is defined as

$$
u(y_1, y_2) = \begin{cases} 1 \; , \; y_1 < 1 \text{ and } y_2 > 2.071 \\ 1 \; , \; y_1 \geq 1 \text{ and } y_2 > 2.1 \\ 0 \; , \text{ otherwise} \end{cases} \quad (35)
$$

Using this feedback control, we can integrate the system with the initial point $y_1 = 1, y_2 = 0$. The optimal state trajectory is plotted as a solid line in Figure 2. In Figure 3 we show the evolution of the state variables and the optimal control.

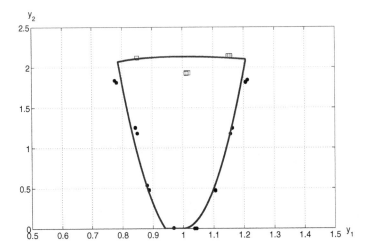

**Fig. 2.** Optimal state trajectory approximation

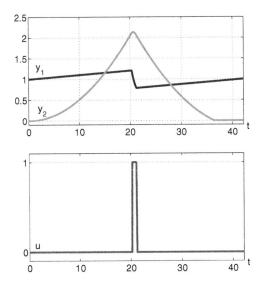

**Fig. 3.** Approximated periodic solution and optimal control

The value of the objective function calculated on this pair is approximately 0.077072. Comparing it with (33), one can conclude that the admissible pair which has been constructed is an approximation to the solution of (2).

Curiously enough, the evolution of the optimal sending rate $y_1$ resembles a "saw-tooth" behavior of the "instantaneous" TCP sending rate. We recall that the variables in our fluid model stand for average values. Hence, the optimal solution suggests that the dynamics of the control in AQM should be much slower than the dynamics of TCP. It is also interesting to see that the use of the sigmoidal utility function results in the recommendation to use the control rarely but intensively. In fact, this has a natural explanation. A user of a streaming application prefers to have a good quality of service most of the time and to experience a bad quality of service seldom and for very short periods of time rather than to have a not satisfactory quality of service all the time.

We have also tested the current objective function for the interaction between AIMD and AQM-non-ECN and for the MIMD congestion control. For those cases, we have also detected similar periodic optimal solutions.

# 5    Optimal Steady State Solution

As in the previous section, let us consider the interaction between AIMD TCP (8) and the AQM-ECN router (10). However, in contrast to the above consideration, let us choose the following objective function

$$g(u, y) = y_1 - y_2. \tag{36}$$

This choice of the utility function corresponds to the throughput maximization. That is, take $\psi(y_1) = y_1$, $\kappa = 1$, and $M = 0$ in (12). Note that, as can be easily verified, in this case, the solution of the steady state optimization problem (4) is

$$\bar{u} = 0.02 , \quad \bar{y}_1 = 1 , \quad \bar{y}_2 = 0 \tag{37}$$

and, in particular, $G_{ss} = \bar{y}_1 - \bar{y}_2 = 1$ . We define the grid of $U \times Y$ as in the previous Section 4.

The approximating LPP (15) in this specific case is of the form

$$\max_{\gamma \in W_N^\Delta} \sum_{i,j,k} \left( y_{1,j}^\Delta - y_{2,k}^\Delta \right) \gamma_{i,j,k} = G_N^\Delta, \tag{38}$$

where $W_N^\Delta$ has exactly the same form as in (32).

**Proposition 1.** *For any $N = 1, 2, ...,$ and any $\Delta > 0$ such that $\frac{0.02}{\Delta} \stackrel{def}{=} i^*$ is integer, there exists a basic optimal solution $\gamma^{N,\Delta} \stackrel{def}{=} \{\gamma_{i,j,k}^{N,\Delta}\}$ of the LPP (38) defined by the equations*

$$\gamma_{i^*,j^*,k^*}^{N,\Delta} = 1 , \quad \gamma_{i,j,k}^{N,\Delta} = 0 \ \forall \ (i,j,k) \neq (i^*,j^*,k^*) , \tag{39}$$

*where $i^*$ is as above and $j^* = \frac{1}{\Delta}$, $k^* = 0$ .*

*Proof* of the proposition is given in [2].

Since, by definition, $u_{i^*}^\Delta = 0.02$ , $y_{1,j^*}^\Delta = 1$ , $y_{2,k^*}^\Delta = 0$ , that is, $(u_{i^*}^\Delta, y_{1,j^*}^\Delta, y_{2,k^*}^\Delta)$ coincides with the optimal steady state regime $(\bar{u}, \bar{y}_1, \bar{y}_2)$ defined in (37), one obtains the following corollary of Proposition 1 (see (26) and (27)).

**Corollary 2.** *The periodic optimization problem (2) has a steady state solution and this steady state solution is defined by (37). In particular,*

$$G_{per} = G_{ss} = \bar{y}_1 - \bar{y}_2 = 1 . \tag{40}$$

We have also checked numerically the other criteria with concave utility functions for the sending rate. It appears that if the utility function for the sending rate is concave, the optimal solution is steady state. The same conclusion holds for the case of interaction between AIMD TCP and AQM-non-ECN and when MIMD is used instead of AIMD.

# 6   Conclusions

We have analyzed the interaction between TCP and AQM using the fluid model approach. The fluid model approach leads to a periodic optimization problem. We have shown that depending on the choice of the utility function for the sending rate, the optimal solution is either periodic or steady state. In particular, we have obtained steady state solution for all concave utility functions and periodic solutions for sigmoidal utility functions. In the case of sigmoidal utility functions, the optimal periodic solution resembles strikingly the "saw-tooth" behavior of the instantaneous TCP sending rate evolution. Moreover, we note that the optimal dynamics of AQM is much slower than the dynamics of TCP. The optimal AQM strategy can be implemented with the help of a simple queue with two thresholds. The choice of the sigmoidal utility functions results in the recommendation to apply the control rarely but intensively. With the help of linear programming approach for periodic optimization we have succeeded to prove that the steady state solution is indeed an optimal solution for the given non-linear periodic optimization problem.

# References

1. K. Avrachenkov, U. Ayesta, and A. Piunovsky, "Optimal Choice of the Buffer Size in the Internet Routers", in Proceedings of IEEE CDC-ECC 2005.
2. K. Avrachenkov, L. Finlay, and V. Gaitsgory, "TCP-AQM Interaction: Periodic Optimization via Linear Programming", in *Control Theory Applications in Financial Engineering and Manufacturing*, Eds.: H. Yan, G. Yin, and Q. Zhang, Springer (Kluwer), Fred Hillier's International Series in Operations Research and Management Sciences, To Appear.
3. M. Allman, V. Paxson and W. Stevens, TCP congestion control, *RFC 2581*, April 1999, available at `http://www.ietf.org/rfc/rfc2581.txt`.
4. M. Christiansen, K. Jeffay, D. Ott and F. Donelson Smith, "Tuning RED for Web Traffic", *IEEE/ACM Trans. on Networking*, v.9, no.3, pp.249-264, June 2001. An earlier version appeared in Proc. of ACM SIGCOMM 2000.
5. F. Colonius, "Optimal Periodic Control", Lecture Notes in Mathematics, Springer-Verlag, Berlin, 1988.
6. M. Fazel and M. Chiang, "Network Utility Maximization with Nonconcave Utilities Using Sum-of-Squares Method", in Proceedings of IEEE CDC-ECC 2005.
7. S. Floyd and V. Jacobson, "Random Early Detection Gateways for Congestion Avoidance", *IEEE/ACM Trans. on Networking*, v.1, no.4, pp.397–413, 1993.
8. V. Gaitsgory, "Suboptimization of Singularly Perturbed Control Problems", *SIAM J. Control and Optimization*, 30 (1992), No. 5, pp. 1228 - 1240.

9. V. Gaitsgory and S. Rossomakhine "Linear Programming Approach to Deterministic Long Run Average Problems of Optimal Control", *SIAM J. Control and Optimization*, 44 (2005/2006), No 6, pp 2006-2037.
10. V. Jacobson, Congestion avoidance and control, *ACM SIGCOMM'88*, August 1988.
11. T. Kelly, "Scalable TCP: Improving performance in highspeed wide area networks", *Computer Comm. Review*, v.33, no.2, pp.83-91, 2003.
12. J.-W. Lee, R. Mazumdar, and N. Shroff, "Non-Convex Optimization and Rate Control for Multi-Class Services in the Internet", *IEEE/ACM Trans. on Networking*, v.13, no.4, 2005.
13. S. Low, F. Paganini and J. Doyle, "Internet Congestion Control", *IEEE Control Systems Magazine*, v.22, no.1, pp.28-43, February 2002.
14. V. Misra, W. Gong and D. Towsley, "A Fluid-based Analysis of a Network of AQM Routers Supporting TCP Flows with an Application to RED", in Proceedings of ACM SIGCOMM 2000.
15. M. May, J. Bolot, C. Diot and B. Lyles, "Reasons Not to Deploy RED", in Proceedings of 7th International Workshop on Quality of Service (IWQoS'99), June 1999, London, UK.
16. J. Postel, User Datagram Protocol, *RFC 768*, August 1980, available at http://www.ietf.org/rfc/rfc0768.txt.
17. K. Ramakrishnan, S. Floyd and D. Black, The Addition of Explicit Congestion Notification (ECN) to IP, *RFC 3168*, September 2001, available at http://www.ietf.org/rfc/rfc3168.txt.
18. S. Shenker, "Fundamental Design Issues for the Future Internet", *IEEE J. Selected Areas Commun.*, v.13, no.7, pp.1176-1188, 1995.
19. R. Srikant, *The Mathematics of Internet Congestion Control*, Birkhaüser, Boston, 2004.
20. ILOG CPLEX http://ilog.com/products/cplex/

# TCP Versus TFRC over Wired and Wireless Internet Scenarios: An Experimental Evaluation[*]

Luca De Cicco and Saverio Mascolo

DEE Politecnico di Bari, Via Orabona 4, 70125 Bari, Italy
{ldecicco, mascolo}@poliba.it

**Abstract.** TCP NewReno is the standard transport protocol originally designed to transport bulk data over the Internet. During the years it has been very successful to provide Internet stability due to its congestion control scheme. However TCP is not very suitable for multimedia streaming applications, that are time sensitive, because of its retransmission and multiplicative decrease mechanisms. The alternative to TCP is the User Datagram Protocol (UDP) which works as a simple packet multiplexer/demultiplexer and does not implement any congestion control scheme or retransmission mechanism. However, it has been pointed out that applications that don't use congestion control schemes are dangerous for the stability of the Internet [1]. The TCP Friendly Rate Control (TFRC) is currently been discussed within the IETF as a possible leading standard for streaming multimedia flows. This paper aims at investigating the performances of TCP and TFRC congestion control schemes in wired public Internet and in mixed wired/wireless Internet using a commercial UMTS card. The experiments carried out have shown that TFRC exhibits smoother rate dynamics in all wired scenarios, whereas in the case of UMTS scenario its burstiness is comparable to that of TCP.

## 1 Introduction

Nowadays Wireless Internet allows users to achieve ubiquitous access to the Internet. Moreover the new standards for broad band wireless networks such as IEEE 802.16, 802.16a, 802.11a/g and the new 3G UMTS networks enable users to access rich audiovisual contents.

TCP NewReno is the standard transport protocol originally designed to transport bulk data over the Internet, which has been very successful to provide Internet stability due to its congestion control scheme. TCP is not very suitable as a transport protocol for multimedia streaming applications because of its retransmission and multiplicative decrease features that are not useful with delay sensitive flows. The alternative to TCP is the User Datagram Protocol (UDP) which works as a simple packet multiplexer/demultiplexer and does not implement any congestion control scheme or retransmission mechanism. However it has been pointed out that applications that don't use congestion control schemes

---

[*] This work was supported by the MIUR-PRIN project no. 2005093971 "FAMOUS Fluid Analytical Models Of aUtonomic Systems".

Y. Koucheryavy, J. Harju, and V.B. Iversen (Eds.): NEW2AN 2006, LNCS 4003, pp. 530–541, 2006.

to adapt their rate in order to avoid congestion collapse, are dangerous for the stability of the Internet [1]. Many efforts have been carried out to the purpose of designing new end-to-end protocols able to efficiently stream multimedia flows over wired/wireless scenarios and to assure network stability such as TFRC [2] and RAP [7]. When a new protocol is proposed it has to satisfy the following requirements: i) the rate of generated flows should be smooth, i.e. rates should exhibit contained oscillations in order to keep the receiver buffer as small as possible; ii) it has to be TCP friendly i.e. competing TCP flows should gain similar long term throughput; iii) it has to be fair i.e. flows using the same congestion control should gain the same long term throughput; iv) it has to be responsive i.e. flows should quickly react to network condition changes.

The TCP Friendly Rate Control (TFRC) is currently discussed within the IETF as a possible leading standard for streaming multimedia flows [2].

This paper aims at investigating the performances of TCP and TFRC congestion control schemes in wired public Internet and in mixed wireless/wired Internet using a commercial UMTS card.

The paper is organized as follows. In section 2 we will briefly describe TCP Reno congestion control and TFRC basics. Section 3 describes the tools we developed and used to collect experimental analysis; moreover we describe the testbed used in our experiments. In section 4 we report results obtained over both wired Internet and using a commercial UMTS card provided by a telecom operator. Section 5 reports burstiness indices measures for each scenario we tested. In the final section we report conclusions and open issues.

## 2    TCP and TFRC Congestion Control Basics

The version of the TCP (TCP NewReno) congestion control algorithm which is currently implemented in TCP/IP stacks is largely based on [6] and on its modifications. TCP congestion control is made of two main different phases: the probing phase and the decreasing phase. In the probing phase the channel is probed by exponentially increasing the congestion window (*slow start phase*) until the slow start threshold `ssthresh` is hit. At this point the congestion window `cwnd` is linearly increased (*Additive Increase* or *congestion avoidance* phase).

The decreasing phase, also called *Multiplicative Decrease*, is instead triggered when a congestion episode is experienced. TCP assumes that a congestion takes place when three duplicate acknowledgment packets (3DUPAK) are received by the sender or a timeout expires. When such an event occurs the congestion window is halved in order to quickly react to the congestion episode.

The pseudo code of TCP according to [3] is the following:

1. On ACK reception:
   - *cwnd* is increased according to the Reno algorithm
2. When 3 DUPACKs are received:
   - `ssthresh = max(FlightSize/2, 2);`
   - `cwnd = ssthresh;`

3. When coarse timeout expires:
   - `ssthresh = 1;`
   - `cwnd = 1;`

One of the main drawbacks of classic TCP congestion control is experienced when accessing lossy links such as 802.11b/g and 2G/3G network. In fact TCP triggers the Multiplicative Decrease even if the loss is due to interference on the wireless channel and not to congestion.

The TCP Friendly Rate Control (TFRC) is a rate based congestion control which aims at obtaining a smooth rate dynamics along with ensuring friendliness towards Reno TCP [5]. To provide friendliness, a TFRC sender emulates the long term behavior of a Reno connection using the equation model of the Reno throughput developed in [4]. In this way, the TFRC sender computes the transmission rate as a function of the average loss rate, which is sent by the receiver to the sender as feedback report.

## 3    Experimental Testbed and Measurement Tools

Figure 1 shows the wired and wireless scenarios we tested. In both the wired and wireless scenarios TCP and TFRC senders were located at University of Uppsala (Sweden), whereas the receivers were located in Bari (Italy) accessing the wired Internet through a host located at Politecnico di Bari or the public wireless Internet on UMTS from a commercial telecom operator. The reverse traffic flows have been generated running TCP or TFRC senders at Politecnico di Bari and TCP or TFRC receivers at Uppsala.

In order to generate TCP flows we used `iperf` which has been modified to incorporate `libnetmeas` (see below) and to automatically produce log files. As for

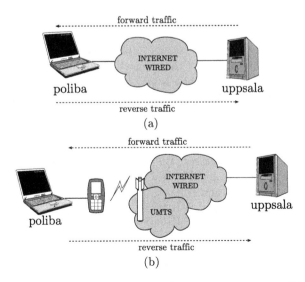

**Fig. 1.** Experimental testbed for wired (a) and wired/wireless (b) scenarios

the TFRC flows we used experimental code at sender and receiver side [12].We have no control of other competing flows on the network as we evaluated protocols over the public Internet so we can't isolate them from other flows. In each scenario we have measured goodputs, fairness indices and burstiness indices we report in the following sections.

When conducting and collecting live Internet experiments one of the most difficult task is to log TCP variables such as *congestion window, slow start threshold, round trip time* and so on. Since TCP is implemented in the kernel of the operating system those variables are kept hidden to user space application making their logging from the user space an impossible task. In order to work around this issue researchers have proposed several solutions: instrumenting the kernel code, developing TCP user space implementation [9] and using packet sniffers along with tcptrace application. Each of these solutions are not well suited because it is difficult to validate instrumented implementations and it is even more difficult to verify a user space TCP implementation. Using a packet sniffer is not suitable as well because sniffed packets don't contain any information about the TCP internal state.

In [8] authors describe a new and less intrusive solution which consists of a kernel patch and a library (libweb100) which exposes to the user space variables of each TCP flow. The interface between the kernel-space and the user-space is the virtual *proc* filesystem where statistics about flows are kept. Each flow is associated to a file in /proc/web100/CID where CID is a number incremented on the establishment of a new TCP flow. In order to log a TCP flow, it is necessary to know the CID and then it is possible to use one of the web100 tools (i.e. readvars) in a loop. This is a really difficult task because the CID is not known and the user has to manually find the CID matching the right connection.

In order to log TCP flows we developed a library which is able to overcome the aforementioned issues [11]. The library depends on libweb100, it is written in C language using glib and it is shipped with a very simple API (Application Program Interface) in order to be easily integrated in existing applications. The library is initialized using a function of the API which starts an internal thread that will automatically log flows matching a string creating or using a specified socket in a file called *tcp_ <CID>_ <timestamp>.txt* where CID is the connection ID and timestamp is the UNIX timestamp of the first data logged. Moreover the API offers a way to select the congestion control algorithm to use. The integration of the library in an existing application is really trivial: the application must call the initialization function on the ports (or on the socket) it wants to listen on and must link the library to the application (see library documentation for further details).

## 4   Experimental Results

### 4.1   Experiments over Wired Internet

In this section we report results obtained testing TCP and TFRC in the following scenarios (see Figure 1 (a)): i) single TCP vs single TFRC flow without reverse traffic; ii) Single TCP vs single TFRC flow with reverse traffic; iii) Single TCP vs single TFRC flow with parallel UDP connection; iv) 3 TCP flows vs 3 TFRC flows without reverse traffic.

In all tests the receiver is located at Politecnico di Bari, Italy and the sender is located at University of Uppsala, Sweden. In each of the considered scenario we will report goodputs, and instantaneous throughput of the most interesting experiments. In each test we run consecutively TCP and TFRC flows for two minutes in order not to have inconsistent results due to different network conditions.

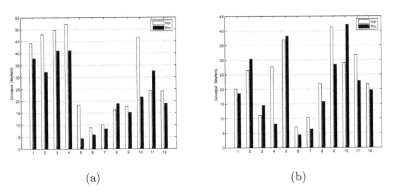

(a)                           (b)

**Fig. 2.** Goodputs of single TCP vs single TFRC flow (a) without reverse traffic (b) in the presence of reverse traffic

**Single TCP vs Single TFRC flow without reverse traffic.** Figure 2 (a) shows goodput achieved by TCP and TFRC flows that were run at different times and in different days. It is worth noticing that in 10 out of the 12 tests TCP achieves higher throughput with respect to TFRC. In some experiments (number 5 and number 10) the gap between TCP and TFRC is noticeable.

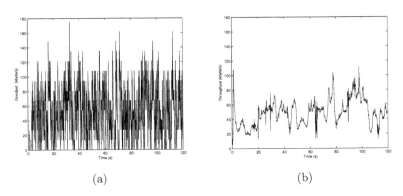

(a)                           (b)

**Fig. 3.** Instantaneous throughputs of (a) TCP; (b) TFRC

Figures 3 (a) and (b) show instantaneous throughput of a TCP and a TFRC connection. By comparing the two figures we can conclude that TFRC flows are smoother than TCP flows in this scenario but TFRC achieves a lower goodput than TCP.

**Single TCP vs single TFRC flow with reverse traffic.** In this scenario we evaluate TCP and TFRC performances in the presence of a TCP flow in the backward path. Figure 2 (b) shows the TCP and TFRC achieved throughputs. By comparing Figure 2 (a) and (b) we can see that the TCP goodput is sensitive to the congestion on the backward path whereas TFRC is not due to the fact that feedback reports are much less frequent than ACK packets (one packet every RTT). In the considered scenario TCP flows achieve a higher throughput with respect to TFRC in 6 tests over 12.

Figures 4 (a) and (b) show instantaneous throughputs for a TCP and a TFRC connection. Also in this scenario TFRC exhibit a smoother throughput dynamics respect to TCP.

**Single TCP vs single TFRC flow with parallel UDP connection.** In this scenario we report results obtained testing TCP and TFRC with one concurrent

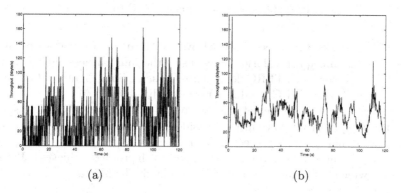

(a)                                (b)

**Fig. 4.** Instantaneous throughputs of: (a) TCP; (b) TFRC (b)

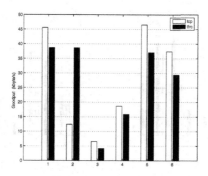

**Fig. 5.** Single TCP vs Single TFRC flow with parallel UDP connection

UDP flow generated by the Uppsala Host in order to force a 20 Kbyte/s bandwidth limitation at the receiver. Figure 5 shows goodputs for the TCP and TFRC
flows: the TCP provides better link utilization.

Figures 6 (a) and (b) depict the dynamics of TCP and TFRC instantaneous
throughputs respectively in the considered scenario. Again, TFRC exhibit a
smoother dynamics with respect to TCP.

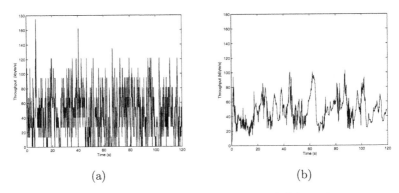

(a)                                    (b)

**Fig. 6.** Instantaneous TCP (a) and TFRC (b) goodput

**Three TCP flows vs three TFRC flows without reverse traffic.** In this
scenario we test intraprotocol friendliness using the Jain Fairness index [13] when
three TCP flows or three TFRC flows share the same link. Figures 7 (a) and (b)
report the TCP and TFRC goodputs respectively.

Figures 8 (a) and (b) depict the behaviour of TCP and TFRC instantaneous throughput of all flows in the considered scenario. Again, TFRC exhibit
a smoother dynamics with respect to TCP.

The Jain Fairness index has been evaluated. Each protocol achieves a JF index
near to 1, which is the maximum possible value for the index.

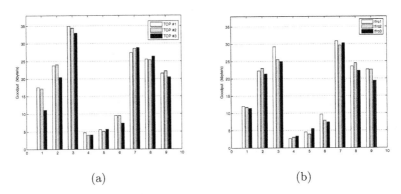

(a)                                    (b)

**Fig. 7.** Goodput for: (a) 3 TCP flows; (b) 3 TFRC flows without reverse traffic

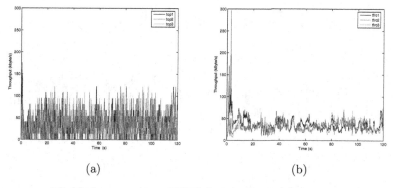

**Fig. 8.** Instantaneous TCP (a) and TFRC (b) goodput

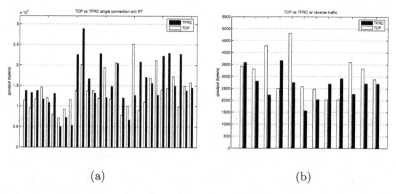

**Fig. 9.** TCP vs TFRC goodputs (a) without reverse traffic (b) with reverse traffic

## 4.2 Experiments over the UMTS Link

**Single connection without reverse traffic.** In this section we describe results obtained when a single TCP or TFRC connection uses an UMTS downlink. Figure 9 (a) depicts throughput as measured at the receiver. It is worth noticing that, nor TCP neither TFRC, reach the nominal downlink capacity of 384 $Kbps$.

Figure 10 depicts the TCP and TFRC received throughput during two consecutive tests. Both protocols show remarkable oscillations in throughput. Moreover it is worth noticing that the TFRC transient is long (approximately $20s$) if compared to the TCP transient time. It seems that using TFRC for video streaming in UMTS scenarios would require a longer buffering phase if compared to TCP behaviour. Moreover, by comparing Figure 3 (a) and Figure 10 (a) we can observe different behaviour of TCP in the wired and wireless scenario respectively. In fact in the UMTS scenario the TCP burstiness is clearly mitigated and it is comparable to that of TFRC. We have obtained similar results by repeating the experiments many times over different days.

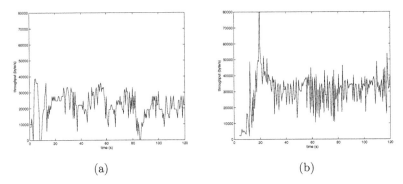

**Fig. 10.** Instantaneous throughput of (a) TCP and (b) TFRC without reverse traffic

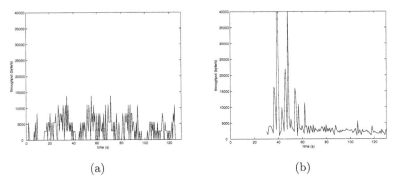

**Fig. 11.** Instantaneous throughput of (a) TCP and (b) TFRC in the presence of reverse traffic

**Single connection with reverse traffic.** In this scenario the TCP and TFRC have been tested in presence of homogeneous reverse traffic in order to evaluate if the protocols are sensitive to congestion on the backward path. For what concerns TCP, we run iperf in bidirectional mode on both the UMTS clients in Bari and the wired host at Uppsala, whereas to test TFRC in this scenario we run both the sender and receiver on UMTS client in Bari and on Uppsala client.

By comparing Figure 9 (b), which shows goodputs in the present scenario, and Figure 9 (a), we can notice that goodputs suffer a dramatic drop when using TCP or TFRC in presence of reverse traffic. This results are quite disappointing if an UMTS connection has to be used in a peer to peer system when a bidirectional communication is set up. Figure 11 reports instantaneous rates of a TCP and a TFRC flow. Both TFRC and TCP provide a very low link utilization in the presence of reverse traffic (around 25 *Kbps* on average).

**One TFRC flow and one TCP flow sharing the downlink.** Here we collect results obtained when one TCP and one TFRC flow share the UMTS

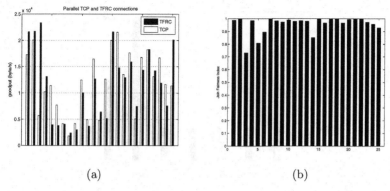

**Fig. 12.** TCP and TFRC accessing the same link; (a) goodputs, (b) Jain Fairness index.

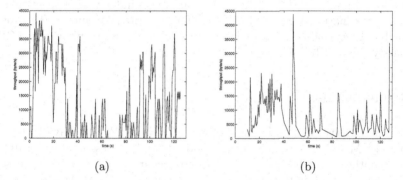

**Fig. 13.** TCP (a) and TFRC (b) instantaneous throughput when simultaneously accessing the link

downlink. Examining Figure 12 (a) we can notice that the throughput of each connection is not affected from the other flow and the downlink capacity is not underutilized. In order to produce a quantitative measurement of the inter-protocol fairness we evaluated the Jain Fairness Index. Figure 12 (b) shows fairness indices which are near to the maximum value of 1 in most of the tests. Figure 13 shows instantaneous throughputs of a TCP flow and a TFRC flow which simultaneously access the UMTS link. It is worth noticing that even if the channel utilization is quite good each flow exhibits pronounced oscillations.

## 5   Burstiness of Received Data

In order to evaluate the smoothness of the data transfer rate we have evaluated the burstiness index of each transfer. The burstiness index is defined as $b = \frac{\sigma(r)}{E[r]}$ where $\sigma(r)$ represents the standard deviation of the received rate $r$ and $E[r]$ is the average value [14].

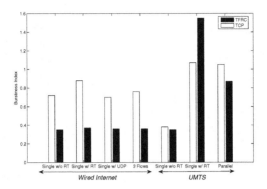

**Fig. 14.** Burstiness Indices

Figure 14 shows that TFRC halves the burstiness index with respect to TCP in all wired scenarios. However in the UMTS scenario TCP and TFRC provide similar burstiness indices, except in the case with reverse traffic where TCP is less bursty than TFRC.

## 6    Conclusions

We conducted several experiments testing TCP and TFRC behaviour both in wired and UMTS networks measuring goodputs and fairness indices. TFRC exhibits smoother rate dynamics in all wired scenarios, whereas in the case of UMTS scenario its burstiness is comparable to that of TCP. Moreover experimental results have shown that when TFRC and TCP flows are accessing an UMTS link they both are not able to provide full link utilization in the presence of reverse traffic, which could be a severe limitation in peer-to-peer applications.

## References

1. S. Floyd and K. Fall, "Promoting the use of end-to-end congestion control in the Internet", *IEEE/ACM Transaction on Networking*, vol. 7, no. 4, pp. 458-472, 1999.
2. M. Handley,S. Floyd, J. Padhye and J. Widmer, "TCP Friendly Rate Control (TFRC): Protocol Specification", *RFC 3448*, January 2003.
3. M. Allman, V. Paxson, W. Stevens, "TCP Congestion Control", *RFC 2581*, April 1999.
4. J. Padhye, V. Firoiu, D. Towsley, and J. Kurose. "Modeling TCP throughput: A simple model and its empirical validation", ACM Sigcomm '98, pages 303-314, Vancouver BC, Canada, 1998.
5. S. Floyd, M. Handley, J. Padhye, and J. Widmer. "Equation-based congestion control for unicast application", ACM SIGCOMM 2000, August 2000
6. V. Jacobson, "Congestion Avoidance and Control", ACM Computer Communications Review, 18(4): 314 - 329, August 1988.

7. R. Rejaie, M. Handley, D. Estrin, "RAP: An End-to-end Rate-based congestion control mechanism for Real-time streams in the Internet", Proc. of INFOCOM, March 1999
8. M. Mathis, J Heffner and R Reddy, "Web100: Extended TCP Instrumentation for Research, Education and Diagnosis", ACM Computer Communications Review, Vol 33, Num 3, July 2003.
9. T. Dunigan, F. Fowler, "A TCP-Over-UDP test Harness", Technical report
10. Iperf, http://dast.nlanr.net/Projects/Iperf/
11. LIBrary for NETwork MEASurements, http://c3lab.poliba.it/index.php/ Libnetmeas
12. TFRC experimental Code, http://www.icir.org/tfrc/code/
13. R. Jain, "The art of Computer Systems Performance Analysis Techniques for Experimental Design, Measurement, Simulation and Modeling", John Wiley and Sons, April 1991.
14. YR. Yang, MS. Kim, S. Lam , "Transient behaviors of TCP- friendly congestion control protocols.", Proc. IEEE INFOCOM 2001, April 2001

# Dynamic Configuration of MAC QoS Mechanisms in 802.11 Access Networks

Simona Acanfora[1], Filippo Cacace[2], Giulio Iannello[1,2], and Luca Vollero[1,2]

[1] Laboratorio ITeM, CINI, Via Diocleziano 328, Napoli, 80124, Italy
simona.acanfora@gmail.com, vollero@ieee.org
[2] Facoltà di Ingegneria, Università Campus Bio-Medico di Roma
Via Emilio Longoni 83, Roma, 00155, Italy
{f.cacace, g.iannello}@unicampus.it

**Abstract.** In this paper, we define a general architecture that integrates micromobility protocols and dynamic management of 802.11 QoS MAC mechanisms and evaluate it through simulation in a scenario where a specific micromobility protocol, Cellular IP, is integrated with a specific QoS mechanism, ACKnowledgement Skipping. Our results demonstrate that seamless, QoS aware connectivity can be provided to mobile users, and identify directions to fully integrate these functionalities in a comprehensive WLAN management system.

## 1 Introduction

The notion of Mobile Internet refers to scenarios where mobile users may have access anytime and anywhere to conventional and emerging web applications requiring multimedia and interactive communications. Wireless local area networks (WLANs) based on the IEEE 802.11 standard are a key element to implement these scenarios, representing a typical access network for mobile users.

A mobile user moving in an area covered by a WLAN negotiate a contract about the quality of a communication session. Once established the communication session, it should be maintained with the negotiated quality as long as possible. The crucial point is that when a handoff takes place, the contract should be still fulfilled, providing that there are enough resources at the new access point.

To meet these requirements both continuous connectivity and resource reservation have to be guaranteed to mobile stations. For what is concerned with the former issue, micromobility protocols have been proposed in recent years to provide low overhead, seamless connectivity in WLANs [1]. As to the resource reservation issue, much work has been done in order to extend the 802.11 MAC protocol with mechanisms supporting traffic differentiation [2], and a new standard has been recently defined [3].

Previous works exist about the integration of mobility and QoS in WLANs [4, 5, 6]. Surprisingly, to the best of our knowledge, none of them considers the integration of mobility and QoS MAC mechanisms proposed in the context of the 802.11 standard, and in a few cases these mechanisms have been considered in dynamic scenarios.

Y. Koucheryavy, J. Harju, and V.B. Iversen (Eds.): NEW2AN 2006, LNCS 4003, pp. 542–553, 2006.

A fundamental point about the integration of 802.11 QoS mechanisms and mobility is the ability to change the configuration of MAC parameters when users move around, i.e. when mobile stations enter or leave wireless cells, and change their traffic patterns.

In this paper, we define a general architecture that integrates micromobility protocols and dynamic management of 802.11 QoS mechanisms and evaluate it through simulation in a scenario where a specific micromobility protocol, Cellular IP [7], is integrated with a specific QoS mechanism, ACKnowledgement Skipping [8]. Our results demonstrate that seamless, QoS aware connectivity can be provided to mobile users, and identify directions to fully integrate these functionalities in a comprehensive WLAN management system.

## 2 Terminology

Recently, a new IETF Working Group, named Control and Provisioning of Wireless Access Points (CAPWAP), started its activity with the goal of finding solutions to problems like configuration, monitoring, control, and management of large scale deployments of IEEE 802.11 networks [9]. The CAPWAP WG has identified a number of functions not included in the 802.11 standard, among which mobility and QoS management are explicitly mentioned, although no further details about their provisioning are given. For this reason we believe that the dynamic management of QoS MAC mechanisms should be considered in the general framework defined by CAPWAP working group. In particular, we make extensive use of the terminology introduced in [9].

In this paper an 802.11 WLAN consists of one or more wireless cells, called Basic Service Sets (BSSs), consisting of an Access Point (AP) and one or more mobile stations (STAs) associated to the AP. The architectural component used to interconnect BSSs is the Distribution System (DS). The term Access Point (AP) refers to the logical entity that controls a BSS and provides a Distribution System Service (DSS) to the associated STAs. We also introduce the term Wireless Termination Point (WTP) to refer to the physical device that includes the RF antenna and implements the 802.11 physical layer, and the term Access Controller (AC) to refer to a logical entity that may be integrated in the WLAN to provide control and other management functions in a centralized fashion.

Traditionally, the AP is a single physical device, including the notion of WTP, and supporting autonomously all the services defined by the IEEE 802.11 standard. Correspondingly the DS is a wired network connecting APs, which may include either L2 (switches), or L3 (routers) devices, or both. This WLAN architecture is called *Autonomous* and it is characterized by the absence of a clear centralized control point. Alternatively, a number of vendors have recently integrated in their WLAN architecture Access Controllers providing a number of control and management functions. These architectures are called *Centralized* since they naturally provide a point of coordination and control of the WLAN.

## 3    Related Work

Much work has been done on mobility management in WLANs with multiple WTPs. A number of micromobility protocols have been proposed to handle local movements of mobile hosts [1]. Micromobility protocols provide several services, including host localization, paging, routing and fast handoffs. Although they use different strategies to deal with these issues, the operational principles that govern the most popular micromobility protocols, i.e. Cellular IP, Hawaii and Hierarchical Mobile IP (HMIP) are largely similar.

A problem related to seamless mobility is the need of an association procedure before any communication can take place between a moving STA and a new WTP. In the 802.11 standard, this procedure includes WTP discovery, authentication and the actual association, and it is performed exchanging special L2 frames between the STA and the network components in charge of these functions. Seamless mobility remarkably depends on this L2 handoff process and much work has been done so far to analyze and improve its performance [10, 11].

In this paper, we do not explicitly consider problems related to L2 handoff performance, but rather we focus our attention primarily on the provisioning of service guarantees *after* handoffs rather than *during* handoffs.

Many extensions to the standard 802.11 MAC protocol have been proposed to introduce traffic differentiation in 802.11 networks [2, 3]. For space reasons, we give here only a few details about ACKnowledgement Skipping (ACKS), the technique used in the simulation tests reported in section 5, and therefore relevant for the following discussion. ACKS [8] is based on the following behavior of the standard 802.11 DCF: after sending a packet, a station waits for an Acknowledgement (ACK) frame, and, if the frame is not received within an assigned timeout, it assumes a collision and increases its Contention Window (CW). Hence, if the AP skips the ACK reply to STAs with some probability, these stations will "see" a collision rate higher than the actual one, and will contend with larger CWs, resulting in a reduced channel access capability. This skipping probability, called *dropping factor*, is the parameter controlling differetiation.

Some work is then available on QoS management in presence of mobility. In [5], efficient integration of the handoff protocol with the RSVP signaling process in the wired portion of the network is considered. Performance evaluation is focused on the reservation phase, since it is assumed that communication flows experiment the QoS requested once reservation is successfully completed.

In [6], a mobility management protocol is extended to deal with handoff requests towards APs that have not enough resources to provide the desired QoS. Again, the proposal considers QoS mechanisms available at routers, providing traffic differentiation only in the wired portion of the WLAN infrastructure.

In [12] a hierarchical QoS architecture that extends the DiffServ QoS model to mobile hosts in an 802.11 environment is presented. The approach provides a quite comprehensive solution to the integration of mobility and QoS. However, the problem of managing QoS requirements inside BSSs is solved through mechanisms working at layer 3 or above. The advantage is that QoS can be managed in WLANs compliant with the current 802.11 standard, but the issue of how new QoS MAC mechanisms may be coupled with mobility is completely neglected.

## 4    Integration of Mobility and QoS MAC Mechanisms

In this section, we describe an architecture that integrates mobility support with dynamic management of QoS MAC mechanisms to provide service guarantees to mobile users. To make discussion concrete, we basically assume an Autonomous WLAN architecture where mobility is managed by a micromobility protocol, such those mentioned in section 3. Nevertheless, the services we identify can be implemented either in a single physical device or across multiple network elements, and their definition is compatible with either Autonomous or Centralized WLAN architectures. In the following, however, to make discussion simpler, we assume an Autonomous architecture and use the traditional terms of Access Point and Router to refer to the WLAN physical components.

As to QoS, we assume that some traffic differentiation mechanisms extending the 802.11 MAC protocol is available, and that service guarantees in the wired part of the WLAN are enforced through overprovisioning, or, alternatively, through admission control and resource reservation based upon solutions similar to those proposed in [5, 6, 12].

According to what required by the 802.11 standard, we assume that *beacon* packets are periodically broadcast by WTPs, and STAs listen to them and initiate handoff based on signal strength measurements. To perform a handoff, a STA tunes its radio transmitter to the new WTP, and enters the authentication and association procedures envisaged by the standard. After the association completes, the new WTP continues the handoff procedure according to the specific micromobility protocol considered.

To implement dynamic QoS management, each AP is controlled by a *QoS manager*, which carries out at least the following minimal set of functions: (i) acquisition of information about traffic patterns and QoS requirements of the moving station when a handoff initiates; (ii) computation of the MAC parameter configuration that best matches traffic offered by the STAs active in the BSS; (iii) actual setting of MAC parameters of the APs and of all active stations in the BSS; (iv) setting of L3 mechanisms to provide QoS guarantees to downstream flows.

In the following, we describe how these functions can be implemented using existing protocols with minimal extensions.

*QoS information acquisition.* We assume that in the WLAN there is a *global session directory* where a traffic pattern description is maintained for every active communication session. Any STA starting a new session with QoS guarantees has to register to the session directory, providing sufficient information to characterize its traffic patterns.

When the new association is started, the MAC layer notifies the QoS manager and informs it about the identity of the moving station. Using this information, the QoS manager accesses the session directory and get the corresponding traffic information.

Note that the proposed scheme is robust, since a default traffic pattern (e.g. best effort traffic) may always be assumed if, for any reasons, the traffic pattern of a moving station is not available (e.g. a STA just entered the WLAN or it is not registered in the session directory).

*Computing MAC configuration.* Once traffic information is acquired, the QoS manager can carry out its second task, i.e. to compute the optimal MAC parameter configuration. The computation can be either *model*-based, if analytical models are available for wireless channel characterization, or *table*-based, if optimal parameter values have been derived off-line through simulation or experimentation. Although this problem, in its general form, is still open for most QoS MAC mechanisms, several useful results are available in the literature, and there is much ongoing research trying to extend these results.

Using either model or off-line characterization, the QoS manager can determine if QoS requirements of handoff requests can be satisfied by the resources available in the BSS. This means that the manager might perform also access control over handoff requests and notify the micromobility protocol when the requested service guarantees cannot be met. To fully exploit this ability, however, the micromobility protocol should be modified introducing the concept of QoS-conditionalized handoff [6]. In this context, we prefer to follow a simpler and conservative approach. In case the request cannot be satisfied, the QoS manager computes the best possible configuration, and the moving station will experiment a service degradation if the handoff is completed successfully.

*Setting MAC configuration.* After having computed a new configuration of MAC parameters for the stations in the BSS, the QoS manager can force BSS configuration by broadcasting configuration information to all active stations. Note that this step is generally needed since most QoS MAC mechanisms require a coordinated configuration of all stations in a BSS.

*Setting L3 mechanisms.* Finally, since downstream flows may have different QoS requirements, they should receive a differentiated treatment, likewise upstream flows transmitted by the mobile hosts. All downstream flows are transmitted by the same 802.11 device (the AP), so, in most cases, differentiation among these flows cannot be satisfactorily managed at layer 2. A reasonable solution is therefore that the MAC parameters of the AP are set in such a way that enough resources are made available to all downstream traffic as a whole. QoS constraints on individual downstream flows can then be enforced through L3 mechanisms, i.e. through mechanisms similar to those implemented in QoS-aware routers, based on traffic conditioning.

The functionalities just described can be implemented by the architecture shown in figure 1, where are reported the functional building blocks and how they interact in the control plane. In the figure functions are mapped onto the two components of the WLAN: the access network, consisting of the the distribution system and the access points, and the mobile stations. This means that in real

WLANs there are multiple instances of some blocks. For instance, the MAC and physical layer on both sides are replicated in each AP and in each STA, as well as, the IP layer is replicated in each STA. Other blocks (e.g. the QoS manager) may or may not be replicated, according to performance considerations and to which WLAN architecture (Autonomous or Centralized) is chosen. A distinctive element of the architecture is the presence of a session layer entity in both the access network and in the STAs that implement the global session directory needed to relate mobility events with the update of MAC parameters.

It is worth noting that most of the interactions reported in the figure rely on existing or forthcoming protocols. Messages between MAC entities on both sides are the control frames defined by the standard to perform L2 handoffs. Communications between MAC/IP blocks and QoS managers in the access network can be implemented using the forthcoming CAPWAP protocol that is still in a proposal stage, but that explicitly includes messages to notify mobility events and set 802.11e parameters [13, 14]. Finally, session registrations in the global session directory can use existing message formats to specify traffic characteristics (see for instance [15]).

Before closing the section, we briefly make some remarks about the actual mapping of the functional elements shown in figure 1 onto physical network devices. A first issue concerns the alternative between Autonomous and Centralized WLAN architectures. In the former case a natural choice is to replicate the QoS manager on each AP, whereas in the latter case it may alternatively be mapped on the AC, especially if the WLAN is a variant of the Centralized architecture called Split-MAC [9, 13]. Similar considerations can be done for the IP block that is located at the gateway connecting the DS with the Internet, but that can be replicated if more routers are present in the DS. As to the global session directory, the natural choice is a centralized implementation, located at the gateway or at the AC, according to the physical topology of the DS. Nevertheless, more sophisticated implementation schemes may be chosen to improve scalability of QoS management functions with the WLAN size. For instance the

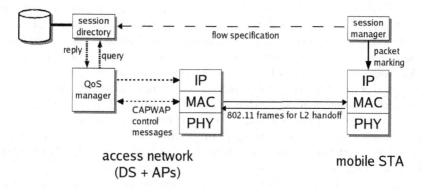

**Fig. 1.** Architecture for the integration of mobility and QoS (control plane)

global session directory can be implemented as a hierarchy of modules to keep local the interactions with STAs and QoS managers.

# 5   Evaluation

In this section, we present a set of simulation results aiming to provide a first evaluation of our architecture.

*Simulation environment.* To setup the simulation environment we needed to choose a specific micromobility protocol and a specific set of QoS mechanisms extending the standard 802.11 MAC protocol. As a micromobility protocol we chose Cellular IP. Indeed, much previous work on the integration of mobility and QoS support assumes Cellular IP as a micromobility protocol, which makes reasonable here its choice to study the integration of mobility and MAC parameter configuration. To limit the complexity of this first analysis, we then chose ACKnowledgement Skipping (ACKS) as a QoS mechanism because it is very simple to configure (just one parameter at the AP when two traffic classes are considered), and it does not introduce any configuration delay (no communication between the AP and the STAs is required).

As a simulation environment we chose the Berkeley Network Simulator, ns-2, that has been widely used in the evaluation of both micromobility protocols and QoS MAC mechanisms. In particular we used the implementation of Cellular IP provided by the Columbia IP Micromobility Suite (CIMS). The choice of using CIMS requires the use of ns-2 ver. 2.1b6, which has a number of minor limitations, including the limit of 2 Mbps for the 802.11 wireless link, since this version of ns-2 does not simulate accurately the 802.11b physical layer.

In our simulations we used multiple APs interconnected to a network of Cellular IP routers. As just stated the radio links have a bitrate of 2 Mbps, whereas the wired links have a bitrate of 10 Mbps. In all experiments, the traffic pattern has the following general characteristics. Multiple mobile hosts belonging either to a best effort (BE) class or to a high priority (HP) class are associated to each AP. All hosts communicate with a correspondent node (CN) located somewhere outside the access network. BE STAs generate an upstream flow towards the correspondent node consisting of constant bitrate data stream of 200 Kbps. HP STAs generate an upstream flow towards the CN and, at the same time receive from it a downstream flow. Both flows are constant bitrate data streams of 200 Kbps each. All flows consist of UDP packets and we use as a main performance index the *goodput* of these flows, i.e. the bitrate corresponding to successfully received packets. During each simulation, one or more mobile hosts (either BEs or HPs) move through the access network performing handoffs. If not stated otherwise, hosts move at a speed of 5 m/s.

It is worth noting that this traffic pattern is quite different than the one usually used to evaluate micromobility protocols or, alternatively, QoS MAC mechanisms. On the one hand, multiple downstream flows are introduced having the same bitrate than the corresponding upstream flows. This roughly models the traffic generated by symmetric multimedia applications like VoIP or videoconferencing. On the other hand, communications of HP hosts (both upstream and downstream) must share the wireless channel with a variable interfering traffic, either because BE hosts leave and join the associated AP during the simulation, or because HP hosts move around and perform handoffs. These rich traffic patterns are needed in our scenarios because we are interested in evaluating how mobility and QoS mechanisms work in a coordinated manner. Indeed, in this scenario, the AP requires much more radio resources than each single STA in the BSS. Hence, the presence of interfering traffic quickly leads to a degradation of the goodput of downstream flows if no differentiation is applied to best effort traffic.

Finally, to test how micromobility protocols and QoS MAC parameter configuration integrate in a dynamically changing scenario, we needed a method to compute the new MAC parameters configuration of the AP when there are changes in the workload because STAs move around. Since we use ACKS to differentiate between two traffic classes, we had to determine the dropping factor to be applied to frames transmitted by the BE STAs in every traffic condition of interest. Even though, a model is available to configure ACKS [16], it can be used only when all the STAs are in saturation conditions. Unfortunately, this is not the case of our traffic scenarios where either no differentiation is needed because the workload does not saturate the channel, or only the BE stations are in saturation conditions because base and HP stations transmit at assigned bitrate. For this reason, we used a table-based approach to MAC parameter configuration, performing off-line simulations to determine the dropping factors to be used in each case.

*Results.* For space reasons, as a representative scenario among the many simulations we performed, we consider here a configuration like that depicted in figure 2: a HP STA moves from an AP where it is the only STA to an AP where there are already two HP and three BE STAs. We focus our attention on the goodputs and delays of the moving HP STA, and consider both the case when no dropping factor is applied at the final AP, and the case when it is properly updated according the table approach mentioned above.

Figures 3 and 4 show the goodputs and delays of the moving HP STA. Upstream delays are not shown because they are always very low. For goodputs, data refer to mean values taken every 10 s, excluding the range 77.5–90 s, where data refer to mean values taken every 0.5 s to point out the transient behavior.

Approximately, at simulation time 79 s the moving STA enters the second BSS, and at time 84.2 s it performs the handoff. For upstream flows (figure 3-a and 3-b), the goodput is essentially constant, with a short break in the communication flow during the handoff[1]. For downstream flows, the goodput is remarkably less than offered traffic when no dropping factor is applied (figure 4-a), whereas it remains substantially constant when the proper dropping factor is applied (figure 4-b). Note that this behavior is expected, since HP downstream

---

[1] This is when the QoS manager of the target AP is notified that the new mobile STA has performed the association.

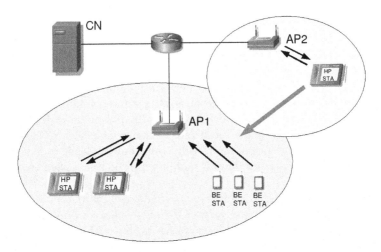

**Fig. 2.** Simulation scenario

flows are much more affected by the interfering traffic than HP upstream flows. The effectiveness of the dynamic reconfiguration of the MAC parameters is also apparent by looking at the downstream delay (figure 4-c and 4-d), which is dramatically reduced when the right dropping factor is applied.

The above data, however, show that there is a short interruption in the communication flows even when the dropping factor is applied, i.e. when resources are reserved to protect HP flows. This is due to the delay induced by the congestion at the target BSS in exchanging the messages used by ns-2 to model STA re-association. We therefore repeated the simulation changing the dropping factor at 79 s, i.e. just when the moving STA enters the coverage area of the second AP. Results are reported in figure 3. With respect to figures 3-b, 4-b, and 4-d, both upstream and downstream goodputs are remarkably improved and downstream delay further reduced, leading to an almost seamless handoff with QoS guarantees.

Although the L2 handoff process is not precisely modeled by the simulator, the observed behavior is coherent with available data on L2 handoffs (see for instance [17]). Our results suggest that an anticipated configuration of QoS MAC parameters is desirable. Although a modification to mobility protocols goes beyond the scope of this paper, it is worth noting that such an anticipated configuration could be not too difficult to implement, e.g. by enabling the moving STA to perform some kind of *pre-association* to the target AP before the actual handoff takes place.

These experiments confirm the possibility to guarantee QoS requirements after the handoff when QOS MAC mechanisms are dynamically configured according to the offered traffic patterns. This result clearly holds at steady state (i.e. before and after handoffs). As to the behavior during handoffs, we observed that an

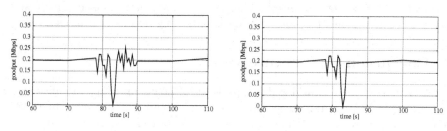

(a) goodput without dropping factor    (b) goodput with variable dropping factor

**Fig. 3.** Upstream flows of the moving HP STA

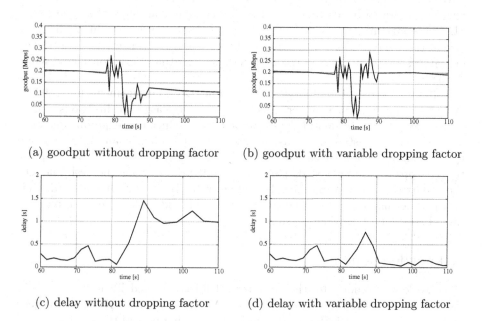

(a) goodput without dropping factor    (b) goodput with variable dropping factor

(c) delay without dropping factor    (d) delay with variable dropping factor

**Fig. 4.** Downstream flows of the moving HP STA

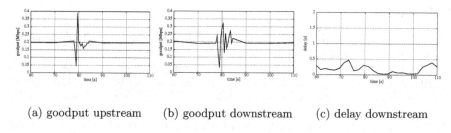

(a) goodput upstream    (b) goodput downstream    (c) delay downstream

**Fig. 5.** When the dropping factor change is anticipated at 79 s

anticipated BSS configuration might improve handoff performance. This result however is only qualitative since handoff is only roughly modeled in ns-2.

# 6   Conclusions and Future Work

In this paper we have considered the problem of how to manage mobility support and QoS MAC parameter configuration in an integrated fashion in 802.11 WLANs. In this respect, we have proposed a cross-layer management architecture characterized by two main functional elements operating at session and MAC layer, respectively. The architecture's control plane basically relies on existing or forthcoming mechanisms and protocols and it may be integrated with CAPWAP functions for WLAN management. We have also presented a preliminary evaluation of benefits deriving from dynamic management of a simple QoS MAC mechanism through simulation. Our results demonstrate that the integration of these mechanisms may provide almost seamless connectivity and service guarantees to mobile users in traffic scenarios richer than those considered so far in the literature.

Full integration of mobility and dynamic QoS management into 802.11 WLANs still requires, however, much work in at least two directions. First, the functions we have identified have to be better refined and experimented on a real testbed. For instance, we have not experimented L3 mechanisms to manage QoS of individual downstream flows (or classes of them). Second, a more sophisticated approach to QoS MAC parameter configuration would be desirable. On the one hand, ACKS should be substituted by the IEEE 802.11e standard, and, on the other hand, model-based configurations should be considered in place of table-based solutions when possible.

## Acknowledgements

This work was funded by the Ministero dell'Istruzione dell'Universit e della Ricerca with the FIRB 2001 project "WEB-MINDS" and the PRIN 2004 project "QUASAR". We thank Hewlet Packard for providing part of the hardware, under the Teaching Grant Initiative – 2005.

## References

1. Campbell, A., J. Gomez, S.K., Turanyi, Z., Valkò, A., Wan, C.Y.: Internet micromobility. Journal of High Speed Networks **11** (2002) 177–198
2. Lindgren, A., Almquist, A., Schelén, O.: Quality of Service Schemes for IEEE 802.11 Wireless LANs: An Evaluation. Special Issue of the Journal on Special Topics in Mobile Networking and Applications (MONET) on Performance Evaluation of Qos Architectures in Mobile Networks **8** (2003) 223–235
3. IEEE 802.11e/D6.0: Part 11: Wireless LAN Medium Access Control (MAC) and Physical Layer (PHY) Specifications: Medium Access Control (MAC) Enhancements for Quality of Service (QoS). Draft Supplement to IEEE 802.11 Standard (2003)

4. Moon, B., Aghvami, A.: Quality-of-service mechanisms in all-ip wireless access networks. IEEE Journal on Selected Areas in Communications **22**(5) (2004)
5. Paskalis, S., Kaloxylos, A., Zervas, E., Merakos, L.: An efficient rsvp-mobile internetworking scheme. Mobile Networks and Applications **8** (2003) 197–207
6. Sroka, S., Karl, H.: Performance evaluation of a qos-aware handover mechanism. In: Proceedings of the 8th IEEE Int. Symp. on Computers and Communication (ISCC'03). (2003)
7. Valkò, A.: Cellular ip: a new approach to internet host mobility. ACM SIGCOMM Computer Communications Review **29**(1) (1999) 50–65
8. Vollero, L., Iannello, G.: ACK Skipping: enabling QoS for multimedia communications in WiFi hot spots. Int. Journal of High Performance Computing and Networking (2006)
9. Yang, L., Zerfos, P., Sadot, E.: Architecture Taxonomy for Control and Provisioning of Wireless Access Points (CAPWAP). RFC 4118 (Informational) (2005)
10. Mishra, A.: An empirical analysis of the ieee 802.11 mac layer handoff process. ACM SIGCOMM Computer Communications Review **33**(2) (2003) 93–102
11. Shin, S., Forte, A., Rawat, A., Schulzrinne, H.: Reducing mac layer handoff latency in ieee 802.11 wireless lans. In: Proceeding of MobiWac'04. (2004)
12. Garcia-Macias, J.A., Rousseau, F., Berger-Sabbatel, G., Toumi, L., Duda, A.: Quality of service and mobility for the wireless internet. Wireless Networks **9** (2003) 341–352
13. Calhoun, P., O'Hara, B., Suri, R., Winget, N.C., Kelly, S., Williams, M., Hares, S.: Light Weight Access Point Protocol. Internet-Draft (2005)
14. Govindan, S., Yao, Z., Zhou, W., Yang, L., Cheng, H.: Objectives for Control and Provisioning of Wireless Access Points (CAPWAP). Internet-Draft (2005)
15. Partridge, C.: A Proposed Flow Specification. RFC 1363 (1992)
16. Vollero, L.: Providing throughput guarantees in wlans using acks. In: Proceeding of the 6th IEEE International Symposium on a World of Wireless Mobile and Multimedia Networks. (2005)
17. Montavont, N., Noel, T.: Analysis and evaluation of mobile ipv6 handovers over wireless lan. Mobile Networks and Applications **8** (2003) 643–653

# SMPCS: Sub-optimal Model Predictive Control Scheduler[*]

Mehran Mahramian[1], Hassan Taheri[1], and Mohammad Haeri[2]

[1] Electrical Engineering Department, Amirkabir University of Technology, Hafez Ave.,
Tehran, Iran
m_mahramian@isc.iranet.net, htaheri@aut.ac.ir
[2] Electrical Engineering Department, Sharif University of Technology, Azadi Ave.,
Tehran, Iran
haeri@sharif.edu

**Abstract.** An approximated quadratic programming algorithm is proposed to
determine a model predictive controller, which is applied to the scheduling
problem. We name the algorithm Sub-optimal Model Predictive Control Scheduler (SMPCS). The SMPCS prepares a platform to guarantee end to end proportional delay in DiffServ architecture. This paper investigates the complication
of SMPCS and its implementation problems in high speed routers. In order to
efficiently implement the controller we admit sub-optimal results against reduction in the complexity of the optimizer. Simulation results show that the proposed approximation improves speed of the controller considerably.

## 1 Introduction

RECENTLY, many researchers are involved in the control theoretic approaches to the
networking problems. They propose design of high performance, stable and robust
controllers that theoretically formulate their algorithms. Although most of these works
utilize various kinds of traditional PID controller [1], the advantages of modern controllers should also be investigated in the new algorithms. The only advanced control
technology which has affected the industrial process control is model predictive control (MPC) [2]. Because of the requirements of the optimization associated with the
MPC, it is computationally too expensive and can be used only for slow dynamic
systems. Furthermore, existing implementations of MPC typically perform numerical
calculations using 64-bit Floating Point (FP) arithmetic, which is too expensive,
power demanding and large in size [3].

If the model is linear and the constraints are in a special form, there are exact solutions for the MPC. However, for the application of MPC in networking and other fast
dynamic systems, the computational complexity becomes bottleneck for the implementation. Many recent works try to increase the performance of MPC [4] at high
speeds by utilizing Field Programmable Gate Arrays (FPGA) [5], modifying MPC in
a way that most of the computationally expensive calculations are calculated off-line

---

[*] This work was supported in part by the Iran Telecommunication Research Center under Grant
500/8916.

Y. Koucheryavy, J. Harju, and V.B. Iversen (Eds.): NEW2AN 2006, LNCS 4003, pp. 554–565, 2006.
© Springer-Verlag Berlin Heidelberg 2006

[6], [2], using new CPU architectures to implement MPC on a "System on a Chip" (SoC) [7] and proposing heuristic methods for optimization [8].

In this paper, we provide a sub-optimal solution to the minimization of the cost function in order to speed up the solution time. The proposed model predictive controller is especially tuned for a scheduler in a router in the DiffServ architecture [9] to guarantee delay constraints on aggregated traffic classes. DiffServ tries to provide scalable QoS by aggregating users with similar requirements. Under the DiffServ architecture, algorithms are required to manage different priorities for different traffic classes. Although current traffic classes in DiffServ provide Premium Service and Assured Service, they are not so accurate and new types of aggregated traffic classes should be defined to establish a reliable and scalable structure for QoS guarantee. Some recent works try to strengthen guarantee of QoS by quantifying the differentiation [1].

When scheduling algorithms may not guarantee all absolute delay constraints, proportional delay and loss probability can be considered [10]. This means one can guarantee that delay of the higher priority service is not more than a fraction of the delay of the lower priority service in addition to the absolute boundaries for delay. A solution for QoS aware scheduling is changing service rates of different traffic classes in a way that QoS requirements of the high priority classes like delay and loss are met while preventing the starvation for the low priority classes. In our previous works, we applied MPC to guarantee proportional and absolute delay for different classes of traffic [11], [12].

The rest of the paper is organized as follows. System formulation and a brief explanation of the Model Predictive Control are investigated in Section 2. SMPCS is presented in Section 3. Section 4 deals with computational complexity of the controller. Simulation results are presented in Section 5 and finally, Section 6 concludes the paper.

## 2 Modeling the System Dynamics

### 2.1 Formulation

As described earlier, our aim is to use model predictive control to provide proportional differentiation on average delay for different traffic classes. Let us assume that there are $Q$ classes in the system. Let the measured average output of class $i$ at event $n$ be $\bar{y}_i(n)$. In a proportional delay service, the proportion of measured outputs between classes $i$ and $i+1$ is assumed to be some predetermined value $k_i$:

$$\frac{\bar{y}_{i+1}(n)}{\bar{y}_i(n)} = k_i, \quad i = 1, \ldots, Q-1 \tag{2.1}$$

The network administrator decides the values of $k_i$. In order to provide guaranteed proportional delay, a controller is applied to the system. Figure 1 shows the block diagram of the system. We introduce a new variable, which is sum of the weighted delays of the queues. This variable can be used as the set point of the system outputs.

$$y(n) = \frac{\sum_{i=1}^{Q} (\prod_{k=1, k \neq i}^{Q} m_k) y_i(n)}{Q} \tag{2.2}$$

where $m_i = \prod_{j=1}^{i-1} k_j$ and $m_1 = 1$.

The system can be assumed a special case of MIMO system. The sum of the manipulated variables ( $u_i(n)$ ) should not exceed the link capacity: Also, The scheduler should not go to the idle mode when any of the queues are in the busy period (nonempty). $C$ is the link capacity.

$$\sum_{i=1}^{Q} u_i(n) = C \qquad (2.3)$$

To approximate the system model, an adaptive queue model based on an ARX structure is employed:

$$H_i(z) = \frac{Y_i(z)}{U_i(z)} = \frac{\sum_{j=0}^{m} b_j z^{-j}}{1 + \sum_{j=1}^{p} a_j z^{-j}} \qquad (2.4)$$

where $Y_i(z)$ and $U_i(z)$ are $z$ transform of $y_i(n)$ (output of the system) and $u_i(n)$ (manipulated variables). $p$ and $m$ are number of poles and zeros of the model. This model is converted to the state space model:

$$y(n+1) = Ay(n) + Bu(n) + \omega(n) \qquad (2.5)$$

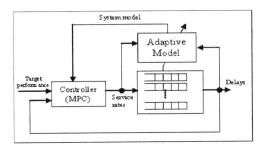

**Fig. 1.** Block diagram of SMPCS

where $y(n)$ and $u(n)$ are $Q$ -dimensional vectors of outputs and inputs. $A$ and $B$ are system dynamic matrices and $\omega(n)$ is the modeling error. The state space model is used in our algorithm to predict the future states of the system. To determine the model parameters, recursive least squares method is implemented.

## 2.2  Model Predictive Control

The first ideas of MPC were mostly developed by industrial professionals. Since late 70's, many publications proposed different types of model predictive control approach. All of these works shared the some basic features, which contained an explicit internal model, the receding horizon idea and optimizing the predicted plant behavior [2].

The main attractions of MPC that motivated us to apply it to the scheduling problem were that it can intrinsically take account of system and controller constraints and naturally handles multivariable control problems. It also allows operation closer to the boundaries of constraints compared to conventional control. Besides, it handles an event based as well as a time based model. The main ideas in model predictive control are [2]:

- The controller uses a model by which the future behavior of the system is predicted. i.e.:

$$y(n+1) = Ay(n) + Bu(n)$$
$$y(n+2) = A^2 y(n) + ABu(n) + Bu(n+1)$$
$$\vdots \qquad\qquad (2.6)$$
$$y(n+H_p) = A^{H_p} y(n) + A^{H_p-1} Bu(n) + ... + Bu(n+H_p - 1)$$

$H_p$ is the receding (prediction) horizon. Receding is one of the brilliant ideas of MPC. It allows the optimization problem to be solved in the open loop, which reduces the complexity. This is discussed in more details in the following paragraphs.

- The controller calculates a set of control inputs in order to optimize an objective function. Thus, it is considered as an optimal control method. The criterion to be minimized is:

$$J = \sum_{i=H_w}^{H_p} \left\| y(n+i) - s(n+i) \right\|_{W(i)}^2 + \sum_{i=0}^{H_u-1} \left\| \Delta u(n+i) \right\|_{R(i)}^2 \qquad (2.7)$$

where $\|x\|_W^2 = x^T W x$, $y$ is the predicted output of the system, $s$ is the set point, $\Delta u(n+i)$ is the difference between manipulated variable $u(n+i)$ and $u(n+i-1)$. $H_w$ is the starting step for penalizing output deviation from the set point, $H_u$ is the control horizon, i.e. number of steps that $u$ is allowed to be changed and $W(i)$ and $R(i)$ are positive weighting matrices to penalize deviations. The optimization problem is:

Determine $U = [u(n), u(n+1), \cdots, u(n+H_u -1)]^T$ subject to constraints, so that $J$ is minimized.

- The mentioned optimization is performed at each sample interval.
- One of the strengths of MPC is that it explicitly handles constraints. We consider the following constraints:

$$L_{\Delta u} \Delta U \le K_{\Delta u}, \quad L_u U \le K_u, \quad L_y Y \le K_y$$
$$\Delta U = [\Delta u(n), \Delta u(n+1), \cdots, \Delta u(n+H_u -1)]^T \qquad (2.8)$$
$$Y = [y(n+H_w), u(n+H_w +1), \cdots, u(n+H_p)]^T$$

where $L_{\Delta u}$, $K_{\Delta u}$, $L_u$, $K_u$, $L_y$ and $K_y$ are matrices with appropriate dimensions. Using the prediction model 2.6, all three inequalities in 2.8 can be rewritten in respect to $\Delta U$ :

$$L\Delta U \leq K \tag{2.9}$$

Detail calculations for obtaining $L$ and $K$ matrices can be found in [2]. The given performance index in 2.7 can be rewritten in general form of the quadratic function. Therefore problem described in the above sentences is a standard Quadratic Programming defined as follows:

$$\Delta U = \arg \min_{\Delta U} J = \Delta U^T H \Delta U - P\Delta U + J_0$$

$$subj\ to\ L\Delta U \leq K \tag{2.10}$$

Fortunately, the defined problem is convex [2] and therefore there is only one global minimum. It is also guaranteed that the minimum is eventually reached.

# 3   Sub-optimal Model Predictive Control Scheduler (SMPCS)

## 3.1  Notation

$Q$ : number of traffic classes in DiffServ architecture.

$n$ : number of packet arrivals since start of the last busy period, i.e. since the time at least one of the queues had packets to be served.

$k_i$ : The proportional guarantee on delay, i.e. the proportion of $\overline{D}_{i+1}$ to $\overline{D}_i$, which are average delays of the related queues.

$C$ : The output link capacity.

$D_i(n)$ :  Delay of the packet in the head of the $i^{th}$ queue.

$B_i(n)$ :  Total backlog packets in the $i^{th}$ queue.

$r_i(n)$ :   Service rate of the $i^{th}$ queue.

$\Delta r_i(n)$ : The difference between service rate of previous and present sample times.

## 3.2  Modeling

The system here consists of $Q$ queues, while the first queue has the highest priority. Packets arrive in their corresponding queues and wait until being serviced. Thus, there are $Q$ systems, in which the manipulated variable of each system is the service rate and the output variable is the delay of the corresponding queue.

The desired value (set point) for the delay of each class should be calculated in respect to delay of other classes. To obtain a uniform objective function, based on the average delays of different queues, results of Theorem 1 is applied to define a new weighted delay:

$$D_i^*(n) = (\prod_{k=1,k\neq i}^{Q} m_k)D_i(n), \quad 1 \leq i \leq Q \tag{3.1}$$

where $m_i = \prod_{j=1}^{i-1} k_j$ are dummy variables to define a uniform objective function for all traffic classes. $m_1$ is assumed to be one. The object is to make all $D_i^*(n)$'s equal, i.e. change service rates of traffic classes, to minimize absolute differences of $D_i^*(n)$'s.

The model is event based. An event is defined as an input or output to each of the queues. Equation 3.2 shows an ARX model i.e. the transfer function between $D_i^*(n)$ (output of the system) and $r_i(n)$ (service rate which is manipulated variable) in the specified model:

$$H_i(z) = \frac{\tilde{D}_i^*(z)}{R_i(z)} = \frac{\sum_{j=0}^{m} b_j z^{-j}}{1 + \sum_{j=1}^{p} a_j z^{-j}} \tag{3.2}$$

where $\tilde{D}_i^*(z)$ and $R_i(z)$ are $z$ transforms of $D_i^*(n)$ and $r_i(n)$, $m$ and $p$ number of zeros and poles of the system, and $b_i$ and $a_i$ time varying coefficients that are determined by a RLS algorithm. $Q$ traffic classes have $Q$ transfer functions as described in Equation 3.2. Thus, $H$ is a multi-input multi-output (MIMO) transfer function, which is a diagonal $Q$ by $Q$ matrix.

$$H(z) = diag[H_1, H_2, \cdots, H_Q] \tag{3.3}$$

The ARX model in fact calculates the output at the next step ($D_i(n)$) using the previous inputs and measured outputs of the system:

$$D_i^*(n) = \theta_i' \varphi_i(n-1) \tag{3.4}$$

where;

$$\varphi_i(n-1) = [-D_i^*(n-1), ..., -D_i^*(n-p), r_i(n), ..., r_i(n-m)]' \tag{3.5}$$

$$\theta_i = [a_1, ..., a_p, b_0, ..., b_m]' \tag{3.6}$$

The RLS algorithm is used for the model adaptation.

## 3.3 Algorithm

At the next step of the SMPCS, constraints on the states and control signals should be considered. The constraints are (3.7) and:

$$\ell_i < r_i < C - \sum_{j=1, j \neq i}^{Q} \ell_i \tag{3.7}$$

$\ell_i$ is the lower band for $i^{th}$ service rate which avoids starvation for the $i^{th}$ queue. The network administrator may decide the values of $\ell_i$. In our experiments, we considered the value of the lower band to be 2% of the output link capacity. It is obvious that $\ell_i$ should be set to zero during the steps that the related $i^{th}$ queue is empty.

One of the most important parts of the SMPCS design is how to map and tune MPC parameters to queue parameters. These parameters are:

- $H_p$ : This parameter is set to 1 to minimize the size of matrices.
- $H_u$ : This parameter is set to 1 to minimize the computational complexity.
- $W(i)$, $R(i)$, $s(n)$ and RLS initial parameters are chosen same as in 12.

To solve the quadratic programming, which is dominant part of the MPC for the calculations, we propose a sub-optimal search within the response space. As a matter of fact, the optimization method should search all of the response space to find the optimum answer. Here, we consider that in each step, each one of the manipulated variables may remain unchanged or change only one predetermined step. The step is a fraction of the input range, e.g. in our problem, it is $1/S$ of the output link capacity $C$ i.e. $C/S$. In this manner, all possible values for the $\Delta r$ are the following 19 states:

$$
\begin{aligned}
\Delta r \in \{ & (0,0,0,0),(0,0,\delta,-\delta),(0,\delta,0,-\delta),(0,0,-\delta,\delta), \\
& (0,-\delta,0,\delta),(0,\delta,-\delta,0),(0,-\delta,\delta,0),(\delta,0,-\delta,0), \\
& (-\delta,0,\delta,0),(\delta,0,0,-\delta),(-\delta,0,0,\delta),(\delta,-\delta,0,0), \\
& (-\delta,\delta,0,0),(\delta,\delta,-\delta,-\delta),(\delta,-\delta,\delta,-\delta),(\delta,-\delta,-\delta,\delta), \\
& (-\delta,\delta,\delta,-\delta),(-\delta,\delta,-\delta,\delta),(-\delta,-\delta,\delta,\delta)\}
\end{aligned}
\tag{3.8}
$$

where $\delta$ is $C/S$. It should be noticed that due to work conservativeness of the system, sum of each element of the mentioned set is zero.

Thus, the value of the object function $J$ has to be calculated only 19 times for each step. Fortunately, all these calculations can be done in parallel utilizing parallel processors or 19 hardware units. For our special case of MIMO system here, $H$ in 3.14 is a diagonal matrix. Therefore, $J$ can be calculated in 12 operations. If each operation takes 3 clock cycles to be performed, for a processor with 2 GHz clock, calculations will place in $12\times3/(2\times10^9)=18ns$. If we assume the minimum IP packet size, which is 40 bytes, SMPCS can run for a $40\times8/(18\times10^{-9})=17.7Gbps$ link capacity.

## 4  Computational Complexity

Form the algorithmic point of view; the computational complexity of the SMPCS is determined by some factors:

Number of queues: $Q$

Number of poles of the model: $p$

Control horizon: $H_u$

The complexity of the MPC is $O(n^3)$ [7], where $n$ is the number of manipulated variables multiplied by control horizon i.e. $QpH_u$. Our simulation results in [12] show that proportional delay error increases for higher number of poles. So $p$ is chosen to be 1. $H_u$ is also considered as 1. Therefore, the complexity of the algorithm is $O(Q^3)$. The dominant part of the complexity is the optimization. The optimization consists mainly of matrix operations that are matrix inversion and matrix–matrix multiplication. SMPCS performs the optimization part in 19 parallel operations. Although the number of parallel processors increases with the 3rd power of the number of traffic classes, but most of the time, the number of traffic classes does not take values more than 4 or 5.

## 5 Simulation Results

We simulated our algorithm in *ns-2* environment over networks with several nodes and background traffic to show its performance. We also compare the algorithm with our previous work (AMPCS) and with one of the latest scheduling algorithms (JoBS). To solve optimum quadratic programming, a Fortran program was attached to *ns-2* [14]. The network topology is shown in Figure 2. There are 4 source and 4 sink nodes These nodes can be considered as edge routers. The source number $n$ produces some traffic from the traffic class $n$. The sources produce two types of traffic. The first

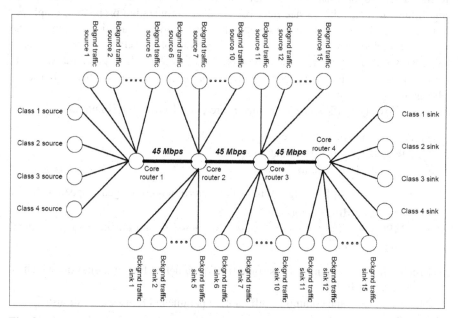

**Fig. 2.** The network topology for the simulation. There are 4 core routers and 4 traffic classes. Background traffic sources produce cross traffic.

type is greedy TCP traffic, we have used the infinite duration FTP of *ns-2* simulator for this type of traffic. The second one is high bursty Pareto ON-OFF traffic. The average traffic rate of this type is 2.5 Mbps, $\alpha$ is 1.9 and the average times for the idle and burst durations are $10\,ms$. So, the average traffic rate for the ON period is 5 Mbps. The number of Pareto ON-OFF user sources is 3.

All core routers 1, 2, 3 and 4 are equipped with SMPCS as scheduler, i.e. SMPCS is implemented as scheduler in all output ports. The core links capacity is 45 Mbps each and the link delay is $3\,ms$. All other links are 100 Mbps and have $1\,ms$ delay.

There are also 15 source and 15 sink nodes that produce background traffic to investigate if our algorithm can handle the background traffic from different sources. Nine nodes produce TCP traffic (1-3, 6-8 and 11-13) and others produce UDP traffic with the Pareto ON-OFF distribution. TCP traffic is of the ON-OFF type and is produced using the FTP application at the *ns-2* simulator. It periodically produces some packets and the window size for the TCP agent is set to 50. Packet sizes are 500 bytes and the number of produced packets during each interval is random uniformly distributed between 300 and 1000. The time interval of OFF period for ON-OFF FTP is considered to be $5\,ms$, so each background source produces 1.0 Mbps of traffic. Traffic in the simulation scenario is highly variable with considerable amount of background traffic.

The JoBS algorithm 1, which is one of the latest scheduling algorithms, is chosen to compare our algorithm to the previous works. JoBS allocates a guaranteed service rate to each class and dynamically shares the remaining bandwidth between classes in a way that proportional and absolute delay constraints are met. Although JoBS algorithm calculates the service rates based on an optimization problem, it employs a heuristic algorithm to reduce the complexity. In the JoBS heuristic algorithm, delay constraint violations are monitored every $N$ arrivals of packets (For maximum accuracy, we set $N$ equal to 1). When constraints violations are less than a threshold, the service rates isn't changed; otherwise, an adaptive proportional controller is used to change the service rates.

The following configuration parameters are considered in the simulation:

1) SMPCS and AMPCS are configured as follows:
- ARX model with 1 pole and no zeros.
- $H_p$ and $H_u$ are set to 1.
- $\lambda$ parameter of RLS algorithm, which determines convergence speed, is 0.99.
- $S$ for the SMPCS algorithm is 5.
- AMPCS uses the primal-dual method of Goldfarb and Idnani to solve the quadratic programming [13].

2) JoBS heuristic algorithm has some fixed parameters for configuration, which are the same as in 1.

There are also some common configuration parameters for two algorithms:
- All of the proportional delay values ($k_i$) are 2.
- 4 different levels of service (traffic class) are considered.

Our metrics for comparison are:

- Average proportional delay error: This error is calculated as:

$$\overline{E_i} = \sum_{n=1, D_{i,n}>0}^{N} \left|(\overline{D}_{i+1,n} / \overline{D}_{i,n}) - k_i\right| / N_{Busy} / k_i \tag{5.1}$$

where $\overline{D}_{i,n}$ is the averaged delay of the $i^{th}$ queue at the $n^{th}$ arrival over a window size of 1000, $N$ is total number of packet arrivals and $k_i$ is the desired value for the proportional delay of $(i+1)^{th}$ to $i^{th}$ queues. The $N_{busy}$ is the number of events that $i^{th}$ queue is not empty.

- Maximum delay ($D_{max,i}$): The maximum delay that a packet tolerates during the simulation time.

Table 1 shows the values of metrics for end to end proportional delay for three algorithms. Each metric has three or four columns. The three columns under proportional metric ($\overline{E_i}$) are related to $i = 1, 2, 3$, respectively and the four columns under maximum delay ($D_{max,i}$) are related to $i = 1, 2, 3, 4$, respectively. It can be seen that for all experiments, SMPCS controls the proportional delay better than JoBS algorithm, but it has lost some accuracy, because of sub-optimality of quadratic programming algorithm. Although it can not control the proportional delay as accurate as AMPCS, its accuracy is acceptable.

**Table 1.** The value of metrics for end to end proportional delay, which evaluate the three algorithms

| Algorithm | Average prop. delay error $\overline{E_i}$ (%) | | | Maximum delay $D_{Max,I}$ (ms) | | | |
|---|---|---|---|---|---|---|---|
| | $i=1$ | $i=2$ | $i=3$ | $i=1$ | $i=2$ | $i=3$ | $i=4$ |
| SMPCS | 19.0 | 21.0 | 29.6 | 131 | 71 | 216 | 294 |
| AMPCS | 12.9 | 17.1 | 24.9 | 131 | 67 | 129 | 256 |
| JoBS | 18.5 | 22.7 | 34.9 | 131 | 70 | 161 | 252 |

The maximum delay ($D_{max,i}$) parameter for the SMPCS is worse than the other two algorithms. This shows that the controller for some short intervals may allocate much less service rate than required for some classes, because of the limitation of the $\Delta r$, which is $C/S$. For example if service rate of some class is near zero and the optimal value for the next step is $C$ (i.e. all of the bandwidth should be allocated to that class), it takes at least $S$ steps that the service rate reaches the optimal value. During these $S$ steps, packets in the queue may receive much more delay than the other algorithms.

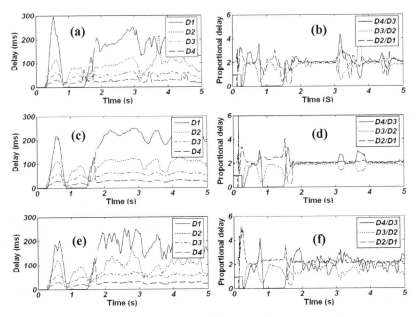

**Fig. 3.** End to end absolute and proportional delay during simulation of SMPCS (a, b), AMPCS (c, d) and JoBS (e, f) algorithms at core router 1. Both SMPCS and AMPCS give better results.

Figure 3 shows the end to end absolute and proportional delay for three algorithms. It can be seen that the variations of the proportional delay for the SMPCS algorithms is less than the variations for the JoBS algorithm and is more than the variations for the AMPCS algorithm. Although there are no negotiations between routers that use any of three algorithms, the end to end proportional delay is guaranteed precisely, i.e. if all routers of the network support any of SMPCS, AMPCS or JoBS algorithms and the administrator of the network configures the network correctly, the end to end proportional requirements can be guaranteed. The results in our previous work [12] show that the AMPCS is able to guarantee also absolute delay as well as proportional delay.

Utilizing a single processor for simulation, SMPCS was approximately 10 times faster than AMPCS.

## 6 Conclusion

Although MPC is a powerful control algorithm, there is not a clear and transparent rule to adjust its parameters for nonlinear systems. We performed some simulations to do the job by trial and error. The RLS algorithm has also run by initial parameters, which were selected in random and forgetting factor that was tuned again by trial and error.

One of the research objectives was to develop a computationally efficient MPC technology for very high sampling rate systems. Real-time implementation of model predictive control is currently only possible for relatively slow systems with low dynamic performance specification. This is due to the high computational load of the

MPC technology. In networking systems, faster dynamics need to be controlled within the constraints of the system. In this research some possibilities are discussed to improve the computational efficiency of model predictive control technology. More specifically, a special sub-optimal solution for the quadratic programming optimization is proposed. The SMPCS can be implemented for links with more than 10 Gbps data rate. The simulation results also show that SMPCS guarantees end to end proportional delay in DiffServ architecture.

# References

1. N. Christin, J. Liebeherr and T. F. Abdelzaher, "A quantitative assured forwarding service", Technical Report, University of Virginia, 2001.
2. J. M. Maciejowski, "Predictive control with constraints", Pearson education, 2002.
3. L. G. Bleris, M. V. Kothare, "Real time implementation of Model Predictive Control", In proceedings on American Control Conference, pp. 4166-4171, USA, June 2005.
4. A. Tyagunov, "High performance model predictive control for process industry", Ph. D. thesis, Eindhoven university of technology, 2004.
5. O. A. Palusinski, S. Vrudhula, L. Znamirowski and D. Humbert, "Process control for microreactors", Chemical Engineering Progress, pp. 60-66, 2001.
6. P. Tondel and A. Johansen, "Complexity reduction explicit linear model prediction control", IFAC world congress, Barcelona, 2002.
7. L. G. Bleris, J. Garcia, M. V. Kothare, M. G. Arnold, , "Towards embedded model predictive control for system-on-a-chip application, Seventh IFAC symposium on dynamics and control of process systems. Boston, 2004.
8. M. Fu, Z. Q. Luo and Y. Ye, "Approximation algorithms for quadratic programming", Journal of combinatorial optimization, Vol. 2, No. 1, pp. 29-50, 1998.
9. S. Blake, D. Black, M. Carlson, E. Davies, Z. Wang and W. Weiss, "An Architecture for Differentiated Services" RFC2475, Dec. 1998.
10. Dovrolis, "Proportional differentiated services for the Internet", PhD thesis, University of Wisconsin-Madison, December 2000.
11. M. Mahramian, H. Taheri, "A model predictive control approach for guaranteed proportional delay in DiffServ architecture", Proceedings of the 10th Asia-Pacific Conference on Communications, pp. 592-597, China, 2004.
12. M. Mahramian, H. Taheri, M. Haeri, "AMPCS: Adaptive Model Predictive Control Scheduler for guaranteed delay in DiffServ architecture", In proceedings of IEEE Conference on Control Applications, pp. 910-915, Canada, August 2005.
13. Goldfarb, A. Idnani, "A numerically stable method for solving strictly convex quadratic programs", Mathematical Programming, Vol. 27, pp. 1-33, 1983.
14. K. Schittkowski, "QL: A Fortran code for convex quadratic programming", http://www.klaus-schittkowski.de, ver 2.1, September 2004.

# Analysis and Simulation of the Signaling Protocols for the DiffServ Framework

A. Sayenko, O. Alanen, O. Karppinen, and T. Hämäläinen

University of Jyväskylä, MIT department
P.O. Box 35, Mattilaniemi 2 (Agora), Jyväskylä, Finland
{sayenko, opalanen, ollkarp, timoh}@cc.jyu.fi

**Abstract.** This paper considers signaling protocols for the DiffServ QoS framework. Originally, DiffServ had no standardized signaling solution, which resulted in the static configuration for the DiffServ domain. However, the dynamic allocation of resources within the domain allows to ensure the per-flow QoS guarantees and achieve better performance. At the moment, several signaling solutions for DiffServ framework are available. Thus, it is crucial to analyse these solutions and interconnections between them. In particular, the RSVP, aggregated RSVP, GIST, COPS, and SIP protocols will be considered. The simulation comprises several scenarios that present that the dynamic allocation of resource within the domain can provide considerably better results when compared to the static configuration.

## 1 Introduction

The Internet Engineering Task Force (IETF) has defined two frameworks to provide Quality-of-Service (QoS) in the Internet: the Integrated Services (IntServ) [1] and Differentiated Services (DiffServ) [2]. While the IntServ QoS framework relies upon the Resources reSerVation Protocol (RSVP) [3] to provide the strict QoS guarantees, the DiffServ approach to service management is more approximate in the nature of its outcome. Though the DiffServ scales good for the large networks, there is no requirement for the network to inform the application that a request cannot be admitted. It is left to the application to determine if the service has been delivered or not. It has been assumed that the configuration for the DiffServ domain should rely upon the Service Level Agreement (SLA) negotiated between a customer and a provider. However, it is often the case that the outcome of such an approach is the static configuration, which does not take account of the fact that the customer applications do not require bandwidth resources all the time.

What appears to be useful within the DiffServ framework is the resource availability signaling. Having a possibility to request resources dynamically, a customer may reduce his expenses and a provider may allocate resources more efficiently. For instance, if an organization pays for the bandwidth, then the signaling will allow to reserve certain amount of bandwidth resources depending on the daytime and/or day of the week. Indeed, the organization would need only

Y. Koucheryavy, J. Harju, and V.B. Iversen (Eds.): NEW2AN 2006, LNCS 4003, pp. 566–579, 2006.

minimal bandwidth in the night or during the holiday. Though these conditions can be a part of the SLA, it is more flexible to let the organization a possibility to decide how much bandwidth it needs and when. At the same time, if there is a bandwidth and delay critical application for which the *usage-based* or even *time-based* pricing is used, then the signaling will allow to reserve resources when they are actually needed. From a provider's point of view, the QoS signaling allows to achieve more optimal functioning due to the allocation of free bandwidth to other customers.

Another impetus for implementing the signaling in the core IP networks is the rapid development of the wireless networks. The architecture of the most local wired networks, such as Ethernet, does not assume the presence of any local coordinator. It means that either a customer application or some node should signal the core network to reserve resources. At the same time, wireless environments, such as WLAN and WiMAX, assume the presence of the access point or the base station, which is connected to the wired medium. This access point can act, either implicitly or explicitly, as the local coordinator capable of tracking the number of active wireless stations and their QoS requirements. Based on this information, the access point can signal the core network and reserve necessary resources. Even if the access point does not possess any information about the number of active stations and their QoS requirements, then it can reserve resources on behalf of the whole cell thus ensuring some level of QoS.

The aim of this research works is to show the need for the signaling within the DiffServ QoS framework. At the moment, there is no standardized signaling solution adopted by all the major manufactures. Thus, several solutions proposed by the IETF and other researchers will be considered and some of them will be simulated. It is also important to mention that the main focus is on the protocols for the *intra-domain* resource management. We will show that it is possible, and actually necessary, to decouple the end-to-end signaling from the intra-domain signaling. However, we will also mention the end-to-end protocols since the intra-domain resource management depends on the QoS requirements of the end-user applications.

The rest of this article is organized as follows. Section II considers the major signaling solutions for the DiffServ QoS framework. Section III presents simulation scenarios to analyse the efficiency of several signaling approaches. Also, this section provides an analysis for other solutions. The final section summarizes the article and draws future research directions.

## 2    Signaling Protocols

### 2.1    RSVP

One approach for the DiffServ signaling is to combine the IntServ and DiffServ architectures into an end-to-end model, using the IntServ as the architecture that allows applications to interact with the network, and DiffServ as the architecture to forward data [4]. Based on the interaction between the IntServ and

DiffServ parts, it is possible to present three major scenarios [5]: a) static allocation in the DiffServ domain, b) dynamic allocation by RSVP in the DiffServ domain, and c) dynamic allocation by other means. The first case assumes that the ingress DiffServ router, which acts as the RSVP-capable node, performs the admission control for the statically allocated bandwidth along some data path within the domain. However, the ingress routers cannot perform efficiently the admission control for the whole DiffServ domain since the router does not know how many active flows at other edge routers are and, as a result, the amount of free bandwidth available in the core. The second case assumes that some of the DiffServ core routers participate in the end-to-end RSVP signaling. Such an approach puts a significant burden on the core router because the number of the reservation states can be huge. Finally, the last case relies upon the intra-domain signaling mechanisms that are out of the scope of the RSVP protocol. We will consider several solutions in the subsequent sections.

## 2.2    Aggregated RSVP

As mentioned above, the core routers cannot participate in the end-to-end RSVP signaling due to the huge number of reservations, each of which requires some amount of message exchange, computation, and memory resources. Thus, a possible solution is to aggregate reservations to a more manageable level to reduce the number of the reservation states. However, a key problem in the design of the RSVP protocol is that it cannot aggregate individual reserved sessions into a common class. One of the reasons is that there is no proper way to classify the aggregate at the forwarding level. The original design of the RSVP protocol assumed that the routers use the multi-field classifiers to select packets belonging to a particular flow from the incoming data. With the development of DiffServ, packets of a particular class can be marked with the required DiffServ codepoint (DSCP) value and so classified.

The aggregated RSVP [6] describes a way to aggregate the individual RSVP reservations. Figure 1 presents the common actions of the DiffServ routers when a new end-to-end reservation state is established. The end-to-end PATH message is sent transparently through the DiffServ domain. When the end-to-end RESV message arrives at the egress router, it maps the flow into one of the aggregates. Then, the router sends the aggregated RESV message so that the core routers can adjust bandwidth for the required aggregate. Having received a confirmation from the ingress router, the egress router adds the DCLASS object [7] with the DSCP value into the original end-to-end RESV message, so that the ingress router can set up the classifier and mapper properly. Note that this simple scenario does not consider how the initial aggregated state is established.

It is interesting to note that in the case of the aggregated RSVP the edge *egress* router is responsible for mapping a flow into one of the DiffServ aggregates and adjusting the aggregated reservation. Such a behaviour is dictated by the RSVP protocol, in which the receiver is responsible for making a reservation. At the same time, while aggregating the RSVP states over the MPLS/DiffServ tunnels [8], all the decisions are made by the ingress router.

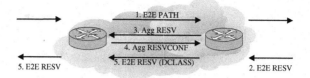

**Fig. 1.** The aggregated RSVP in the DiffServ domain

It is also worth mentioning that the aggregated RSVP reservations can be established even if the RSVP protocol is not used end-to-end. Other signaling protocols, which are recognized by the edge routers, can trigger and alter the aggregated reservation states. In the absence of the end-to-end signaling, a provider may set the aggregated reservations based on some static rules.

## 2.3   GIST

To overcome some limitations of the RSVP protocol [9], the IETF has established the Next Steps In Signaling (NSIS) group, purpose of which is to develop the General Internet Signaling Transport (GIST) protocol [10]. The requirements for such a protocol were outlined in [11]. GIST is a general signaling protocol that is capable of carrying the QoS data in various scenarios, such end-to-end, end-to-edge, and edge-to-edge, supporting the receiver- and sender-initiated reservations, as well as resource querying. Unlike RSVP, GIST provides several layers of abstraction: transport, signaling [12], and signaling for a particular QoS model. It makes GIST considerably more flexible than RSVP. Since this article analyses the signaling protocols for the DiffServ framework, we will consider the case when the GIST protocol is used for the intra-domain, i.e. edge-to-edge, signaling.

The usage of the GIST protocol for the DiffServ QoS framework is described in [13]. Conceptually, it is quite similar to the aggregated RSVP. Only the edge routers keep the per-flow state information, while the core routers act as the reduced state nodes that keep the QoS requirements at the aggregate level. Every time a new flow appears or an existent one stops functioning, the edge router updates the resource reservations within the domain.

Figure 2 presents the common usage of the GIST protocol in the DiffServ domain. We assume that customer applications use the RSVP protocol end-to-end, however, other protocols can be also deployed. As the RESV message arrives at the ingress edge router, the latter maps a flow to one of the aggregates and sends the REQUEST message along the data path to inform the network about new requirements. The REQUEST message contains the QSpec object [12] that specifies the amount of resources to reserve. If the router receives a positive response, then the RESV message is sent towards the sender. Otherwise, the RESVERROR message can be generated and sent to the receiver.

The difference between this solution and the aggregated RSVP is that the edge *ingress* router makes the mapping decision and reserves resources within the domain by using the GIST protocol. Furthermore, the exchange of the signaling messages is somewhat simpler when compared to the aggregated RSVP.

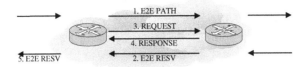

**Fig. 2.** The GIST protocol in the DiffServ domain

## 2.4   COPS

Unlike RSVP and GIST, the Common Open Policy Protocol (COPS) [14] is used
for the vertical signaling. Its framework assumes the presence of two types of
nodes: the Policy Decision Point (PDP), which is eligible for making the policy
and/or admission control decisions, and the Policy Enforcement Point (PEP),
which asks the PDP for a decision and enforces it. Since the PDP is responsible
for making admission control decisions, its functionality is very similar to the
Bandwidth Broker (BB). Hence, we will refer to this entity as PDP/BB, though
physically it can comprise two different systems.

At the moment, the IETF has standardized several client types for the COPS
protocol: COPS-RSVP [15] and COPS-PR [16]. The COPS-RSVP client defines
how the RSVP messages are sent to the PDP/BB over the COPS protocol. This
model of interaction is often referred to as *outsourcing* because the policy and
admission control decisions are not made locally, but outsourced to the PDP/BB.
The COPS-RSVP client presents a particular interest for the DiffServ networks.
If the edge routers outsource the RSVP messages to the PDP/BB, then the
latter has the information about all the reservation states within the domain
and can allocate bandwidth in the core routers appropriately. Then, the COPS-
PR client type is used to configure the DiffServ routers. This model is referred
to as *provisioning* because the PDP/BB provides or pushes the configuration to
the routers. The configuration data is sent in the form of the Policy Information
Base (PIB) that is standardized for the DiffServ routers in [17].

Figure 3 presents the usage of the COPS protocol and the standardized clients
within the DiffServ domain. As the RESV message arrives to the ingress edge
router, it is sent to the PDP/BB by the means of the COPS-RSVP client. The
PDP/BB makes the admission control decisions and if there is enough bandwidth, then it answers positively to the COPS-RSVP client. After that, the edge
router sends the initial RESV message to the sender. At the same time, the
PDP/BB can update configuration of the routers in the domain, including the
edge router, by using the COPS-PR client.

Among other client types proposed for the COPS protocol, it bears mentioning
the "Outsourcing DiffServ Resource Allocation" (ODRA) [18] client. Like COPS-
RSVP, this client type works at the edge routers and outsources the resource
allocation requests to the PDP/BB. However, the COPS-ODRA client may be
triggered by any end-to-end signaling protocol, which provides a unified way for
managing resources within the domain. Furthermore, the COPS-ODRA client

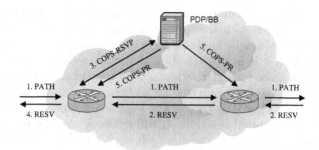

**Fig. 3.** COPS-RSVP and COPS-PR clients in the DiffServ domain

assumes that the edge router maps the incoming flow to one of the DiffServ aggregates, which allows to send a bandwidth request for an aggregate, not for a distinctive flow like the COPS-RSVP client does. It reduces greatly the number of request states within the PDP/BB because only aggregate state information is kept allowing for higher scalability.

It is also worth mentioning the COPS-SIP client [19] for the Session Initiation Protocol (SIP) [20]. However, it may be useful only for the policy control decisions. As the incoming SIP flow is mapped to one of the DiffServ aggregates at the edge router, it is better to use the COPS-ODRA client type to reserve the necessary amount of bandwidth resources or send the GIST REQUEST message along the data path.

## 3   Simulation

The simulation of the signaling protocols for the DiffServ QoS framework was done in the NS-2 simulator. For these purposes, the RSVP and COPS protocols were implemented.[1] To provide an exhaustive analysis, the following simulation scenarios are considered:

1. Static provisioning of the DiffServ domain
2. Dynamic provisioning of the DiffServ domain by the means of RSVP
3. Dynamic provisioning of the DiffServ domain by the means of the RSVP and COPS protocols

We will also show that other signaling protocols and interactions between them fit into these basic scenarios.

Figure 4 presents the simulation environment. It consists of two client networks, two edge ingress routers, one core router, and one egress edge router. The edge ingress routers implement the Weighted Fair Queuing (WFQ) scheduling

---

[1] The implementation of the RSVP protocol is available at the following address: http://www.cc.jyu.fi/~sayenko/src/ns-2.28-rsvp2.diff.gz. The implementation of the COPS protocol for NS-2 is available at http://www.cc.jyu.fi/~sayenko/src/ns2cops-0.7.tar.gz

discipline, while the core router relies upon the simpler Weighted Round Robin (WRR) scheduler. Such a choice corresponds to the DiffServ framework that mandates the usage of simple mechanisms in the core by pushing the complexity into the edge routers. It is worth noting that the presented environment has several bottleneck links: each ingress router is the bottleneck for the appropriate client network, and the core router is the bottleneck for all traffic coming from the ingress routers. Thus, the task of the signaling protocols is to inform the network about the required resources. It will enable a provider to configure the routers in such a way, that all the flow QoS requirements are ensured. To find the configuration for the given scheduler type based on the QoS requirements, we will use algorithms presented in [21].

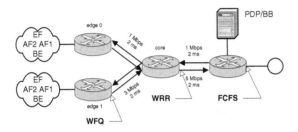

**Fig. 4.** Simulation environment

The last simulation scenario also uses the PDP/BB that is attached to the *egress* edge router. The reason it is placed there is that we want to test the efficiency of the signaling protocols when a) it takes some time before the signaling packets reach the PDP/BB and b) the signaling packets have to compete with the user packets for the available resources.

Each client network consists of customer applications that belong to different aggregates, details of which are presented in Table 1. The EF aggregate corresponds to the real-time audio services. It is simulated by the G.711 audio codec that transmits data over the Real Time Protocol (RTP). The AF2 and AF1 aggregates represent general purpose services. They are simulated by the FTP-like applications that generate bulk data transferred over the TCP protocol. The reason for choosing such a type of application is that it always tries to send data thus behaving very aggressively. The bandwidth in Table 1 specifies the minimum rate that a flow must obtain within its aggregate. The BE aggregate represents non-critical user data, such as the mail system, that has no particular requirements at all. Since the BE flows do not have any QoS requirements, the bandwidth of 100 Kbps is reserved for the whole aggregate and all the data streams compete for the available resources. The Telnet application is chosen to generate general purpose BE data that is carried over the TCP protocol.

The maximum flow number in Table 1 specifies the maximum number of flows within an aggregate. However, during the simulation, the number of active flows varies all the time thus creating the variable load on the network and variable requirements for the traffic aggregates.

**Table 1.** Parameters of aggregates

| Aggregate | Max flows | Price for 1MB | Band. (Kbps) | P.size (bytes) |
|-----------|-----------|---------------|--------------|----------------|
| EF  | 10 | 16.8 | 78 | 300 |
| AF2 | 15 | 8.4  | 50 | 840 |
| AF1 | 25 | 4.2  | 10 | 640 |
| BE  | 40 | –    | –  | 340 |

It is worth mentioning that there is also another aggregate, purpose of which is to carry the intra- and inter-domain signaling data. Though, some documents propose to carry the signaling data from the customer applications within an aggregate the application belongs to [22], it is not efficient since the signaling packets can be delayed in queues of the bottleneck routers. As a result, other routers do not receive notifications about required or freed resources immediately. Furthermore, due to the finite size of buffers, the signaling packets can be even dropped. Thus, we will introduce a special aggregate, which will be referred to as SIG.

Each simulation scenario will be evaluated in terms of parameters, such as obtained QoS guarantees and the amount of the signaling information. We will also use the total revenue to analyse the efficiency of the functioning of the DiffServ domain from the economical point of view. The total revenue will be calculated by using a simple usage-based pricing, i.e. the amount of sent data times the price for one unit of data. To a provide a fair comparison of results, each scenario will last exactly 90 seconds. Furthermore, the same behaviour of flows will be submitted in each case.

### 3.1   Static Provisioning

This simulation scenario considers the case when all the DiffServ routers rely upon the static configuration. Table 2 presents the static weight values assigned to the schedulers at the edge and the core routers. Since the network does not participate in the signaling, there is no need for the customer applications to use any signaling protocol. Furthermore, since no signaling data is sent during the simulation, resources for the SIG aggregate are not allocated.

**Table 2.** Static configuration for schedulers

| Router | Discipline | EF | AF2 | AF1 | BE |
|--------|------------|------|-------|-------|------|
| edge | WFQ | 0.45 | 0.275 | 0.175 | 0.05 |
| core | WRR | 5 | 2 | 1 | 1 |

Table 3 provides the results for this simulation case that include the amount of sent data and the mean per-flow rate within each aggregate, as well as the total revenue. These results were gathered at the egress router, i.e. when packets leave the domain. As follows from the results, if a provider has enough bandwidth resources, then the static configuration is capable of ensuring the QoS guarantees.

By comparing per-flow rate in Table 3 with the requirements in Table 1, it is possible to arrive at the conclusion that all the requirements are met.

**Table 3.** Simulation results

| Quantity | EF | AF2 | AF1 | BE |
|---|---|---|---|---|
| Data (MB) | 13.00 | 28.97 | 11.67 | 1.93 |
| Rate (Kbps) | 76.89 | 99.02 | 28.40 | 2.64 |
| Total revenue | | 510.087 | | |

However, it is significant to note that the provision of the QoS guarantees in this scenario was possible only because we knew the maximum number of data flows within each aggregate and their behaviour. Based on this, we could find such a static configuration for the routers, that all the aggregates have enough bandwidth regardless of the number of active flows. In the real life, the task of finding the static configuration for all the routers along the data path is much more complicated and sometimes even impossible. Furthermore, if a provider has scarce bandwidth resources, then the static configuration fails to ensure any QoS guarantees.

### 3.2   Dynamic Provisioning (RSVP)

This simulation scenario considers the case when the customer applications use the RSVP protocol to signal the network about the required resources. However, only the edge ingress routers participate in the RSVP signaling. Such an approach allows the ingress routers to allocate resources on demand based on the number of active flows and their QoS requirements. Every time the RESV message is supposed to leave the DiffServ domain, the correspondent edge router maps the flow to one of the DiffServ aggregates based on the FLOWSPEC object stored in the RESV message. Then, a new configuration for the scheduler is calculated based on the new bandwidth requirements of the aggregates.

The core and egress routers transmit the RSVP packets as ordinary IP datagrams without analysing them. Their configuration remains static because they do not participate in the RSVP signaling. Table 4 presents the configuration for the WRR scheduler at the core router. It is noticeable that the core router also allocates some resources for the SIG aggregate. Furthermore, each edge ingress router reserves the bandwidth of 56 Kbps for this aggregate.

**Table 4.** Static configuration for the core router

| Router | Discipline | EF | AF2 | AF1 | BE | SIG |
|---|---|---|---|---|---|---|
| core | WRR | 5 | 2 | 1 | 1 | 1 |

Table 5 presents the results for this simulation case. As in the previous simulation scenario, they were gathered at the egress router. When compared to the results presented in Table 3, it is possible to notice that they are almost the same. However, the total revenue is a little bit less. It is explained by the fact that some bandwidth resources were taken from the user aggregates and allocated for the signaling data. At the same time, the static configuration of the core router does not allow to utilize the dynamic allocation of resources at the edge routers. Since the core router is the bottleneck one, the resulting allocation of resources depends more on its configuration.

**Table 5.** Simulation results

| Quantity | EF | AF2 | AF1 | BE | SIG |
|---|---|---|---|---|---|
| Data (MB) | 12.81 | 29.13 | 11.70 | 1.97 | 0.09 |
| Rate (Kbps) | 76.87 | 98.53 | 29.86 | 2.67 | 8.78 |
| Total revenue | | | 509.13 | | – |

Based on these results, it is possible to arrive at the conclusion that it is not enough that only the edge routers participate in signaling. To achieve flexible and efficient allocation of resources within the DiffServ domain, the core routers should also allocate resources in accordance with the current bandwidth requirements for each aggregate. Since it is not likely that the core routers will participate in the per-flow end-to-end signaling, other protocols are necessary that will carry the QoS requirements between the edge and core routers.

At the same time, such a signaling scenario is viable if a provider allocates some fixed bandwidth within the domain for packets coming from each edge router. For instance, the MPLS trunk can be created for these purposes. Then, the edge routers can perform the admission control for the whole trunk. However, it is possible if a provider has enough bandwidth in the core.

### 3.3 Dynamic Provisioning (RSVP and COPS)

This simulation case analyses the signaling scenario presented in section 2.4. The customer applications use the RSVP protocol to signal the network about the required resources. The edge routers act as the COPS-RSVP clients and *outsource* the RSVP messages to the PDP that implements the functionality of the BB. The edge routers outsource only the RSVP RESV messages, while the PATH messages are processed locally. Since the PDP/BB possesses information about the QoS requirements at the edge routers, it can update bandwidth requirements at the core router, by the means of the COPS-PR client. The edge routers also run the COPS-PR client so that the PDP/BB can configure them.

This simulation case presents the completely adaptive DiffServ domain. Neither edge routers nor the core router rely upon the static configuration. Instead, the routers update the scheduler configuration based on the current QoS requirements of the customer applications.

Figure 5 presents the sequence of signaling actions (the size of packets *including* the network/transport layer headers is given in brackets). Every time the edge router receives the RESV message, it delegates the admission control decision to the PDP/BB. Having received a positive response from the PDP/BB, the edge router forwards the original RESV message to the sender. At the same time, the PDP/BB constructs and sends a new configuration for the core router and edge routers over the COPS-PR client. When the PATHTEAR message arrives at the ingress edge router, the latter deletes the appropriate state at the PDP/BB that, in turn, updates the configuration of all the routers.

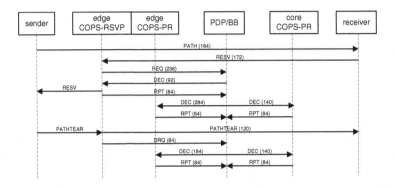

**Fig. 5.** Sequence of the signaling actions

It is important to note that the PDP/BB does not send the whole configuration to the core and edge routers, but rather updates only the necessary entries within the routers' PIB. That's why there are different decisions for the core and edge routers. In the case of the core router, when a flow appears or disappears, the PDP/BB has to adjust the bandwidth requirements of an aggregate this flow belongs to, while the configuration of other aggregates and components remains the same. As a result, the decision object includes only one `dsMinRateEntry` from the DiffServ PIB. In the case of the edge router, the PDP/BB has to set up also the classifier so that the incoming data is mapped to the correct aggregate. Thus, the decision object for the edge router consists of the following PIB entries: `dsMinRateEntry`, `dsClfrElementEntry`, and `FrwkIpFilterEntry`.

**Table 6.** Simulation results

| Quantity | EF | AF2 | AF1 | BE | SIG |
|---|---|---|---|---|---|
| Data (MB) | 12.80 | 35.15 | 5.66 | 2.00 | 0.09 |
| Rate (Kbps) | 76.88 | 118.07 | 15.13 | 2.70 | 9.64 |
| Total revenue | | 531.50 | | | − |

Table 6 presents the results for this simulation scenario. As follows from the results, considerably better resource allocation can be achieved. All the QoS requirements are ensured and the total revenue is bigger when compared to the previous simulation scenarios.

The amount of the signaling data presented in Table 6 contains only the RSVP packets because the COPS packets do not leave the DiffServ domain. Thus, Table 7 presents the amount of the COPS signaling data sent to and from the PDP/BB. It is noticeable that the overall amount of the signaling data is larger when compared to the previous simulation case. However, since the COPS-RSVP clients carry the QoS requirements to the PDP/BB that configures the core router, more efficient resource allocation is achieved.

**Table 7.** Amount of the COPS signaling data (in KB)

| to PDP/BB from | | | from PDP/BB to | | |
|---|---|---|---|---|---|
| edge 0 | edge 1 | core | edge 0 | edge 1 | core |
| 49.8 | 52.1 | 69.4 | 59 | 61.9 | 57.3 |
| | 171.3 | | | 178.2 | |

It is possible to reduce the amount of signaling information generated by the COPS-RSVP clients by deploying a new client type, such as COPS-ODRA, to the edge ingress routers. According to [18], the typical COPS-ODRA REQ message size is 128 bytes, while the minimum size of the REQ message of the COPS-RSVP client is 236 bytes (see Fig. 5). Since the COPS-RSVP and COPS-ODRA requests are triggered by the same events, it is quite easy to approximate that the COPS-ODRA client can reduce the amount of the signaling information sent *to* the PDP/BB almost by two times. The free bandwidth can be allotted to the user aggregates thus increasing the total revenue. Furthermore, the COPS-ODRA client eliminates the need to run the COPS-PR client at the edge router since the mapping decision is done locally. It reduces the amount of the signaling data carried *from* the PDP/BB to the edge router.

## 3.4   Other Scenarios

The simulation results from the considered scenarios can assist in analysing the behaviour of the DiffServ domain when other signaling protocols are used. For instance, if the DiffServ domain uses the GIST protocol to signal the core routers about the necessary resources, then the results will be similar to those obtained in the case of the dynamic provisioning with the RSVP and COPS protocols. However, it is anticipated that the results will better because the GIST RESERVE message is smaller than the RSVP RESV message encapsulated in the COPS-RSVP REQ message. Furthermore, no PDP/BB is necessary. As a result, packets will be destined to the egress router and the core routers will not have to wait for the PDP/BB decision, but rather participate in the edge-to-edge GIST signaling.

Since the aggregated RSVP exchanges QoS data in the horizontal plane, it is anticipated that it will provide results quite similar to GIST. However, since the GIST messages are smaller when compared to the aggregated RSVP, the GIST protocol should provide a little bit better results. In the similar way we can show that if the SIP protocol, or even GIST, communicates the QoS requirements end-to-end, then the results will be similar to those scenarios when the RSVP protocol is used in the horizontal plane. The small differences in results will be explained by different packet sizes of the RSVP, SIP, and GIST protocols.

## 4    Conclusions

In this paper, we have presented several signaling solutions for the DiffServ QoS framework. Having analysed them, it is possible to arrive to the important conclusion. The current state of the signaling protocols allows to decouple the end-to-end signaling from the core signaling. While customers can choose protocols to signal the provider about the required resources, a provider is free to choose mechanisms to allocate resources within the domain, regardless of mechanisms used in domains of other providers. The only requirement is that the edge routers must recognize the required end-to-end protocols. It is a significant breakthrough when compared to the IntServ framework, which mandates the usage of the RSVP protocol both end-to-end and in the core routers.

According to the simulation results, the dynamic bandwidth management provides better results when compared to the static one. Though the static configuration can ensure the QoS guarantees, a provider has to overprovide the DiffServ domain to meet all the QoS requirements. The simulation results have also revealed that it is not enough that only the edge routers use the signaling protocols. The best results are achieved when all the DiffServ routers, including the core ones, participate in the signaling. Resource are allocated dynamically in the domain thus ensuring the QoS requirements of all the data flows.

Further investigation is necessary to decide which solution provides the best results. A solution that is based on COPS and PDP/BB requires that the core routers run only minimal software. However, the response from the PDP/BB takes some time, which may not be acceptable for large networks. The solutions based on GIST and the aggregated RSVP do not have this problem, but the core routers have to take part in the edge-to-edge signaling. Thus, our future works aim to evaluate the efficiency of the GIST protocol and compare it to the aggregated RSVP and its extensions for the traffic engineering networks.

## References

1. Braden, R., Clark, D., Shenker, S.: Integrated Services in the Internet Architecture: an Overview. IETF RFC 1633 (1994)
2. Blake, S., Black, D., Carlson, M., Davies, E., Wang, Z., Weiss, W.: An Architecture for Differentiated Services. IETF RFC 2475 (1998)
3. Braden, R., Zhang, L., Berson, S., Herzog, S., Jasmin, S.: Resource reservation protocol (RSVP) – version 1 functional specification. IETF RFC 2205 (1997)
4. Bernet, Y.: The complementary roles of RSVP and Differentiated Services in the full-service QoS network. IEEE Communications 38 (2000) 154–162
5. Bernet, Y., Ford, P., Yavatkar, R., Baker, F., Zhang, L., Speer, M., Braden, R., Davie, B., Wroclawski, J., Felstaine, E.: A framework for Integrated Services operation over DiffServ networks. IETF RFC 2998 (2000)

6. Baker, F., Iturralde, C., Faucher, F.L., Davie, B.: Aggregation of RSVP for IPv4 and IPv6 reservations. IETF RFC 3175 (2001)
7. Bernet, Y.: Format of the RSVP DCLASS object. IETF RFC 2996 (2000)
8. Faucheur, F.L., Dibiasio, M., Davie, B., Davenport, M., Christou, C., Hamilton, B., Ash, J., Goode, B.: Aggregation of RSVP reservations over MPLS TE/DS-TE tunnels. INTERNET DRAFT (work in progress) (2006) draft-ietf-tsvwg-rsvp-dste-01.txt.
9. Manner, J., Fu, X.: Analysis of existing quality-of-service signaling protocols. IETF RFC 4094 (2005)
10. Schulzrinne, H., Hancock, R.: GIST: General Internet signalling transport. INTERNET DRAFT (work in progress) (2005) draft-ietf-nsis-ntlp-08.txt.
11. Brunner, M.: Requirements for signaling protocols. IETF RFC 3726 (2004)
12. Manner, J., Karagiannis, G., McDonald, A., den Bosch, S.V.: NSLP for Quality-of-Service signalling. INTERNET DRAFT (2006) draft-ietf-nsis-qos-nslp-09.txt.
13. Bader, A., Westberg, L., Karagiannis, G., Kappler, C., Phelan, T.: RMD-QOSM - the resource management in DiffServ QoS model. INTERNET DRAFT (work in progress) (2006) draft-ietf-nsis-rmd-05.txt.
14. Durham, D., Boyle, J., Cohen, R., Herzog, S., Rajan, R., Sastry, A.: The COPS (common open policy service) protocol. IETF RFC 2748 (2000)
15. Herzog, S., Boyle, J., Cohen, R., Durham, D., Rajan, R., Satry, A.: COPS usage for RSVP. IETF RFC 2749 (2000)
16. Chan, K., Seligson, J., Durham, D., Gai, S., McCloghrie, K., Herzog, S., Reichmeyer, F., Yavatkar, R., Smith, A.: COPS usage for policy provisioning (COPS-PR). IETF RFC 3084 (2001)
17. Chan, K., Sahita, R., Hahn, S., McCloghrie, K.: Differentiated services Quality of Service policy information base. IETF RFC 3317 (2003)
18. Salsano, S.: COPS usage for outsourcing DiffServ resource allocation. INTERNET DRAFT (expired) (2001)
19. Gross, G., Rawlins, D., Sinnreich, H., Thomas, S.: COPS usage for SIP. INTERNET DRAFT (expired) (2001)
20. Rosenberg, J., Schulzrinne, H., Camarillo, G., Johnston, A., Peterson, J., Sparks, R., Handley, M., Schooler, E.: SIP: Session initiation protocol. IETF RFC 3261 (2002)
21. Sayenko, A.: Adaptive scheduling for the QoS supported networks. PhD thesis, University of Jyväskylä, Finland (2005)
22. Babiartz, J., Chan, K., Baker, F.: Configuration guidelines for DiffServ service classes. INTERNET DRAFT (work in progress) (2006) draft-ietf-tsvwg-diffserv-classes-02.txt.

# Author Index

# Lecture Notes in Computer Science

For information about Vols. 1–3904

please contact your bookseller or Springer

Vol. 3956: G. Barthe, B. Gregoire, M. Huisman, J.-L. Lanet (Eds.), Construction and Analysis of Safe, Secure, and Interoperable Smart Devices. IX, 175 pages. 2006.

Vol. 3955: G. Antoniou, G. Potamias, C. Spyropoulos, D. Plexousakis (Eds.), Advances in Artificial Intelligence. XVII, 611 pages. 2006. (Sublibrary LNAI).

Vol. 3954: A. Leonardis, H. Bischof, A. Pinz (Eds.), Computer Vision – ECCV 2006, Part IV. XVII, 613 pages. 2006.

Vol. 3953: A. Leonardis, H. Bischof, A. Pinz (Eds.), Computer Vision – ECCV 2006, Part III. XVII, 649 pages. 2006.

Vol. 3952: A. Leonardis, H. Bischof, A. Pinz (Eds.), Computer Vision – ECCV 2006, Part II. XVII, 661 pages. 2006.

Vol. 3951: A. Leonardis, H. Bischof, A. Pinz (Eds.), Computer Vision – ECCV 2006, Part I. XXXV, 639 pages. 2006.

Vol. 3950: J.P. Müller, F. Zambonelli (Eds.), Agent-Oriented Software Engineering VI. XVI, 249 pages. 2006.

Vol. 3947: Y.-C. Chung, J.E. Moreira (Eds.), Advances in Grid and Pervasive Computing. XXI, 667 pages. 2006.

Vol. 3946: T.R. Roth-Berghofer, S. Schulz, D.B. Leake (Eds.), Modeling and Retrieval of Context. XI, 149 pages. 2006. (Sublibrary LNAI).

Vol. 3945: M. Hagiya, P. Wadler (Eds.), Functional and Logic Programming. X, 295 pages. 2006.

Vol. 3944: J. Quiñonero-Candela, I. Dagan, B. Magnini, F. d'Alché-Buc (Eds.), Machine Learning Challenges. XIII, 462 pages. 2006. (Sublibrary LNAI).

Vol. 3943: N. Guelfi, A. Savidis (Eds.), Rapid Integration of Software Engineering Techniques. X, 289 pages. 2006.

Vol. 3942: Z. Pan, R. Aylett, H. Diener, X. Jin, S. Göbel, L. Li (Eds.), Technologies for E-Learning and Digital Entertainment. XXV, 1396 pages. 2006.

Vol. 3941: S.W. Gilroy, M.D. Harrison (Eds.), Interactive Systems. XI, 267 pages. 2006.

Vol. 3940: C. Saunders, M. Grobelnik, S. Gunn, J. Shawe-Taylor (Eds.), Subspace, Latent Structure and Feature Selection. X, 209 pages. 2006.

Vol. 3939: C. Priami, L. Cardelli, S. Emmott (Eds.), Transactions on Computational Systems Biology IV. VII, 141 pages. 2006. (Sublibrary LNBI).

Vol. 3936: M. Lalmas, A. MacFarlane, S. Rüger, A. Tombros, T. Tsikrika, A. Yavlinsky (Eds.), Advances in Information Retrieval. XIX, 584 pages. 2006.

Vol. 3935: D. Won, S. Kim (Eds.), Information Security and Cryptology - ICISC 2005. XIV, 458 pages. 2006.

Vol. 3934: J.A. Clark, R.F. Paige, F.A. C. Polack, P.J. Brooke (Eds.), Security in Pervasive Computing. X, 243 pages. 2006.

Vol. 3933: F. Bonchi, J.-F. Boulicaut (Eds.), Knowledge Discovery in Inductive Databases. VIII, 251 pages. 2006.

Vol. 3931: B. Apolloni, M. Marinaro, G. Nicosia, R. Tagliaferri (Eds.), Neural Nets. XIII, 370 pages. 2006.

Vol. 3930: D.S. Yeung, Z.-Q. Liu, X.-Z. Wang, H. Yan (Eds.), Advances in Machine Learning and Cybernetics. XXI, 1110 pages. 2006. (Sublibrary LNAI).

Vol. 3929: W. MacCaull, M. Winter, I. Düntsch (Eds.), Relational Methods in Computer Science. VIII, 263 pages. 2006.

Vol. 3928: J. Domingo-Ferrer, J. Posegga, D. Schreckling (Eds.), Smart Card Research and Advanced Applications. XI, 359 pages. 2006.

Vol. 3927: J. Hespanha, A. Tiwari (Eds.), Hybrid Systems: Computation and Control. XII, 584 pages. 2006.

Vol. 3925: A. Valmari (Ed.), Model Checking Software. X, 307 pages. 2006.

Vol. 3924: P. Sestoft (Ed.), Programming Languages and Systems. XII, 343 pages. 2006.

Vol. 3923: A. Mycroft, A. Zeller (Eds.), Compiler Construction. XIII, 277 pages. 2006.

Vol. 3922: L. Baresi, R. Heckel (Eds.), Fundamental Approaches to Software Engineering. XIII, 427 pages. 2006.

Vol. 3921: L. Aceto, A. Ingólfsdóttir (Eds.), Foundations of Software Science and Computation Structures. XV, 447 pages. 2006.

Vol. 3920: H. Hermanns, J. Palsberg (Eds.), Tools and Algorithms for the Construction and Analysis of Systems. XIV, 506 pages. 2006.

Vol. 3918: W.K. Ng, M. Kitsuregawa, J. Li, K. Chang (Eds.), Advances in Knowledge Discovery and Data Mining. XXIV, 879 pages. 2006. (Sublibrary LNAI).

Vol. 3917: H. Chen, F.Y. Wang, C.C. Yang, D. Zeng, M. Chau, K. Chang (Eds.), Intelligence and Security Informatics. XII, 186 pages. 2006.

Vol. 3916: J. Li, Q. Yang, A.-H. Tan (Eds.), Data Mining for Biomedical Applications. VIII, 155 pages. 2006. (Sublibrary LNBI).

Vol. 3915: R. Nayak, M.J. Zaki (Eds.), Knowledge Discovery from XML Documents. VIII, 105 pages. 2006.

Vol. 3914: A. Garcia, R. Choren, C. Lucena, P. Giorgini, T. Holvoet, A. Romanovsky (Eds.), Software Engineering for Multi-Agent Systems IV. XIV, 255 pages. 2006.

Vol. 3911: R. Wyrzykowski, J. Dongarra, N. Meyer, J. Waśniewski (Eds.), Parallel Processing and Applied Mathematics. XXIII, 1126 pages. 2006.

Vol. 3910: S.A. Brueckner, G.D.M. Serugendo, D. Hales, F. Zambonelli (Eds.), Engineering Self-Organising Systems. XII, 245 pages. 2006. (Sublibrary LNAI).

Vol. 3909: A. Apostolico, C. Guerra, S. Istrail, P. Pevzner, M. Waterman (Eds.), Research in Computational Molecular Biology. XVII, 612 pages. 2006. (Sublibrary LNBI).

Vol. 3908: A. Bui, M. Bui, T. Böhme, H. Unger (Eds.), Innovative Internet Community Systems. VIII, 207 pages. 2006.

Vol. 3907: F. Rothlauf, J. Branke, S. Cagnoni, E. Costa, C. Cotta, R. Drechsler, E. Lutton, P. Machado, J.H. Moore, J. Romero, G.D. Smith, G. Squillero, H. Takagi (Eds.), Applications of Evolutionary Computing. XXIV, 813 pages. 2006.

Vol. 3906: J. Gottlieb, G.R. Raidl (Eds.), Evolutionary Computation in Combinatorial Optimization. XI, 293 pages. 2006.

Vol. 3905: P. Collet, M. Tomassini, M. Ebner, S. Gustafson, A. Ekárt (Eds.), Genetic Programming. XI, 361 pages. 2006.